Power Systems Handbook

Power Systems Protective Relaying
Volume 4

Power Systems Handbook

Series Author
J.C. Das
Power System Studies, Inc., Snellville, Georgia, USA

Volume 1: Short-Circuits in AC and DC Systems:
ANSI, IEEE, and IEC Standards

Volume 2: Load Flow Optimization and Optimal Power Flow

Volume 3: Harmonic Generation Effects Propagation and Control

Volume 4: Power Systems Protective Relaying

Power Systems Protective Relaying
Volume 4

J.C. Das

CRC Press
Taylor & Francis Group
Boca Raton London New York

CRC Press is an imprint of the
Taylor & Francis Group, an **informa** business

CRC Press
Taylor & Francis Group
6000 Broken Sound Parkway NW, Suite 300
Boca Raton, FL 33487-2742

© 2018 by Taylor & Francis Group, LLC
CRC Press is an imprint of Taylor & Francis Group, an Informa business

No claim to original U.S. Government works

Printed on acid-free paper

International Standard Book Number-13: 978-1-4987-4550-5 (Hardback)

This book contains information obtained from authentic and highly regarded sources. Reasonable efforts have been made to publish reliable data and information, but the author and publisher cannot assume responsibility for the validity of all materials or the consequences of their use. The authors and publishers have attempted to trace the copyright holders of all material reproduced in this publication and apologize to copyright holders if permission to publish in this form has not been obtained. If any copyright material has not been acknowledged please write and let us know so we may rectify in any future reprint.

Except as permitted under U.S. Copyright Law, no part of this book may be reprinted, reproduced, transmitted, or utilized in any form by any electronic, mechanical, or other means, now known or hereafter invented, including photocopying, microfilming, and recording, or in any information storage or retrieval system, without written permission from the publishers.

For permission to photocopy or use material electronically from this work, please access www.copyright.com (http://www.copyright.com/) or contact the Copyright Clearance Center, Inc. (CCC), 222 Rosewood Drive, Danvers, MA 01923, 978-750-8400. CCC is a not-for-profit organization that provides licenses and registration for a variety of users. For organizations that have been granted a photocopy license by the CCC, a separate system of payment has been arranged.

Trademark Notice: Product or corporate names may be trademarks or registered trademarks, and are used only for identification and explanation without intent to infringe.

Library of Congress Cataloging-in-Publication Data

Names: Das, J. C., 1934- author.
Title: Power systems protective relaying / J.C. Das.
Description: Boca Raton : Taylor & Francis, a CRC title, part of the Taylor & Francis imprint, a member of the Taylor & Francis Group, the academic division of T&F Informa, plc, [2018] | Includes bibliographical references and index.
Identifiers: LCCN 2017017946 | ISBN 9781498745505 (hardback : acid-free paper) | ISBN 9781498745512 (ebook)
Subjects: LCSH: Protective relays. | Electric power systems--Protection.
Classification: LCC TK2861 .D33 2018 | DDC 621.31/7--dc23
LC record available at https://lccn.loc.gov/2017017946

Visit the Taylor & Francis Web site at
http://www.taylorandfrancis.com

and the CRC Press Web site at
http://www.crcpress.com

Contents

Series Preface xix
Preface to Volume 4: Power Systems Protective Relaying xxi
Author xxiii

1. **Modern Protective Relaying: An Overview** 1
   1.1 Design Aspects and Reliability 1
   1.2 Fundamental Power System Knowledge 2
   1.3 Design Criteria of Protective Systems 3
       1.3.1 Selectivity 3
       1.3.2 Speed 3
       1.3.3 Reliability 4
   1.4 Equipment and System Protection 4
   1.5 Unit Protection Systems 5
       1.5.1 Back-Up Protection 7
   1.6 Smart Grids 7
       1.6.1 Framework for the Smart Grids 7
       1.6.2 Fundamental Layer 9
           1.6.2.1 Foundational Infrastructure and Resources 9
           1.6.2.2 Organization and Process 10
           1.6.2.3 Standards and Models 10
           1.6.2.4 Business and Regulatory 10
       1.6.3 Enabling Layer 11
           1.6.3.1 Enabling Infrastructure 11
           1.6.3.2 Incremental Intelligence 11
       1.6.4 Application Layer 11
           1.6.4.1 Grid and Customer Analysis 11
           1.6.4.2 Real-Time Awareness and Control 11
           1.6.4.3 Customer Interaction 11
       1.6.5 Innovation Layer 12
           1.6.5.1 Research and Development 12
           1.6.5.2 Research and Demonstration Projects 12
   1.7 Load Profiles: Var–Volt Control 12
   1.8 Some Modern Technologies Leading to Smart Grids 13
       1.8.1 WAMSs and PMUs 14
       1.8.2 System Integrity Protection Schemes 16
       1.8.3 Adaptive Protection 17
   1.9 Cyber Security 18
   1.10 NERC and CIP Requirements 20
   References 22

2. **Protective Relays** 23
   2.1 Classification of Relay Types 23
       2.1.1 Input 23
       2.1.2 Operating Principle 23

		2.1.3	Performance	23
		2.1.4	Construction	24
	2.2	Electromechanical Relays		24
	2.3	Overcurrent Relays		26
		2.3.1	ANSI Curves	26
		2.3.2	IEC Curves	28
	2.4	Differential Relays		28
		2.4.1	Overcurrent Differential Protection	29
		2.4.2	Partial Differential Schemes	34
		2.4.3	Overlapping the Zones of Protection	34
		2.4.4	Percent Differential Relays	34
	2.5	Pilot Wire Protection		36
	2.6	Directional Overcurrent Relays		37
	2.7	Voltage Relays		41
	2.8	Reclosing Relays		45
	2.9	Breaker Failure Relay		45
	2.10	Machine Field Ground Fault Relay		46
	2.11	Frequency Relays		47
	2.12	Distance Relays		50
	2.13	Other Relay Types		52
References				52
3. Instrument Transformers				**53**
	3.1	Accuracy Classification of CTs		53
		3.1.1	Metering Accuracies	53
		3.1.2	Relaying Accuracies	53
		3.1.3	Relaying Accuracy Classification X	55
		3.1.4	Accuracy Classification T	55
	3.2	Constructional Features of CTs		56
	3.3	Secondary Terminal Voltage Rating		56
		3.3.1	Saturation Voltage	57
		3.3.2	Saturation Factor	58
	3.4	CT Ratio and Phase Angle Errors		59
	3.5	Interrelation of CT Ratio and Class C Accuracy		62
	3.6	Polarity of Instrument Transformers		64
	3.7	Application Considerations		66
	3.8	Series and Parallel Connections of CTs		70
	3.9	Transient Performance of the CTs		70
		3.9.1	CT Saturation Calculations	73
		3.9.2	Effect of Remanence	74
	3.10	Practicality of CT Applications		75
	3.11	CTs for Low-Resistance Grounded Medium-Voltage Systems		76
	3.12	Future Directions in CT Applications		77
	3.13	Voltage Transformers		79
		3.13.1	Rated Primary Voltage and Ratios	79
		3.13.2	Accuracy Rating	82
		3.13.3	Thermal Burdens	83
		3.13.4	PT Connections	83
		3.13.5	Ferroresonance Damping	84

Contents

    3.14 Capacitor-Coupled Voltage Transformers 84
        3.14.1 Transient Performance 85
        3.14.2 Applications to Distance Relay Protection 88
    3.15 Line (Wave) Traps 88
    3.16 Transducers 91
    References 91

4. **Microprocessor-Based Multifunction Relays** 93
    4.1 Functionality 93
        4.1.1 Protection Features 93
        4.1.2 Voltage-Based Protections 93
        4.1.3 Monitoring Features 94
        4.1.4 Communications and Controls 94
    4.2 Front Panel 94
    4.3 Environmental Compatibility 95
    4.4 Dimensions 95
    4.5 Specifications 95
    4.6 Settings 101
        4.6.1 The Setting Groups 101
    4.7 Relay Bit Words 106
    4.8 Time Delay Overcurrent Protection 106
    4.9 Voltage-Based Elements 111
    4.10 Power Elements 113
    4.11 Loss of Potential 114
    4.12 Frequency Settings 114
    4.13 Trip and Close Logic 114
        4.13.1 Trip Logic 114
        4.13.2 Close Logic 118
        4.13.3 Reclose Logic and Supervision 119
    4.14 Demand Metering 120
    4.15 Logical Settings 122
    4.16 Latch Bits: Nonvolatile State 123
    4.17 Global Settings 124
    4.18 Port Settings 125
    4.19 Breaker Monitor 125
    4.20 Front Panel Operations 127
        4.20.1 Rotating Display 130
    4.21 Analyzing Events 130
        4.21.1 Sequential Event Recorder 131
        4.21.2 Triggering 131
        4.21.3 Aliases 131
    4.22 Setting the Relay 132
    Reference 133

5. **Current Interruption Devices and Battery Systems** 135
    5.1 High-Voltage Circuit Breakers 135
        5.1.1 DC Control Schematics 137
    5.2 Battery Systems 138
        5.2.1 Battery Types 138

	5.2.2	Plante Batteries	139
	5.2.3	Pasted Plate Batteries	140
	5.2.4	Tubular Plate Batteries	140
	5.2.5	Sealed (Valve-Regulated) Lead Acid Batteries	140
	5.2.6	Battery Monitoring System	141
	5.2.7	Nickel–Cadmium Batteries	142
	5.2.8	Pocket Plate Nickel–Cadmium Batteries	142
5.3	Sizing the Batteries		143
	5.3.1	Standards for Sizing the Batteries	144
	5.3.2	System Configurations for Batteries	144
	5.3.3	Automatic Transfer Switches	145
	5.3.4	Battery Chargers	147
		5.3.4.1 Floating Operation	147
		5.3.4.2 Equalizing Charge	148
		5.3.4.3 Switch Mode Operation	148
	5.3.5	Battery Charger as a Battery Eliminator	148
	5.3.6	Short-Circuit and Coordination Considerations	148
5.4	Capacitive Trip Devices		149
5.5	Lockout Relays		149
5.6	Remote Trips		150
5.7	CT and PT Test Switches		151
5.8	Fuses		151
	5.8.1	Medium-Voltage Fuses	155
		5.8.1.1 Variations in the Fuse Time–Current Characteristics	159
	5.8.2	Selection of Fuse Types and Ratings	159
	5.8.3	Semiconductor Fuses	161
5.9	Low-Voltage Circuit Breakers		162
	5.9.1	Molded Case Circuit Breakers	162
	5.9.2	Current-Limiting MCCBs	164
	5.9.3	Insulated Case Circuit Breakers	168
	5.9.4	Low-Voltage Power Circuit Breakers	168
	5.9.5	Short-Time Bands of LVPCBs' Trip Programmers	169
	5.9.6	Motor Circuit Protectors	171
	5.9.7	Other Pertinent Data of Low-Voltage Circuit Breakers	172
5.10	Selective Zone Interlocking		173
5.11	Electronic Power Fuses		175
5.12	Low- and Medium-Voltage Contactors		176
References			179

6. Overcurrent Protection: Ideal and Practical 181
  6.1 Fundamental Considerations 181
  6.2 Data for the Coordination Study 182
  6.3 Computer-Based Coordination 183
  6.4 Initial Analysis 184
  6.5 Coordinating Time Interval 184
    6.5.1 Relay Overtravel 184
  6.6 Fundamental Considerations for Coordination 185
    6.6.1 Settings on Bends of Coordination Curves 186
  6.7 Some Examples of Coordination 186

		6.7.1	Low-Voltage Distribution System	186
		6.7.2	2.4 kV Distribution	192
		6.7.3	Ground Fault Protection	196
		6.7.4	Coordination in a Cogeneration System	197
	6.8	Coordination on Instantaneous Basis		199
		6.8.1	Selectivity between Two Series-Connected Current-Limiting Fuses	200
		6.8.2	Selectivity of a Current-Limiting Fuse Downstream of Noncurrent-Limiting Circuit Breaker	203
		6.8.3	Selectivity of Current-Limiting Devices in Series	207
	6.9	NEC Requirements of Selectivity		211
		6.9.1	Fully Selective Systems	212
		6.9.2	Selection of Equipment Ratings and Trip Devices	214
	6.10	The Art of Compromise		215
	6.11	Zone Selective Interlocking		221
	6.12	Protection and Coordination of UPS Systems		227
	References			230

7. System Grounding 231

7.1	Study of Grounding Systems	231
7.2	Solidly Grounded Systems	232
	7.2.1 Hazards in Solidly Grounded Systems	235
7.3	Low-Resistance Grounded Systems	236
7.4	High-Resistance Grounded Systems	237
7.5	Ungrounded Systems	239
7.6	Reactance Grounding	241
7.7	Resonant Grounding	242
7.8	Corner of Delta Grounded Systems	242
7.9	Artificially Derived Neutrals	243
7.10	Multiple Grounded Systems	245
	7.10.1 Equivalent Circuit of Multiple Grounded Systems	246
7.11	NEC and NESC Requirements	247
7.12	Hybrid Grounding System for Industrial Bus-Connected Generators	248
7.13	Grounding of ASDs	249
7.14	Grounding in Mine Installations	253
References		254

8. Ground Fault Protection 257

8.1	Protection and Coordination in Solidly Grounded Systems	257
	8.1.1 NEC Requirements	257
	8.1.2 Self-Extinguishing Ground Faults	263
	8.1.3 Improving Coordination in Solidly Grounded Low-Voltage Systems	263
8.2	Ground Fault Coordination in Low-Resistance Grounded Medium-Voltage Systems	266
8.3	Remote Tripping	268
8.4	Ground Fault Protection in Ungrounded Systems	268
	8.4.1 Nondiscriminatory Alarms and Trips	270
8.5	Ground Fault Protection in High-Resistance Grounded Systems	271

		8.5.1	Nondiscriminatory Alarms and Trips..271

       8.5.1   Nondiscriminatory Alarms and Trips 271
       8.5.2   Selective Ground Fault Clearance 271
       8.5.3   Pulsing-Type Ground Fault Detection System 272
       8.5.4   Protection of Motors 273
       8.5.5   Protection against Second Ground Fault 274
       8.5.6   Insulation Stresses and Cable Selection for HR Grounded Systems 275
  8.6 Ground Fault Protection in Resonant Grounded Systems 276
  8.7 Studies of Protection and Coordination in Practical Systems 277
       8.7.1   Ground Fault Protection of Industrial Bus-Connected Generators 277
       8.7.2   Directional Ground Fault Relays 280
       8.7.3   Operating Logic Selection for Directional Elements 285
                 8.7.3.1   Single-Line-to-Ground Fault 285
                 8.7.3.2   Double-Line-to-Ground Fault 285
  8.8 Selective High-Resistance Grounding Systems 285
       8.8.1   EMTP Simulation of a HRG 287
       8.8.2   Generator 100% Stator Winding Protection 294
       8.8.3   Accuracy of Low Pickup Settings in MMPR 299
  8.9 Monitoring of Grounding Resistors 299
References 300

9. Bus-Bar Protection and Autotransfer of Loads 301
  9.1 Bus Faults 301
  9.2 Bus Differential Relays 301
       9.2.1   Low-Voltage Bus Bars 301
  9.3 High-Impedance Differential Relays 301
       9.3.1   Sensitivity for Internal Faults 305
       9.3.2   High-Impedance MMPRs 305
       9.3.3   Open-Circuited CT 308
  9.4 Low-Impedance Current Differential Relays 308
       9.4.1   CT Saturation 312
       9.4.2   Dynamic Bus Replica 314
       9.4.3   The Differential Settings 315
       9.4.4   Comparison with High-Impedance Relays 316
  9.5 Direction Comparison Bus Protection 316
  9.6 Bus Protection Using Linear Couplers 317
       9.6.1   Linear Couplers 317
  9.7 Differential Protection of Common Bus Configurations 318
       9.7.1   Single Bus 318
       9.7.2   Sectionalized Bus 318
       9.7.3   Double Bus Double Breaker 319
       9.7.4   Main and Transfer Bus 319
       9.7.5   Double Bus Single Breaker with Bus Tie 319
       9.7.6   Breaker and a Half Scheme 322
       9.7.7   Ring Bus 322
       9.7.8   Combined Bus Differential Zones 322
       9.7.9   Ground Fault Bus Differential Protection 325
  9.8 Reclosing 326
  9.9 Bus Transfer Schemes 326
       9.9.1   Fast Bus Transfer 327

		9.9.2 Residual Voltage Transfer	327

        9.9.2   Residual Voltage Transfer 327
        9.9.3   In-Phase Transfer 329
    9.10 Momentary Paralleling 330
        9.10.1   Fault Conditions 330
        9.10.2   Dropout of Motor Contactors 330
        9.10.3   Autotransfer of Synchronous Motors 331
    References 334

10. Motor Protection 335
    10.1 Motor Characteristics 335
    10.2 Motor Protection 335
        10.2.1   Medium-Voltage Induction Motors 335
        10.2.2   Medium-Voltage Synchronous Motors 336
        10.2.3   Low-Voltage Motors 336
    10.3 Motor Protection and Coordination Study 337
    10.4 Coordination with Motor Thermal Damage Curve 338
    10.5 RTD Biasing 347
        10.5.1   Locked Rotor Protection Using Device 21 347
    10.6 Medium-Voltage Motor Starters 348
        10.6.1   Class E1 349
        10.6.2   Class E2 349
        10.6.3   Low-Voltage Magnetic Contactors up to 600 V 350
    10.7 Two-Wire and Three-Wire Controls 352
        10.7.1   Schematic Control Circuit of Krondroffer Starter 353
    10.8 Undervoltage Protection of Motors 354
    10.9 NEC and OSHA Requirements 354
    10.10 Motor Insulation Classes and Temperature Limits 357
        10.10.1 NEMA Standards for Insulation Temperature Rise 357
        10.10.2 Embedded Temperature Detectors 359
                 10.10.2.1 Polling of RTDs 360
    10.11 Bearing Protection 361
        10.11.1 Antifriction Ball or Roller Bearings 361
        10.11.2 Sleeve Bearings 361
        10.11.3 Bearing Failures 361
                 10.11.3.1 Lubricant Problems 361
                 10.11.3.2 Mechanical Problems 361
                 10.11.3.3 Excessive Axial or Thrust Loading 362
        10.11.4 Bearing Protection Devices 362
        10.11.5 End Play and Rotor Float for Coupled Sleeve Bearings Horizontal Motors 362
    10.12 Vibrations 363
        10.12.1 Relative Shaft Vibrations 363
                 10.12.1.1 Standard Machines 363
                 10.12.1.2 Special Machines 364
        10.12.2 Axial Vibrations 364
        10.12.3 Limits of Relative Shaft Vibrations 365
    10.13 Motor Enclosure 365
        10.13.1 IP Designations 366
        10.13.2 IC Designations 366

- 10.14 Effect of Negative-Sequence Currents 367
  - 10.14.1 Protection for Negative-Sequence Currents 369
    - 10.14.1.1 Phase Balance Protection 369
    - 10.14.1.2 Time Delay and Instantaneous Negative-Sequence Protection 370
- 10.15 Differential Protection 371
  - 10.15.1 Flux Balancing Current Differential 371
  - 10.15.2 Full Differential Protection with Ground Fault Differential 372
- 10.16 Ground Fault Protection 372
- 10.17 Variable-Speed Motor Protection 373
- 10.18 Synchronous Motors Starting and Synchronization 373
  - 10.18.1 Brush-Type Controllers 376
  - 10.18.2 Brushless-Type Excitation System 377
  - 10.18.3 Types of Field Controllers 378
- 10.19 Stability Concepts of Synchronous Motors 379
  - 10.19.1 Calculation of PO Power Factor 383
- 10.20 Rotor Ground Fault Protection, Slip Ring-Type Synchronous Motors 385
- References 385

11. Generator Protection 387
- 11.1 Ratings of Synchronous Generators 387
- 11.2 Protection of Industrial Generators 387
- 11.3 Functionality of a Modern MMPR for Generator Protection 388
- 11.4 Voltage-Controlled and Voltage-Restraint Protection (51V) 390
- 11.5 Negative-Sequence Protection: Function 46 397
- 11.6 Loss of Excitation. Protection: Function 40 400
  - 11.6.1 Steam Turbine Generators 400
  - 11.6.2 Hydro Generators 400
  - 11.6.3 Protection 400
- 11.7 Generator Thermal Overload 404
- 11.8 Differential Protection 407
  - 11.8.1 Generator Winding Connections 407
  - 11.8.2 Protection Provided by Differential Relays 408
  - 11.8.3 Self-Balancing Differential Scheme 409
  - 11.8.4 Application Depending on Winding Connection 409
  - 11.8.5 Backup Protection Using Differential Relays 410
  - 11.8.6 Characteristics 411
  - 11.8.7 The Tripping Modes 415
- 11.9 Generator Stator Ground Fault Protection 415
  - 11.9.1 HR Grounded through Distribution Transformer Unit Generators 415
    - 11.9.1.1 Third Harmonic Differential Scheme 415
    - 11.9.1.2 Subharmonic Voltage Injection 416
  - 11.9.2 Generator Ground Fault Differential 418
- 11.10 Rotor Ground Fault Protection 419
  - 11.10.1 Slip Ring Machines 419
  - 11.10.2 Brushless Type of Generators 419
- 11.11 Volts per Hertz Protection 420
- 11.12 Over- and Underfrequency Protection 423

Contents

        11.12.1 Load Shedding 424
        11.12.2 Protection 424
        11.12.3 Rate of Change of Frequency 426
   11.13 Out-of-Step Protection 427
   11.14 Inadvertent Energization of Generator 429
   11.15 Generator Breaker Failure Protection 431
   11.16 Antimotoring Protection 433
   11.17 Loss of Potential 434
   11.18 Under- and Overvoltage Protection 436
   11.19 Synchronization 436
        11.19.1 Manual Synchronizing 436
        11.19.2 Automatic Synchronizing 437
   11.20 Tripping Schemes 437
        11.20.1 Sequential Tripping 438
   References 440

12. Transformer Reactor and Shunt Capacitor Bank Protection 443
   12.1 Transformer Faults 443
        12.1.1 External Faults 443
            12.1.1.1 Short-Circuit Faults 443
            12.1.1.2 Overloads 443
            12.1.1.3 Overvoltage and Underfrequency 444
        12.1.2 Internal Faults 444
            12.1.2.1 Short Circuit in the Windings 444
            12.1.2.2 The Interturn Winding Faults 444
            12.1.2.3 Faults in ULTC and Off–Load Tap Changing Equipment 445
            12.1.2.4 Bushing Flashovers 445
            12.1.2.5 Part-Winding Resonance 445
   12.2 NEC Requirements 445
   12.3 System Configurations of Transformer Connections 446
        12.3.1 Radial System of Distribution 446
        12.3.2 Primary Selective System 448
        12.3.3 Group Feed System 448
        12.3.4 Dedicated Circuit Breakers 448
        12.3.5 Fixed Mounted Primary Circuit Breaker 448
        12.3.6 Secondary Selective System 448
   12.4 Through-Fault Current Withstand Capability 449
        12.4.1 Category I 450
        12.4.2 Category II 451
        12.4.3 Categories III and IV 452
            12.4.3.1 Category III 452
            12.4.3.2 Category IV 453
        12.4.4 Observation on Faults during Life Expectancy of a Transformer 454
        12.4.5 Dry-Type Transformers 454
   12.5 Construction of Through-Fault Curve Analytically 455
   12.6 Protection with Respect to Through-Fault Curves 458
        12.6.1 Withstand Ratings of UT (GSU) and UAT 458
            12.6.1.1 UAT 459
   12.7 Transformer Primary Fuse Protection 460

	12.7.1 Variations in the Fuse Characteristics	460
	12.7.2 Single Phasing and Ferroresonance	461
	12.7.3 More Considerations of Fuse Protection	461
12.8	Overcurrent Relays for Transformer Primary Protection	463
12.9	Listing Requirements	466
12.10	Effect of Transformer Winding Connections	468
12.11	Requirements of Ground Fault Protection	469
12.12	Through-Fault Protection	470
	12.12.1 Primary Fuse Protection	470
	12.12.2 Primary Relay Protection	471
12.13	Overall Transformer Protection	471
	12.13.1 System Configuration	471
	12.13.2 Coordination Study and Observations	474
	12.13.3 Addition of Secondary Relay	477
12.14	Differential Protection	477
	12.14.1 Electromechanical Transformer Differential Relays	480
	12.14.2 Harmonic Restraint	483
	12.14.3 Protection with Grounding Transformer Inside Main Transformer Protection Zone	483
	12.14.4 Protection of an Autotransformer	483
	12.14.5 Protection of Three-Winding Transformers	483
	12.14.6 Microprocessor-Based Transformer Differential Relays	483
	12.14.6.1 CT Connections and Phase Angle Compensation	486
	12.14.6.2 Dynamic CT Ratio Corrections	488
	12.14.6.3 Security under Transformer Magnetizing Currents	488
12.15	Sensitive Ground Fault Differential Protection	489
	12.15.1 Protection of Zigzag Grounding Transformer	491
12.16	Protection of Parallel Running Transformers	491
12.17	Volts per Hertz Protection	493
12.18	Shunt Reactor Protection	494
	12.18.1 Oil-Immersed Reactors	497
12.19	Transformer Enclosures	498
	12.19.1 Dry-Type and Cast Coil Transformers	499
	12.19.2 Liquid-Filled Transformers	500
	12.19.2.1 Sealed Tank Construction	500
	12.19.2.2 Positive Pressure Inert Gas	500
	12.19.2.3 Conservator Tank Design	501
12.20	Transformer Accessories	502
	12.20.1 Pressure–Vacuum Gauge and Bleeder Valve	502
	12.20.2 Liquid Level Gauge	502
	12.20.3 Pressure-Relief Device	502
	12.20.4 Rapid Pressure Rise Relay	502
	12.20.5 Liquid Temperature Indicator	503
	12.20.6 Winding Temperature Indicator (Thermal Relays)	503
	12.20.7 Combustible Gas Relay	504
	12.20.8 Underload Tap Changing Equipment	504
	12.20.9 Surge Protection	506
12.21	Shunt Capacitor Bank Protection	507
References		507

13. Protection of Lines 509
- 13.1 Distribution Lines 509
- 13.2 Transmission and Subtransmission Lines 509
- 13.3 Protective Relays 510
  - 13.3.1 Application of Directional Overcurrent Relays 517
  - 13.3.2 Loop System with One Source of Fault Current 520
- 13.4 Distance Protection 522
  - 13.4.1 Zoned Distance Relays 523
  - 13.4.2 Distance Relay Characteristics 524
  - 13.4.3 Operating Time in the First Zone 528
  - 13.4.4 Effect of Arc Fault Resistance 528
- 13.5 Load Encroachment Logic 530
  - 13.5.1 Communication-Assisted Tripping 530
- 13.6 Ground Fault Protection 531
  - 13.6.1 Zero-Sequence Overcurrent 531
  - 13.6.2 Quadrilateral Ground Distance and Mho Ground Distance Characteristics 533
  - 13.6.3 Effect of Nonhomogeneous System on Reactance Elements 534
  - 13.6.4 Zero-Sequence Mutual Coupling 535
  - 13.6.5 High-Resistance Fault Coverage and Remote Infeed 536
- 13.7 Protection of Tapped 345 kV Transmission Line 538
- 13.8 Series Compensated Lines 546
  - 13.8.1 Subsynchronous Resonance 547
  - 13.8.2 Steady-State Excitation 547
  - 13.8.3 Models 547
  - 13.8.4 Analysis 548
  - 13.8.5 An Example with EMTP Simulations 549
- 13.9 Mitigation of Subsynchronous Resonance in HV Transmission Lines 551
  - 13.9.1 NGH-SSR 551
  - 13.9.2 Thyristor-Controlled Series Capacitor 552
  - 13.9.3 Supplemental Excitation Damping Control 554
  - 13.9.4 Torsional Relay 555
- References 556

14. Pilot Protection 559
- 14.1 Pilot Systems 559
- 14.2 Signal Frequencies 561
- 14.3 Metallic Pilot Wire Protection Using Electromechanical Relays 562
- 14.4 Modern Line Current Differential Protection 564
  - 14.4.1 Differential Protection of Two-Terminal and Three-Terminal Lines ... 564
  - 14.4.2 The Alpha Plane 565
  - 14.4.3 CT Saturation 567
  - 14.4.4 Three-Terminal Protection 568
  - 14.4.5 Enhanced Current Differential Characteristics 569
- 14.5 Direct Underreaching Transfer Trip 571
- 14.6 Permissive Underreaching Transfer Trip 571
- 14.7 Direct Overreaching Transfer Trip 572
- 14.8 Blocking and Unblocking Pilot Protection 574
  - 14.8.1 Direct Blocking Scheme 574

	14.8.2	Directional Comparison Blocking Scheme 575
	14.8.3	Directional Comparison Unblocking Scheme 575
14.9	Phase Comparison Schemes 577
	14.9.1	Single-Phase Comparison Blocking 578
	14.9.2	Dual-Phase Comparison Blocking 579
	14.9.3	Segregated Phase Comparison 579
14.10	Power Line Carrier 583
	14.10.1	Coupling Capacitor and Drain Coils 584
	14.10.2	Line Tuner 584
	14.10.3	Coaxial Cable 585
	14.10.4	Transmitter/Receiver 585
		14.10.4.1 On–Off Carrier 586
		14.10.4.2 Frequency Shift Carrier: TCF and FSK 587
	14.10.5	Audio Tone Channels 588
	14.10.6	Microwave Channels 589
14.11	Modal Analysis 589
References 591

15. Power System Stability 593
15.1 Classification of Power System Stability 593
 15.1.1 Rotor Angle Stability 594
 15.1.2 Voltage Instability 595
  15.1.2.1 Large Disturbance Instability 595
  15.1.2.2 Small Disturbance Voltage Instability 595
 15.1.3 Static Stability 596
15.2 Equal Area Concept of Stability 597
 15.2.1 Critical Clearing Angle 599
15.3 Factors Affecting Stability 600
15.4 Swing Equation of a Generator 601
15.5 Classical Stability Model 604
15.6 Modern Transient Stability Methods 606
15.7 Excitation Systems 606
 15.7.1 Fast Response Systems 606
 15.7.2 Types of Excitation Systems 607
15.8 Transient Stability in a Simple Cogeneration System 608
15.9 A System Illustrating Application of PSS 618
Bibliography 618

16. Substation Automation and Communication Protocols Including IEC 61850 621
16.1 Substation Automation 621
16.2 System Functions 621
16.3 Control Functions 621
16.4 Wire Line Networks 623
 16.4.1 Point-to-Point Networks 623
 16.4.2 Point-to-Multipoint Networks 623
 16.4.3 Peer-to-Peer Network 623
16.5 System Architecture 624
 16.5.1 Level 1: Field Devices 624
 16.5.2 Level 2: Substation Data Concentrator 624

Contents   xvii

        16.5.3   Level 3: SCADA Systems 624
        16.5.4   LAN Protocols 624
        16.5.5   SCADA Communication Requirements 624
                16.5.5.1   Distributed Network Protocol 625
                16.5.5.2   IEEE 802.3 (Ethernet) 625
  16.6   IEC 61850 Protocol 626
  16.7   Modern IEDs 627
  16.8   Substation Architecture 628
  16.9   IEC 61850 Communication Structure 628
  16.10 Logical Nodes 629
  16.11 Ethernet Connection 630
  16.12 Networking Media 634
        16.12.1 Copper Twisted Shielded and Unshielded 635
        16.12.2 Fiber-Optic Cable 635
  16.13 Network Topologies 636
        16.13.1 Prioritizing GOOSE Messages 636
        16.13.2 Techno-Economical Justifications 638
  16.14 A Sample Application 638
  References 640

17. Protective Relaying for Arc-Flash Reduction 641
  17.1   Arc-Flash Hazard 641
        17.1.1   Arc Blast 642
        17.1.2   Fire Hazard and Electrical Shock 642
        17.1.3   Time Motion Studies 642
  17.2   Arc-Flash Hazard Analysis 643
        17.2.1   Ralph Lee's Equations 644
        17.2.2   IEEE 1584 Equations 644
  17.3   Hazard/Risk Categories 647
        17.3.1   Hazard Boundaries 648
  17.4   System Grounding: Impact on Incident Energy 649
  17.5   Maximum Duration of an Arc-Flash Event 651
        17.5.1   Equipment Labeling 652
  17.6   Protective Relaying and Coordination 652
  17.7   Arc Protection Relays 652
        17.7.1   Principle of Operation 652
        17.7.2   Light Sensor Types 654
        17.7.3   Other Hardware 657
        17.7.4   Selective Tripping 657
        17.7.5   Supervision with Current Elements 659
        17.7.6   Applications 659
        17.7.7   Self-Testing of Sensors 660
  17.8   Accounting for Decay in Short-Circuit Currents 660
        17.8.1   Reducing Short-Circuit Currents 664
  17.9   Arc-Resistant Switchgear 664
  17.10 Arc-Flash Calculations 665
        17.10.1 Reduction of HRC through a Maintenance Mode Switch 666
  17.11 System Configuration for Study 668
        17.11.1 Coordination Study and Observations 668

       17.11.2 Arc-Flash Calculations in Figure 17.12 672
       17.11.3 Reducing HRC Levels with Main Secondary Circuit Breakers 674
       17.11.4 Maintenance Mode Switches on Low-Voltage Trip Programmers 679
  References 682

Appendix A: Device Numbers according to IEEE C37.2 683

Index 687

Series Preface

This handbook on power systems consists of four volumes. These are carefully planned and designed to provide state-of-the-art material on the major aspects of electrical power systems, short-circuit currents, load flow, harmonics, and protective relaying.

An effort has been made to provide a comprehensive coverage, with practical applications, case studies, examples, problems, extensive references, and bibliography.

The material is organized with sound theoretical base and its practical applications. The objective of creating this series is to provide the reader with a comprehensive treatise that could serve as a reference and day-to-day application guide for solving the real-world problem. It is written for plasticizing engineers and academia at the level of upper-undergraduate and graduate degrees.

Though there are published texts on similar subjects, this series provides a unique approach to the practical problems that an application engineer or consultant may face in conducting system studies and applying it to varied system problems.

Some parts of the work are fairly advanced on a postgraduate level and get into higher mathematics. Yet the continuity of the thought process and basic conceptual base are maintained. A beginner and advanced reader will equally benefit from the material covered. An underground level of education is assumed, with a fundamental knowledge of electrical circuit theory, rotating machines, and matrices.

Currently, power systems, large or small, are analyzed on digital computers with appropriate software. However, it is necessary to understand the theory and basis of these calculations to debug and decipher the results.

A reader may be interested only in one aspect of power systems and may choose to purchase only one of the volumes. Many aspects of power systems are transparent between different types of studies and analyses—for example, knowledge of short-circuit currents and symmetrical component is required for protective relaying and fundamental frequency load flow is required for harmonic analysis. Though appropriate references are provided, the material is not repeated from one volume to another.

The series is a culmination of the vast experience of the author in solving real-world problems in the industrial and utility power systems for more than 40 years.

Another key point is that the solutions to the problems are provided in Appendix D. Readers should be able to independently solve these problems after perusing the contents of a chapter and then look back to the solutions provided as a secondary help. The problems are organized so these can be solved with manual manipulations, without the help of any digital computer power system software.

It is hoped the series will be a welcome addition to the current technical literature.

The author thanks CRC Press editor Nora Konopka for her help and cooperation throughout the publication effort.

— J.C. Das

Preface to Volume 4: Power Systems Protective Relaying

Modern Protective Relaying

Protective relaying is an indispensable part of electrical power systems. In modern times, the advancement in protective relaying is being dictated by microprocessor-based multifunction relays (MMPRs); Chapter 4 is devoted to their functionality and capabilities. Furthermore, smart grids, integration of wind and solar generation, and microgrids are also the driving aspects of fast changes and innovations in protective relaying. The modern framework of smart grids and the current technologies driving smart grids are discussed in Chapter 1. Cybersecurity is a growing concern for the grid systems.

Chapter 3 is devoted to instrument transformers. Here, the CT saturation has taken a step further, documenting the practical limitation of providing required accuracy CTs based on analytical calculations. The advancement made in the digital relays to accommodate CT saturation and inaccuracies and ensure required performance is the recent trend through digital technology.

Chapter 6 forms the background of overcurrent coordination. Coordination on instantaneous basis is a new frontier discussed in this chapter with an example for a fully coordinated system based on NEC requirements. The coordination may require change of device ratings, or other system modifications may be necessary.

Chapters 7 and 8 discuss system grounding and ground fault protection. The problems of full coordination in solidly grounded low-voltage systems are described and their transition to high-resistance grounded systems. A detailed example of selective coordination in interconnected medium-voltage high-resistance grounded systems, with generators, is a new trend, amply discussed with an example. This is compared with recent IEEE recommendations of hybrid grounded systems for bus-connected generators.

All the relaying systems as applicable to generator, transformer, motor, bus, and transmission lines are covered in dedicated chapters. The emphasis is to exploit the functionality of the MMPRs. In each of these, some recent trends and innovations are described, for example, calculations of pullout power factor of a synchronous motor and plotting its negative-sequence withstand capability and coordinating it with an inverse overcurrent device.

The clarity of the reading is enhanced with many illustrations, examples, and practical study cases.

Chapter 15 is added to provide basic concepts of power system stability, power system stabilizer (PSS) and the impact of protective relaying on stability. It has a study case for the transient stability in a cogeneration system and another for application of PSS.

The substation automation and communication protocols including IEC 61950 are important in the current relaying environment, which are covered in Chapter 16. Finally, special considerations for arc-flash reduction and protective relaying are discussed in Chapter 17. This volume provides the complete coverage of the protective relaying in power systems.

— J.C. Das

Author

J.C. Das is an independent consultant, Power System Studies, Inc. Snellville, Georgia. Earlier, he headed the electrical power systems department at AMEC Foster Wheeler for 30 years. He has varied experience in the utility industry, industrial establishments, hydro-electric generation, and atomic energy. He is responsible for power system studies, including short circuit, load flow, harmonics, stability, arc flash hazard, grounding, switching transients, and protective relaying. He conducts courses for continuing education in power systems and is the author or coauthor of about 70 technical publications nationally and internationally. He is the author of the following books:

- *Arc Flash Hazard Analysis and Mitigation*, IEEE Press, 2012.
- *Power System Harmonics and Passive Filter Designs*, IEEE Press, 2015.
- *Transients in Electrical Systems: Analysis Recognition and Mitigation*, McGraw-Hill, 2010.
- *Power System Analysis: Short-Circuit Load Flow and Harmonics*, Second Edition, CRC Press 2011.
- *Understanding Symmetrical Components for Power System Modeling*, IEEE Press, 2017.

These books provide extensive converge, running into more than 3000 pages, and are well received in the technical circles. His interests include power system transients, EMTP simulations, harmonics, passive filter designs, power quality, protection, and relaying. He has published more than 200 electrical power system study reports for his clients.

He has published more than 200 study reports of power systems analysis addressing one problem or the other.

Das is a Life Fellow of the Institute of Electrical and Electronics Engineers, IEEE, (USA), Member of the IEEE Industry Applications and IEEE Power Engineering societies, a Fellow of the Institution of Engineering Technology (UK), a Life Fellow of the Institution of Engineers (India), a Member of the Federation of European Engineers (France), a Member of CIGRE (France), etc. He is registered Professional Engineer in the states of Georgia and Oklahoma, a Chartered Engineer (CEng) in the UK, and a European Engineer (EurIng) in Europe. He received a meritorious award in engineering, IEEE Pulp and Paper Industry in 2005.

He earned a PhD in electrical engineering at Atlantic International University, Honolulu, an MSEE at Tulsa University, Tulsa, Oklahoma, and a BA in advanced mathematics and a BEE at Panjab University, India.

1
Modern Protective Relaying: An Overview

Protective relaying has been called an "art" and also a "science." This is so because there is a judgment involved in making selections, which require compromises between conflicting objectives, such as maximum protection, reliability, fast fault clearance times, economics, and selectivity. A fault in the system should be detected fast, and only the faulty section isolated without impacting the unfaulted system. Protective relaying is an essential feature of the electrical system which is considered concurrently with the system design. Protection is not a substitute for poorly designed systems; that is, protecting a poorly designed system will be more complex and less satisfactory than a properly designed system.

In many continuous processes industrial plant distribution systems, a single nuisance trip can result in colossal loss of revenue and it may take many hours to days to restore the processes to full-stream production.

In terms of modern technology, a revolution has taken place in the development and application of microprocessor-based multifunction relays (MMPRs). The single-function electromechanical relays are now outdated, and these are being replaced with MMPRs in many industrial and utility systems. Chapter 4 describes the functionality of a feeder relay.

1.1 Design Aspects and Reliability

Protective relaying must be considered alongside the design of power systems, large or small. It is difficult and even unsatisfactory to protect a badly designed system, and protective relaying, which sometimes comes in last, cannot cover the lapses of the inadequate system designs. See Chapter 1 of Volume 1 for the fundamental concepts of design and planning of electrical power systems.

The safety and reliability of a power system cannot be considered based on only one aspect. It is a chain where the weakest link can jeopardize the reliability and security. The basic concepts of reliability are discussed in Chapter 1 of Volume 1 and are not repeated here.

Many utilities establish standards of quality of service based on a number and duration of outages on a given type of circuit on a yearly basis. In continuous process plants, a single loss of critical equipment due to nuisance trip may result in colossal loss of revenue. This returns us to system designs, redundant sources of power for the critical equipment, standby generators or tie lines, UPS, etc.

A number of ANSI/IEEE standards have been developed for protection and relaying and are continuously updated. These standards cover the application of protective devices, that is, the manner in which these need to be applied for specific protection systems. Protection standards for bulk power facilities require that redundancy exists within protection system designs. Redundancy requires that the failure of a protection component, protective relay, circuit breaker, or communication channel will not result in failure to detect and isolate faults.

The North American Electric Reliability Corporation (NERC) has the statuary responsibility to regulate bulk power system users and producers through adoption and enforcement of their standards. In 2007, the Federal Energy Regulatory Commission (FERC) which is the U.S. federal agency granted NERC the legal authority to enforce reliability standards for bulk power systems in the United States and made compliance with these standards mandatory and enforceable.

1.2 Fundamental Power System Knowledge

A number of power system design concepts are covered in Volumes 1–3 of this series and some data are transparent with respect to protective relaying. The following underlying concepts are not repeated in this volume:

- Nature of modern power systems, generation, distributed generation, transmission and subtransmission systems, industrial and commercial systems (see Volume 1)
- Renewable energy sources, solar and wind power plants (see Volume 1)
- Short-circuit calculations, three-phase and unsymmetrical faults like single-line-to ground faults, double-line-to-ground faults, line-to-line faults, open conductor faults, and 30-cycle faults for protective relaying (see Volume 1)
- Symmetrical components theory and its applications (see Volume 1)
- Rating structures of high-voltage circuit breakers, fundamental characteristics of high- and low-voltage power fuses, low-voltage power circuit breakers, molde case and insulated case circuit breakers (see Volume 1)
- Calculations of transmission line and cable parameters (see Volume 1)
- Fundamental concepts of AC current interruption, rating structure of circuit breakers, fuses, low-voltage circuit breakers, MCCBs (Molded Case Circuit Breakers), etc. (see Volume 1)
- Shunt reactors, their switching, application considerations, and transients (see Volume 1)
- ANSI/IEEE standards of system voltages (see Volume 2)
- Power transformers, synchronous generators, and motors, their models, and their operations (see Volumes 1 and 2)
- Load flow and reactive power compensation (see Volume 2)
- Starting of motors (see Volume 2)
- FACTs devices (see Volume 2)
- Effect of harmonics on power system equipment (see Volume 3)
- Protection of shunt capacitor banks is covered in Volume 3 and not repeated in this volume

In addition, familiarity with phasors, vectors, per unit system, electrical circuit concepts, and matrix algebra are required.

In particular, the protective relaying demands knowledge of calculations of short-circuit current in the systems and also symmetrical components. See [1–3] for Volumes 1–3 referred here. Research works [4–10] list some popular books on protective relaying.

1.3 Design Criteria of Protective Systems

The logic of protective relaying looks at a complex distribution system as an integration of subsystems. In all cases, some common criteria are applicable. These are as follows:

- Selectivity
- Speed
- Reliability
- Simplicity
- Economics
- Maintainability (sometimes)

1.3.1 Selectivity

A protection system must operate so as to isolate the faulty section only. In a radial distribution system, which is a common system configuration in the industrial power distribution systems, inverse time overcurrent relays are used as the primary protection devices. The desired selectivity is attained by coordinating upstream relays with the downstream relays, so that the upstream relay is slower than the downstream relay. A proper time delay should be selected between two overcurrent relays in series by providing either a certain appropriate time delay, called coordinating time interval, or variations of the inverse time–current characteristics, not forgetting the definite time–current characteristics. This coordination is discussed in Chapter 6. This increases the time delay for fault clearance toward the source, which is not desirable from arc flash hazard limitation and equipment damage. Separate zones of protection can be established around each equipment, which are called unit protection systems (Section 1.5). The unit protection systems are discussed throughout this volume; for example, a differential system is a unit protection system.

1.3.2 Speed

Fault damage to the system components and the stability between synchronous machines and interconnected systems are related to the speed of operation of the protective systems. In case all faults could be cleared instantaneously, the *equipment damage as well as the arc flash hazard* will be a minimum. Thus, there is a direct relation between limiting the arc flash hazard and equipment damage. Unit protection systems, with overlapping zones of protection, can limit equipment damage and reduce arc flash hazard.

Practically, unit protection systems are not applied throughout an industrial distribution, primarily because of cost. However, the concepts are changing; for example, commercial low-voltage switchgear is available with differential and zone interlocking protection (see Chapter 6).

From transient stability considerations, there is a critical fault clearing time and even a slight delay of one-fourth of cycle exceeding this time can result in system separation, see Chapter 15. Single-pole closing, fast load shedding, bundle conductors, fast excitation systems, power system stabilizers, series and shunt compensation of transmission lines, Static Var Conpensator (SVC) and STATCOM, and FACT controllers can enhance the stability limits of a power system. In industrial plants having cogeneration facilities, fast fault clearance times and system separation for a fault close to the generator become of importance.

1.3.3 Reliability

Dependability and security are the measures of reliability. The protection must be dependable and operate in response to system faults within its required area and be secure against incorrect trips from all other conditions, for example, voltage regulation due to load demand changes, high magnitude of through-fault currents, inrush currents, etc. Thus, these two objectives of reliability mutually oppose each other.

Designing more flexibility into system designs, for example, double-ended substations, duplicate feeders, auto-switching, and bus transfer schemes, will increase the complexity and hence reduce the security of the protective systems.

1.4 Equipment and System Protection

Protective relaying can be distinctively classified into two categories:

- Equipment protection
- System protection

Equipment protection narrows down the protection to individual equipment, that is, generator, transformer, bus, cable, transmission line, and motor protection.

The system protection involves protecting a system, with all its components and power equipment, for example, industrial distribution systems, which may consist of a number of substations, main power distribution at medium-voltage, high-voltage utility substation, and plant generators operating in cogeneration mode. The distribution, subtransmission, and transmission systems may entail a number of power system components, substation protection, automation, etc.

These qualifications make it abundantly clear that:

1. A protection engineer must understand and apply all the protective features demanded by equipment and system protection for an efficient and effective protection.
2. Protective relaying is a vast subject. The short-circuit currents in a system cannot be easily reduced, especially in an existing distribution system. Manipulation, coordination, selection, and application of protection devices can impact the results. It is recognized that the descriptive literature can only lay down rules and guidelines with specimen examples, yet it requires a good deal of experience and practice to apply these to real-world situations.

3. Special knowledge and experience are required to apply proper protection philosophies depending upon the power system; for example, utility systems, industrial systems, and commercial systems each have their specific characteristics and requirements. Standards and industry practices have been established over the course of years and coupled with that many innovations are occurring in the protective relaying on account of microprocessor relaying technology, see Chapter 4.
4. System Integrity Protection Schemes (SIPSs) and adaptive protection can be cited as two modern trends. According to the Power System Relaying Committee of the IEEE, SIPSs embrace a wide range of measures such as underfrequency and undervoltage load shedding, adaptive load mitigation, out-of-step tripping, voltage and angular instability, advanced warning schemes, overload and congestion mitigation, system separation, shunt capacitor switching, tap changer control, SVC/STATCOM control, HVDC controls, etc.

1.5 Unit Protection Systems

Unit protection systems are of special significance. In Chapter 6, we will study the time–current coordination systems and their limitations; that starting from downstream, as higher sources of power upstream are coordinated, the fault clearance times go on increasing. For mitigating the damage to the power system components, the short-circuit current must be cleared fast. The higher short-circuit current upstream, coupled with higher fault clearance times, will give rise to increased equipment damage and a possibility of loss of stability.

A separate zone of protection can be established around each system element so that any fault within that zone will cause tripping of the circuit breakers to isolate the fault quickly, without looking at the coordination or protective devices in the rest of the system. If a fault occurs outside the protective zone, the protective system will not operate; that is, it is stable for all faults outside the protective zone. Such zones of protection constitute unit protection systems.

Unit protection systems can be applied to any individual element or sometimes to a group of elements in power distribution systems. That is, separate zones of protection can be created around:

- Generators
- Transformers
- Motors
- Bus bars
- Cables
- Overhead lines

Sometimes more than one equipment to be protected can be covered in a single zone of protection. An example is unit-connected generator, where the generator and the transformer are protected as one unit and there is no generator circuit breaker.

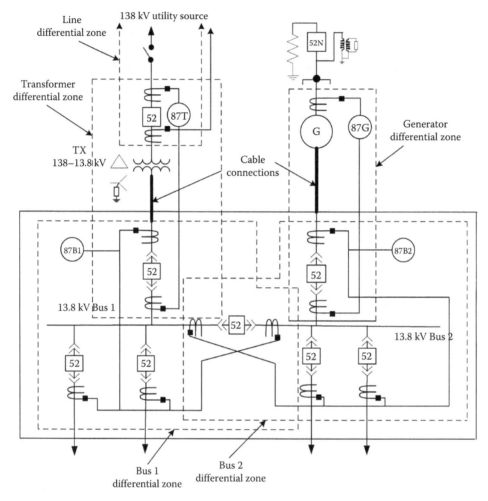

FIGURE 1.1
Overlapping zones of differential protection.

Figure 1.1 shows a single-line diagram of two interconnected 13.8 kV buses for primary distribution of power in an industrial plant, with cogeneration facility. Bus 1 receives utility source power through 40/64 MVA transformer and a 50 MVA 13.8 kV generator is connected to bus 2; it operates in synchronism with the utility source. The plant loads are served from both buses 1 and 2.

This figure shows distinct zones of protection created to protect the utility tie transformer TX and 13.8 kV buses 1 and 2. There are two zones of bus differential protection and the bus section breaker is covered in both the zones. Thus, for a bus fault, only the faulty bus is isolated. The secondary cables from transformer TX and generator G are included in the transformer and generator differential zones. For the utility incoming line to the transformer, a separate zone of protection is provided by the utility company. This figure also shows the location of current transformers and the circuit breakers. The cables from the feeder circuit breakers are not in any differential zone of protection.

1.5.1 Back-Up Protection

In a protective system design, the protection system is backed up in the sense that if the primary protection fails to trip, the second protective device in line must trip. The back-up protection considers failure of the relaying scheme, a breaker, or control supply failure. Relaying for a mesh or ring-connected bus configuration will be different from that for a radial system, even though these systems may interconnect the same size of transformers, feeders, and generators. In a time–current coordinated system, the back-up protection is inherent. If the intended relay or circuit breaker fails to trip, the next upstream breaker will trip with a greater time delay, which will increase the fault damage. As a general practice a unit protection system, for example, differential relaying, is backed up with time overcurrent protection.

The breaker failure schemes are discussed in Chapters 4 and 11.

1.6 Smart Grids

The electrical utility industry is evolving at a pace that can be compared to the advancements that occurred in the beginning of the 20th century. This evolution is being driven by distributed energy resources (DERs) and reduction of greenhouse emissions. It relies heavily on new communications, controls, automation, and power electronics technologies and unprecedented growth in the computing power. Governments around the world have played a vital role by funding landmark initiatives to trigger innovation and providing financial and tax incentives to the interested parties. In the United States, this effort has been described under various names: smart grids, grid of the future, grid modernization, or utility of the future. This implies a grid that can be controlled in real time to allow for providing a reliable, safe, and secure service and empower customers to actively participate and benefit.

In the United States and other countries, utilities are currently engaged in addressing the aforesaid elements, some of which are shown in Figure 1.2. A reader may peruse Chapters 1–3 of Volume 1 to appreciate this figure. It involves not only technical solutions, but also policy matters and planning; see also [11].

1.6.1 Framework for the Smart Grids

Key layers and components of the framework are shown in Figure 1.3 and explained further:

Foundational layer

- Foundational infrastructure and resources
- Organization and processes
- Standards and models
- Business and regulatory

Enabling layer

- Enabling infrastructure
- Incremental intelligence

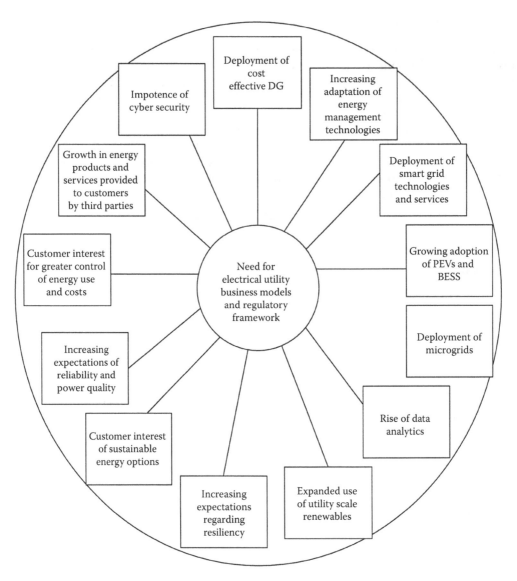

FIGURE 1.2
The drivers for new utility business models and regulatory framework. (Romero et al., *IEEE Power Energy*, 14(5) 29–37, 2016.)

Application layer

- Grid and customer analytics
- Real-time awareness and control
- Customer interaction

Innovation layer

- Pilot and demonstration projects
- Research and development

Modern Protective Relaying: An Overview

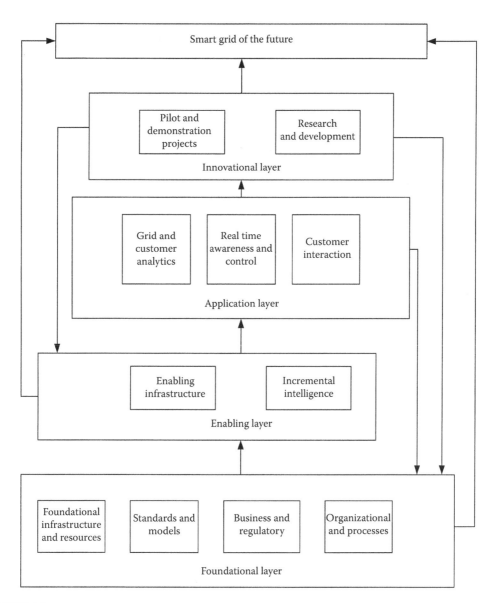

FIGURE 1.3
The layers and components of the conceptual framework of the utility of the future.

1.6.2 Fundamental Layer

1.6.2.1 Foundational Infrastructure and Resources

This is the basic layer of utility structure and the future of the foundational layer. This includes generation facilities, transmission and distribution lines, substations, and distribution transformers. This means infrastructure that enables power production, delivery, and utilization by the end users. This includes basic controls, existing communication facilities, operation's engineering, planning, and maintenance of equipment. This comprises the most essential components of the electrical utility, largely responsible for a reliable service, on the top of which all remaining layers are built.

This importance is well recognized by utilities, which have made significant investments to replace aging infrastructure. FERC Order 1000 has introduced new opportunities and challenges regarding transmission grid planning. The order focuses on three main issues:

1. The removal of federal right of first refusal (ROFR) utility will not have the ROFR to build, own, and operate large transmission projects within their service territories. This introduces competition.
2. All transmission providers must participate in a regional transmission planning process, which also addresses large transmission projects spanning multiple regions.
3. Regions must allocate and collaborate to achieve equitable and economical cost allocation for multiple regions projects.

Additional challenges include issues related to DER. See Chapter 3 of Volume 1 for microgrids, with transmission and distribution systems serving as an integration platform. Note that the PV and wind power generation are located mostly away from the load centers. Their intermittent operation and nature of generation input additional burdens on the already taxed transmission and distribution networks.

1.6.2.2 Organization and Process

The visions of the future require engineering, planning, and empowering the users to interactively act with the service providers giving them options to participate in market transactions.

An example is the formation of new organizations within utilities to processes dictated to planning engineering pertaining to DERs, distribution management systems, and advance metering technologies. The benefits derived from DERs need to be correctly estimated. An example is California's distribution resource planning, which requires state-owned utilities to incorporate distributed generation, energy storage, demand response, electrical vehicles, and energy efficiency measures in their multidisciplinary planning.

1.6.2.3 Standards and Models

This has prompted an increasing need to review, update existing standards, and generate new standards. Examples of ongoing developments in this area include update of IEEE 1547 (standard for interconnecting distributed resources). This leads to activities regarding development of interoperability standards to ensure that equipment and solutions developed by different manufacturers and deployed for automation and control such as voltage and current sensors, smart meters, and inverters which can communicate and exchange data flawlessly. Yet another area of importance is joint modeling and studies of the long- and short-term dynamics.

1.6.2.4 Business and Regulatory

The technology and customer expectations will have an important impact on regulatory and policy aspects of electrical utility. Additional challenges are asset ownership, investment recovery, and related legal matters. The bidirectional data and services exchange between utilities and customers require a flexible, progressive, and adaptive framework. Note that the end users may not only be the consumers, but also producers of electricity (prosumers).

1.6.3 Enabling Layer

1.6.3.1 Enabling Infrastructure

This component comprises intelligent assets, communications, and information systems that have not been used extensively in transmission and distribution systems. There is a plethora of new devices connected to the grid such as microprocessor-based intelligent electronic devices (IEDs), phasor measurement units (PMUs), and inverters and controllers for implementation of substation and feeder automation solutions, such as fault location, service restoration, volt–var optimization, advanced enclosures, and FACTS (see Volume 2).

This infrastructure enables the real-time monitoring, protection, automation, and control of grids. There is ongoing work on standards that govern the implementation of smart inverter functions and var–volt control. The ongoing work on standards to define the functionalities of microgrid controllers, IEEE Project P2030.7, standard for the specifications of microgrid controllers, is currently addressing this topic.

1.6.3.2 Incremental Intelligence

To an extent, the existing systems are intelligent. The purpose of this component is to bridge the gap, which can be implemented in building new systems or enhancing existing systems. An example is informational systems that can share data between utilities seamlessly. RTs (Remote Terminals) include management systems that are responsible for processing and analyzing the data collected by enabling systems. On the load side, this covers systems and platforms required to collect and process data that are needed to manage the relationship and interaction with customers.

1.6.4 Application Layer

1.6.4.1 Grid and Customer Analysis

This component includes data applications to internal utility processes, which can be either utility or grid related. These include planning, load forecasting, reliability, engineering designs and standards, energy consumption, and billing analytics. Based on these data, meaningful information is provided for future improvements.

1.6.4.2 Real-Time Awareness and Control

This component is made of all applications needed for real-time operation, automation, and control. Also it embraces seamless integration of conventional and variable renewable generation sources and market participation.

1.6.4.3 Customer Interaction

These applications are intended to lead to "customer-centric" vision. The data collected from customer assets can be processed to enhance overall customer experience, for example, web and mobile applications, time-of-use rated, and market participation.

1.6.5 Innovation Layer

1.6.5.1 Research and Development

Research and development is necessary to keep pace with the multifunction scenarios described above. These identify the market changes, and reference inputs to update the selected utility of future strategy. There can be widespread changes that have far-reaching impacts and can change the techno-economical environment. An example is the oil production from shale, which makes the United States a world leader in oil production. Since 1999, energy use per square foot has dropped by about 18% in the U.S. commercial business sector.

1.6.5.2 Research and Demonstration Projects

These are important instruments for utilities to evaluate in a controlled environment the potential benefits, costs, and implementations for deploying new technologies and solutions, without undue risks or sharing resources.

1.7 Load Profiles: Var–Volt Control

For panning and pricing, it has been conventional to divide consumer loads into commercial, residential, and industrial categories; see Chapter 1 of Volume 2. Its validity is eroded by proliferation of distributed energy. Figure 1.4 shows a consumer load with and without solar panels. Thus, in terms of load profiles the conventional concepts are changing. Rather than classifying into commercial and industrial, the load is simply watts and vars requiring a certain power quality, irrespective of its ultimate consumption.

Figure 1.5 shows the voltage regulation in a new environment. The converter and technology vendors are regulating voltage, minimizing the need for mechanical regulation and providing distributed support and distributive intelligence to the grid.

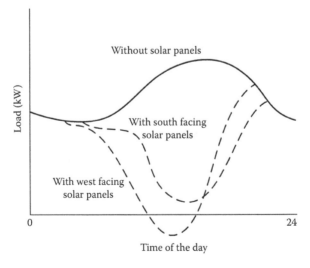

FIGURE 1.4
A consumer load profile without and with solar generation.

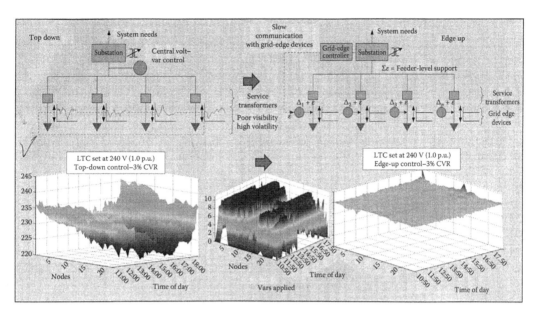

FIGURE 1.5
A modern concept of voltage–var control, smart grids.

1.8 Some Modern Technologies Leading to Smart Grids

In the future, the power generation, transmission, and distribution will undergo profound changes, need based—environmental compatibility, reliability, improved operational efficiencies, integration of renewable energy technologies such as wind and solar power, and distributed generation. The modern grid systems are being controlled and will be controlled and operated *so that the dynamic state of the grid is known in terms of:*

1. Rotor angle stability and voltage stability.
2. Increase/decrease in transmission capability that can take place in real time over transmission systems.
3. Control and regulation of power flow to maintain grid parameters.
4. Remedial action schemes (RAS) and SIPSs.
5. Identifying what remedial measures should be taken to avoid an extreme contingency, i.e., cascading and blackouts.
6. How these corrective actions can be physically implemented?.

The technologies driving the self-healing smart grid are as follows: wide-area measurement systems (WAMSs), SIPSs, PMUs, energy management systems, FACTS, communication systems, and dynamic contingency analysis (DCA); all are somewhat related.

It is amply clear that the stability of a system is not a fixed identity and varies with the operating and switching conditions. Some, not so common, contingencies in a system can cascade and bring about a shutdown of a vital section. Historically, the Great North East Blackout of November 9–10, 1965, and more recently the 2003 East Coast Blackout can be mentioned.

1.8.1 WAMSs and PMUs

Wide-area network measurements have been around for the last 60 years, and have been used in economic dispatch, generation control, and real-time measurements of power flows. Supervisory control and data acquisition systems (SCADAs) are of late 1960s origin and provided real-time state estimates of power systems. The measurement comprised a data window of several seconds, without regard to the instant at which the precise measurement was made, while the system may have drifted meanwhile from the instant of measurement. Developments in microprocessor-based relays got an impetus in the 1970s, based upon the requirements that symmetrical components of currents and voltages at relay locations be estimated from synchronized sampled data on a system-wide basis. In the 1980s, global positioning systems (GPSs) began to be deployed and the prospect of synchronizing sampled data on a system-wide basis became a reality.

This requirement led to the concept of PMU. The concept is simple; a sinusoidal waveform can be represented by a magnitude and a phase angle (Figure 1.6). The magnitude is the peak or rms value of the sinusoid and the phase angle is given by the frequency and the time reference. The synchrophasor representation X of a signal $x(t)$ is the complex value given by

$$\begin{aligned} X &= X_r + jX_i \\ &= \left(X_m/\sqrt{2}\right)e^{j\phi} \\ &= \frac{X_m}{\sqrt{2}}(\cos\phi + j\sin\phi) \end{aligned} \quad (1.1)$$

where

$X_m/\sqrt{2}$ is the rms value of the signal $x(t)$ and φ is the phase angle relative to cosine function at normal system frequency synchronized to universal time coordinated (UTC). This angle is 0° when maximum of $x(t)$ occurs at the UTC second roll over (one pulse per second time signal) and −90° when the positive zero crossing occurs at the UTC second roll over. Synchrophasors are, thus, phasor values that represent power system sinusoidal waveforms referenced to nominal system frequency and UTC. The phase angle is uniquely determined by the time of measurement, waveform, and system frequency. If a sinusoid is observed at intervals $(0, T_0, 2T_0, \ldots, nT_0, \ldots)$ leading to phasor representations $(X_0, X_1, X_2, \ldots)$ and observation time interval T_0 is an integer multiple of sinusoid $T = 1/f$, then a constant phasor is obtained at each observation. If observation time T_0 is not an integer multiple of

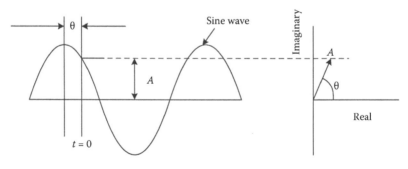

FIGURE 1.6
Concept of phasor representation of a sinusoidal waveform.

T, the observed phasor has a constant magnitude, but angles of the phasors ($X_0, X_1, X_2,...$) will change uniformly at a rate $2\pi(f - f_0)T_0$, where $f_0 = 1/T_0$.

System frequencies are not rock steady and vary. An interconnected system runs at the same frequency and all phase angles rotate together, one way or the other. Because of this rotation, the phase angle measurements should be made exactly at the same rate. As an example, state estimators run at intervals ranging from a few seconds to tens of minutes, and a phasor system running at 6+ samples cannot directly feed into the slower system. A solution will be to use synchronized samples drawn (periodically) from the full data set. This led to the development of IEEE Standard C37.118 [12] revised in 2005. The basic measurement requirements, including angle–time relationship, are detailed in this standard. The accuracy of phasor estimate is compared with a mathematically predicted value using a total vector error. This can be defined as root square difference of the values and compliance with the standard [5] requires a difference within 1% under various conditions:

$$\text{TVE} = \sqrt{\frac{(X_r(n) - X_r)^2 + (X_i(n) - X_i)^2}{X_r^2 + X_i^2}} \quad (1.2)$$

where $X_r(n)$ and $X_i(n)$ are the measured values and X_r and X_i are the theoretical values of the input signal at the instant of time of measurement.

Obtaining a phasor equivalent of an arbitrary sinusoidal signal requires a sample of waveform taken at appropriate frequency—the quality of phase estimate has to be ensured.

The discrete Fourier transform (DFT) is the most commonly used method of phase estimation. This technique uses the standard Fourier estimate applied over one or more cycles at nominal system frequency. At a sufficient sample rate and at accurate synchronization with UTC, it produces an accurate and usable phasor value for most system conditions. Problems with DFT response like roll off can occur with varying frequency and must be corrected, for example, by centering the measurement window. GPS is universally used for the UTC time reference. Other PMUs may rely upon time signal, such as IRIG-B from an external GPS receiver.

The PMU functions are built into MMPRs, (see Chapter 4), and digital fault recorders. These may have variable capabilities. Based upon analog inputs, three-phase quantities as well as positive-, negative-, and zero-sequence phasors can be outputted. The sensing elements, that is, the accuracy of potential and current transformers, becomes a question mark and so far ANSI/IEEE relaying class accuracies are found adequate; further work is being done by the North American Synchrophasor Initiative project.

Figure 1.7 shows a typical hierarchical system of the PMUs feeding into phasor data concentrators (PDCs) at a control center. PDCs are produced which interface with other products, such as monitor/control platforms and a data historian. PDCs connect to multiple PMUs, and receive, parse, and sort incoming data. Due to sheer amount of data, it is an overwhelming computer processing task. IEEE standard [12] establishes PMU data protocols. Over the course of years, CPU processing power has increased. The number of incoming PDC devices that may be deployed is limited by CPU processing power.

Originally WAMSs were limited to single utilities. The interutility data exchange enables wide area visibility. When interfacing with SCADA, the data must be reduced to match SCADA data rates and interface with protocols used by SCADA. (Most estimators draw data from SCADA.) The widely used IEEE COMTRADE (*Com*mon format for *Tra*nsient *D*ata *E*xchange), IEEE Standard C37.111 [13], developed for time sequence data supports

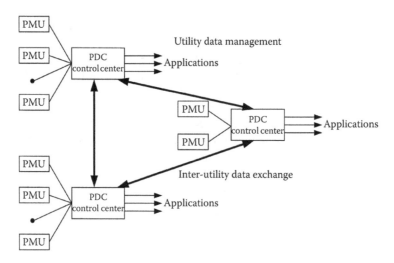

FIGURE 1.7
Typical star hierarchical system of PMUs feeding into PDCs that feed local and remote applications. These can extend to share data between utilities.

binary and floating point formats. Most control center applications use a data historian for analysis and trending. And generally, these will accept data at full rate.

The phasor estimation is primarily developed for steady-state signals, and the next step will be to apply it to system dynamics. Most power system dynamics are slower compared with the speed of phasor systems. An IEEE working group is formed to revise C37.118 to include dynamic performance requirements. DCA will make stability assessment and issue real-time control signals. The stability functional requirements will dictate the system performance.

1.8.2 System Integrity Protection Schemes

SIPSs are automated systems that protect the grid against system contingencies and minimize the potential for wide outages. Without SIPS, it may not be possible to provide for many contingencies, address transmission paths, alternate routes, corrective measures, and prior warnings. A SIPS design is based upon system studies of predefined contingencies for a variety of conditions. According to the IEEE Power System Relaying Committee (PSRC), following is the list of SIPS measures:

- Generator and load rejection
- Underfrequency and undervoltage load shedding
- Adaptive load mitigation
- Out of step tripping
- Voltage instability and angular instability advance warning schemes
- Overload and congestion mitigation
- System separation
- Shunt capacitor switching
- Tap changer control
- SVC/STATCOM control

- Turbine valve control
- HVDC (High Voltage DC) controls
- Power system stabilizer control
- Discrete excitation
- Dynamic breaking
- Generator runback
- Bypassing series capacitor
- Black-start or gas turbine start-up
- AGC (Automatic Generation Control) actions
- Bus bar splitting

We have discussed the basic concepts of many of these items in this book. Applied to complex grid systems, SIPS is the last line of defense to protect the integrity of the power system, and propagation of disturbances for severe system emergencies caused by unplanned operating conditions.

1.8.3 Adaptive Protection

The objective of adaptive relaying is to adjust relay performance or settings to changing system conditions. It can be defined as follows: "A protection philosophy that permits and seeks to make adjustments automatically in various protection functions in order to make them more attuned to prevailing system conditions." This is being achieved by phenomenal advancements in the microprocessor-based technology applied to protective relaying. As an example, current differential schemes with high-speed communications can be applied to transmission lines.

As stated before, dependability and security are measures of reliability, which mutually oppose. To be dependable the protection must always trip, even if there is a nuisance trip. Consider that a system is robust and a dependable protection is applied to it—if the system changes, whether due to planned or unplanned outages, the strength of the system to withstand the same amount of trips becomes questionable. With WAMSs and digital devices, it is possible to reorganize to reorientate the relay performance from dependable to secure. Figure 1.8 shows a scheme, where three protective schemes can trip independently, without supervision to security, while tripping decisions are connected so that out of three at least two schemes should operate correctly. After the 2003 East Coast Blackout, NERC recommended to remove all unnecessary zone 3 distance tripping to avoid "over-tripping" by these elements.

Other smart grid issues can be itemized as follows:

- Requirements for renewable portfolio standards, limits on greenhouse gases, and demand response.
- Advanced metering structures at consumer loads and development of smart meters.
- Integration of solar, wind, nuclear, and geothermal facilities that pose their own challenges. For example, the large-scale solar plants or wind generation may be located in areas distant from existing transmission facilities. New protection and control strategies, interconnection standards, for example, low-voltage ride through capabilities, forecasting, and scheduling are required.

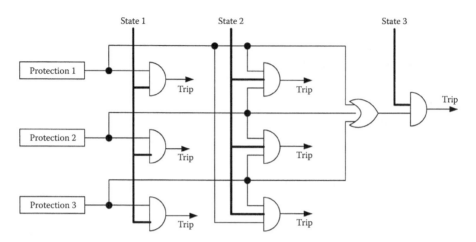

FIGURE 1.8
Schematic of adaptive protection and redundancy.

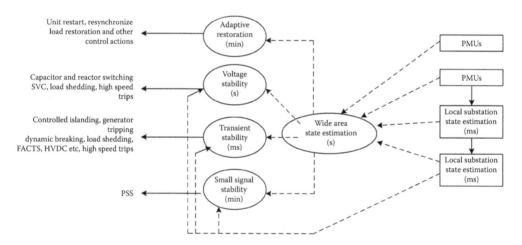

FIGURE 1.9
Wide-area control framework.

- Circuit congestion and managing distribution system overloads.
- Role of information and automation technologies.

Figure 1.9 shows the wide-area control framework. Although it may not be entirely possible to avoid multiple contingencies blackouts, the probability, size, and impact of widespread outages can be reduced.

1.9 Cyber Security

The grids of the future monitored and controlled by devices are cyber vulnerable. Consider the hacking of a personal ID that can result in lots of expense and embarrassment to the

TABLE 1.1

A Comparison of Traditional IT and ICS Cyber Security Priorities

Attribute	Traditional IT Focus	CS Focus
Maintain confidentiality	High	Low
Message integrity	Low–medium	Very high
System availability	Low–medium	Very high
Authentication	Medium–high	High
Nonrepudiation	High	Low–medium
Safety	Low	Very high
Time criticality	Delays tolerated	Critical
System downtime	Tolerated	Not acceptable
Security skills/awareness	Usually good	Usually poor
System life cycle	3–5 years	15–25 years
Interoperability	Not critical	Critical
Computing resources	Unlimited	Very limited
Standards	ISO27000	ISA/IEC 62443

CS: Cyber Security.

user. Compare this with data breach with the potential of physical destruction of the assets operated by a utility. A cyberattack on essential services such as the financial system or utilities of a country can bring it down to its knees.

It is necessary to expand the conception of cyber security beyond Internet-connected devices and servers. The National Institute of Standards and Technology published "Federal Information Processing Standard Publication 200," defining a cyber incident as an occurrence that actually or potentially jeopardizes the confidentiality, integrity, or availability of an information system or the information that the system processes, stores, or transmits or that constitutes a violation or imminent threat of violation of security policies, security procedures, and acceptable use policies; for more details refer [14, 15].

The electric utility has become more automated and connected to the Internet through smart grid and other industrial Internet initiatives, making power production, transmission, and distribution more operationally dynamic.

Traditional IT security breaches have targeted data and personal information. The CS intrusions target physical processes and a comparison between IT and CS cyber security approaches is presented in Table 1.1.

The control systems and the smart devices connected to the grid include IEDs used for controlling circuit breakers, transformers, capacitor banks and power system equipment, PLCs (Power Line Carrier), voltage/frequency/power control monitors, sensors, valves, motors, or device status indicators. These devices are critical to the safe operation of power generation and delivery systems. A malicious attack perpetuated internally by a malicious employee or externally by a hostile party has the potential for large-scale disruption, see Table 1.2.

A challenge facing the utilities is the scarcity of resources with both IT and CS expertise. Few CS experts are also IT experts and vice versa.

In a recent article of Associated Press published in March 2016, Duke Energy CEO said: "If I were to share with you the number of attacks that came into Duke Network everyday, you will be astounded. It is from nation-states that are trying to penetrate systems." In fact, there has been 800 total CS cyber incidents globally that have been documented since 1980.

Table 1.3 is issued by the DHS (Department of Homeland Security).

TABLE 1.2

A Comparison of Cyber-Related Events and What Future Events May Look Like

	Northeast Outage, 2003	Arizona Public Service Outage, 2007	Florida Power and Light Outage, 2008	Ukraine Cyber Attack, 2015	The Next Cyber Caused Event?
Cyber	Yes	Yes	Yes	Yes	Yes
Intentional or unintentional	Unintentional	Unintentional	Unintentional	Intentional	Could be either
Load lost	61,800 MW	400 MW	4,300 MW	Unconfirmed[a]	—

[a] Reports range, with 225,000 customers experiencing outage.

TABLE 1.3

Aurora Vulnerability (Department of Homeland Security)

Elements necessary for an attack	Programmable digital relay or other devices that control a breaker
	High speed breakers
	Access (front panel, modem, Internet, wireless, or SCADA)
Necessary knowledge	Power engineering skills (attack planning and device setting skills)
	Hacking skills (exploit the relay and conduct an attack)
Time required to conduct an attack after gaining access	Less than 1 min
	No additional software introduced
	Uses the Internet settings of the embedded relay software

1.10 NERC and CIP Requirements

Current NERC CIP standards are shown in Figure 1.10. While the utilities are working for the compliance, some important devices are not included under standards provisions. For instance, SCADA and substations for subtransmission and distribution facilities are not specified under the standards. However, compliance with the standards is a good starting point.

The NERC has developed standard PRC-004 titled "Analysis and Mitigation of Transmission and Generation. Protection System Misoperations." Related standards are:

NERC Standard PRC-016—Special protection system misoperations

NERC Standard PRC-022—Undervoltage load shedding program performance

NERC Standard PRC-009—Underfrequency load shedding performance following an underfrequency event

The NERC works with eight regional entities in carrying out its mission. A protection engineer needs to be aware of these standards.

The NERC Standard PRC-004 lays guidelines that misoperations of protective systems that relate to reliability of bulk electric system (BES) are analyzed and mitigated. The BES is identified as power system facilities that are part of electric network and operated at

Modern Protective Relaying: An Overview

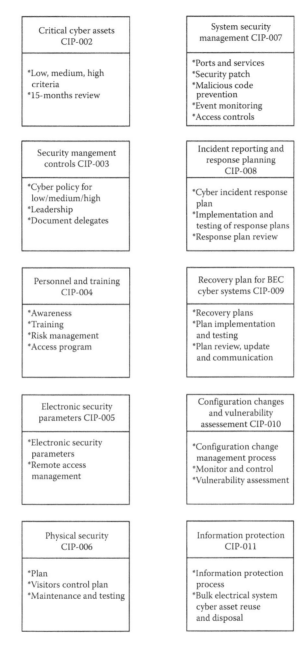

FIGURE 1.10
Current NERC CIP standards as of March 2016.

100 kV or higher voltages. The standard applies to transmission owners, distribution providers, and owners of generating facilities. According to the NERC, a protection system includes the following:

- Protective relays
- Communication systems

- Voltage and current sensing devices
- Station DC battery systems
- Control circuits associated with protective functions

The protective system misoperations are described as follows:

- Any failure of protective system element to operate as intended.
- Any operation for a fault not within a zone of protection, that is, lack of coordination or nuisance trips.
- An unintended operation without a fault or any abnormal condition, unrelated to onsite maintenance and testing activity.

References

1. JC Das. *Power System Handbook, Short-Circuit Calculations in AC and DC Systems*, Vol. 1. CRC Press, Boca Raton, FL, 2018.
2. JC Das. *Power System Handbook, Load Flow and Optimal Load Flow*, Vol. 2. CRC Press, Boca Raton, FL, 2018.
3. JC Das. *Power System Handbook, Generation Effects and Control of Harmonics*, Vol. 3. CRC Press, Boca Raton, FL, 2018.
4. CR Mason. *The Art and Science of Protective Relaying*. John Wiley & Sons, New York, 1956.
5. AR Washington. *Protective Relays. Their Theory and Practice*, Vol. 1. John Wiley & Sons, New York, 1962. Vol. 2, Chapman and Hall, London, 1974.
6. WA Elmore (Ed.). *Applied Protective Relaying*. Westinghouse Electric Corporation, Coral Springs, FL, 1979.
7. SH Horowitz. *Protective Relaying for Power Systems*. IEEE Press, Piscataway, NJ, 1980.
8. JL Blackburn, TJ Domin. *Protective Relaying: Principles and Applications*. CRC Press, Boca Raton, FL, 2014.
9. PM Anderson. *Power System Protection*. Wiley-IEEE Press, New York, 2015.
10. A Wright, C Christopoulos. *Electrical Power System Protection*. Chapman and Hall, London, 1993.
11. J Romero, A Khodaei, R Masiello. The utility and grid of the future. *IEEE Power and Energy*, 14(5), 29–37, 2016.
12. IEEE Std. C37.118. IEEE Standard for Synchrophasors for Power Systems, 2005.
13. IEEE Std. C37.111. IEEE Standard for Common Format for Transient Data Exchange (COMTRADE) for Power Systems.
14. J Weiss. *Protecting Industrial Control Systems from Electronic Threats*. Momentum Press, New York, 2010.
15. E Smith, S Corzine, D Racey, P Dunne, C Hassett, J Weiss. Going beyond cybersecurity compliance. *IEEE Power and Energy*, 14(5), 48–56, 2016.

2
Protective Relays

2.1 Classification of Relay Types

The protective relays can be regulating, controlling, alarming, restoring, synchronizing, load shedding, out of step, rate of pressure rise, inadvertent energization, temperature, vibration sensing, relays specific for ground fault protection, etc. The specific applications and characteristics of these relays are covered throughout this volume. The purpose of this chapter is to provide an overview of the major relay types and their characteristics. Throughout this chapter, references are made to other chapters in this volume, which contain the settings, characteristics, and applications of relay types covered in this chapter, *and not covered in this chapter.*

The protective relays have also been classified based upon the following.

2.1.1 Input

- Current
- Voltage
- Power
- Temperature
- Vibration
- Frequency

2.1.2 Operating Principle

- Percentage differential
- Product type
- Pressure sensing
- Thermal

2.1.3 Performance

- Distance
- Mho
- Ground or phase
- Undervoltage
- Directional comparison

2.1.4 Construction

With respect to construction, the protective relays are loosely classified as follows:

Electromechanical: These are single function relays and no longer in use. The construction of an overcurrent time delay and instantaneous relay is described next.

Static: It is a loose term, and does not specifically point out to the inner circuit used in the relay for the desired functionality. Even electromechanical relays may use some static components, for example, varistors in differential relays. However, a static relay may have all solid-state components.

Digital: This signifies that digital technology has been used in the relay. There may be some solid-state components such as integrated chips, transistors, or other devices.

Numerical: This is a term which is more popular in Europe. Numerical relays are, generally, microprocessor based. These can be applied for multifunctions. For example, a numerical relay may be equally suitable for the differential protection of a transformer, generator, or motor.

MMPRs: Microprocessor-based multifunction relays (MMPRs) are the most popular relays today and practically used for all power system relaying applications. The capabilities and functionality of MMPRs are discussed in respective sections in this book. Chapter 4 is solely devoted to the functionality and capability of a feeder MMPR.

2.2 Electromechanical Relays

As stated before, the electromechanical relays are no longer in use. Yet in most texts, these are described first—it may be easier to understand the operating principles better than the logic circuit diagrams of an equivalent MMPR.

The construction of an induction pattern relay, out of its draw-out case, is shown in Figure 2.1. The pulling out of the relay element from the case short circuits the CT secondary winding circuit and prevents open circuit of the CT secondary. The disk is mounted on a rotating shaft, restrained by a spring. The moving contact is fastened to the shaft, and the operating torque is produced by an electromagnet having a main and lag coil, which produce out-of-phase magnetic flux, see Figure 2.2. A damping magnet provides the restraint after the disk starts to move, resulting in the desired time–current characteristics. There are discrete taps that determine the current pickup setting, and the time dial setting determines the initial position of the moving contact. Different current characteristics are obtained by modification of the electromagnetic design. Figure 2.3 shows the comparison of typical ANSI curve shapes. An instantaneous unit is mounted in the same case. The relays are of single-phase type; that is, three units are required for three-phase and an additional single-phase unit for the ground fault protection. There are two adjustments: the pickup current tap and the time dial. The pickup is determined by several taps on the coil that are furnished in different current ranges. The time dial determines the initial position of the contacts when the current is below pickup setting. It is calibrated, say, from 1 to 10. An auxiliary seal-in relay, is incorporated as a target indicator, and also to relieve the current carrying duty of the moving contacts.

It is worthwhile to note that the taps are available only in certain ranges. The time–current characteristics of the relay are fixed; that is, it can be extremely inverse, inverse, very inverse, etc. To apply a different characteristic, the complete relay must be replaced. Definite time characteristics are not available. Thus, to cover a setting range, different relay

Protective Relays

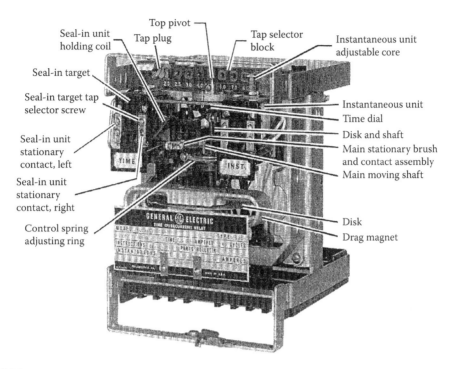

FIGURE 2.1
Construction of an electromechanical phase overcurrent time delay and instantaneous relay. (ANSI/IEEE Std. 242. IEEE Recommended Practice for Protection and Coordination of Industrial and Commercial Power Systems, 1986.)

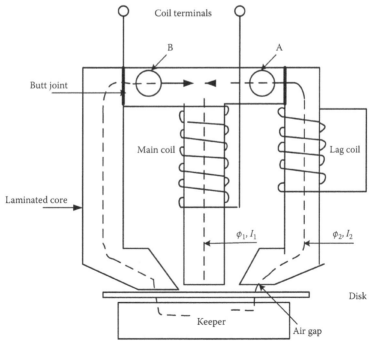

FIGURE 2.2
Operating principle of an electromechanical induction type overcurrent relay.

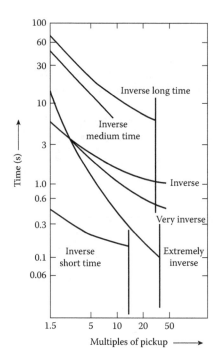

FIGURE 2.3
ANSI inverse time overcurrent current shapes.

TABLE 2.1

Typical Tap Settings of Electromechanical Overcurrent Relays

Tap Range	Tap Settings
0.5–2.5 or 0.5–2.0	0.5, 0.6, 0.8, 1.0, 1.2, 1.5, 2.0, 2.5
0.5–4	0.5, 0.6, 0.7, 0.8, 1.0, 1.2, 1.5, 2.0, 2.5, 3.0, 4.0
1.5–6 (or 2–6)	1.5, 2.2.5, 3, 3.5, 4, 5, 6
4–16 (or 4–12)	4, 5, 6, 7, 8, 10, 12, 16
1–12	1.0, 1.2, 1.5, 2.0, 2.5, 3.0, 3.5, 4, 5, 6, 7, 8, 10, 12

types need to be selected. Table 2.1 shows some typical time ranges. This table does not show the instantaneous attachment ranges which also vary. The lower setting ranges are applicable to ground fault protection.

2.3 Overcurrent Relays

2.3.1 ANSI Curves

The overcurrent protection described in Section 2.2 is most widely used in the protective relaying. Different inverse current shapes are shown in Figure 2.3. Mathematically, ANSI curve shapes are given by the following equation:

TABLE 2.2

Constants for Overcurrent Relay ANSI/IEEE Curves

ANSI Curve	Constants				
	A	B	C	D	E
Extremely inverse	0.0399	0.2294	0.5000	3.0094	0.7222
Very inverse	0.0615	0.7989	0.3400	−0.2840	4.0505
Normally inverse	0.0274	2.2614	0.3000	−4.1899	9.1272
Moderately inverse	0.1735	0.6791	0.8000	−0.0800	0.1271

$$T = M \times \left[A + \frac{B}{(1.03-C)} + \frac{D}{(1.03-C)^2} + \frac{E}{(1.03-C)^3} \right] \quad 1 \leq \frac{I}{I_{PKP}} < 1.03$$

$$= M \times \left[A + \frac{B}{(1/I_{PKP}-C)} + \frac{D}{(1/I_{PKP}-C)^2} + \frac{E}{(1/I_{PKP}-C)^3} \right] \quad 1.03 \leq \frac{I}{I_{PKP}} < 20 \quad (2.1)$$

$$= M \times \left[A + \frac{B}{(20-C)} + \frac{D}{(20-C)^2} + \frac{E}{(20-C)^3} \right] \quad 20.0 \leq \frac{I}{I_{PKP}}$$

where
  T = operate time (s)
  M = multiplier set point, commonly called time dial
  I = input current
  I_{PKP} = pickup current set point, commonly called "pickup"
  A, B, C, and D are constants given in Table 2.2.

A typical setting range of an electromechanical relay for phase fault protection of a certain characteristic can be as follows:

- Time overcurrent tap settings: 1–12 A; tap adjustments 1, 1.5, 2, 3, 4, 5, 6, 7, 8, 10, 12 A
- Overcurrent instantaneous settings: 40–160 A, in terms of CT secondary current
- Time dial: 0.5–10
- Indicator contactor switch settings: 0.2 or 2 A

The indicator contactor switch indicates operation of the relay and is hand reset. Two indicator contactor switches, one for time delay and the other for instantaneous function, are provided, see Figure 2.1. These indicator contactor switches have target coils in series with the main trip circuit and when these drop out for indication, the coils and springs are bypassed. The time–current characteristic is preset and cannot be changed. Thus, a protection engineer has to decide in advance what type of characteristics will be appropriate in a certain application. More often, it is rather difficult to decide the time–current characteristics unless a rigorous coordination study is undertaken in the design stage. This is one major limitation. This simple and rugged construction held its field for the past 50 years, but it is not recommended for modern projects.

TABLE 2.3

Constants for IEC Overcurrent Relay Curves

IEC (BS) Curve Shape	Constants	
	K	E
IEC curve A	0.140	0.020
IEC curve B	13.50	1.000
IEC curve C	80.00	2.000
IEC short inverse	0.050	0.040

2.3.2 IEC Curves

The equation for IEC curves is

$$T = M \times \left[\frac{K}{(1.03)^E - 1} \right] \quad 1 \leq \frac{I}{I_{PKP}} < 1.03$$

$$= M \times \left[A + \frac{K}{(1/I_{PKP})^E - 1} \right] \quad 1.03 \leq \frac{I}{I_{PKP}} < 20 \quad (2.2)$$

$$= M \times \left[A + \frac{K}{(20)^E} \right] \quad 20 \leq \frac{I}{I_{PKP}}$$

The relevant constants in Equation 2.2 are given in Table 2.3. In addition to the curve shapes in Tables 2.2 and 2.3, additional curve shapes are available. Figure 2.4 shows the relative time–current curves of seven inverse time characteristic relays. All these curves are plotted for a pickup current of 1200 A and a time dial setting of 3.

Figure 2.5a–e shows the characteristics of U.S. and IEC overcurrent protection relays. (The ANSI/IEEE and IEC characteristics each are labeled (a) through (e).) Considering IEC and U.S. curves together, there are 10 distinct curve shapes. With overlapping setting zones for phase and ground faults, this amounts to at least 100 different relay types. All the characteristics with extended setting ranges and much closer pickup settings in small increments for phase and ground fault protection are available in one single MMPR. Furthermore, as the electromechanical relays are of single-phase type, four relays are required for phase faults and one relay for the ground fault. *All this functionality is available in one single MMPR.*

A functional diagram of a three-phase and neutral static relay is shown in Figure 2.6. Any of five ANSI characteristics can be selected.

2.4 Differential Relays

A differential protection operates on the principle that the current entering a zone of protection is equal to the current leaving that zone of protection. This current balance does not hold if there is a fault within the protected zone. The protection should operate fast even for low magnitudes of currents for a fault within the protected zone and should be stable for a large magnitude of through-fault currents outside the protected zone. This

Protective Relays

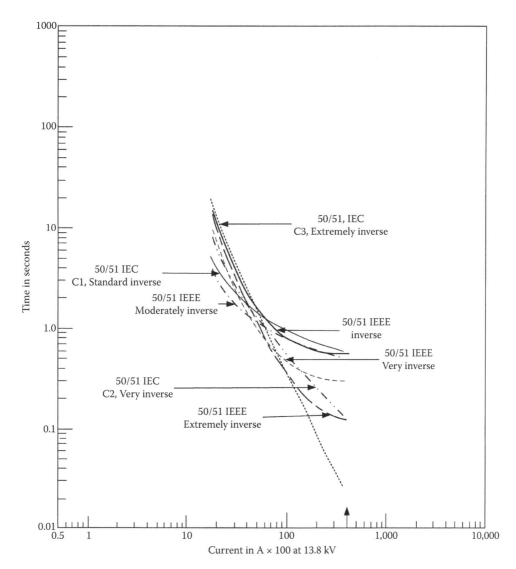

FIGURE 2.4
Time overcurrent characteristics of various overcurrent relay types drawn for the same pickup of 1200 A and a time dial of 3.

chapter describes the basic concepts. The differential protection for motors, generators, bus bars, transmission lines, etc., is discussed in the respective chapters.

2.4.1 Overcurrent Differential Protection

Figure 2.7a shows the basic principle of differential relaying and stability for an external fault of 20 kA, with CT ratios of 2000/5. Only one phase is shown and the differential protection is provided by a simple overcurrent relay. Note the polarity of the current transformers shown in this figure. (The polarities of CTs are discussed in Chapter 3.) A current of 50 A circulates through the CT secondary leads and none flows in the relay. This is an ideal situation.

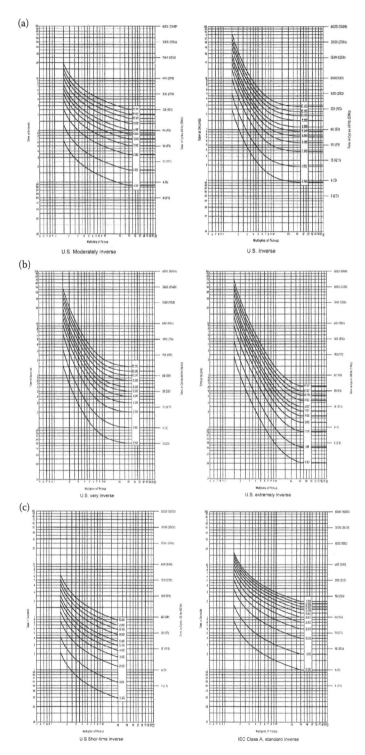

FIGURE 2.5
(a-e) Time–current characteristics of overcurrent relays (ANSI and IEC).

(Continued)

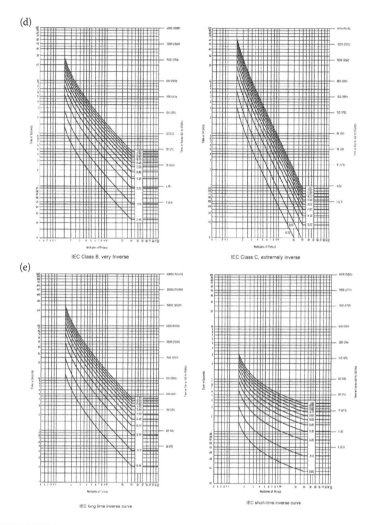

FIGURE 2.5 (CONTINUED)
(a-e) Time–current characteristics of overcurrent relays (ANSI and IEC).

Figure 2.7b shows that for an internal fault of 500 A, a secondary current of 1.25 A should flow in the overcurrent relay and if this relay is set to pickup at this level, it will trip for an internal fault.

This situation is not realized in practice because of the following:

- The CT leads may be of different lengths, imposing different burdens on the CTs.
- Even if the CTs are exactly of the same ratio, there can be variations and CTs do not perform exactly according to their ratios. This difference is caused by variations in manufacture, and a difference in secondary loading and magnetic history. Residual magnetic flux of varying magnitude may be trapped in the CT core, see Chapter 3.
- Though CTs may be selected to avoid saturation, yet under high magnitudes of through-fault currents these may have different saturation characteristics.
- The CT accuracies and errors are discussed in Chapter 3.

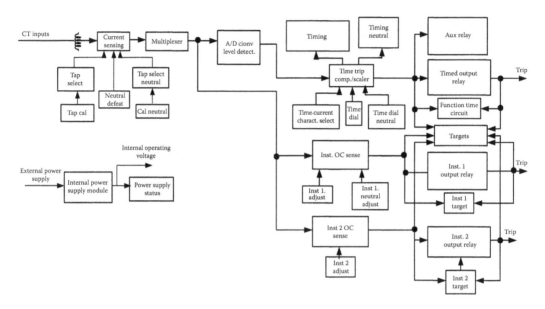

FIGURE 2.6
Functional diagram of a static overcurrent relay.

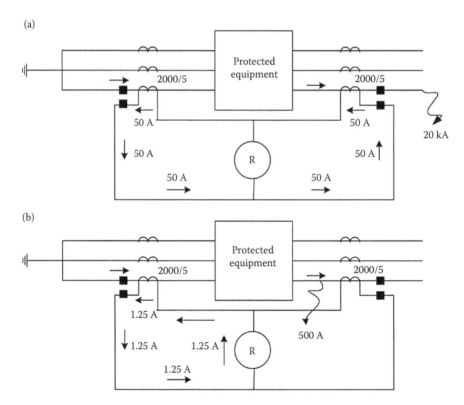

FIGURE 2.7
Concepts of differential protection: (a) a fault external to the protected equipment zone and (b) a fault within the differential protected zone.

Protective Relays

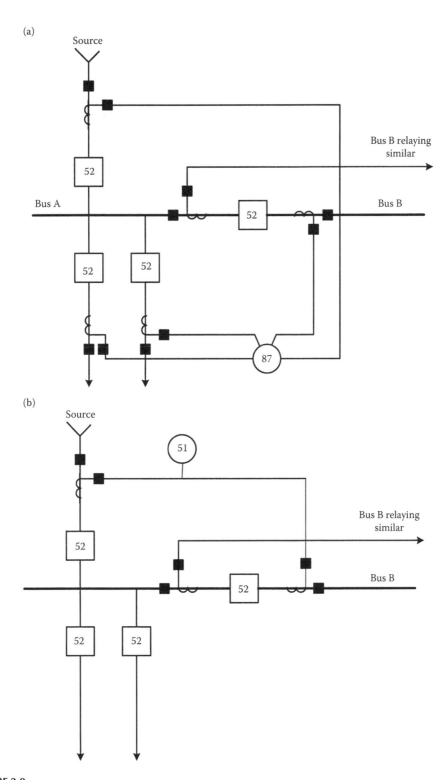

FIGURE 2.8
(a) A differential protection using a simple overcurrent relay and (b) partial differential protection.

Yet, simple overcurrent relays have been applied for differential schemes. Figure 2.8a shows such a scheme for two buses with a bus section switch. This will have poor sensitivity; the overcurrent element must be set high to override the CT spill currents. Also it will be more susceptible to nuisance operation due to CT saturation [1].

2.4.2 Partial Differential Schemes

Figure 2.8b shows a further simplification of the differential protection called a *partial differential scheme*. This is the slowest and least sensitive of any differential scheme, though sometimes used. There are no differential CTs on the feeder circuit breakers. The relay must be set high enough to coordinate with feeder relaying.

2.4.3 Overlapping the Zones of Protection

It is noteworthy that the overlapping of the differential zones of protection is achieved by proper location of the current transformers, and constructing a zone of protection so that one zone overlaps the other and no area is left unprotected. Consider that a differential protection is provided only for the 13.8 kV bus in Figure 2.9. It is a metal-clad switchgear, and the CTs are located on the circuit breaker spouts. This is further illustrated in Figure 2.10, which is a cross section through two-high metal-clad switchgear. Generally, four CTs can be located in one draw-out circuit breaker: two on the source side and two on the load side. Figure 2.9 shows that a bus differential zone is created by a CT located on the source side of the main incoming circuit breaker and the CTs located on the load side of the feeder circuit breakers. In this diagram, the area shown with thick lines is not in the differential zone of protection. The same area enclosed within dotted lines is also shown in Figure 2.10. A fault at location F2 in Figure 2.9 will be cleared by the overcurrent relays 50/51, which are connected to CTs on the source side of the circuit breakers. A fault at location F1 in the incoming cable compartment of the main circuit breaker must be cleared by an upstream protective device. Therefore, the cable terminations and cable compartments remain outside the zone of differential protection. This concept is important because the differential zone of protection is dictated by the location of the CTs. A single-bus differential zone in a metal-clad switchgear leaves the areas shown unprotected.

Figure 2.11a shows the normal location of CTs in a metal-clad switchgear. In some earlier vintage of switchgear, the CTs could not be provided on the source side, and were located only on the load side of the circuit breaker. Yet, overlapping zones of protection can be achieved by proper CT connections as shown in Figure 2.11b. The connections of CTs shown in Figure 2.11c will be inappropriate, the zones do not overlap, and the areas shown dotted remain unprotected.

2.4.4 Percent Differential Relays

These relays have the advantage of high stability for external faults when errors are more likely to produce erroneous differential currents, and good sensitivity to faults in the protected zone is required. These are commonly applied to generators, transformers, and motor differential protection. There are three types of restraint characteristics:

- Fixed percentage
- Variable percentage
- Harmonic-restraint percentage

Protective Relays 35

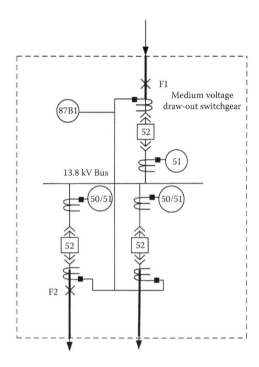

FIGURE 2.9
Differential protection for a bus, showing areas outside the protected zone.

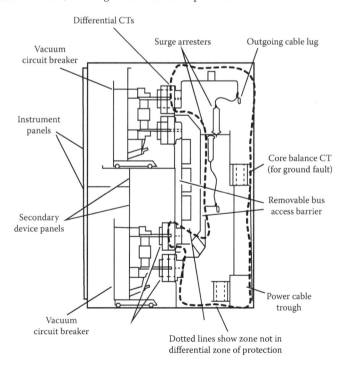

FIGURE 2.10
Cross section through metal-clad switchgear showing the location of CTs and the area in dotted lines which remains unprotected outside the differential zone.

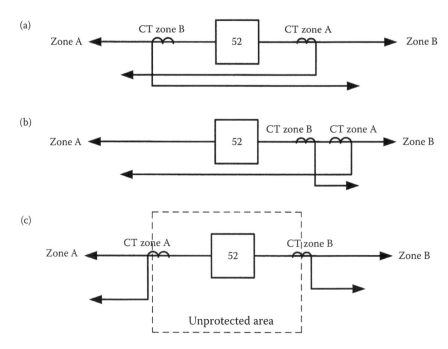

FIGURE 2.11
Creating overlapping zones of protection: (a) CTs located on the source side and load side in the switchgear, (b) CTs located on the load side only, and (c) unacceptable configuration leaving the area between the CTs unprotected.

Variable percentage relays are used for generator protection and are less likely to have nuisance tripping. Harmonic restraint blocks tripping on transformer inrush currents.

The principle of an electromechanical percentage differential relay can be explained with reference to Figure 2.12a. The relay uses an induction principle. The current in the restraining coils, R_1 and R_2, produce a restraining or contact opening torque. An internal fault in the protected zone will unbalance the secondary currents, forcing an operating current I_o in the relay operating coil, O. For a fixed-percentage differential relay, the operating current required to overcome restraining torque is a fixed percentage of restraining current, see Figure 2.12b. For example for a fixed 10% percentage differential current, the relay will trip if the operating current was greater than 10% of restraint current.

In a variable-percentage differential relay, the amount of differential current required to overcome restraining torque is a variable percentage of the restraining current, having a higher percentage at higher currents (Figure 2.12c). The modern differential relays have much better two slope characteristics, see Chapters 9 through 11.

2.5 Pilot Wire Protection

With hardwired differential schemes, a limitation of the distance to which the protection zone can be covered is soon reached depending upon the fault currents, CT burdens, and

Protective Relays

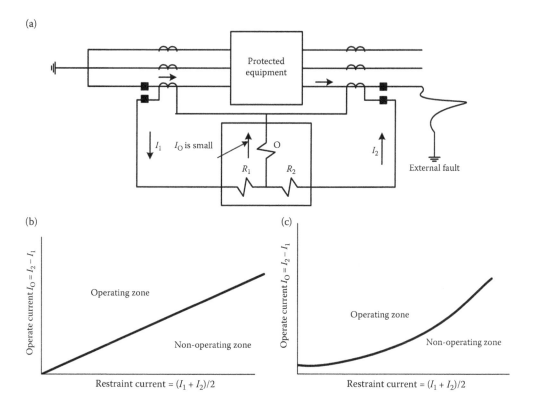

FIGURE 2.12
(a) Principle of a percent differential relay, (b) fixed-restraint characteristics, and (c) variable-restraint characteristics.

the long CT leads. Also intertripping is required. Generally, the schemes can be used for distances no more than approximately 200–350 ft.

The earlier pilot wire scheme, using electromechanical relays and metallic pilot wires for short lines up to approximately 20–30 km, provided fast clearance times for 100% of the line length. This protection is not discussed here; the whole of Chapter 14 is devoted to pilot protection.

2.6 Directional Overcurrent Relays

Directional overcurrent relays will provide tripping in one selected direction, and block the fault currents in the opposite direction. Some applications are as follows:

- Detection of uncleared faults in the utility line, where the fault current can be back-fed from the industrial system, for example, an industrial system with cogeneration, see Figure 2.13. The fault current fed by the in-plant generators and motors will be less than that from the utility source. Furthermore, this fault contribution will be of decaying nature. Thus, sensitive directional settings will be required. With proper settings, there is a possibility that the generator will not trip out, and

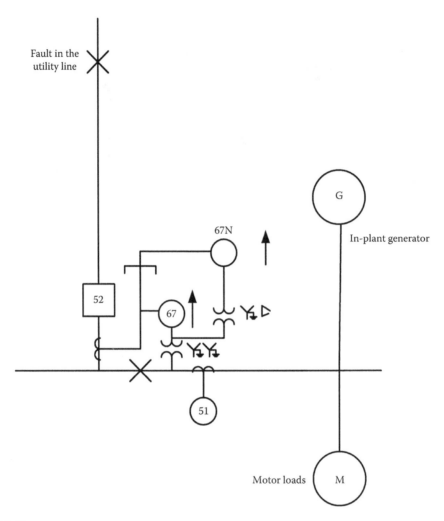

FIGURE 2.13
Application of phase overcurrent and ground overcurrent directional relays.

the total power loss can be avoided. This figure shows the phase directional and ground directional overcurrent protection. Note that the directional relays must have a current or voltage or both current and voltage polarizing input. The current polarizing is obtained by a CT in the neutral grounding conductor of a generator or transformer. Here, the polarizing input for phase directional relay is shown through wye–wye connected PTs and for the ground fault a polarizing voltage derived from an open delta secondary winding. This is discussed in more detail in Chapter 8.
- Network protection, say for parallel running transformers, or for paralleled feeders for selective tripping. This is shown in Figure 2.14a. Here, the system is operated with bus section breaker closed (sources 1 and 2 have same phase shifts and voltage levels and are suitable for paralleling). For a fault at F1, the directional relays should operate fast, tripping breaker A, and the entire load is now served from source 2, without any loss of power to the loads. Note that phase as well as

ground faults should have directional features; the relays 67 (for phase faults) and 67N for ground faults are shown. Also faults at F2, on the bus, and fault at F3, on a feeder, can be selectively cleared with overcurrent relays and ground fault relays. This is done with time–current coordination of the phase and ground relays R1, R2, R3, R4, and R5. Generally, in such systems, bus differential relays are added.

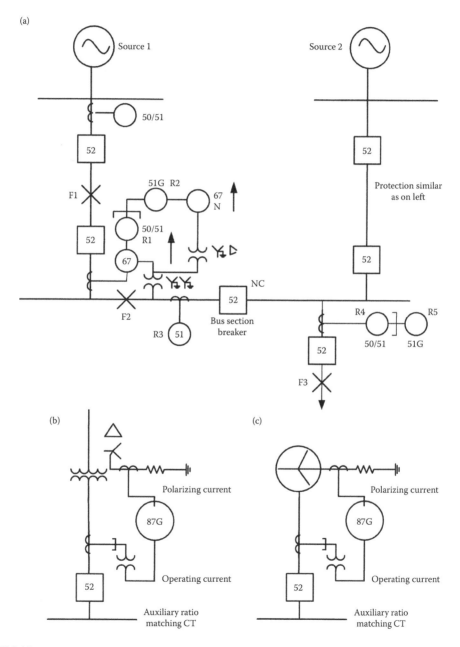

FIGURE 2.14
(a) Selective fault clearance in two parallel running sources with overcurrent nondirectional and directional relays. (b and c) Restricted ground fault protection with 87G devices which may be considered as directional devices.

Examples of coordination of directional relays are provided with illustrative examples in appropriate chapters.

- For providing sensitive high-speed ground fault protection of transformers and generators. The directional control gives the relay the characteristics of a differential relay. This is shown in Figure 2.14b,c for ground fault relays of the product type for generators and transformers. See Chapter 11 for the description of product type of differential relays.

These applications are further discussed and continued with examples in Chapters 6 and 8.

The principle of an electromechanical directional relay is illustrated in Figure 2.15. It consists of a conventional overcurrent disk element and an instantaneous power directional element. The characteristics of nondirectional overcurrent relays as discussed before are applicable. When the current is flowing in the tripping direction, the directional contacts close, which are in the lag coil circuit. This enables the overcurrent element to operate. The relay does not start operating till the directional contacts close. And off course, the

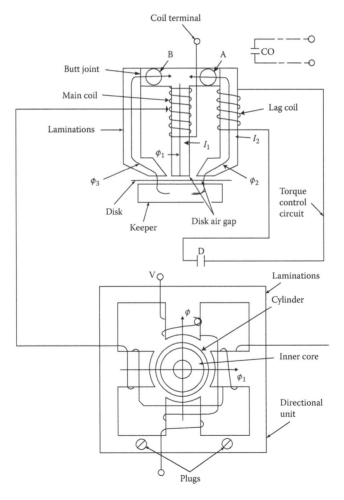

FIGURE 2.15
Construction of an electromechanical directional overcurrent relay. (IEEE Std. 242-1986.)

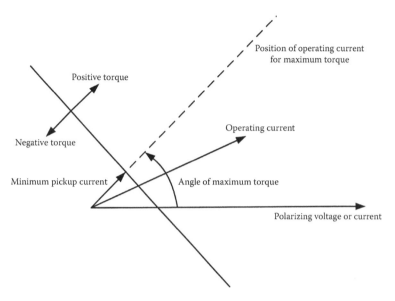

FIGURE 2.16
General operating characteristics of directional overcurrent element.

operating current is above the pickup setting. The directional element has an operating coil and a polarizing coil. The latter is energized by either the current or voltage in order to provide the directional feature. Some units may be dual polarized. Maximum tripping torque is produced in the tripping direction, when the angle between the operating coil current and the polarizing coil current is equal to the maximum torque angle of the relay, see Figure 2.16. As an example, the maximum torque may be produced when the operating current leads the voltage by 45°. In the current polarized relay, the maximum torque may be produced when two currents are in phase. These angles will vary.

An instantaneous electromagnetic directional relay has an instantaneous induction cup element that is controlled by an instantaneous power directional element.

2.7 Voltage Relays

Voltage relays are of undervoltage or overvoltage types, instantaneous or time delay type with inverse characteristics, much alike overcurrent relays. In addition, we have voltage-controlled or voltage-restraint relays, special relays, that is, neutral displacement relays for ground fault detection in ungrounded systems, and voltage balance relays. Some applications are as follows:

1. Undervoltage and overvoltage protection on feeder circuits, generators, and motors. Overvoltage protection is normally provided on machines such as hydro-genates where excessive terminal voltages may be produced following load rejection, without necessarily exceeding V/Hz limits of the machine.
2. Voltage-controlled and voltage-restraint relays are used for generator backup protection, see Chapter 11. When an external fault to the generator occurs, the

voltage collapses to a relatively low value, while on an overload the voltage drop is relatively small. The relays modify the time–current characteristics so that the relay will ride through permissible power swings, but respond to fault currents. In the voltage-controlled relay, an auxiliary overcurrent element controls the operation of the induction disk element. When applied voltage is below a predetermined level, an undervoltage contact is closed in a shaded pole circuit, permitting relay to develop torque and operate like a conventional overcurrent relay. In voltage-restraint relay, a voltage element provides restraining torque proportional to the voltage and thus, shifts the relay pickup current. The larger the voltage dip, the more sensitive the relay on pickup. The relay is set so that it rides through permissible power swings at nominal voltage. The characteristics and settings on these two types of relays for generator protection are illustrated in Chapter 11 and also in Chapter 6.

3. Overvoltage and undervoltage bus protection.
4. Ground fault protection in high-resistance grounded generators, see Chapter 8.
5. An instantaneous dc undervoltage relay is used for protection of dc control circuits of the breakers.
6. An instantaneous undervoltage relay is used for bus transfer between two sources. Figure 2.17 shows that the undervoltage is detected on two sources, normally the utility source is in vice. On failure of the utility source, the standby generator can be automatically started and the generator breaker closed, provided there is no fault tripping. On restoration of normal source, the generator is taken out of service and the normal power restored.
7. A special rectifier type undervoltage relay is used for the detection of a ground fault on the load side of a bridge rectifier or converter. Figure 2.18 shows a single pole rectifier type of relay and Figure 2.19 shows its application. For a ground fault on the converter side, the current flows in unidirectional pulses, and cannot be detected by a conventional pulsing type of ground fault detection equipment described in Chapter 8. However, a rectifier type of relay will respond to these pulses and can be set to detect the ground fault on the converter side.

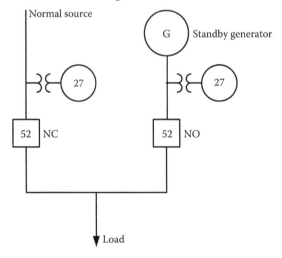

FIGURE 2.17
Application of undervoltage relays for sensing of sources and auto-switching from normal source to standby generator and then back to normal source on restoration of power.

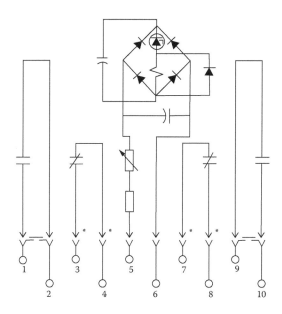

FIGURE 2.18
A rectifier type of voltage relay.

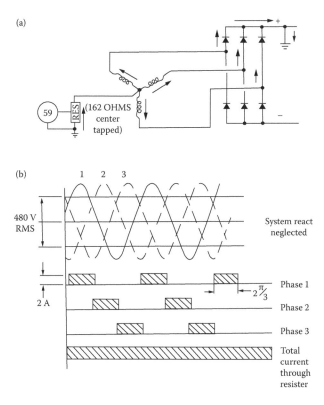

FIGURE 2.19
Ground fault on the converter side of the rectifier bridge, and the flow of ground fault current is in pulses. Device 59 for ground fault detection should be a rectifier type of relay, as shown in Figure 2.18.

8. Voltage unbalance relay is used to protect three-phase motors from damage which may be caused by single phase operation. On operation of a fuse in a three-phase circuit, the motor will continue to operate and its back EMF will maintain the full voltage across the open phase. The voltage unbalance relay is fundamentally a negative-sequence voltage relay. The operating time is of the order of 10–35 ms, depending on the tap setting. In modern motor protection technology, a separate relay is rarely used and the protection is provided by device 46, via current unbalance settings (see Chapter 10).

9. In generator circuits, a voltage balance relay is used to continuously monitor the PT voltage and to block operation of the protective relays and control devices that will operate incorrectly when a PT fuse operates. Two sets of PTs (PT1 and PT2) are provided in the generator circuit: one set supplies potential for the backup overcurrent relays, or distance relays, directional power and loss of excitation relay, and the other set supplies potential for the synchronizing relays, metering and voltage regulator. The PTs can be a bus connected PT on the top of the generator breaker. If dead bus start-up of a generator is not required, then the two sets of PTs have identical outputs, the relay is balanced, and both the right and left contacts are open. When a fuse operates in any phase of PT1, the unbalance will cause the left contact to close, which can be used for alarm and block operation of 51 V, 32, 40 devices. When a fuse opens in any phase of PT2, the unbalance will cause the right contact to close, which operates an alarm and switches the voltage regulator to manual mode of operation to prevent it from rising to the ceiling voltage, see Figure 2.20.

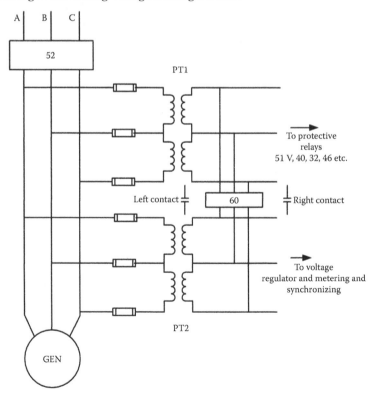

FIGURE 2.20
Application of a voltage balance relay, device 60, to a generator.

2.8 Reclosing Relays

Reclosing relays are applied in distribution circuits, to reclose a circuit breaker, one, two, or three times after the circuit has been tripped by the protective relays. Many a time, the faults on the overhead distribution and transmission systems are of temporary nature and the power can be restored through reclosing. The relay has a timer setting, the first reclosing can be instantaneous or time delay, and one complete cycle may be adjustable between some timing limits such as 15–150 s. The relay can be adjusted to reset after a successful recloser. On unsuccessful reclosing, the close and trip circuits are locked out and an alarm is sounded. With high-speed reclosing, the circuit breaker closing time is compared with the drop out time of the protective relays that initiate a trip out. The protective relays that tripped the breaker in the first instance must open their contacts before the circuit breaker recloses; otherwise, the breaker will be tripped again even if the fault is cleared. The breaker closing circuit must be pump-free, see Chapter 5.

In modern protective schemes, discrete relays will be rarely used. This feature can be programmed based on the timers and the trip and close logic in MMPRs. A detailed application and description of reclosing logic with an example is included in Chapter 4.

2.9 Breaker Failure Relay

One configuration of breaker failure logic is shown in Chapter 11; this can be implemented in an MMPR using inbuilt timers and other functions, though discrete relays are also available. Figure 2.21 illustrates a timing circuit with successful and inoperative breaker. The input to the breaker failure timer is the fault detector which receives each phase current

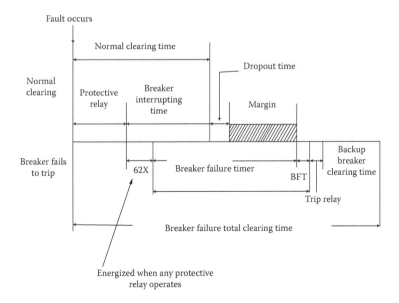

FIGURE 2.21
Timing circuit diagram of a breaker failure relay.

and $3I_0$, with level detectors which produce an output when the instantaneous magnitude exceeds its sensitivity. 62X is actuated by any protective relay functions intended to trip the breaker. Note that this input is applied to the BFT when the breaker is in the normal tripping mode. The IEEE Relaying Committee recommends at least three cycle of margin. For further discussions and applications, see Chapter 11.

2.10 Machine Field Ground Fault Relay

The ground fault detection of the fault in the *rotor circuit* of the machines, brush type and brushless type, has been discussed in Chapters 10 and 11.

For the machines (generators or motors) field-to-ground fault protection, see Figure 2.22. The machine field to ground resistance is sensed, and if this is low enough, the relay senses this fault condition. A fault on the field winding is the connection of the field winding to ground through-fault resistance R_G. This completes a circuit so that the sensitivity current I_S is flowing from a rectifier to positive to negative through relay coil marked A in Figure 2.22. This coil is calibrated to pickup at a fixed value of I_S. The sequence of operation is given as follows:

- I_S flows through from relay rectifier positive through coil A, normally closed contact AX, diode D1, machine field winding, and R_G, and returns to the relay rectifier negative terminal through stud 5.
- Thermal time delay element T is energized through A contact and time delay link.

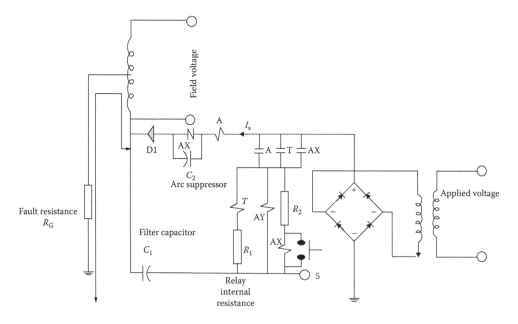

FIGURE 2.22
Field-to-ground fault protection of synchronous generators or motors—functional diagram of a relay.

Protective Relays

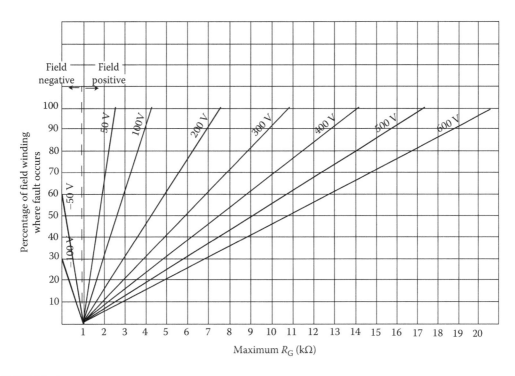

FIGURE 2.23
Percentage of field windings protected with respect to ground fault resistance.

- After a delay of 2 s, thermal timer contact T closes energizing AY coil and AX coil through R2.
- AX seals itself and AY through its A contact, and simultaneously it interrupts I_S, which reenergizes units A and T.
- AY, sealed in through AX contacts, sets target and closes alarm or trip contacts.
- Unit AY cannot be reset unless unit T has cooled off.

Figure 2.23 illustrates ground fault sensitivity for 100% voltage for fault on negative and positive fields.

2.11 Frequency Relays

The frequency relays are of underfrequency and overfrequency types with adjustable time delays. Their main applications are as follows:

- These relays are used in protection of steam turbines which have resonant frequencies not far removed from 60 Hz. See Figure 11.31 which is generalized in IEEE standard, based on some manufacturer's data. A specific curve for safe cooperation of steam turbines for off frequency operation can be obtained from the manufacturer. Serious damage can occur if the underfrequency and overfrequency settings exceed the specified limits.

- Underfrequency relays are applied when the loads are supplied exclusively by local generators or by a combination of local generators and utility ties. When a major generator drops offline in a system exclusively supplied by the plant generators, the frequency relays will open the specific programmed breaker so that the total load is less than the generation. Otherwise, the overloads can plunge the system into a total shutdown. The same situation applies when the utility system and the plant generators are operated in synchronism. On loss of the utility source, appropriate loads should be tripped, so that the total load is somewhat less than the generation. When there is an overload, the generators will tend to slow down and frequency drops. Enough load should be tripped to arrest this frequency drop.
- Underfrequency relays are used for load shedding as described above and can enhance the stability of the system, see Chapter 15. Generally, the loads are shed in steps; for example, the first load shedding occurs at 59.6 Hz, and if the frequency drop is not arrested, the next load shedding will occur at a lower frequency of 59 Hz.
- Overfrequency relays are used for protection of generators against over speed during startup or when the unit is suddenly separated from the system with little or no load.

Three types of frequency relays are available: induction disk relays, induction cup relays, and static or microprocessor-based relays. The induction disk relay is subjected to two fluxes whose phase relationship changes with frequency to produce contact opening torque above the frequency setting and closing torque below it. A time dial, like an overcurrent relay, is used to adjust initial contact separation. The operating principle of the induction cup relay is the same. These types of relays are no longer in use.

Figure 2.24 shows the block circuit diagram of a static frequency relay of a manufacturer. A load conservation scheme can be designed for the following:

- To trip off blocks of loads in several steps with progressively lower frequency settings;
- To trip off blocks of loads in several steps on a time basis at one level of frequency, so that as each time step is reached additional load is shed;
- Any combination of the above two bullet points.

Figure 2.25 shows the frequency versus time to open the breaker after the disturbance starts. The underfrequency conditions must persist for 4.5 cycles to a maximum of 80 cycles before an output is produced. The relay has a cut-off feature adjustable between 20% and 90% of the rated ac voltage, and blocks all outputs when voltage is below the set point value.

The relay has also overfrequency restore function, which works in conjunction with underfrequency trip function. In general, substantial time delay of the order of seconds is used in the automatic restoration scheme. The restoration is set up in small load blockswith sufficient time between each step. In this way when each block is restored, the system has a chance to absorb this load and settle at its new frequency. If this new frequency is not below the rated frequency, the next load block is restored after a time delay. If at any step the system frequency is not restored, the load restoration step will be blocked. The restoration timers are set for long time delays, and these should not reset after a system transient disturbance. Auxiliary relays will be required.

Protective Relays 49

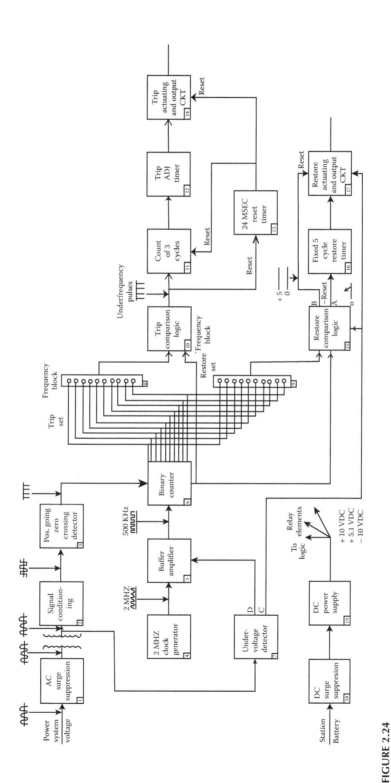

FIGURE 2.24
Operational diagram of an underfrequency and overfrequency static relay.

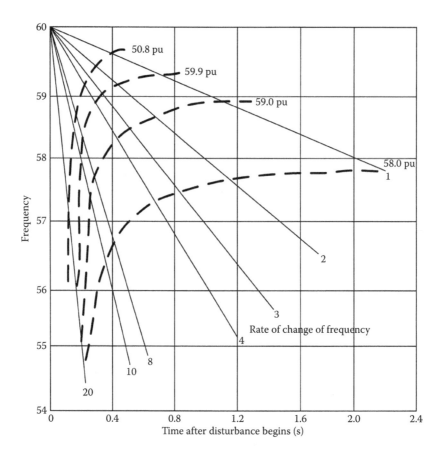

FIGURE 2.25
Rate of change of frequency after the disturbance.

2.12 Distance Relays

The distance relays (impedance relays) are extensively applied for transmission line protection. Also, these are applied as a backup protection of a generator, and loss of excitation of a generator.

The principles can be explained by considering a line of impedance Z_L, and a fault at a distance of nL from the source. The voltage at the fault point is zero (bolted fault). Then at the relay, referring to Figure 2.26a, the impedance measured to the fault point is

$$Z_n = \frac{I \times nZ_L}{I} = nZ_L \qquad (2.3)$$

This is independent of current for three-phase faults, phase-to-phase faults, and two-phase-to-ground faults, as the current values get canceled in Equation 2.3. Providing such fixed reach is not possible with overcurrent relays. The settings and operation are a function of impedance from the source end.

Protective Relays

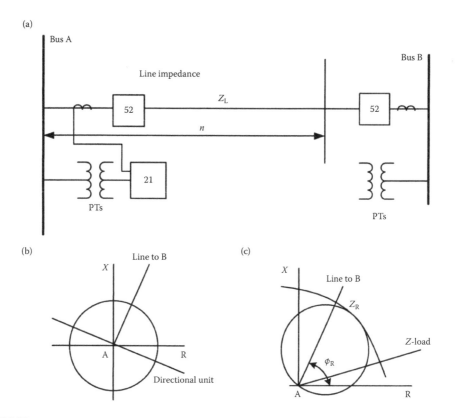

FIGURE 2.26
(a) Application of a distance protection to a transmission line, (b) impedance characterizes, and (c) mho characteristics.

The characteristics are plotted on an R–X diagram. Typical characteristic of a distance unit is shown in Figure 2.26b. The origin is the relay location. If the set Z_n falls within the circle, the relay operates.

In Figure 2.26c, the circle through the origin is called an mho unit, which is widely used for line protection. It shows a directional operating characteristic. The load currents will be at angles 0°–30° while the fault currents will be more lagging. For heavy loads, the impedance Z_R phasor moves toward the origin, while for low loads, it moves away from the origin. Operations for fault currents less than the load currents are, therefore, possible.

On the R–X diagram, the lagging load current from A to B will be phasor in the first quadrant (Figure 2.26c) while for a lagging load current from Bus B to Bus A it will be in the second quadrant.

Example 2.1

Consider a 115 kV line, and that the distance relay has a CT ratio of 800/5 and a PT ratio of 120 kV/120 V. The load current is 4 A (secondary). Then, the load impedance is

$$Z_{\text{load}} = \frac{120}{\sqrt{3} \times 4} = 17.32 \ \Omega$$

Consider that the load angle is 30°. Then, with reference to Figure 2.26c, the relay reach Z_R at angle ϕ_r can be written as

$$17.32 = Z_R \cos(75° - 30°)$$

This considers a typical angle of 75° for the mho unit.

Then, Z_R in terms of primary resistance is equal to 153.1 Ω. If the line impedance is of the order of 1 Ω per mile (approximate), the reach extends to 153 miles.

While the above forms an introduction, the detailed application and discussions and settings of distance relays are covered in Chapter 13.

2.13 Other Relay Types

In order to avoid repetitions, see respective chapters for the protective relay applications to electrical power systems. The objective of this chapter is to provide a background for the latter chapters, though some applications discussed in this chapter, such as overcurrent relay characteristics, are quite comprehensive and complete.

The cited references [3–11] provide further reading.

References

1. AR Washington. *Protective Relays. Their Theory and Practice*, Vol. 1. John Wiley & Sons, New York, 1962. Vol. 2, Chapman and Hall, London, 1974.
2. ANSI/IEEE Std. 242. *IEEE Recommended Practice for Protection and Coordination of Industrial and Commercial Power Systems*, 1986.
3. CR Mason. *The Art and Science of Protective Relaying*. John Wiley & Sons, New York, 1956.
4. WA Elmore (Ed.). *Applied Protective Relaying*. Westinghouse Electric Corporation, Coral Springs, FL, 1979.
5. SH Horowitz (Ed.). *Protective Relaying for Power Systems*. IEEE Press, Piscataway, NJ, 1980.
6. JL Blackburn, TJ Domin. *Protective Relaying Principles and Applications*. CRC Press, Boca Raton, FL, 2014.
7. PM Anderson. *Power System Protection*. Wiley-IEEE Press, New York, 2015.
8. A Wright, C Christopoulos. *Electrical Power System Protection*. Chapman and Hall, London, 1993.
9. Collection of IEEE Standards: IEEE Guides and Standards for Protective Relaying Systems-1989 Edition.
10. Silent Sentinels, Relay Pointer Letter (RPL) 65-3 through 86-1 and PRSC-1 through PRSC-8, Westinghouse Electric Corporation, Relay and Telecommunication Division, Coral Springs, FL, 1967–1989.
11. Bibliography of Relay Literature, 1994. *IEEE Trans Power Delivery*, 11(3), 1251–1262, 1996.

3
Instrument Transformers

For the relaying applications, the high system voltages and currents are reduced through the potential and current transformers (called instrument transformers). These also protect and insulate the personnel and relays from high voltages and currents. The performance of instrument transformers is critical in protective relaying, as the reduced secondary currents and voltages should be an exact replica of the primary quantities both under steady-state and transient conditions; for example, under fault conditions when large magnitudes of currents flow. These primary currents and voltages should be applied to the relays reduced in exact proportion and without any waveform distortion. We will see that this ideal situation is not always practically possible and saturation of the current transformers (CTs) has to be allowed for. Generally, the reduced voltages are 120 and 69.3 V and the CT secondary current ratings are 5 and 1 A. We alluded to current transformer saturation and their accuracies in earlier chapters. This chapter provides a conceptual base for proper applications of CTs for relaying quantities, considering steady-state and transient behavior.

3.1 Accuracy Classification of CTs

3.1.1 Metering Accuracies

The instrument transformers have two distinct accuracy classifications: metering and relaying. Table 3.1 from Reference [1] shows metering class accuracies for the voltage and current transformers. A metering CT must reproduce faithfully the primary quantities, and it is immaterial if it saturates on heavy fault currents. In fact, this saturation means that there will be little output from the CT and the meter connected to the secondary of the CT will not be subjected to high magnitudes of secondary currents. On the contrary, a relaying class CT must reproduce the high primary fault currents accurately without saturation for proper relay operation. Referring to Table 3.1, for revenue metering, sometimes, even better accuracies than 0.3 metering class are demanded, where large power supplies to a consumer are to be metered. Generally, such revenue metering instrument transformers do not have any other secondary burden, except the meter to which these are connected.

A single set of current transformers is often used for both relaying and metering in industrial relaying applications. *While relaying class accuracy is acceptable for metering, the metering class accuracy is not acceptable for relaying.*

3.1.2 Relaying Accuracies

Relaying CTs have C, X, or T classifications [1,2].

C classification covers transformers in which the leakage flux in the core does not have a considerable effect on the ratio, and the ratio correction can be calculated. The excitation curve is plotted on a log–log paper between secondary exciting current and

TABLE 3.1

Standard Accuracy Class for Metering Service and Corresponding Limits of Transformer Correction Factor (0.6–1.0 Lagging Power Factor of Metered Load)

Metering Accuracy Class	Voltage Transformers (at 90%–110% of Rated Voltage)		Current Transformers			
			At 100% Rated Current		At 10% Rated Current	
	Minimum	Maximum	Minimum	Maximum	Minimum	Maximum
0.3	0.997	1.003	0.997	1.003	0.994	1.006
0.6	0.994	1.006	0.994	1.006	0.998	1.012
1.2	0.998	1.012	0.998	1.012	0.976	1.024

Source: ANSI/IEEE, Standard, C57.13, *Requirements for Instrument Transformers*, 1993 (R 2008).

voltage. A typical excitation curve for class C transformers with nongapped cores is shown in Figure 3.1. The data represent secondary exciting rms currents by applying rms voltage to the current transformer secondary windings, with primary open circuiting. This gives the approximate exciting current requirements for a secondary voltage. The *knee point* is defined as the point where tangent is 45° to the abscissa. For CTs with gapped cores, this angle is 30°. The maximum excitation values above knee are shown in Figure 3.1.

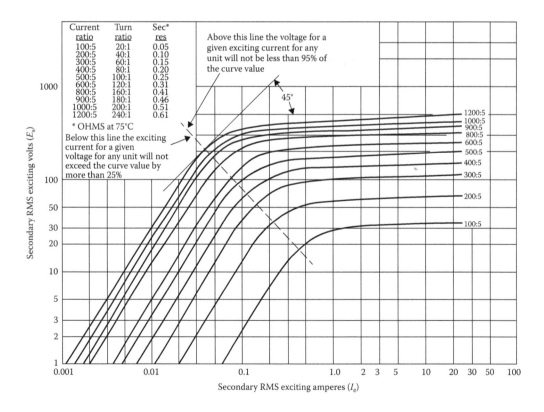

FIGURE 3.1
Excitation characteristics of C type current transformers, showing knee point voltage.

Instrument Transformers

The knee point voltage, V_k, is defined as the sinusoidal voltage applied to the secondary terminals; all other windings are open circuited, which, when increased by 10%, causes excitation current to increase by 50%.

3.1.3 Relaying Accuracy Classification X

The accuracy classification X is user defined for a specific condition, where the requirements are given as follows:

V_k = minimum knee point voltage

I_k = maximum exciting current at V_k

R_{ct} = maximum allowed secondary winding resistance, measured with DC current, corrected to 75°C

3.1.4 Accuracy Classification T

Class T transformers have considerable leakage flux and the ratio correction must be determined by test. Typical ratio curves for class T transformers are plotted over the range from 1 to 22 times normal primary current for all standard burdens which cause a ratio correction of 50% [1]. Wound-type transformers are sometimes used as auxiliary CTs for ratio matching or these can be separate CTs in themselves, available for low-voltage to high-voltage applications. Figure 3.2 shows the excitation curves of a class T current transformer.

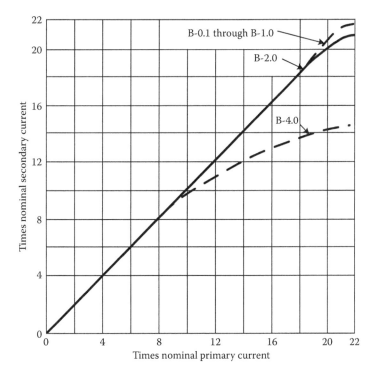

FIGURE 3.2
Excitation characteristics of T type current transformers.

3.2 Constructional Features of CTs

Figure 3.3 shows constructional features of some current transformers. In the window-type CT, the primary conductor consists of just one single conductor passing through the window (see Figure 3.3a). The CTs located in indoor metal-clad switchgears, and those located on transformers and outdoor high-voltage circuit breaker bushings are window type. These do not have high basic insulation levels (BILs). A fully insulated window-type CT is shown in Figure 3.3b. For high-voltage outdoor substation applications, a window-type CT mounted in an oil-filled tank and installed on a separate steel framework with in and out high-voltage connections can be applied (see Figure 3.3c). Figure 3.3d shows a CT of large window diameter and ratio that can be mounted on generator iso-phase bus ducts. The window-type CTs, generally, meet the above definition of knee point voltage.

The bushing current transformers are most widely used for relaying because their installation cost is low. It consists of a tapped secondary winding on an annular magnetic core and is referred as multiratio bushing current transformer. (Note that window-type current transformers can also be of multiratio.) The core encircles a high-voltage insulating bushing used on circuit breakers, power transformers, and generators, through which the primary conductor passes to form a primary turn. The secondary turns of the bushing current transformers should be distributed to minimize leakage reactance. This is achieved by distributing each section of the winding completely around the circumference of the core.

A wound-type current transformer has a primary winding of one or more turns and a secondary winding on a common core, similar to a power transformer.

The auxiliary transformers are sometimes used in the secondary circuits of the main CTs to change the ratio or phase angles. There characteristics can seriously impact the performance, and their use should be avoided as much as possible. Modern microprocessor-based multifunction relays (MMPRs) can tolerate a greater CT mismatch and have more sensitive settings to obviate the necessity of auxiliary CTs.

3.3 Secondary Terminal Voltage Rating

A class C or T transformer will deliver to the standard burden a secondary terminal voltage at 20 times the secondary current, without exceeding 10% ratio correction. Furthermore, the ratio correction should be limited; 3%–10% from rated secondary current to 20 times the rated secondary current at the specified rated burden [1]. For example, for a C200 CT the ratio correction should not exceed 10% at 20 times the rated secondary current at a standard 2 Ω burden, at 0.5 power factor (2.0 Ω multiplied by 100 A equals 200 V). If the relaying class accuracy is C200 and the CT secondary current is 1 A, then the burden to develop secondary terminal voltage will be 200 V/(1 A × 20) = 10 Ω.

When multiratio CTs are applied, care has to be taken that the accuracy quoted by the manufacturer applies to full secondary winding.

Table 3.2 shows the limits of ratio errors for relaying class CTs and Table 3.3 shows the standard burdens and secondary terminal voltages.

Instrument Transformers

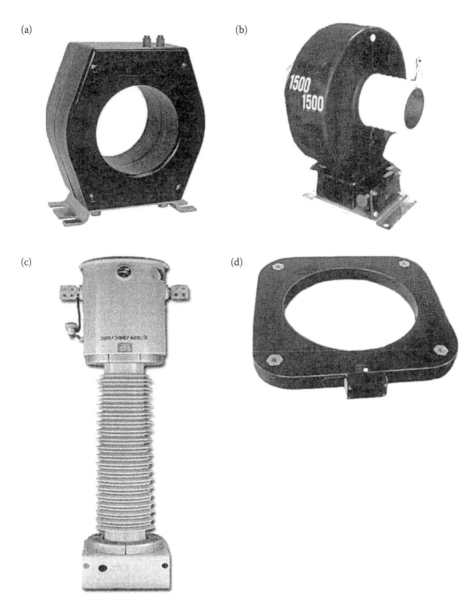

FIGURE 3.3
Construction of various CT types: (a) window-type CT window diameter up to 8", ratios up to 5000:5, accuracy up to C800. Single ratio, double ratio, or multi-ratio, (b) a fully insulated window-type CT up to 34.5 kV, ratios up to 4000:5, accuracy up to C800. (Say BIL=110 kV for 13.8 kV.) Single ratio, double ratio, or multi-ratio, (c) oil-filled head type outdoor CT, 25–161 kV, accuracy up to C800. Single ratio, double ratio, or multi-ratio and (d) an encapsulated generator CT, window size up to 35", ratio 40,000:5, accuracy up to C800. Single ratio, double ratio or multi-ratio for mounting on iso-phase bus.

3.3.1 Saturation Voltage

Saturation voltage is defined as the voltage across the secondary winding at which peak induction just exceeds the saturation flux density. *It is not the same as the knee point voltage.* This difference can be illustrated graphically. Consider the point on the excitation curve

TABLE 3.2

Ratio Errors for Relaying Accuracy CTs

Relaying Accuracy Class	@ Rated Current (%)	@ 20 Times Rated Current
C and T classifications	3	10%
X classification	1	User defined

Source: ANSI/IEEE, Standard, C57.13, *Requirements for Instrument Transformers*, 1993 (R 2008).

TABLE 3.3

Class C or T Relaying Accuracy

Secondary Terminal Voltage	Burden Designation	Resistance (Ω)	Inductance (mH)	Impedance (Ω)	Total Power (VA at 5 A)	Power Factor
100	B-1.0	0.5	2.30	1.0	25	0.5
200	B-2.0	1.0	4.60	2.0	50	0.5
400	B-4.0	2.0	9.20	4.0	100	0.5
800	B-8.0	4.0	18.40	8.0	200	0.5

Source: ANSI/IEEE, Standard, C57.13, *Requirements for Instrument Transformers*, 1993 (R 2008).
Note: Secondary voltage and relaying burden.

where the straight line relation just starts to deviate. It is found graphically by locating the intersection of the straight portions of the excitation curve on log–log axes. This is clearly shown in Figure 3.4, where V_k is the knee point voltage and V_{sat} is the saturation voltage.

3.3.2 Saturation Factor

Saturation factor is the ratio of the saturation voltage to the excitation voltage. It is an index of how close to saturation a CT is applied in a given application.

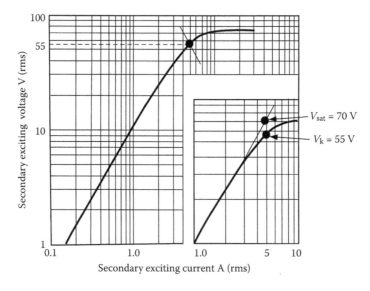

FIGURE 3.4
Excitation characteristics of a 100/5 core-balance CT. V_k is the knee point voltage and V_{sat} is the saturation voltage.

Instrument Transformers

These definitions and concepts are important for the analysis of the saturation of CT discussed further.

3.4 CT Ratio and Phase Angle Errors

The phasor diagram of a current transformer can be constructed much alike a power transformer; Figure 3.5 is applicable for class C CTs. The primary current in a CT is not determined by the secondary load and there is no considerable voltage across the primary terminals of the CT. To magnetize the core, a small current I_0 flows, which can be resolved into two components: I_m, the magnetizing current, which is in phase with the magnetic flux and I_e, the eddy current, which is in quadrature. I_e is required to meet the hysteresis and eddy current loss, given by the following equations:

$$P_h = K_h f B^s$$
$$P_e = K_e f^2 B^2$$
(3.1)

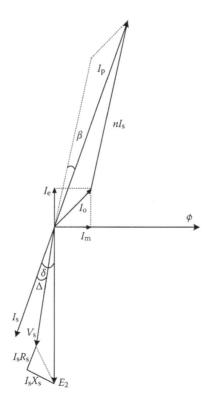

FIGURE 3.5
Phase diagram of a class C current transformer.

where P_h and P_e are the hysteresis and eddy current loss, respectively, f is the frequency, B is the flux density, K_h and K_e are constants, and s is the Steinmetz exponent, which varies from 1.5 to 2.5 depending upon the core material. Generally, the value of Steinmetz exponent is 1.6.

The magnetizing impedance of the CT, obtained by dividing the excitation voltage and excitation current (Figure 3.1), is not constant. It is highly nonlinear, changing from a high value at low excitations to a low value at high excitations.

This gives rise to a ratio error. The ratio correction factor, RCF, is defined as

$$\mathrm{RCF} = \frac{I_p}{nI_s} \quad (3.2)$$

where I_s is the secondary current and n denotes the number of secondary turns. The flux φ produces a voltage E_2, in the secondary. The voltage at the secondary terminals of the CT is V_s, which is obtained by subtracting the vectors of voltage drops $I_s R_s$ and $I_s X_s$, where R_s and X_s are the resistance and reactance of the secondary burden plus CT resistance and reactance. The secondary winding leakage reactance of the CT is small. It is shown exaggerated in Figure 3.5 for clarity. This gives rise to a phase angle error β, which is given by

$$\beta = \tan^{-1} \frac{I_m}{nI_s} \quad (3.3)$$

The relaying standard burdens are at a power factor of 0.5 (see Table 3.2). Figure 3.5 shows that if the actual burden is resistive, the CT error will be much less.

The IEC [3] defines a composite error, which is given by the following equation:

$$\frac{100}{I_p} \sqrt{\frac{1}{\tau} \int_0^{\tau} (nI_s - I_p)^2 \, d\tau} \quad (3.4)$$

This may be considered to take into account ratio and phase angle errors as well as waveform distortion (see Figure 3.6). The IEC designations of the CTs are based upon the composite error. For example, a 5P30 CT means a protection class CT with composite error no more than 5% and VA burden of 30 VA.

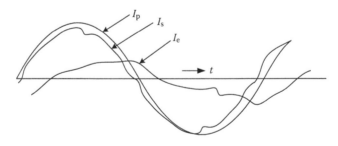

FIGURE 3.6
Composite error according to IEC standards [3].

ANSI definition of total correction factor is

$$\text{TCF} = \text{RCF} - \frac{\beta}{2600} \qquad (3.5)$$

where RCF is the ratio correction factor and β is the phase angle in minutes. This is of importance for metering class CTs. Similarly for voltage transformers

$$\text{TCF} = \text{RCF} - \frac{\gamma}{2600} \qquad (3.6)$$

where γ is the phase angle in minutes for the voltage transformers. These relations are shown in Figure 3.7 for the current transformers and Figure 3.8 for the voltage transformers.

The ANSI/IEEE Standard C57.13 [1] gives the following expression for RCF:

$$\text{RCF} = \frac{I_{st}}{I_s} \qquad (3.7)$$

where $I_{st} = I_s + I_e$.

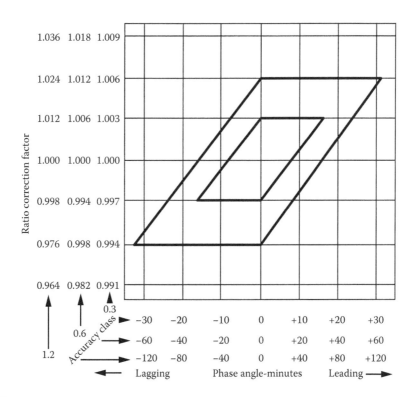

FIGURE 3.7
Limits of accuracy classes for current transformers for metering service.

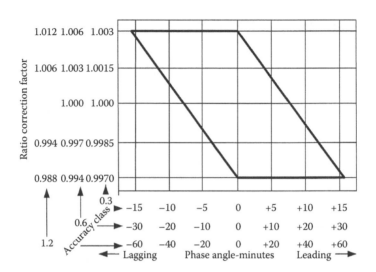

FIGURE 3.8
Limits of accuracy classes for potential transformers (PTs) for metering service.

3.5 Interrelation of CT Ratio and Class C Accuracy

The following considerations can be postulated.

1. When the CT ratio is low, the ampere-turns (denoted as $\widehat{A}$) are low. In the window-type CT, the primary consists of a single turn, i.e., the ampere-turns is equal to primary amperes. (The use of the term $\widehat{A}$ is no longer in use; instead the term magnetomotive force (MMF) is used. However, for illustrative purposes, the use of ampere-turns is retained.)
2. The low $\widehat{A}$ makes the design of high-accuracy CT difficult as the flux produced is low.
3. The larger the dimension of the window, the lesser the flux that will be produced by the same primary current.
4. A larger core cross section is required as the flux is equal to flux density multiplied by the area of the cross section of the core.
5. A larger core length means that the magnetization and core-loss components of the current will increase, which in turn increases ratio and phase angle errors.
6. The $\widehat{A}$ required to establish a certain amount of flux in an air gap is higher by a factor of hundreds. Thus, the flux must be increased to account for even a small air gap in the current transformer of a split-core construction.
7. The introduction of an air gap into the core of the CT leads to fringing and leakage of flux, which needs to be controlled. The winding reactance is a function of the leakage flux.
8. The higher the accuracy, the greater the voltage across the secondary windings which should be produced at a minimum error; that is, the saturation of the magnetic material should not be reached.

Considerations 6 and 7 apply to split-core CTs. The split-core construction allows ease of installation without dismantling, say, around a bus bar. Thus, there are limitations in obtaining high accuracy with low ratio CTs.

Analytically, the MMF is written as

$$\text{MMF} = \oint H\, dl \tag{3.8}$$

where H is the magnetic intensity and $\oint$ means that the integration is taken all the way around the closed circuit. For a solenoid, MMF = NI, where N is the number of turns and I is the current. The flux ϕ is given by

$$\phi = \frac{\text{MMF}}{S} = \frac{\text{MMF}}{l/\mu\mu_0 a} \tag{3.9}$$

where S is the reluctance of the magnetic circuit, l is the length of the circuit in meters, a is the area of cross section in m^2, μ is the relative permeability of the CT core magnetic material, and μ_0 is the permeability of air, which is given as follows:

$$\mu_0 = 4\pi 10^{-7} \tag{3.10}$$

This gives flux in webers. Also

$$\phi = Ba \tag{3.11}$$

where B is the flux density in Wb/m².

The secondary voltage induced in the CT core is given by

$$E_2 = -N \frac{d\phi}{dt} \tag{3.12}$$

Rewriting and adjusting for the units, the steady-state secondary rms voltage is

$$E_2 = 4.44 N \phi f 10^{-8} = 4.44 NBaf \times 10^{-8}\, \text{V} \tag{3.13}$$

The magnetic field produced in air due to an infinitely long straight conductor carrying a current I_p consists of concentric circles, which lie in a plane perpendicular to the axis of the current carrying conductor, and have its center on its axis. The magnetic field intensity, H, at a distance r from the conductor is given by

$$H \propto \frac{I_p}{r} \tag{3.14}$$

Also

$$\phi = Ba = \frac{Ha}{\mu\mu_0} \tag{3.15}$$

When a CT of *lower current ratio is* to be designed, the primary ampere-turns are small. Thus, the magnetic flux produced in the core by the primary single-turn conductor is

TABLE 3.4

C Class Accuracy Window-Type CTs Normally Provided on Medium-Voltage Switchgear and Also Higher Accuracy CTs That Can Be Provided

Ratio	Accuracy Class Normally Provided	Accuracy Class That Can Be Provided
50:5	C10	C20
75:5	C10	C20
100:5	C10	C20
150:5	C20	C50
200:5	C20	C50
300:5	C50	C100
400:5	C50	C100
600:5	C100	C200
800:5	C100	C200
1200:5	C200	C400
1500:5	C200	C400
2000:5	C200	C400
3000:5	C200	C400
4000:5	C200	C400

small. A linearity of the excitation curve (Figure 3.1) should be obtained for certain accuracy. The secondary turns or the cross-sectional area of the core must be increased. The turns are fixed by the transformation ratio, i.e., for a 100:5 CT, $n = 20$. This limits the secondary voltage and the class C accuracy that can be obtained. From Equation 3.11, a core of higher cross-sectional area and lower reluctance is required.

Thus, the maximum secondary VA burden and accuracy is controlled mainly by the primary ampere-turns.

For split-core CTs, introducing an air gap into the core makes the design all the more difficult. The permeability of iron is approximately 600 times that of the air and more ampere-turns (MMF) are required for the flux to cross even a small air gap. The fringing and leakage effect must be considered, so that leakage reactance is small. Thus, the design is an optimization of a number of conflicting parameters.

Table 3.4 shows the normal class C accuracies of window-type transformers that are normally provided for medium-voltage switchgear. In metal-clad switchgear, the CTs are located on the circuit breaker spouts (see Figure 3.9). Two sets of CTs of normal accuracy can be provided on the source and load sides of the circuit breaker. With high-accuracy CTs, the width of the CT increases, and it may not be possible to accommodate two CTs on the source and load sides.

3.6 Polarity of Instrument Transformers

We have shown the polarity of CTs in many figures in the previous chapters by a square dot (■); it is also denoted by an "X" sometimes. The polarity indications shown in Figure 3.10 are applicable to both current and voltage transformers. Figure 3.10a shows *subtractive* polarity. This signifies that the current flowing out at the polarity-marked terminal on the

Instrument Transformers

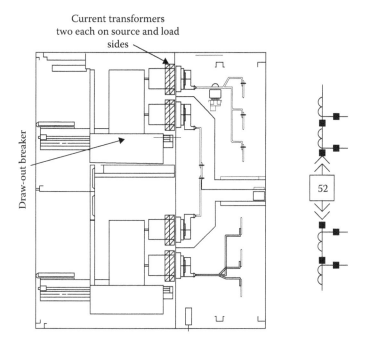

FIGURE 3.9
Location of window-type CTs in a metal-clad draw-out switchgear. Two CTs each on source and load sides can normally be located on breaker spouts.

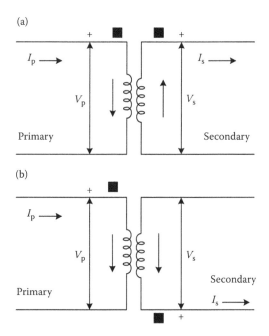

FIGURE 3.10
(a) Subtractive polarity and (b) additive polarity, shown by darkened square blocks.

secondary side is substantially in phase with the current flowing in the polarity-marked terminal on the primary side. This subtractive polarity is in common use.

Figure 3.10b shows the additive polarity. This signifies that the voltage drop from the polarity marked terminal to the nonpolarity marked terminal on the primary side is substantially in phase with the voltage drop from the polarity marked to nonpolarity marked terminals on the secondary side.

According to ANSI standards, all power and distribution transformers and also dry-type power transformers have subtractive polarity.

3.7 Application Considerations

The following generalities apply to the selection of class C accuracy CTs.

1. *Select a CT ratio*

    This is dependent upon the desired sensitivity of relaying and also on ratio balancing that is required in some relaying applications, such as ground fault differential relaying. For example, many MMPRs will correct the CT mismatch in differential scheme up to a certain level only.

    Another important consideration for the selection of ratio is the accuracy of the CT. By definition, class C accuracy holds for only 20 times its primary current, at rated burden (thereafter nonlinearity can set in and the errors are unpredictable). Now consider a switchgear phase CT of 600:5 and the primary symmetrical three-phase short-circuit current of 40 kA. The maximum short-circuit current to which 600/5 CT should be applied is 12 kA and the minimum ratio that should be selected for a 40 kA primary current application is 2000:5. A calculation can be made when the actual burden is other than the standard burden. It is not unusual to see this misapplication in almost every switchgear installation. A switchgear manufacturer will provide a CT ratio as demanded by the customer, and this qualification of 20 times the maximum current is ignored.

    As an example, consider a metal-clad switchgear rated at 50 kA interrupting, $K = 1$ (Volume 1), according to ANSI/IEEE Standard C37.04. A manufacturer should not install a phase CT of any ratio less than 2500:5. Now consider that the load to be served from a feeder circuit breaker is only 200 A. It may be necessary to provide two sets of CTs: one set for the overload conditions and the second set for the fault conditions. This may resolve the concern for relay maloperation during fault conditions, but requires additional metering device. This practice is not adopted though recommended in Reference [2].

    Further considerations of ratio selection are applicable; for example, for generators a CT ratio of at least 150% of the generator current rating is chosen. Ratio selection also impacts saturation [4].

    An exception is that the ratio selection of a core-balance CT is not dependent upon the phase load current. *For a given application, select the maximum permissible ratio.*

2. *Make a single-line diagram of the CT connections*

    This should show all the protective and other devices connected in the CT circuit, the secondary cable sizes and lengths, the VA burdens of the relays, and instruments at the calculated settings. It is usual to combine protection and metering functions in the same CTs in industrial distribution systems. This practice needs to be reviewed in terms of transient overloads that relaying class CTs can impose on the metering circuits. Furthermore, the applications and accuracies for relaying and metering class CTs are different.

3. *CT burden*

   Accurately calculate the CT secondary burden as reflected on the CT terminals, including the CT secondary windings. Though, the power factor of the burden is often neglected, it is desirable to include it for accuracy. The burden is usually expressed in VA at a certain power factor or it can be expressed as impedance in $R + jX$ format.

4. *Short-circuit currents and asymmetry*

   Accurately calculate the short-circuit currents on the primary side and their asymmetry and the fault point X/R ratio. For time delayed devices, it is necessary to calculate 30-cycle currents.

5. *Calculate steady-state performance*

   The excitation curve of the CT, similar to Figure 3.1, is required from the manufacturer. Consider a 600:5 CT in Figure 3.1. Its knee point voltage can be read approximately 90 V. Corresponding to 10 A excitation current, the voltage is 200 V. The knee point voltage is approximately 46% of the excitation voltage corresponding to 10 A excitation, and the excitation voltage at 10 A current is also the C rating. Thus, for the steady-state performance, *as a rule of thumb*, the calculated secondary impedance including CT winding resistance when multiplied by the CT secondary current under maximum fault condition should not exceed the CT C rating. Vectorial calculation is required.

6. *Calculate steady-state errors*

   This is illustrated in Example 3.1 and also examples in References [5,6]. Once the secondary voltage is known, the excitation current can be read from it and the error calculated.

Example 3.1

A CT ratio of 600:5 has the following devices connected in the secondary circuit:

- A very inverse electromechanical relay, set at a tap of 4 A time delay overcurrent, no instantaneous. According to the manufacturer, the relay burden at the 4 A tap is 2.38 VA, power factor = 68%.
- CT secondary loop impedance corrected to 75°C = $0.0820 + j0.0062$ Ω.
- An ammeter VA burden = 0.05 VA at unity power factor.
- A wattmeter VA burden = 077W at 4 A and power factor = 0.54.
- CT secondary winding resistance (reactance ignored) = 0.31 Ω (see Figure 3.1).

The total secondary impedance is $0.5663 + j0.1262$ Ω = $|0.5802|$ Ω (see Table 3.5 for the calculations).

The primary short-circuit current is 18 kA rms. Thus, the maximum secondary current is 150 A.

Therefore, the secondary voltage developed across CT windings is 84.9 V, ignoring reactance, and 87.02 V if the reactance is considered. Table 3.5 also calculates the burden

TABLE 3.5

Calculation of CT Secondary Resistance and VA Burden, Example 3.1

Device	Specified Burden Data	Burden, VA at 150 A	Impedance $R + jX$ (Ω)
Overcurrent relay	2.38 VA at 4.0 A setting, 68% PF	3347	$0.101 + j0.1090$
CT secondary leads	$0.0820 + j0.0062$ Ω	1850	$0.0820 + j0.0062$
CT secondary resistance	0.31 Ω	6975	$0.31 + j0$
Ammeter	1.05 VA, 5 A at unity PF	945	$0.042 + j0$
Wattmeter	0.82 VA at 5 A and 0.94 PF	740	$0.031 + j0.011$

at the maximum current of 150 A. The secondary impedance based on the calculated burden is

$$\frac{13,857}{(150)^2} = 0.616 \, \Omega$$

which gives an excitation voltage of 92.38 V.

The excitation current read from Figure 3.1 for 600:5 CT at this excitation voltage is approximately 0.08 A.

Therefore, the percentage ratio error is

$$\left(\frac{0.08}{150}\right) \times 100 = 0.053\%$$

This is acceptable. The excitation current will increase steeply, if the secondary voltage increases.

Example 3.2

In Example 3.1, the overcurrent relay is provided with an instantaneous function, which is set at 50 A. Then, in the above calculation of CT secondary connected resistance, we add the extra burden of the instantaneous element, e.g., 50 VA at 50 A pickup and at 80°. The impedance of $0.003 + j0.197$ is added to the calculations of impedance in Example 3.1, and the total secondary impedance becomes $0.5693 + j0.3232 \, \Omega = 0.6546 \, \Omega$.

However, to calculate the secondary voltage, we do not consider the maximum fault current of 150 A. We consider the maximum setting of the instantaneous overcurrent element, which is 50 A. Then the secondary excitation voltage is 32.73 V. The ratio error will be even smaller as compared to Example 3.1.

Example 3.3

This example addresses a major problem of CT saturation, when an auxiliary CT is used to step up the current for ratio matching. Figure 3.11 shows the configuration. Figure 3.11a illustrates a residual connection, and an auxiliary CT of ratio 1:30 is interposed to increase the sensitivity of ground relay pickup, shown as GR in this figure. The effective CT ratio as seen by the relay GR becomes 100:5. Compare this with Figure 3.11b which shows a zero-sequence CT of ratio 100:5, connected to GR, as an alternative connection. Thus, the relay GR should see the same effective secondary current in either of the two connections, but this is not true.

Instrument Transformers

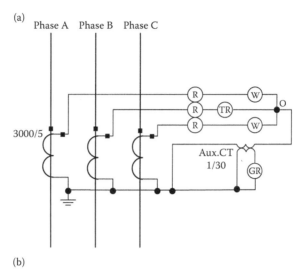

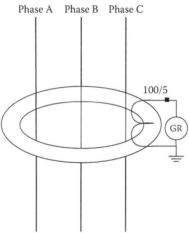

FIGURE 3.11
(a) Residual connection of a GR fault relay, through auxiliary wound-type step-up CT of ratio 1:30 and (b) equivalent connection through a 100/5 core-balance CT.

First consider the connection in Figure 3.11b. The ground fault current is 400 A and the excitation characteristics of 100/5 core-balance CT are shown in Figure 3.4. Then, the CT secondary current $I_s = 20$ A. Assuming a CT secondary winding resistance $R_s = 0.5\,\Omega$ and a secondary burden $R_b = 0.5\,\Omega$, the total burden is 1.0 Ω. The burden will be at a certain power factor, which we ignore in this example. The CT secondary voltage V_s is given by

$$V_s = I_s(R_s + R_b + jX_b) \tag{3.16}$$

where R_s is the CT secondary resistance and R_b is the CT secondary burden. Ignoring the power factor, V_s is equal to 20 V. Thus, even a C20 rating is adequate. From Figure 3.4, the excitation current is 1.0 A, which gives a ratio error of 5%.

Again considering 400 A ground fault current and residual connection of the CTs in Figure 3.11a, the secondary current in auxiliary CT windings is 20 A. This being a wound-type CT, its secondary resistance will be higher. Considering the same burden and the same secondary resistance for comparison, a total of 1 Ω, when reflected on the primary side, becomes

$$R_p = n^2 \times (R_s + R_b) = 900 \, \Omega \qquad (3.17)$$

The phase CTs should not saturate for a voltage of 900 × 0.67 = 600 V, ignoring all other burdens. This is rather a large secondary voltage and the phase CTs must be designed for C800 or better accuracy. Figure 3.11b illustrates the connection of a core-balance CT for the same application.

The IEEE Standard C37.110 [7] recommends that the auxiliary CTs should be avoided in a step-up configuration, though it has been an industry practice. An example of step-up connection for differential ground fault protection is shown in Figure 3.12. Note the connections of auxiliary ratio balancing CT and the flow of the currents in the differential product type relay for a 200 A external fault. The MMPR relays for this application internally balance the CT mismatch and provide more sensitive and stable differential protection.

It is not unusual to see metering class or class T CTs applied as step-up CTs, and the resultant nuisance trips.

3.8 Series and Parallel Connections of CTs

The current transformers can be connected in series and parallel. Figure 3.13a shows a single CT of ratio 600:5, a secondary resistance of 0.31 Ω, and the lead and devices connected to a CT having a burden of 1.0 Ω. The CT secondary voltage is 162.5 V, which gives high excitation currents. If two similar CTs are connected in series with the same secondary burden (Figure 3.13b), the secondary voltage across each CT winding reduces to 100.6 V.

The CT burdens can also be reduced by using the following:

- Multifunction microprocessor relays (MMPRs) can be used to reduce burden. As an example, the relay burden is only 0.1–0.2 VA, see Chapter 4.
- A larger size CT secondary lead can be used to reduce their burden.
- The electronic meters have a small requirement of CT burden, typically 0.1–0.2 VA.

A CT of higher ratio, considering the sensitivity of the pickup desired and the available relay setting range, may be possible to be selected. The CT of higher ratio will have higher winding resistance, but the CT secondary current will be reduced, which has a more pronounced effect in reducing the secondary voltage. Also, a CT of higher ratio can have higher class C accuracy.

The other CT specifications are continuous thermal rating factor, short-time rating, and BIL, which are not discussed here.

3.9 Transient Performance of the CTs

The above steady-state analysis of calculation is not adequate. The transient performance and saturation on short-circuit asymmetry should be considered. This may even result in

Instrument Transformers

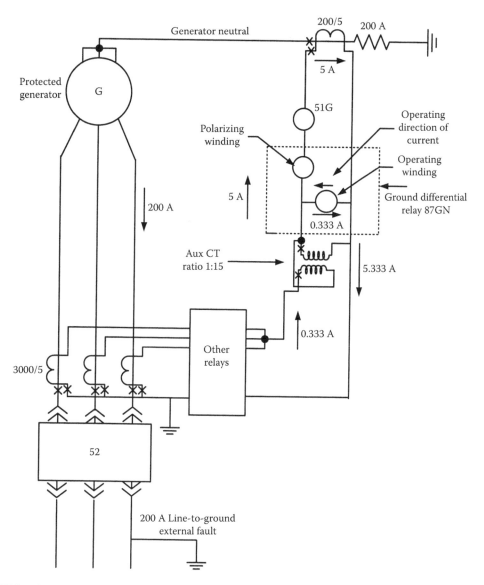

FIGURE 3.12
Connections of a product type electromechanical ground fault relay, showing auxiliary ratio matching CT of ratio 1:15 (see text).

nonoperation of the instantaneous devices [4,8]. This is of special importance for differential and instantaneous relaying.

The CT saturation is addressed in References [4–14]. The integrity of protection can be seriously jeopardized and nuisance trips can occur, if the CTs are not properly selected for the application and saturation characteristics are not accounted for. Recommendations of the manufacturers for a certain application must be followed.

Figure 3.14 [8] shows the saturation of a CT. This is reproduced from Chapter 1, Volume 3. The top curve shows the asymmetrical fault current and the bottom curves show progressive saturation.

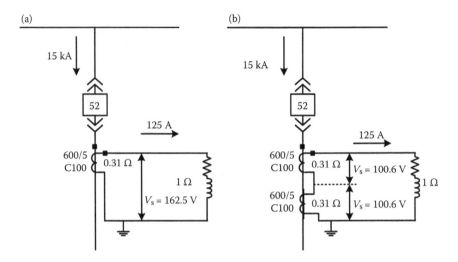

FIGURE 3.13
(a) Connection of a single 600:5 ratio CT (secondary voltage, 162.5 V) and (b) connection of two identical 600/5 CTs in series (the secondary voltage across each CT reduced to 100.6 V).

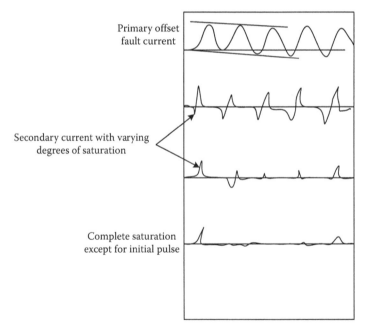

FIGURE 3.14
Oscillogram showing progressive saturation of a CT [8].

A completely saturated CT does not produce an output except during the first pulse, as there is a finite time to saturate and desaturate. The transient performance must consider DC component, as it has more pronounced effect in producing severe saturation of the CT compared to AC component.

An equivalent circuit of the CT is shown in Figure 3.15. Note that the CT magnetizing impedance is nonlinear and can vary over large values. As the excitation current increases,

Instrument Transformers

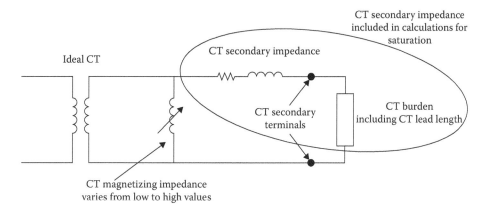

FIGURE 3.15
Equivalent circuit of a CT.

the impedance falls. *Ideally, if the saturation characteristics in Figure 3.1 were a straight line, the saturation would not occur.*

3.9.1 CT Saturation Calculations

In the calculation in above examples in order to avoid saturation, the CT secondary saturation voltage, V_x, was calculated based upon the following equation:

$$V_x \geq I_s Z_s \tag{3.18}$$

where Z_s is the total secondary burden including CT resistance. When the offset waveform concept was introduced, it resulted in the following equation:

$$V_x \geq K I_s Z_s \tag{3.19}$$

The following values of K have been used:

$$K = 1.6$$

$$K = 2$$

$$K = 2\sqrt{2}$$

In the mid-1980s, Zocholl and Kotheimer published their papers [9–11]. Later, the IEEE Power engineering Society Relay Committee addressed this topic, and in 1996 the IEEE standard 110-1996 formalized the work of these publications [7]. This standard recommends the (1 + X/R) method. To avoid saturation with DC component of the fault current, the following equation holds:

$$V_x > I_s Z_s \left(1 + \frac{X}{R} \times \frac{R_s + R_b}{\sqrt{(R_s + R_b)^2 + (X_b)^2}} \right) \tag{3.20}$$

where X/R is calculated at the fault point. Equation 3.20 takes into account the inductive component of the CT burden. If this is ignored, the simplified equation is

$$V_x > I_s Z_s \left(1 + \frac{X}{R}\right) \tag{3.21}$$

The short-circuit current equations for a three-phase fault are derived in Volume 1 and not repeated here. The concept of exponentially decaying components is discussed. Even if the fault occurs when the voltage peaks in one phase, the DC component is zero; but in a three-phase system, as the phases are displaced by 120 electrical degrees the DC component will be present in at least two phases.

Equation 3.23 accounts for this asymmetry in the short-circuit current.

As the CT saturation increases, so does the secondary harmonics, before the CT goes completely into saturated mode. Harmonics of the order of 50% 3rd, 30% 5th, 18% 7th, 15% 9th, or higher may be produced. These can cause improper operation of the protective devices. Thus, the following two interrelated issues need to be addressed:

- Evaluation of the saturation of the CT
- Effect of distorted waves and harmonics on the operation of the protective devices

3.9.2 Effect of Remanence

When a magnetic material is subjected to an alternating exciting current, the hysteresis loop is traced (see Figure 3.16). The relation is different with increasing and decreasing values of magnetic intensity. This is due to an irreversible process which results in energy dissipation, produced as heat. The first time the magnetic core is excited, neutral or virgin

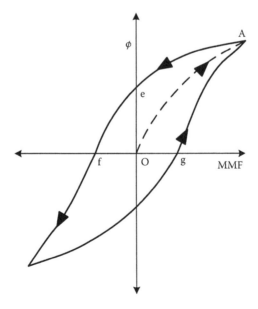

FIGURE 3.16
Hysteresis loop on magnetization of a magnetic material.

Instrument Transformers 75

curve OA is produced, but it cannot be reproduced in the reverse direction. Some magnetism is left as the MMF drops to zero, given by *oe*, which is the residual magnetism. To bring it to zero, a reverse MMF (*of* or *og*) must be applied, which is called the coercive force. The area under the curve, and *oe*, *of*, and *og* depend upon the magnetic material.

This is very akin to power transformers, where the residual magnetism can give higher inrush currents and saturation on switching.

In a CT to avoid saturation due to remanence:

$$V_x > \frac{I_s Z_s \left(1 + \frac{X}{R} \times \frac{R_s + R_b}{\sqrt{(R_s + R_b)^2 + (X_b)^2}}\right)}{1 - \text{per unit remanence}} \quad (3.22)$$

3.10 Practicality of CT Applications

Consider that in a 13.8 kV system the available three-phase fault current is 35 kA and the X/R ratio is 30. In fact, the fault point X/R at the primary voltage of distribution in an industrial power system can be as high as 80, when the distribution system contains large generators. The X/R ratio of a generator of 80 MVA can be close to 100. Also short-circuit current limiting reactors have high X/R ratios to limit the fundamental frequency copper loss.

For 35 kA of fault current, X/R is equal to 30; a CT ratio of 1500:5 meets the criterion of 20 times the current to limit the ratio error to 10%. Let us consider a CT of ratio of 2000:5, a secondary resistance of 1.15 Ω, a secondary burden of 0.550 Ω, and a total secondary burden of 1.7 Ω. Then, the secondary voltage calculated from various equations cited above is given as follows:

- $V_s = 102$ V (Equation 3.18)
- $V_s = 163.2$ V, $K = 1.6$ (Equation 3.19)
- $V_s = 204$, $K = 2$ (Equation 3.19)
- $V_s = 288$ V $K = 2\sqrt{2}$ (Equation 3.19)
- $V_s = 4611$ V (Equation 3.21)
- $V_s = 7685$ V, with 0.4 per unit remanence (Equation 3.22)

The (1 + X/R) factor gives high CT secondary voltages that are impractical to meet in real-world applications. The maximum class C accuracy described in the standards is 800. *This clearly demonstrates that the calculation of saturation using 1 + X/R method is not a real-world situation* [14]. Further, Table 3.6 shows the calculated secondary voltages for various primary fault currents and X/R ratios, ignoring all secondary burden except the CT winding resistance itself. The secondary voltages are too high for practical selection of an accuracy class, even for the X/R = 15. Table 3.7 shows similar calculations, with standard CT burden of C200 accuracy CTs. The secondary calculated voltages are even higher, though no remanence is considered.

The maximum ANSI rating, i.e., C800, can be selected for all CTs. The IEEE Standard C37.110 [7] suggests to use identical CTs and match the knee point voltages in differential

TABLE 3.6

Calculated CT Secondary Voltage (Volts) Using Factor (1 + X/R), and Ignoring Remanence and All CT Secondary Burden Except the CT Secondary Winding Resistance

System Short-Circuit Current in kA rms sym.	Minimum Required CT Ratio	CT Winding Resistance Ω	Secondary Voltage in Volts			
			$X/R = 15$	$X/R = 20$	$X/R = 30$	$X/R = 50$
20	1000:5	0.51	816	1071	1581	2601
30	1500:5	0.84	1344	1764	2604	4284
40	2000:5	1.15	1840	2415	3565	5865
50	2500:5	1.50	2400	3150	4650	7650

TABLE 3.7

Calculated CT Secondary Voltage (Volts) Using Factor (1 + X/R), Ignoring Remanence and Standard CT Secondary Burden

System Short-Circuit Current in kA rms sym	Minimum Required CT Ratio	Standard CT Burden, Ω All CTs of C200 Accuracy	CT Winding Resistance Ω	Secondary Voltage in Volts			
				$X/R = 15$	$X/R = 20$	$X/R = 30$	$X/R = 50$
20	1,000:5	2	0.51	4,016	5,271	7,781	12,801
30	1,500:5	2	0.84	4,544	5,964	8,804	14,484
40	2,000:5	2	1.15	5,040	6,615	9,765	16,065
50	2,500:5	2	1.50	5,600	7,350	10,850	17,850

relaying application, so that these have the same saturation characteristics. The authors of [9–14] further investigate the impact of CT saturation on protective relaying.

An MMPR has internal matching auxiliary CTs, filters, and AD converters with scaling of output. Apart from rigorous CT saturation simulations using electromagnetic transient program, efforts are directed toward development of software and other programs to calculate the impact of CT saturation on protective relaying.

From the above discussions, we come to the following conclusion:

- Use a CT ratio as high as practically possible. The limitation can be the pickup settings in the protective relays. Modern MMPRs have a large band of pickup settings; for example, the minimum setting on overcurrent pickup can be 0.5A. Even if a 4000:5 CT is used, it allows a primary pickup setting of 400 A.
- Reduce the burden. Again modern MMPRs have negligible burdens. Thus, the burden will be mainly of the CT secondary impedance and the CT leads. A conductor size of #10 AWG can be used instead of #14.
- Use high class C accuracy.

3.11 CTs for Low-Resistance Grounded Medium-Voltage Systems

In industrial medium-voltage low-resistance grounded systems, the fault point X/R ratio for a ground fault is low; that is, the zero-sequence reactance is small and the resistance predominates. The ground fault current can be calculated using symmetrical component method. As a specimen calculation, the following data are presented for an industrial system, maximum ground fault current is equal to 400 A, using rigorous calculations:

$$Z_0 = 2.5116 + j0.0290$$

$$Z_1 = 0.0005 + j0.0233$$

$$Z_2 = 0.0008 + j0.0221$$

Based upon these sequence impedances, the X/R ratio of a single line-to-ground fault is 0.029, i.e., power factor is 99.95%, and the fault current is practically in phase with the voltage. For a three-phase fault, the X/R from above data is 46.6. For the short-circuit calculations, according to ANSI/IEEE standards, the X/R ratio is calculated from separate R and X networks. For protective relaying, it is appropriate to calculate it from complex impedance. Note the low X/R ratio and low magnitude of line-to-ground fault current. Thus, the requirements of CT accuracy are minimal. For example, a 400:5 CT, the maximum fault current is limited to 400 A and with 1.0 Ω total burden, including CT resistance and lead length resistance need to have an accuracy of C10 only.

3.12 Future Directions in CT Applications

In Chapter 9, we will discuss the CT saturation and its impact upon low impedance bus differential relaying. It is shown that even with heavy CT saturation a proper operation is obtained with the algorithms built into the relay to account for CT saturation.

This is the direction in which the relaying technology is progressing. CT saturation cannot be avoided in all applications. The CT saturation reduces the apparent current seen by the relay. This can delay operation of time overcurrent elements and the instantaneous elements may not operate at all.

MMPRs employ digital filtering [13] to obtain phasors that eliminate DC component and harmonics. It is important to employ instantaneous elements that operate on the fundamental in the absence of saturation, but respond to peak currents during saturation.

Figure 3.17 shows that the fundamental is severely reduced in a severely saturated waveform; a 100:5 ratio C50 CT, with 40 kA fault current. The magnitude of the fundamental frequency current in severely saturated CT waveform is a poor representation of the actual fault current. The digital filters cannot make accurate measurements once the saturation sets in. The improved response of the rms, peak and cosine filters, with the same fault current but CT of ratio 200:5 and C100 is shown in Figure 3.18. Both peak and rms filters respond quickly to a fast rising signal and exhibit a high transient response as these respond to dc component in the asymmetrically offset waveform. The cosine filter responds to the fundamental frequency component of the signal and is slower, but has admirable characteristics with respect to dc offset and removal of harmonics. Combining a bipolar peak detector with a cosine filter provides an efficient solution to instantaneous element. Figure 3.19 shows instantaneous function logic of a modern MMPR using cosine peak adaptive filter. The cosine filter supplies magnitude of normal sine wave operation and bipolar peak detector for the saturated waveforms. By incorporating a method to determine when the

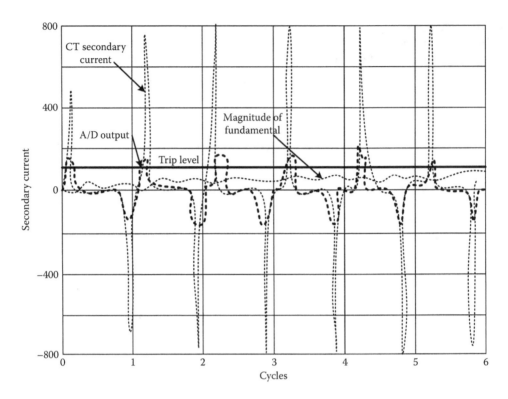

FIGURE 3.17
Output of a 100:5 ratio C50 CT for a 40 kA fault current (severely saturated CT) [13].

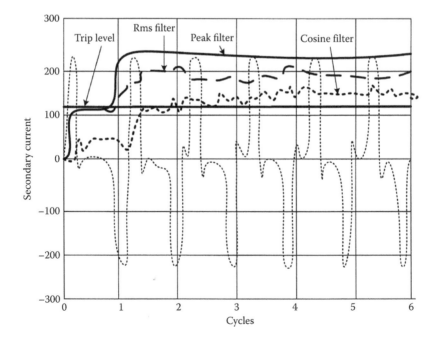

FIGURE 3.18
Improved response for a 40 kA fault current, 200:5 ratio, C100 CT with filters [13].

Instrument Transformers

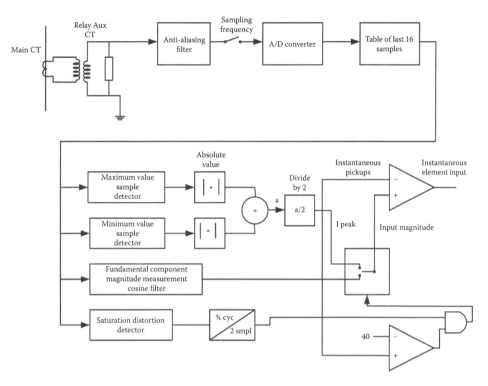

FIGURE 3.19
Adaptive overcurrent element block circuit diagram showing bipolar and cosine filters with CT saturation distortion detector [15].

CT is saturated, the adaptive overcurrent element can switch the instantaneous elements such that they operate on the input of the cosine filter or on the output of the bipolar detector as appropriate. The adaptive overcurrent element determines which filter to use by means of saturation detector, which operates when the harmonic content of A/D converter output exceeds a threshold called the harmonic distortion index. Also, the bipolar detector is enabled only when phase pickup setting is greater than 40 A secondary (5 A relay). This ensures that a bipolar peak detector is active in conditions where CT saturation is likely to affect overcurrent operation [15].

It should not, however, be implied that a CT can be randomly chosen for the MMPRs. The relays are tested for the required performance, and the recommendations of the manufacturer for a particular relay type and application should be followed.

3.13 Voltage Transformers

3.13.1 Rated Primary Voltage and Ratios

Rated primary voltage and ratios are depicted in Figure 3.20.

- Group 1 voltage transformers are shown in Figure 3.20a,b. These are meant for applications with 100% of rated voltage across primary windings when connected line-to-line or line-to-ground. Transformers shall be capable of operation at 125%

of the rated voltage on an emergency basis, provided VA burden does not exceed 64% of the thermal burden rating, without exceeding the specified temperature rises.

- Group 2 voltage transformers are primarily for line-to-line service, and may be applied line-to-ground or line-to-line at a winding voltage equal to primary voltage rating divided by $\sqrt{3}$ (see Figure 3.20c,d). These are capable of continuous operation at 110% rated voltage, provided the burden at this voltage does not exceed the thermal burden rating.
- Group 3 voltage transformers are for line-to-ground connection only (see Figure 3.20e). These may be insulted neutral or grounded-neutral types. Ratings through

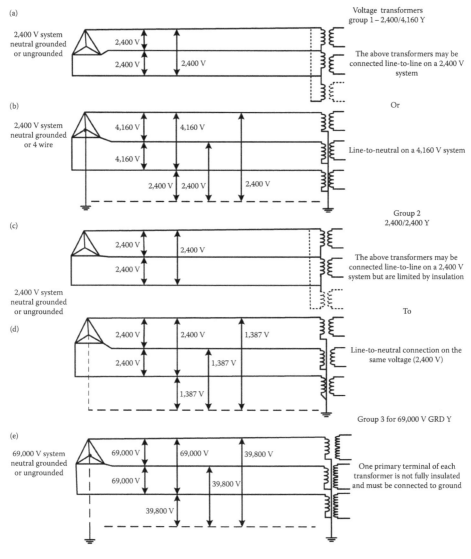

FIGURE 3.20
(a–e) Voltage transformers groups.

(*Continued*)

Instrument Transformers

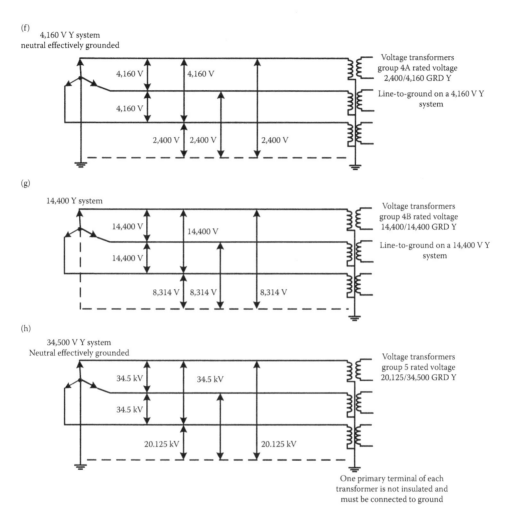

FIGURE 3.20 (CONTINUED)
(f–h) Voltage transformers groups.

92 kV for 161 kV shall be capable for $\sqrt{3}$ times the rated voltage for 1 min without exceeding 175°C temperature rise for copper conductors. Ratings 138 kV for 230 kV Gnd Y and above shall be capable of operation at 140% of rated voltage with the same temperature and time limitations.

- Group 4 transformers (Figure 3.20f,g) are for line-to-ground connection only. These shall be capable of continuous operation at 110% of rated voltage, provided VA burden at this voltage does not exceed the thermal burden rating.
- Group 4A (Figure 3.20f) transformers shall be capable of operation at 125% of rated voltage on emergency (8 h) basis, provided the burden at rated voltage does not exceed 64% of the thermal rating without exceeding temperature rises as specified in the standard. Figure 3.20g is applicable to group 4B.
- Group 5 transformers are for line-to-ground connection only, and these are for use on grounded systems. They may be insulated neutral or grounded-neutral type. These shall be capable of operation at 140% of rated voltage for 1 min without

exceeding a 175°C rise for copper conductors or a 125°C rise for EC aluminum conductors. Group 5 transformers shall be suitable for continuous operation at 110% of rated voltage, provided VA on this voltage does not exceed thermal burden ratings (Figure 3.20h).
- Table 3.8 shows the voltage transformer groups according to IEEE standard [1].

3.13.2 Accuracy Rating

Standard metering class accuracies for voltage transformers establish limits from 90% to 110% of the rated voltage ratings, which correspond to 120 or 115 V secondary. When a voltage transformer is operated at 58% of rated voltage, the accuracy will be different than at 100%. The standard burdens at 120 and 69.3 V in Table 3.9 have different impedances. Therefore, a transformer will have much different errors at 69.3 and 120 V.

TABLE 3.8

Standard Burdens for Voltage Transformers

	Characteristics on Standard Burdens		Characteristics on 120 V Basis			Characteristics on 69.3 V Basis		
Designation	VA	Power Factor	Resistance (Ω)	Inductance (H)	Impedance (Ω)	Resistance (Ω)	Inductance (H)	Impedance (Ω)
W	12.5	0.10	115.2	3.04	1152	38.4	1.01	384
X	25	0.70	403.2	1.09	576	134.4	0.364	192
Y	75	0.85	163.2	0.268	92	54.4	0.0894	64
Z	200	0.85	61.2	0.101	72	20.4	0.0335	24
ZZ	400	0.85	30.6	0.0503	36	10.2	0.0168	12
M	35	0.20	82.3	1.07	411	27.4	0.356	137

TABLE 3.9

Summary of IEEE Voltage Transformer Groups

Group	Number of Bushings	Connection Method	Neutral Grounding	Notes
1	2	Open delta Wye–wye possible	Any	Withstand 25% above rated voltage on an emergency basis
2	2	Open delta Wye–wye possible	Any	Withstand 10% above rated voltage continuously. Primary rated for line-to-line voltage
3	1	Wye–wye–wye	Any	Outdoor, two secondary windings. Withstand 10% over rated voltage continuously
4A	1	Wye–wye	Effectively	Withstand 10% over rated voltage continuously and 25% on emergency basis. For operation at 100% rated voltage
4B	1	Wye–wye Wye-broken corner delta	Noneffectively	Withstand 10% overvoltage continuously. For operation at 58% rated voltage
5	1	Wye–wye	Effectively	Outdoor applications. Withstand 40% over rated voltage for 1 minute and 10% over rated voltage continuously

3.13.3 Thermal Burdens

The standard burdens are based on two secondary voltages, 120 and 69.3 V, as listed in Table 3.9. Thermal burdens are specified in terms of maximum burden in VA that a transformer can carry at rated secondary voltage without exceeding the temperature limits specified in IEEE standard [1].

A voltage transformer is assigned an accuracy rating for each of the standard burdens for which it is designed. For example, it may have an accuracy rating of 0.3W and X, 0.6Y, 1.2Z. When a voltage transformer has two secondaries, the burden on one secondary affects the voltage on the other. The error limits apply to both secondaries and the burden in VA for a given accuracy can be divided between the two secondaries in any desired manner.

3.13.4 PT Connections

Two most common connections of voltage transformers (PTs) for a metal-clad switchgear in draw-out construction protected with primary and secondary current limiting fuses are shown in Figure 3. 21a,b. Figure 3.21a is open delta connection, which is most popular. Note that the 120 V secondary voltages shown between the phases and the middle phase b are grounded. The voltage between phases a and b is also 120 V, but the PT loads are connected from phase a to b and from c to b, and none from phase a to c. Figure 3.21b shows a wye–wye (grounded) connection. There is not much advantage in having this connection. The PT loads are connected between phases a, b, c to grounded neutral.

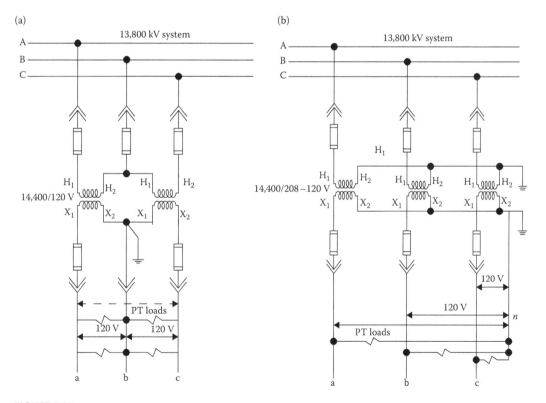

FIGURE 3.21
(a) Open delta connections of PT and (b) wye–wye (grounded) connection of PTs.

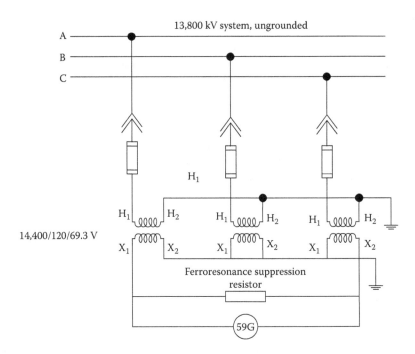

FIGURE 3.22
A wye-open delta connection of PTs for neutral displacement relay and for derivation of zero-sequence polarization voltage in directional ground fault schemes.

3.13.5 Ferroresonance Damping

A wye–wye connection of the PTs is required when a zero-sequence voltage polarization is required for directional ground fault relaying (see Chapter 8). In this case, a broken corner delta PT connection can be made (see Figure 3.22). This will sense ground fault on ungrounded systems through a neutral displacement relay (device 59G in this figure). The configuration is prone to ferroresonance and a damping resistor of 65 Ω is connected across the relay coil (see Chapter 8 for further details).

3.14 Capacitor-Coupled Voltage Transformers

Wire wound transformers above 138 kV are not economical. Capacitor voltage transformers (CVTs) consist of capacitors and inductors that can be applied from 72.5 to 1100 kV. These can be used for relaying and metering applications and can be designed with high accuracy of 0.1 metering. Though, the ferroresonant conditions must be accounted for and transient response carefully considered. The schematic diagram of a capacitor coupled voltage transformer (also sometimes called only CVT) is shown in Figure 3.23. The CVT consists of two main components: the high-voltage capacitor divider stack and the electromagnetic unit. The capacitor stack may comprise of one or more sections, consisting of serially connected capacitor elements housed in hermetically sealed porcelain housing. The capacitor's polypropylene/kraft paper insulation system is impregnated with specially processed oil. Each hermetically sealed section utilizes a stainless-steel expansion chamber to allow for

Instrument Transformers

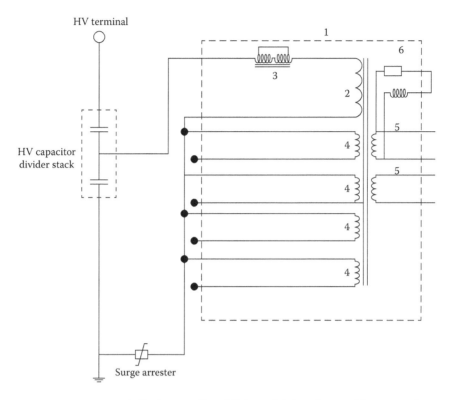

1. Electromagnetic unit (intermediate transformer with compensating reactor)
2. Primary windings or the intermediate voltage transformer
3. Compensating windings
4. Adjustment windings
5. Secondary windings
6. Ferroresonance damping circuit

FIGURE 3.23
Schematic representation of a circuit of CVT.

oil expansion and contraction due to temperature variations. The capacitor divider provides between 5 and 20 kV to the intermediate step-down transformer which has low-voltage output windings. The technical data of a 550 kV CVT are shown in Table 3.10.

3.14.1 Transient Performance

Three considerations are as follows:

- Ability to reproduce rapid changes in the primary voltage.
- The remaining secondary voltage due to short circuit at the primary voltage.
- Transient oscillations in the secondary voltage. These can be classed into two categories: (1) high-frequency components, frequency range 600–4000 Hz, which damp out in about 10 ms and (2) low-frequency components, frequency range 2–15 Hz, which last for a longer period.

TABLE 3.10

Technical Data of a 550 kV CVT

Parameter	Description	
Maximum voltage	550 kV	550 kV
BIL	1,550 kV	1,550 kV
Rated capacitance (pf)	5,000	10,000
Ratio	5,000:1	5,000:1
Secondary voltage	$100/\sqrt{3}$	$100/\sqrt{3}$
Accuracy class, maximum burden (two windings) Class 0.2 Class 0.5	300	250
Thermal VA	1,500	1,500
Creepage distance (mm/kV)	18,700	18,700
$L \times W \times H$ (mm)	855 × 740 × 6910	855 × 740 × 6910
Number of coupling capacitors	3	3
Weight (kg)	2,040	2,220

The transient response is a function of the following:

- Equivalent capacitance: See Figure 3.24. The higher the capacitance, the lower the residual voltage.
- Voltage taps: Higher capacitance and higher tap can work together to reduce the residual voltage.
- Ferroresonance suppression: See Figure 3.25. More effective ferroresonant circuits can suppress the transient voltage.
- Connected burden: See Figure 3.26. The lower the connected burden, the lower the residual voltage.

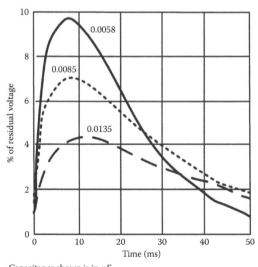

Capacitance shown is in μF

FIGURE 3.24
Effect of capacitance on residual voltage in a CVT.

Instrument Transformers

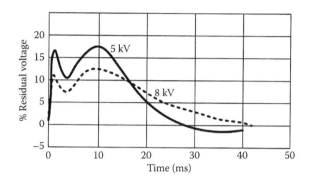

FIGURE 3.25
Effect of ferroresonance suppression on the residual voltage in a CVT.

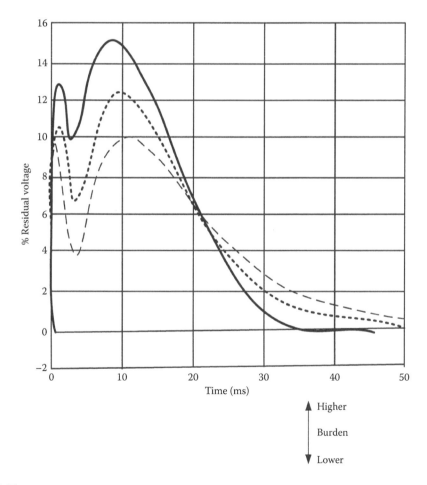

FIGURE 3.26
Effect of burden on residual voltage in a CVT.

Based on the above observations, the modern trend in the design of CVT is to use higher capacitance values, higher tap voltage up to 13 kV, and designs of better ferroresonant circuits. The stack capacitance, tap voltage, and ferroresonance circuits are the parameters of the design of CVT, while burden is under the control of user and applications [16].

3.14.2 Applications to Distance Relay Protection

CVT transients can cause incorrect information to be presented to the relay; however, as the transients last for a short period of time, this is not of consideration for zone 2 or zone 3 protection. The zone 1 protection without intentional time delay can be impacted. The worst case scenario will be for phase-to-phase faults, heavily loaded line, load flowing toward the relay, and the fault occurring at instant voltage wave crossing the zero axis. The maximum reach, depending upon loading, may have to be limited. Modern MMPRs use low-pass filters and counters to mitigate effects of CVT transients; the relay operation may have to be delayed up to 10 ms.

3.15 Line (Wave) Traps

Line (wave) traps are required for power line carrier (PLC) for teleprotection (voice and data communications). It is one of the most economical and reliable forms of communication and versatile in operation. It has three main components:

- Signal carrying medium, which is the high-voltage transmission line itself
- Communication apparatus, transmitter's receivers, and associated components
- Line traps and coupling capacitors

Line traps are connected in series with the HV transmission line and, therefore, must be rated to carry the full-load current of the transmission line, withstanding high mechanical stresses due to short circuit. It presents high impedance to the carrier frequency band and negligible impedance to the power frequency. The high frequency is, thus, confined to the line section between two substations. The carrier signal should be prevented from being dissipated in the substation itself, being grounded in case of a fault outside the transmission path, or being dissipated in a tap line or a branch of the transmission path.

The line traps are connected between the sections of a transmission line as shown in Figure 3.27a and generally suspension mounted, with single-point or multipoint brackets, in Figure 3.27b, but these can be pedestal mounted directly onto coupling capacitors, CVTs, or station post insulators. Several types of mounting pedestals are available.

The line traps are designed to meet the requirements of IEEE Standard C93.3 [17] or IEC Standard 60353 [18]. Table 3.11 shows the rated current and rated inductance according to IEC 60353 as a specimen.

The major components are the main coil, tuning device, and protective device (surge arrester), with optional bird barrier. The main coil is an air-core dry-type reactor. The winding is insulated and terminated at both ends for electrical connections to the transmission line.

Instrument Transformers

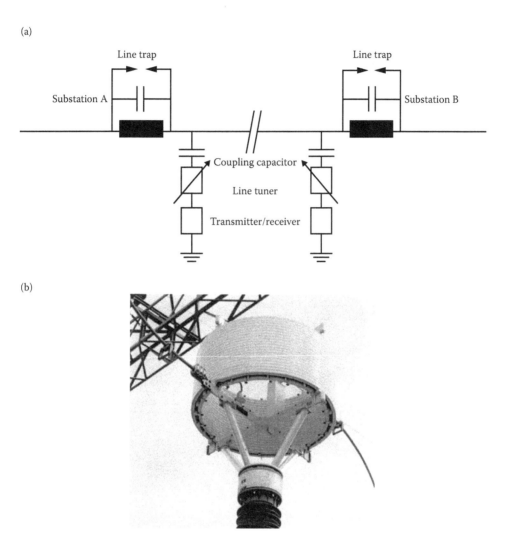

FIGURE 3.27
Typical mounting of a line trap.

TABLE 3.11

Rated Current and Rated Inductance according to IEC 60353 (at 50 Hz)

Rated Current Continuous	Short Time Series 1 kA/1 s	Rated Inductance, mH at 100 kHz						
2000	40	0.2	0.25	0.315	0.4	0.5	1.0	2.0
4000	63	0.2	0.25	0.315	0.4	0.5	1.0	2.0
Rated Current Continuous	Short Time Series 2 kA/1 s	Rated Inductance, mH at 100 kHz						
2000	50	0.2	0.25	0.315	0.4	0.5	1.0	2.0
4000	80	0.2	0.25	0.315	0.4	0.5	1.0	2.0

The tuning device connected across the main coil forms a blocking circuit which provided high impedance over a specified PLC frequency range. The tuning device may consist of capacitors, inductors, and resistors, all having relatively low power ratings. For environment projection, the components may be mounted in a fiberglass enclosure, and installed inside the main coil. The bandwidth of the line trap is the frequency range over which the trap provides a certain minimum blocking impedance/resistance. The blocking impedance will suppress the resonance that may occur with the line trap impedance resonating with the substation impedance.

Figure 3.28 shows the following type of tunings:

Single-frequency tuning
Single-frequency tuning gives narrow blocking bands. Within this band, high blocking impedance can be provided, which results in good PLC signal isolation (Figure 3.28a).

Double-frequency tuning
Double-frequency tuning blocks two relatively narrow bands of frequency. For proper operation, a minimum separation must be maintained between the peak tuning frequencies (see Figure 3.28b).

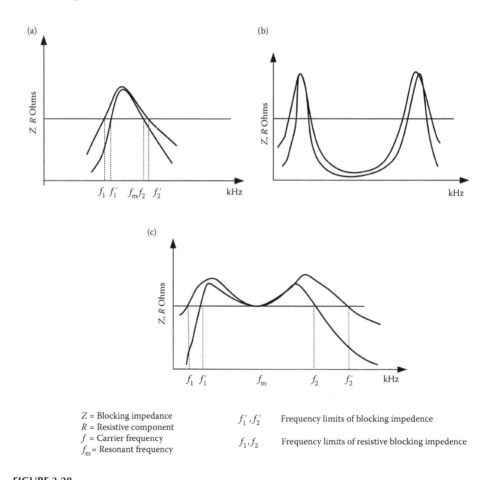

Z = Blocking impedance
R = Resistive component
f = Carrier frequency
f_m = Resonant frequency

f_1', f_2' Frequency limits of blocking impedence

f_1, f_2 Frequency limits of resistive blocking impedence

FIGURE 3.28
(a–c) Various tuning methods of line traps (see text).

Wideband tuning
This is the most common type of tuning, and the traps thus tuned are suitable for multichannel applications. This is so because practically a constant impedance is obtained over a broad frequency range (see Figure 3.28c).

Self-tuned line traps
The self-tuned line traps utilize the self-capacitance of the main coil winding. The inductance of a self-tuned line trap is higher than that of a tuned line trap.

Surge arrester
A surge arrester is connected as shown in Figure 3.23. It protects the tuning device and the main coil by limiting the transient overvoltage levels. The insulation of the main coil and the tuning device is coordinated with the surge arrester protective characteristics.

3.16 Transducers

The transducers can measure the metering quantities such as A, V, W, Var, PF, etc. These provide a scaled low-energy signal that represents the measured quantity. These buffer the control systems from power system, and smaller control cables can be run over relatively longer distance. The inputs to the transducer current or voltage are compatible with standard CT and PT secondaries. Many transducers will operate at levels above their normal operating ranges with little degradation or error. A transducer can output 4–20 or 0–1 ma output proportionate and scaled to the input parameter into a load burden of 10 kΩ. For voltage measurements, scaling is 0–10 V. The transducers have their ground references and double ground, that is, one at the transducer and another at the control panel can cause reliability problems and error. The shielded cables normally used have their shields grounded at the transducers.

References

1. ANSI/IEEE, Standard, C57.13. *Requirements for Instrument Transformers*, 1993 (R 2008).
2. ANSI/IEEE Standard, C57.13.2. *Conformance Test Procedures for Instrument Transformers*, 2005.
3. IEC 60044-1. *Instrument Transformers, Part 1*, (Current Transformers), 1997.
4. JR Linders. Relay performance considerations with low ratio CT's and high fault currents. *IEEE Trans Ind Appl*, 31(2), 392–405, 1995.
5. *Westinghouse Applied Protective Relaying*. Westinghouse Electric Corporation, Newark, 1982.
6. ANSI/IEEE Standard, 242. *IEEE Recommended Practice for Protection and Coordination of Industrial and Commercial Power Systems.*, 1986.
7. IEEE Standard, C37.110. *IEEE Guide for Application of Current Transformers used for Protective Relaying Purposes*, 1996.
8. JC Das, JR Linders. Power system relaying. *Wiley's Encyclopedia of Electronic and Electrical Engineering*, 17, 71–84. John Wiley & Sons, New York, 1999.
9. SE Zocholl, WC Kotheimer, FY Tajaddodi. An analytical approach to the application of current transformers for protective relaying, *43rd Annual Georgia Technology Protective Relaying Conference*, May 3–5, 1989, pp. 1–21, Atlanta.

10. SE Zocholl, WC Kotheimer. CT performance in critical relay application, *44th Annual Georgia Technology Protective Relaying Conference*, 1991, pp. 1–14, Atlanta.
11. SE Zocholl, DW Smaha. Current transformer concepts, *45th Annual Georgia Technology Protective Relaying Conference*, 1992, pp. 1–16, Atlanta.
12. JC Das, R Mullikin. Design and application of low ratio high accuracy split-core, core-balance current transformer. *IEEE Trans Ind Appl,* 46(5), 1856–1865, 2010.
13. J Hill, K Behrendt. Upgrading power system protection to improve safety, monitoring, protection and control, *Conference Record, IEEE Pulp and Paper Industry Technical Conference*, pp. 77–87, Seattle, Washington, 2008.
14. RE Cossé, DG Dunn, M Spiewak. CT saturation calculations: Are they applicable in the modern world?—Part I: The question. *IEEE Trans Ind Appl*, 43(2), 444–452, 2007.
15. SEL. Current transformer selection criteria for relays with adaptive overcurrent elements. *Application Guide* vol. 3, Publication AG2005-04.
16. I Sula, UO Aliya, GK Venoyagamoorthy. Simulation model for assessment of performance of CVT, *IEEE PES Summer Meeting*, 2006.
17. ANSI C93.3. *Requirements for Power-Line Carrier Line Traps*, 2013.
18. IEC 60353. *Line Traps for AC Power System*, 1989.

4

Microprocessor-Based Multifunction Relays

As stated in Chapter 1, the age of discrete electromechanical relays, single function, static relays, and digital relays is over. The industry demands protective relays that can host a number of programmable protective functions, control and metering, programmable inputs and outputs, pre- and postdata fault capture, and a variety of communication protocols with self-diagnostics of software and hardware failures. The redundancy over a period of time is avoided and a variety of applications can be made depending on the programming capability of the application engineer. It is possible to add-on protective and other functions by additional cards that can be inserted at the back of the relay. To provide a conceptual base, a feeder protection relay is described in this chapter.

4.1 Functionality

The standard and optional features put together give the flexibility and functionality as described.

4.1.1 Protection Features

- Phase instantaneous (50P)
- Ground (residual) instantaneous overcurrent (50G)
- Negative-sequence overcurrent (50Q)
- Phase time overcurrent (51)
- Ground residual time overcurrent (51G)
- Neutral time overcurrent (51N)
- Negative-sequence time overcurrent (51Q)
- Frequency, under and over (81)
- Breaker failure protection
- Breaker wear monitor
- Autoclosing control (79)

4.1.2 Voltage-Based Protections

- Undervoltage (27)
- Overvoltage (59)
- Negative-sequence overvoltage (59Q)

- Residual (zero-sequence) overvoltage (59G)
- Power element (32)
- Power factor (55)
- Loss of potential (60LOP)
- Rate of change of frequency (81R)
- Fast rate of change of frequency for Aurora mitigation (81RF)
- Arc-flash protection
- Demand and peak demand metering
- Synchronizing check with under- and overvoltage
- Station battery monitor
- Resistance temperature detector (RTD)-based protection: up to 10 RTDs can be monitored with trip and alarm settings for each RTD

4.1.3 Monitoring Features

- Event summaries that contain relay ID, date and time, trip cause, and current/voltage magnitudes
- Event reports including filtered and raw analog data
- Sequence event report (SER)
- A complete suite of accurate metering functions

4.1.4 Communications and Controls

- EIA-232 front panel port
- EIA-232, EIA-485, single or dual, copper or fiber-optic Ethernet, and fiber-optic rear panel ports
- Modbus RTU slave, Modbus TCP/IP, DNP3, LAN/WAN, Ethernet FTP, Telnet, Mirrored Bits, IEC 61850, DeviceNet, File Transfer Protocols, and Synchrophasors with C37.118 Protocol
- ASCII, compressed ASCII, fast meter, fast operate, fast SER, and fast message protocols
- Programmable Boolean and Math operators, logic functions, and analog comparisons

4.2 Front Panel

It contains four programmable pushbuttons with eight programmable LEDs:

- Eight target LEDs (six programmable).
- Operator control interface.

TABLE 4.1

Environmental and Voltage Information

Condition	Range/Description
Indoor/outdoor use	Indoor
Altitude	Up to 2000 m
Temperature	
IEC performance rating (per IEC/EN 60068-2-1 and IEC/EN 60068-2-2)	−40 to +85°C
Relative humidity	5%–95%
Main supply voltage fluctuations	As high as ±10% of nominal voltage
Overvoltage	Category II
Pollution	Degree 2
Atmospheric pressure	80–110 kPa

- Two digital inputs and three digital outputs. More digital inputs and outputs can be added with additional cards.
- Expansion slots for receiving additional cards.
- Configurable labels for LEDs and pushbuttons.

4.3 Environmental Compatibility

See Table 4.1 for the environmental and voltage information.

4.4 Dimensions

The relay has dimensions of 5.67" width, 7.56" height, and 6.92" depth. It can be panel mounted. Functionally, this single relay is equivalent to a hundred or more of discrete relays. Consider the immense amount of interwiring that is altogether eliminated and immense space saving for an application. This adds to the reliability.

Figure 4.1 shows a number of slots at the rear of the relay for additional cards that contribute to the protective functions, inputs, outputs, RTD inputs, and communication protocols. Figure 4.2 shows the input and output connections, in general; no terminal connections are shown.

4.5 Specifications

The compliance to UL and other standard specifications are shown in Table 4.2. Table 4.3 shows general specifications. The following points are noteworthy:

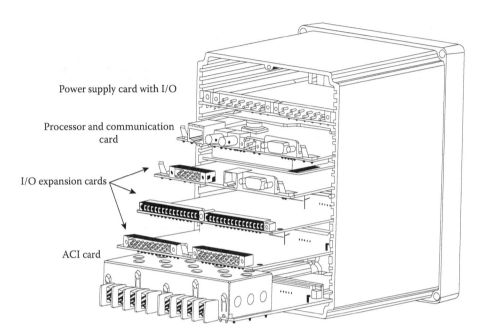

FIGURE 4.1
A rear view of the relay with expansion slots.

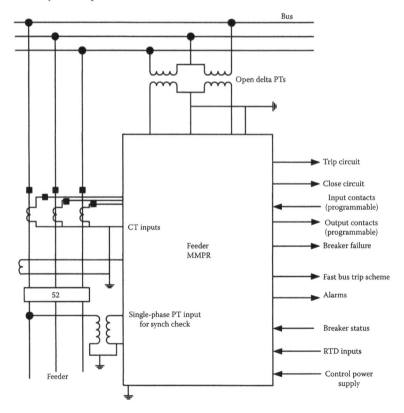

FIGURE 4.2
Input/output connection schematic—terminal numbers not shown.

TABLE 4.2

Compliance with Standards

Standard	Details
UL	Protective relay category NRGU, NRGU7 per UL 508, C.22.2 No. 14
CSA	C22.2 No. 61010-1
CE	CE Mark: CE directive
	Low-voltage directive
	IEC 61010-1:2001
	IEC 60947-1
	IEC 60947-1
	IEC 60947-5-1
Hazardous locations	UL 1604, ISA
Approvals	12,12.01, CSA 22.2 No. 213 and EN60079-15 (Class 1, Division 2)

TABLE 4.3

General Specifications

Parameter	Value	Ratings
Input current	5 A	Continuous rating: 15 A linear to 100 A symmetrical
		1 s thermal: 500 A
		Burden <0.1 VA@ 5 A
	1 A	Continuous rating: 3 A linear to 20 A symmetrical
		1 s thermal: 100 A
		Burden <0.1 VA@ 1 A
	50 mA	Continuous rating: 3 A linear to 1 A symmetrical
		1 s thermal: 100 A
		Burden <2 mVA@ 1 A
	2.5 mA	Continuous rating: 3 A linear to 12.5 mA symmetrical
		1 s thermal: 100 A
		Burden <0.1 mVA@ 1 A
Input voltage	20–250 V	Open delta PTs
	20–440 V	Wye–wye PTs
		Rated continuous voltage 300 Vac
		10 s thermal; 600 V
		Burden <0.1 VA
		Input impedance: 10 MΩ
Power supply	110–240 Vac 50/60 Hz	Input voltage range: 65–264 Vac, 85–300 Vdc, 24–48 Vdc
	110–250 Vdc	Power consumption: <40 Vac (ac), <20 W (dc 110–250 V), <20 W (24–48 Vdc)
	24–48 Vdc	Interruptions: 50ms@ 125 Vac (dc); 100 ms @ 250 Vac/dc; 50 ms @ 48 Vdc
Output contacts		Mechanical operations: 100,000 no load
		Pickup/dropout time ≤8 ms
	DC output ratings	Rated continuous voltage 250 Vdc
		Rated insulation voltage 300 Vdc
		Make: 30 A@ 250 Vdc as per IEEE C37.90
		Continues carry 6a@ 70°C
		Thermal: 50 A
		Contact protection: 360 Vdc, 40 J MOV protection across contacts
		Breaking capacity: 0.20 A L/r 40 ms (shown only for 25 Vdc)
	AC output ratings	Maximum operational voltage = 240 Vac
		Insulation voltage: 300 Vac
		Rated operational current: 3 A@ 120 Vac, 1.5 A @ 240 Vac

(Continued)

TABLE 4.3 (*Continued*)

General Specifications

Parameter	Value	Ratings
	High-speed high-current interrupting	Make 30 A Carry 6 A continuously at 70°C 1 s rating = 50 A Break capacity, 10,000 operations: 125 Vdc = 10.0 A L/R = 40 ms Pickup time : <50 µs resistive load Dropout time <8 µs
Opto-isolated control inputs	When used with DC control signals	250 Vdc: on for 200–312.5 Vdc, off below 106 Vac 125 V: on for 100–156.2 Vdc, off blow 75 Vdc Other voltage ranges not shown
	When used with AC control signals	250 Vac: on for 170.6–312.5 Vac, off below 150 Vdc 125 V: on for 85–156.2 Vac, off blow 53 Vac Other voltage ranges not shown
Analog output	4–20 mA	Load at 20 mA: 0–300 Ω Refresh rate: 100 ms % error full scale at 25°C ≤ ± 1%
	Voltage ±10 V	Load at 10 V: ≥2000 Ω Refresh rate: 100 ms % error full scale at 25°C ≤ ±0.55%
Analog inputs	±20 ma ±10 V	Input impedance: 200 Ω (current mode), >10 kΩ (voltage mode) Accuracy with user's calibration at 25°C = 0.05% of full scale (current mode) 0.025% of full scale (voltage mode)
Arc-flash detectors	Fiber type	1000 µm diameter, 640 nm wavelength, plastic, clear jacketed, or black jacketed Connector type: V-pin
Frequency		50, 60 Hz Phase rotation: ABC or ACB Frequency tracking 15–70 Hz
Time code input		Demodulated IRIG-B On(1) state Vih ≥2.3 V Off (0) state Vil ≤0.8 V Input impedance = 2 kΩ Synchronization accuracy internal clock ± 1 µs
Communication ports		Standard EIA-232 two ports, located front and rear Data speed 300–38,400 bps EIA-485 port, located rear Data speed 300–19,200 bps Ethernet port Single/dual 10/100BASE-T copper (RJ45 connector) Single/dual 10/100BASE-FX (LC connector) Multimode fiber optic Data speed 300–38,400 bps
Port characteristics		Port 1, Ethernet: Wavelength = 1300 nm Optical connector = LC, fiber-type multimode Fiber size: 62.5/125 µm Approximate range: 6.4 km Data rate: 100 Mb Port 2 serial: Wavelength = 820 nm Optical connector = ST, fiber-type multimode Fiber size: 62.5/125 µm Approximate range: 1 km Data rate: 5 Mb Channels 1–4 arc-flash detectors, wavelength = 640 nm, V pin connector, multimode, 1000 µm

(*Continued*)

TABLE 4.3 (*Continued*)

General Specifications

Parameter	Value	Ratings
Communication protocols		Mirror bits, Modbus, DNP3, FTP, TCP/IP, Telnet, SNTP, IEC 61850, EVMSG, C37.118 (Synchrophasors) and DeviceNet
Processing	AC current and voltage inputs	16 samples per system cycle
	Frequency tracking range	15–70 Hz
	Digital filtering	One cycle cosine after low-pass analog filtering. Rejects DC and all harmonics greater than fundamental
	Protection and control processing	Processing interval is four times per power system cycle, except for math variables and analog quantities, which are processed every 100 ms
	Arc-flash processing	Arc-flash light is processed 32 times per cycle; arc-flash current, light, and two fast hybrid outputs are processed 16 times per cycle
Oscillography	Length	15 or 64 cycles
	Sampling rate	16 samples per cycle, unfiltered
		4 samples per cycle filtered
	Trigger	Programmable using Boolean expressions
	Format	ASCII and compressed ASCII
	Time stamp resolution	1 ms

- Note that the relay burden is very small. For example, for a 5 A nominal input it is <0.1 VA. It is so small that it can even be ignored. Correspondingly, for an electromechanical relay, based on vendor's data:
- Phase overcurrent relay, time unit, setting range 4–12 A, has a burden of 146 VA at 40 A at 0.6 power factor. Instantaneous unit will have a burden of 40 A at 0.20 power factor. A wattmeter has a burden of 1.5 A and an ammeter, a burden of 1.4 VA. These high burdens impact the CT saturation. With microprocessor-based multifunction relays (MMPRs), the CT burden will consist of CT resistance and the CT leads. Furthermore, the MMPRs have built-in digital filters and algorithms so that even on CT saturation accurate operation can be obtained. As an example see Chapter 9 for a bus differential relay and also see Chapter 3 for CT operation.
- The relay will accept 5 A and 1 A secondary CT inputs; in addition, inputs of 50 mA and 2.5 mA are also specified. Note that for HRG systems, sometimes, a 2000:1 CT ratio is used. This translates to 50 mA secondary. In the same relay, it is possible to select an input of 5 A for phase overcurrent protection and 1 A for neutral ground fault protection. Note the high thermal ratings. For increased sensitivity of settings, a lower CT ratio can be selected which will increase the CT secondary current and allow sensitive settings, say, for restricted ground fault protection of a transformer.
- Note the ratings of the output contacts. These have a making capacity of 30 A; thus, the application of a lockout relay can be eliminated. This trend is noted, specially, for the relays in applications at utility electrical systems. It is opined that the elimination of a lockout simplifies the circuit and eliminated dependence on one more device to trip. Of course, where a number of breakers have to be tripped,

like in a bus differential protection, a lockout is still required. The pickup time is <8 ms, which is added to the total clearing time. The breaking capability of the contacts is limited, and these should not be used for any breaking applications like opening the close circuit of a breaker.
- The fast hybrid high-speed current interrupting contacts consist of a solid-state circuitry in parallel with a metallic contact. The pickup time is reduced from 8 ms to <50 μs.
- The opto-isolated contact inputs signify that the input contacts are controlled with an LED/light-dependent resistor. The resistance value is of the order of 35 kΩ. Thus, a very small current is needed for change of state of the input contact. One application is to continuously monitor the trip and close coils of a circuit breaker (see Figure 4.3). An input contact is simply wired in the circuit as shown. Most of the voltage is developed across the high resistance of the contact and a very small monitoring current flows through the circuit. On an open circuit of the trip coil, the resistance is infinite and the contact changes state. Note that all input contacts are "wetted" type.
- The power supply specifications show that it is possible to power the relay from 120/240 VAC. However a caution is required. If the AC power source is the same as that being protected on operation of a protective function, the AC control power supply will be lost within a short time delay. Thus, if the relay is powered from an AC power source, it should be from a UPS system and not through an auxiliary

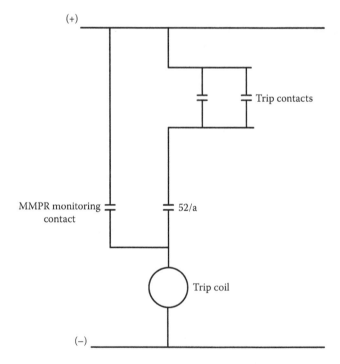

FIGURE 4.3
Continuous monitoring of the trip coil with an input contact.

control power transformer (CPT) connected to the same source which is being tripped. Thus, battery power supply is required, and in remote locations, the relay cannot be powered from a capacitive voltage trip (CVT) device as it can give rise to high voltages and will damage the relay. An electromechanical relay has an advantage that it is CT powered. Some static version of relays having limited functions, for example, overcurrent protection only, can be CT powered.

- A relation between temperature and RTD resistance can be developed.

4.6 Settings

The major protective functions and their accuracy and setting ranges are shown in Table 4.4.

The extended setting ranges of the protective functions are noteworthy. Many electromechanical relays will be required to serve the overlapping setting ranges shown in this table.

Also the accuracy of the settings is much higher than that can be achieved with electromechanical relays with the same pickup/dropout times.

The specifications for RTD protection show that the open circuit and short circuit of the RTDs are detected based on temperature. When an open-circuit or short-circuited RTD is detected, it is removed from the system. A small monitoring current of about 4 mA is continuously circulated through the RTD. A short circuit means that the resistance is zero, while an open circuit means that the resistance is infinite.

4.6.1 The Setting Groups

Table 4.5 illustrates the instantaneous phase overcurrent, neutral overcurrent, residual overcurrent, and negative-sequence overcurrent settings. *It shows three groups of settings.*

- In each group *all the protective functions shown in the table can be replicated.* Considering phase instantaneous overcurrent settings, there are four independent settings. Considering all three groups together, there are a total of 12 instantaneous phase overcurrent settings.
- At any one time, the settings in one selected group are active. Similarly for instantaneous settings of neutral, residual, and negative sequence; one group can be selected.
- On a system input, the settings can be changed online with bump control from one group to another, e.g., from group 1 to group 2 or group 3 and then back to group 1 or group 2.
- A power system is dynamic and can be operated in more than one mode, which means that protective settings in one mode are not applicable when the system mode of operation is changed. This change can be due to a portion of the system being out of service or a tie line brought in or out.
- This forms an example of *"adaptive relaying,"* where the protection settings can be changed depending on the system operation.

TABLE 4.4

Relay Elements

Element	Details	Setting Range
Instantaneous definite time overcurrent	5 A model	0.50–100 A, 0.01 A steps
	1 A model	0.1–20 A, 0.01 A steps
	50 mA model	5–1000 mA, 0.1 mA steps
	2.5 mA model	0.13–12.50 mA, 0.01 mA steps
	Accuracy	±5% of setting range plus ±0.02′ I nom A secondary (steady-state pickup)
	Time delay	0.00–5 s, 0.01 s steps
	Pickup/dropout time	<1.5 cycles
Arc-flash instantaneous current	5 A model	0.50–100 A, 0.01 A steps
	1 A model	0.1–20 A, 0.01 A steps
	Accuracy	0–10% of setting range plus ±0.02′ I nom A secondary (steady-state pickup)
	Pickup/dropout time	2–5 ms/1 cycle
Arc-flash time-overlight	Pickup setting range % of full scale	3.0%–20%, point sensor 0.6%–4%, fiber sensor
	Pickup/dropout time	2–5 ms/1 cycle
Inverse time overcurrent	5 A model	0.50–16 A, 0.01 A steps
	1 A model	0.1–3.2 A, 0.01 A steps
	50 mA model	5–160 mA, 0.1 mA steps
	2.5 mA model	0.13–2.0 mA, 0.01 mA steps
	Accuracy	±5% of setting range plus ±0.02′ I nom A secondary (steady-state pickup)
	Time dial	US: 0.50–15.00, 0.01 step IEC: 0.05–1.0, 0.01 step
	Accuracy	±1.5 cycles plus ±4% between 2 and 30 multiples of pickup
Undervoltage	Setting range	Off, 0.02–1.0 Vnom
	Accuracy	±1% of setting range plus ±0.5 VA secondary (steady-state pickup)
	Pickup/dropout time	<1.5 cycles
Overvoltage	Setting range	Off, 0.02–1.20 Vnom
	Accuracy	±1% of setting range plus ±0.5 VA secondary (steady-state pickup)
	Pickup/dropout time	<1.5 cycles

(*Continued*)

TABLE 4.4 (*Continued*)

Relay Elements

Element	Details	Setting Range
Power Element 32	Instantaneous three-phase element type	+w, −W, +VAR, −VAR
	Pickup setting, 5 A model	1.0–6500 VA, 0.1 VA steps
	Pickup setting, 1 A model	0.2–1300 VA, 0.1 VA steps
	Accuracy	±0.10 A × (L-L voltage secondary) and ±5% of setting at unity power factor for power elements and zero power factor for reactive power elements (5 A nominal) ±0.02 A × (L-L voltage secondary) and ±5% of setting at unity power factor for power elements and zero power factor for reactive power elements (1 A nominal)
	Pickup/dropout time	<10 cycles
Power factor	Setting range	Off, 0.05–0.99
	Accuracy	±5% of full scale for current ≥ 0.5 Inom
Frequency	Setting range	Off, 20.0–70 Hz
	Accuracy	±0.01 Hz (VI> 60 V) with voltage tracking ±0.05 Hz (I1> 0.8 Inom) with voltage tracking
	Pickup/dropout time	<4 cycles
Rate of change of frequency	Setting range	Off, 0.10–15 Hz/s
	Accuracy	±100 mHz/s plus ±3.33% of pickup
Synch check	Pickup range sec voltage	0.00–300 V
	Pickup accuracy sec voltage	±1% plus ±0.5 V over the range of 12.5–300 V
	Slip frequency pickup range	0.05–0.50 Hz
	Slip frequency pickup accuracy	±0.05 Hz
	Phase angle range	0°–80°
	Phase angel accuracy	±4°
Synch check undervoltage	Setting range	Off, 2–300 V
	Accuracy	±1% of setting ±0.5 V
	Pickup/dropout time	<1.5 cycles
Synch check overvoltage	Setting range	Off, 2–300 V
	Accuracy	±1% of setting ±0.5 V
	Pickup/dropout time	<1.5 cycles
Station battery voltage monitor	Operating range	0–350 Vdc
	Pickup range	20–300 Vdc
	Pickup accuracy	±2% of setting value ±2 Vdc

(*Continued*)

TABLE 4.4 (*Continued*)

Relay Elements

Element	Details	Setting Range
Timers	Setting range	As many as 15 timers with various setting ranges
	Accuracy	±0.5% of setting range plus ±1/4 cycle
RTD protection	Setting range	Off, 1°C–250°C
	Accuracy	±2°C
	RTD open-circuit detection	>250°C
	RTD short-circuit detection	<–25°C
	RTD types	PT100, NT100, Ni120, CU10
	RTD lead resistance	25 Ω maximum per lead
	Update rate	<3 s
	Trip/alarm delay	~6 s

Metering Functions

Function	Accuracy
Phase currents	±2% of reading, ±2°
Three-phase average current	±2% of reading
Current imbalance	±2% of reading
Residual current	±2% of reading, ±2°
Neutral current	±2% of reading, ±2°
Negative-sequence current	±3% of reading
System frequency	±0.01 Hz of reading within frequencies 20–70 Hz VI > 60 V with voltage tracking ±0.05 Hz of reading within frequencies 20–70 Hz I1 > 0.8 Inom with current tracking
Line-to-line voltages	±1% of reading, ±1° for voltages between 24 and 264 V
Three-phase average line-to-line voltages	±1% of reading, ±1° for voltages between 24 and 264 V
Line-to-ground voltages	±1% of reading, ±1° for voltages between 24 and 264 V
Three-phase average line-to-ground voltages	±1% of reading, ±1° for voltages between 24 and 264 V
Voltage unbalance %	±1% of reading, ±1° for voltages between 24 and 264 V
Negative-sequence voltage	±3% of reading, ±1° for voltages between 24 and 264 V
Real three-phase power	±5% of reading, ±1° for 0.10<pf<1.0
Reactive three-phase power	±5% of reading, ±1° for 0.10<pf<1.0
Apparent three-phase power	±2% of reading
Power factor	±2% of reading

Note: Any function can be set 'off' and is no longer active.

TABLE 4.5

Instantaneous Overcurrent Settings

Function	Setting Range	Group 1	Group 2	Group 3
Instantaneous phase current	Off, 0.5–100.0 A 0.00–5 s Torque controlled	Setting 1, 50P1P Setting 2, 50P2P Setting 3, 50P3P Setting 4, 50P4P	Setting 1, 50P1P Setting 2, 50P2P Setting 3, 50P3P Setting 4, 50P4P	Setting 1, 50P1P Setting 2, 50P2P Setting 3, 50P3P Setting 4, 50P4P
Instantaneous neutral overcurrent	Off, 0.5–100.0 A 0.00–5 s Torque controlled	Setting 1, 50N1P Setting 2, 50N2P Setting 3, 50N3P Setting 4, 50N4P	Setting 1, 50N1P Setting 2, 50N2P Setting 3, 50N3P Setting 4, 50N4P	Setting 1, 50N1P Setting 2, 50N2P Setting 3, 50N3P Setting 4, 50N4P
Instantaneous residual overcurrent	Off, 0.5–100.0 A 0.00–5 s Torque controlled	Setting 1, 50G1P Setting 2, 50G2P Setting 3, 50G3P Setting 4, 50G4P	Setting 1, 50G1P Setting 2, 50G2P Setting 3, 50G3P Setting 4, 50G4P	Setting 1, 50G1P Setting 2, 50G2P Setting 3, 50G3P Setting 4, 50G4P
Instantaneous negative-sequence overcurrent	Off, 0.5–100.0 A 0.00–5 s Torque controlled	Setting 1, 50Q1P Setting 2, 50Q2P Setting 3, 50Q3P Setting 4, 50Q4P	Setting 1, 50Q1P Setting 2, 50Q2P Setting 3, 50Q3P Setting 4, 50Q4P	Setting 1, 50Q1P Setting 2, 50Q2P Setting 3, 50Q3P Setting 4, 50Q4P

Note: The setting range is shown for 5 A secondary CT; for other CT inputs, see Table 4.4 specifications.

Each group setting consists of the following:

- *General*: Where even the CT ratios, ID of the operation, PT ratios, PT types, neutral CTs, synchronizing voltage inputs, etc., can be changed.
- *Protective functions*: As explained earlier, independent settings on phase and time overcurrent, under/overvoltage, synchronizing, under/overfrequency, metering functions, etc., can be changed.
- Trip and close logic can be changed.
- Reclosing control can be changed.
- Demand metering can be changed.

The global settings that remain common in the three groups are as follows:

- Event messenger
- Synchronized phasor measurements
- Time and date measurement settings
- Breaker failure
- Arc-flash protection
- Analog inputs and outputs
- Station battery monitor
- Input debounce
- Breaker monitor
- Access control
- Time synchronization

The other common settings are as follows:

- Front panel settings
- Report
- Port settings
- Modbus map settings
- DNP map settings

4.7 Relay Bit Words

The relay is set based on the specific relay bit words that are *unique for each function and noninterchangeable*. The Boolean logic accepts these assigned bit words. For example, 50P1P, 50N1P, 50G1P, and 50Q1P are shown for the pickup settings of these functions in Table 4.5. For the time delay setting, the bit words are 50P1D, 50N1D, 50G1D, and 50Q1D, respectively. The torque control bit words in the same sequence are 50P1TC, 50N1TC, 50G1TC, and 50Q1TC. Finally, the trip outputs to be used in the trip equation and other controls and indication logic are 50P1T, 50N1T, 50G1T, and 50Q1T.

Figure 4.4 shows a partial list of relay bit words. There may be hundreds of bit words to complete the entire functionality. These bit words are also used in the block control circuit logic diagram. As an example, a logic control circuit diagram of trip output 50P1T is shown in Figure 4.5. Note the torque control switch 50P1TC. It is closed when the corresponding control bit is asserted. It can be used to block the function in one direction, that is, to impart directional overcurrent features to the settings. Also in this chapter, the residual current refers to the current obtained by summation of the three-phase currents in the phase CTs (see Figure 3.11a). The neutral current is the direct input from the neutral connected CT or core balance CT (Figure 3.11b). This convention is not universal.

4.8 Time Delay Overcurrent Protection

The time delay overcurrent protection is shown in Table 4.6 and Figure 4.6. Note that for phase protection, individual settings for phases A, B, and C are available. This is applicable to unbalanced systems, where the unbalance can be limited by proper settings. Also, there is a maximum phase overcurrent setting: two independent settings per group. Then, there are unbalanced settings, one per group, and residual current and neutral current time delay settings, two per group. The setting ranges are identical for all the settings.

- Note the extended range of pickup settings, which determine the trip level, and the much closer pickup settings in smaller increments. In an electromechanical relay, the settings are limited by the taps on the coils, see Chapter 2.
- Any of the ten curve shapes, five US and five IEC, can be independently selected for each setting.

Bit	Definition
25A1	Synchronism-check element level 1 pickup.
25A2	Synchronism-check element level 2 pickup.
27P1	Phase undervoltage trip 1 pickup (see Figure 4.20).
27P1T	Phase undervoltage trip 1 output (see Figure 4.20).
27P2	Phase undervoltage trip 2 pickup (see Figure 4.20).
27P2T	Phase undervoltage trip 2 output (see Figure 4.20).
27S1	Level 1 vs channel undervoltage element pickup
27S1T	Level 1 vs channel undervoltage element with time delay
27S2	Level 2 vs channel undervoltage element pickup
27S2T	Level 2 vs channel undervoltage element with time delay
50G1P	Definite-time residual overcurrent trip 1 pickup
50G1T	Definite-time residual overcurrent trip 1 output
50G2P	Definite-time residual overcurrent trip 2 pickup
50G2T	Definite-time residual overcurrent trip 2 output
50G3P	Definite-time residual overcurrent trip 3 pickup
50G3T	Definite-time residual overcurrent trip 3 output
50G4P	Definite-time residual overcurrent trip 4 pickup
50G4T	Definite-time residual overcurrent trip 4 output
50N1P	Definite-time neutral overcurrent trip 1 pickup
50N1T	Definite-time neutral overcurrent trip 1 output
50N2P	Definite-time neutral overcurrent trip 2 pickup
50N2T	Definite-time neutral overcurrent trip 2 output
50N3P	Definite-time neutral overcurrent trip 3 pickup
50N3T	Definite-time neutral overcurrent trip 3 output
50N4P	Definite-time neutral overcurrent trip 4 pickup
50N4T	Definite-time neutral overcurrent trip 4 output
50NAF	Sample-based neutral overcurrent element (Arc-flash protection)
50P1P	Definite-time phase overcurrent trip 1 pickup
50P1T	Definite-time phase overcurrent trip 1 output
50P2P	Definite-time phase overcurrent trip 2 pickup
50P2T	Definite-time phase overcurrent trip 2 output
50P3P	Definite-time phase overcurrent trip 3 pickup
50P3T	Definite-time phase overcurrent trip 3 output
50P4P	Definite-time phase overcurrent trip 4 pickup
50P4T	Definite-time phase overcurrent trip 4 output
50PAF	Sample-based phase overcurrent element (Arc-flash protection)
50Q1P	Definite-time negative-sequence overcurrent trip 1 pickup
50Q1T	Definite-time negative-sequence overcurrent trip 1 output
50Q2P	Definite-time negative-sequence overcurrent trip 2 pickup
50Q2T	Definite-time negative-sequence overcurrent trip 2 output
50Q3P	Definite-time negative-sequence overcurrent trip 3 pickup

FIGURE 4.4
A partial list of relay word bits.

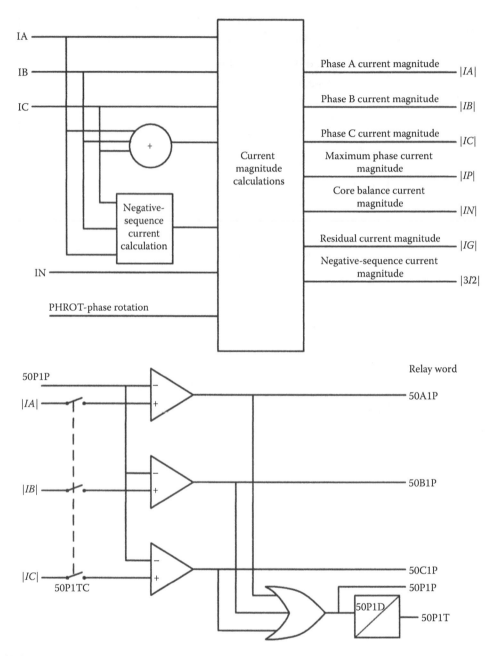

FIGURE 4.5
Instantaneous element logic.

- The time dial settings range from 0.5 to 15.
- EM reset delay is applicable when there are electromechanical relays in the same circuit. These have a higher reset time compared to MMPRs. The selection is "yes" or "no."
- Constant time can be added to any setting, range 0–1.0 s.

TABLE 4.6
Time Delay Settings

Function	Setting Range	Group 1	Group 2	Group 3
Phase A, B, and C time overcurrent	Off, 0.5–16.0 A Curve ANSI/IEC Time dial EM reset delay Const time adder Minimum response time Torque controlled	Phase A setting 51AP Phase B setting 51BP Phase C setting 51CP	Phase A setting 51AP Phase B setting 51BP Phase C setting 51CP	Phase A setting 51AP Phase B setting 51BP Phase C setting 51CP
Maximum phase time overcurrent	Off, 0.5–16.0 A Curve ANSI/IEC Time dial EM reset delay Const time adder Minimum response time Torque controlled	Setting 51P1P Setting 51P2P	Setting 51P1P Setting 51P2P	Setting 51P1P Setting 51P2P
Negative-sequence time overcurrent	Off, 0.5–16.0 A Curve ANSI/IEC Time dial EM reset delay Const time adder Minimum response time Torque controlled	Setting 51QP		
Neutral time overcurrent	Off, 0.5–16.0 A Curve ANSI/IEC Time dial EM reset delay Const time adder Minimum response time Torque controlled	Setting 51N1P Setting 51N2P	Setting 51N1P Setting 51N2P	Setting 51N1P Setting 51N2P
Residual time overcurrent	Off, 0.5–16.0 A Curve ANSI/IEC Time dial EM reset delay Const time adder Minimum response time Torque controlled	Setting 51G1P Setting 51G2P	Setting 51G1P Setting 51G2P	Setting 51G1P Setting 51G2P

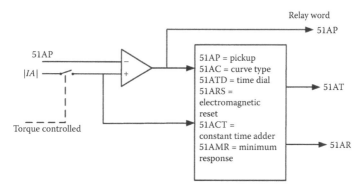

FIGURE 4.6
Phase time overcurrent elements logic.

- Minimum response time can be added between range 0–1.0 s. The selected curves do not account for the time adder and the minimum response time.
- Torque control can be implemented with logic equations.

This provides much desirable flexibility and closer time–current coordination in a given situation. Also, lack of coordination can be avoided.

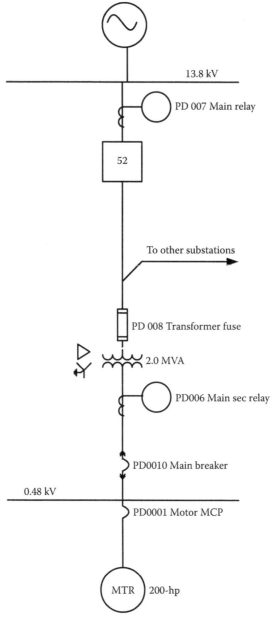

FIGURE 4.7
A system configuration for overcurrent coordination.

Example 4.1

Consider a system configuration with protective devices as shown in Figure 4.7. For arc-flash reduction, the main secondary relay PD006 is added, which trips the main breaker. The coordination of these six devices is shown in Figure 4.8. There is a lack of coordination of the secondary and primary relays, as illustrated, even with the best selection of characteristics. This figure shows three areas of lack of coordination. With MMPRs, the time curves can be combined with instantaneous settings with appropriate delays to achieve better coordination. Figure 4.9 shows that two areas of lack of coordination are eliminated, while the third area cannot be.

4.9 Voltage-Based Elements

The settings are shown in Table 4.7. There are two undervoltage settings and two synchronizing voltage settings for undervoltage per group. Also, there are similar overvoltage settings and also settings on negative-sequence and zero-sequence overvoltage.

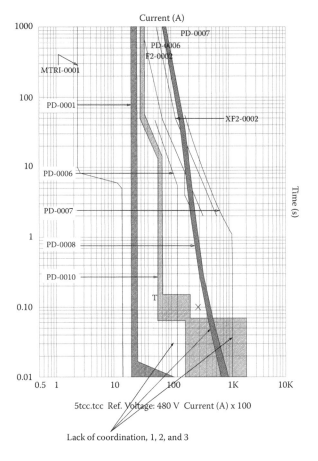

FIGURE 4.8
Time–current coordination with conventional electromagnetic relays, showing three areas of lack of coordination.

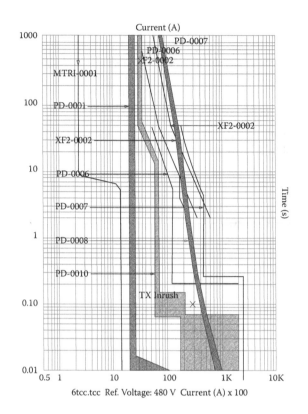

FIGURE 4.9
Time–current coordination with MMPR, showing that two areas of lack of coordination are eliminated.

TABLE 4.7

Under- and Overvoltage Settings

Function	Setting Range	Group 1	Group 2	Group 3
Undervoltage	Off, 0.02–1.0 Vnom	27P1P	27P1P	27P1P
	Time delay = 0.02120 s	27P2P	27P2P	27P2P
Synchronizing undervoltage	Off, 2.0–300 V	27S1P	27S1P	27S1P
	Time delay = 0.02–120 s	27S2P	27S2P	27S2P
Overvoltage	Off, 0.02–1.2 Vnom	59P1P	59P1P	59P1P
	Time delay = 0.02120 s	59P2P	59P2P	59P2P
Synchronizing overvoltage	Off, 2.0–300 V	59S1P	59S1P	59S1P
	Time delay = 0.02–120 s	59S2P	59S2P	59S2P
Overvoltage negative sequence	Off, 0.02–1.2 Vnom	59Q1P	59Q1P	59Q1P
	Time delay = 0.02120 s			
Overvoltage zero sequence	Off, 0.02–1.2 Vnom	S59G1P	S59G1P	S59G1P
	Time delay = 0.02–120 s	59G2P	59G2P	59G2P

The synchronizing settings are shown in Table 4.8. The high and low synchronizing windows can be selected. Also, there is no need for inserting a voltage-matching auxiliary PT—the ratio correction factor can take care of voltage mismatch. The slip and synchronizing angles are adjustable over a wide limit. There is no need for auxiliary PTs for phase angle correction between the two voltages. The angles can be adjusted with respect to line-to-neutral or line-to-line voltages. Finally, the breaker close time input determines the change in angle and slip. In connection with generator protection, the synchronizing settings are further discussed in Chapter 11.

4.10 Power Elements

The power elements +WATTS, −WATTS, +VARS, and −VARS are positive or forward and reverse or negative reactive power. As a result, the operating window (adjustable) is as shown in Figure 4.10. The intended primary WATTS or VARS should be converted to secondary side by application of appropriate CT and PT ratios as inputted in the global settings.

TABLE 4.8

Synchronizing Settings

Function	Setting Range	Group 1	Group 2	Group 3
Synchronizing	VS voltage window 0.00–300 V	25VLO	25VLO	25VLO
	VS synchronizing window high	25VHI	25VHI	25VHI
	Voltage ratio corrections factor 0.05–2.0	25RCF	25RCF	25RCF
	Maximum angle 0°–80°	25RCF	25RCF	25RCF
		25RCF	25RCF	25RCF
	Synchronizing phase VAB, VBC, VCA, or, 30°, 60°, 90°, 120°, 150°, 180°, 210°, 240°, 270°, 300°, 330° lag VAB	SYNPH = VAB	SYNPH = VAB	SYNPH = VAB
	Synchronizing phase VA, VB, VC, or, 30,60,90,120,150,180,210,240, 270,300,330° lag VA	SYNPH = VA	SYNPH = VA	SYNPH = VA
	Breaker close time off, 1–1000 ms	TCLOSD	TCLOSD	TCLOSD

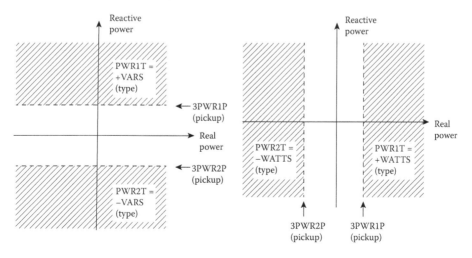

FIGURE 4.10
Adjustable setting windows of power elements.

4.11 Loss of Potential

A loss of potential (LOP) can occur due to a fuse failure or operation of a molded case circuit breaker in the PT circuits. The proper potential is essential for operation of some relay functions, such as over- and underfrequency or under- and overvoltage.

The relay declares LOP when there is more than 25% drop in the measured positive-sequence voltage with no corresponding change in the positive-, negative-, and zero-sequence currents. If this condition persists for more than 60 cycles, LOP is declared. The relay will reset when the voltage returns. There are no settings on LOP, but the required protective functions should be blocked on LOP. For example, the Boolean equation for undervoltage supervision with LOP will be simply (27PiT or 27P2T) and NOT LOP.

The LOP logic diagram is not shown.

4.12 Frequency Settings

The frequency settings are shown in Table 4.9, and the logic diagram for the rate of change of frequency is illustrated in Figure 4.11. As discussed in Chapter 2, the rate of change of frequency and frequency settings can be used to detect and initiate a remedial action.

Referring to Figure 4.11, two frequencies mf1 and mf2 (after a time window dt) are measured, which is determined by the trip level setting 81R1TP. The pickup is 100% of the setting and the dropout is 95%.

The trend can be set as INC or DEC to limit the element to increasing or decreasing frequency, respectively. When the trend is set at INC or DEC, the element receives supervision from nominal frequency, FNOM. If the element is set to ABC, it disregards the frequency trend.

A minimum of positive-sequence voltage and/or current is necessary for the operation of the 81R element. The levels set are specified by the settings 81RISUP (current supervision) and 81RVSUP (voltage supervision). These can be set off. In any case, the element receives supervision from relay word FREQTRK, which ensures that the relay is tracking the measured system frequency.

There are also other settings, e.g., fast rate of change of frequency, which provide a faster response compared to frequency and rate of change of frequency. The fast operating speed makes it suitable for detecting islanding conditions. The element uses a characteristic based on deviation from the nominal frequency and rate of change of frequency to detect islanding conditions. A three-cycle window is used for the calculation. During normal condition, the operating point is close to the origin, whereas during islanding conditions the operating point enters the trip region, depending upon acceleration or deceleration of the isolated system (see Figure 4.12). The logic diagram is not shown.

4.13 Trip and Close Logic

4.13.1 Trip Logic

The trip logic is shown in Figure 4.13, allowing the conditions to be inputted that cause a trip, the conditions that unlatch the trip, and the performance of the relay output contact.

TABLE 4.9

Frequency Settings

Function	Setting Range	Group 1	Group 2	Group 3
Frequency	Enable off, 1–6[a] Off, 20–70 Hz 0–240 s delay	Frequency 1 81D1TP	Frequency 1 81D1TP	Frequency 1 81D1TP
	Off, 20–70 Hz 0–240 s delay	Frequency 2 81D2TP	Frequency 2 81D2TP	Frequency 2 81D2TP
	Off, 20–70 Hz 0–240 s delay	Frequency 3 81D3TP	Frequency 3 81D3TP	Frequency 3 81D3TP
	Off, 20–70 Hz 0–240 s delay	Frequency 4 81D4TP	Frequency 4 81D4TP	Frequency 4 81D4TP
	Off, 20–70 Hz 0–240 s delay	Frequency 5 81D5TP	Frequency 5 81D5TP	Frequency 5 81D5TP
	Off, 20–70 Hz 0–240 s delay	Frequency 6 81D6TP	Frequency 6 81D6TP	Frequency 6 81D6TP
Rate of change of frequency	Enable off, 1–4 81R Voltage SUP off, 0.1–1.3 Vnom 81R Current SUP. off, 0.1–2.0 Inom	General settings	General settings	General settings
	Trip level, off, 010–15.00 Trend Trip delay DO relay	81R1TP	81R1TP	81R1TP
	Trip level, off, 010–15.00 Trend Trip delay 0.1–60 s DO relay 0.1–60 s	81R2TP	81R2TP	81R2TP
	Trip level, off, 010–15.00 Trend Trip delay 0.1–60 s DO relay 0.1–60 s	81R3TP	81R3TP	81R3TP
	Trip level, off, 010–15.00 Trend Trip delay 0.1–60 s DO relay 0.1–60 s	81R4TP	81R4TP	81R4TP

Note: Shows that the functions can be entirely disabled, or from1 through 6 number of settings can be selected.

TDURD defines the minimum trip time, for which the relay word TRIP asserts. This is a rising-edge initiated timer. Trips initiated by the TR relay bit word (included open commands from the front panel of the relay and serial ports) are maintained for at least the duration of TDURD setting.

Any or all intended trip functions can be included in the trip equations using relay bit words. Any trip logic can be designed using Boolean logic. The trips can be initiated also

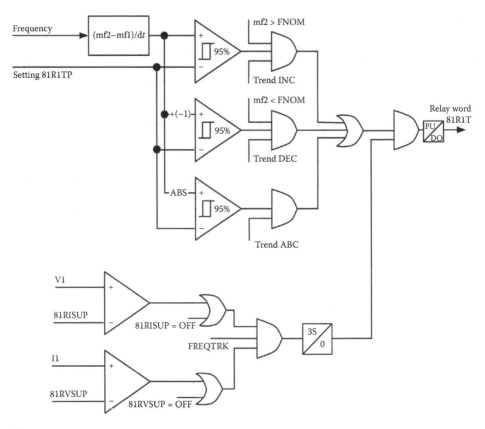

FIGURE 4.11
Frequency rate-of-change logic.

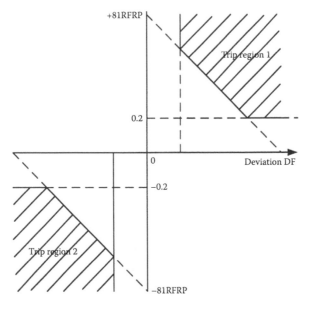

FIGURE 4.12
Schematic of fast rate of frequency characteristics.

Microprocessor-Based Multifunction Relays

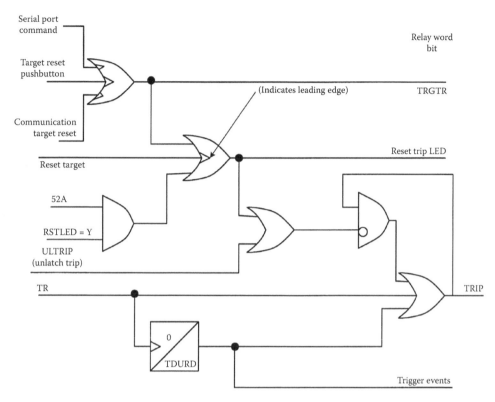

FIGURE 4.13
Trip logic.

from the front panel or serial port (including Modbus and DeviceNet) and remote trips (REMTRIP). The remote trip must be wired to an input contact. Any control input can be mapped to the REMTRIP.

The trip conditions trigger an event report (see Figure 4.13). Also the TRIP relay bit word is wired to output contacts which can be fail-safe or non-fail-safe (user selectable). The trip conditions can be divided and routed to two or more output contacts to distinguish the trips or trip different circuit breakers.

Following a trip, the trip signal is maintained until all the following conditions are satisfied:

- Minimum trip duration time TDURD elapses.
- The TR control equation result deasserts to logical zero. This can be done by negating all the trip functions that initiated the trip or by the breaker status or by an input contact of the status of the breaker.

The trip logic is unlatched, if one of the following occurs:

- Unlatch trip control equation setting ULTRP asserts to logical 1.
- Target reset relay word TRGT asserts, when the TRGT RESET pushbutton of the front panel is pressed.
- Target reset control equation setting RSTTRGT asserts to logical 1.

4.13.2 Close Logic

The close logic is shown in Figure 4.14. The circuit breaker will close for the following:

- For all conditions mapped to CL
- Front panel or serial port (including Modbus and DeviceNet) CLOSE command
- Automatic reclosing when open interval times out, qualified by control equation setting 79CLS

Once close bit is asserted it is sealed in, until any of the following conditions occur:

- Unlatch close logic control equation by setting ULCL asserts to logical 1.
- Relay word 52A asserts to logical 1.
- Close failure relay word bit asserts to logical 1.

The close failure relay is set to highest breaker closing time with some safety margin. If the breaker fails to close, the relay close failure bit word CF asserts.

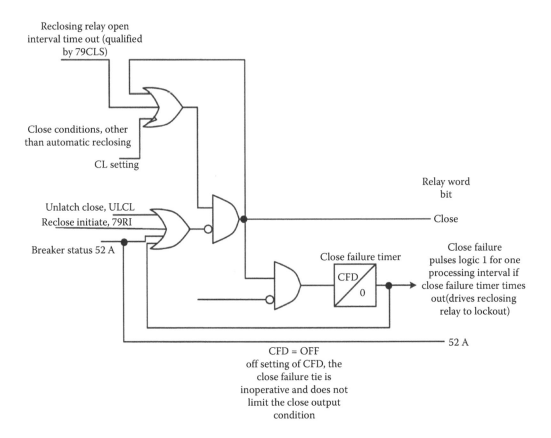

FIGURE 4.14
Close logic.

4.13.3 Reclose Logic and Supervision

For most applications, the reclose supervision limit time is set to 79CLSD = 0 s. For this condition, the logic shown in Figure 4.15a is operative. When an open interval times out, the relay will check reclose supervision setting 79RI (CLS only once). If 79CLS is asserted to logical 1, at the instant of an open interval time out, then logic in Figure 4.15a will automatically reclose the breaker. Consider the configuration in Figure 4.15b. The relays are installed at both ends of a transmission line in a high-speed reclose scheme. When both circuit breakers open after a line fault, relay 1 recloses circuit breaker 52/1 first, which was followed by relay 2 reclosing circuit breaker 52/2, after a synchronizing check across circuit breaker 52/2. Before allowing circuit breaker 52/1 to be reclosed after an open time interval timeout, relay 1 checks that bus 1 voltage is hot and the transmission line voltage is dead. This requires reclose supervision setting 79CLSD = 0 s (only once) and CLS setting to ensure that bus 1 is hot and the transmission line voltage is dead.

Relay 2 checks that bus 2 voltage is hot, the transmission line voltage is hot, and it is in synchronism after the reclosing relay open interval times out, before allowing breaker 52/2 to be reclosed. This requires 79CLSD = 0 s and 79CLS to set that synchronization as okay.

As many as four reclosing shots are possible. The different states of the reclosing relay and its operation are shown in Figure 4.16.

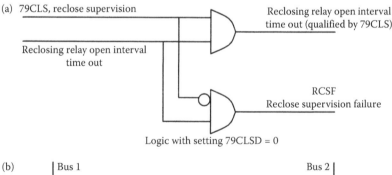

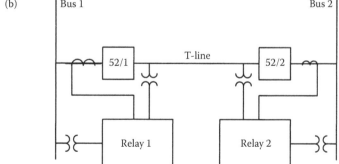

FIGURE 4.15
(a) Operation of reclose supervision logic with timer setting equal to zero and (b) example of reclosing settings on a transmission line.

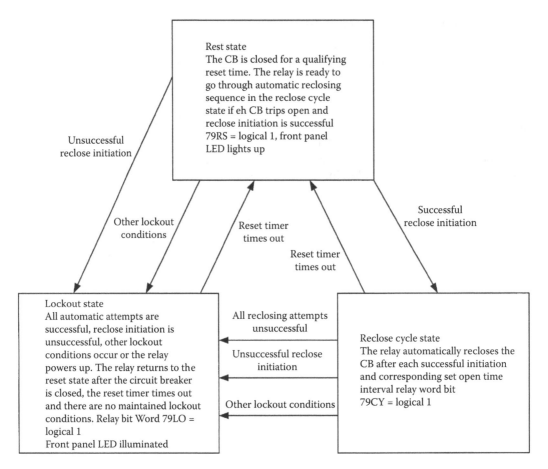

FIGURE 4.16
Reclosing relay states and general operation.

Example 4.2

Consider the configuration in Figure 4.17a. The sequence coordination is shown in Figure 4.17b and the operation in Figure 4.17c. Figure 4.17a shows a relay and a recloser at the far end of the line. The recloser has two curves that are fast and slow. The recloser is set to operate twice on the fast curve and then twice on the slow curve. The slow curve is allowed to operate after the two fast curve operations, because then the fast curves are inoperative for tripping.

The relay phase time overcurrent element (three-phase) 51PT is coordinated with the line recloser fast curve, and the relay single-phase overcurrent elements 51AT, 51BT, and 51CT are coordinated with the recloser slow curve. The sequence is shown in Figure 4.17c.

4.14 Demand Metering

The demand and peak demand metering, which is selectable between thermal and rolling demand types, is provided for the following values:

Microprocessor-Based Multifunction Relays

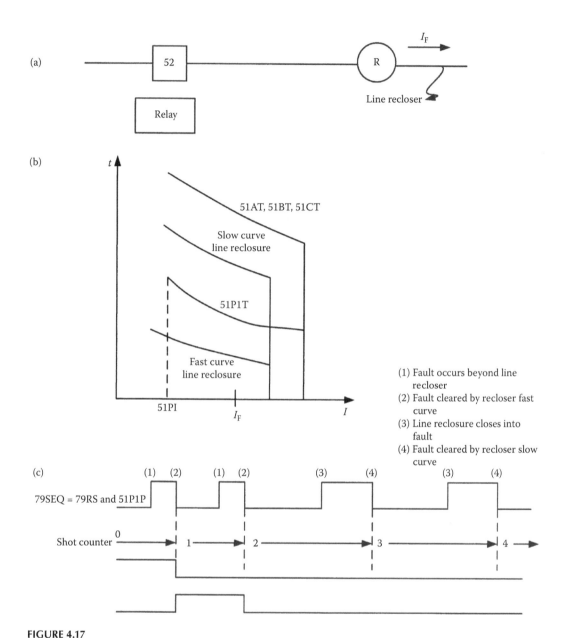

FIGURE 4.17
(a) A single-line diagram, with fault at the far end of the recloser, (b) sequence coordination between relay and reclosure, and (c) operation of relay shot counter with line recloser.

- IA, IB, IC, primary phase currents
- IG residual ground current
- Negative-sequence current

The response of the thermal and rolling demand metering to a step change of current is shown in Figure 4.18. The response of thermal demand metering is analogous to a series RC circuit. In Figure 4.18, the response of the thermal demand meter is 90% of the full

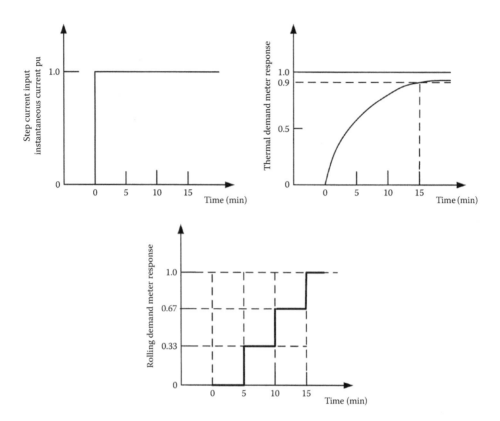

FIGURE 4.18
Response of thermal and rolling demand metering to a step input.

applied value. The rolling demand meter calculates the response with a sliding time-window arithmetic average calculation. It shows that the demand meter integrates the applied signal in 5 min intervals. The average value of an integrated 5 min interval is derived and stored as a 5 min total. The rolling demand meter then averages a number of 5 min totals to produce the rolling meter response. The relay updates the rolling demand meter response every 5 min, after it is calculated in 5 min total.

4.15 Logical Settings

The relay has a maximum of the following:

- 32 logical latches
- 32 SV timers
- 32 counters and
- 32 math variables

None of these need to be selected, or from 1 through 32 any number can be selected. These are available for each group of settings. That is, each of the 32 functions is multiplied by 3.

4.16 Latch Bits: Nonvolatile State

Latch control switches (latch bits are outputs of these switches) replace traditional latching devices, the state is retained even when the power to the relay is lost, and the state of the latch switch is stored in nonvolatile memory, but the relay de-energizes the output contact. When the power is restored, the programmable output contact will go back to the state of latch control switch after device initialization. The tradition latching device output contacts are changed by pulsing the latching device inputs (see Figure 4.19a). Pulse the set input to close and pulse the reset input to open the output contact. The contacts wired to latching device are from remote control equipment such as SCADA and RTU.

In Figure 4.19b if the set input asserts to logical 1, the latch bit LT1 also asserts to logical 1. If reset asserts to logical 1, the LT1 deserts to logical 0. The state of the latch bits is retained if the power to the device is lost, and then restored. As the latch bit states are stored in the nonvolatile memory, these can be retained during power loss or setting change.

Each of the 32 timers has the following settings:

- SV input: any relay word bit in Boolean logic can be used, e.g., 27P1T OR 27P2T and NOT LOP.
- Pickup setting range: 0–3000 cycles.
- Dropout range: 0–3000 cycles.

The counters are up- or down-counting elements, updated every processing interval, and confirm to standard counter function block #3 in IEC 1131-3.

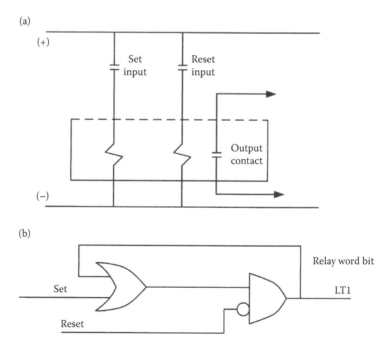

FIGURE 4.19
(a) Schematic diagram of a traditional latching device and (b) logic diagram of a latching switch.

A number of output contacts are available, which can be selected fail-safe or non-fail-safe. Any control equation can be mapped to any output contact. If the fail-safe is enabled, the relay output is held in its energized position when relay control power is applied. The output fails to its de-energized position when control power is removed. When TRIP output fail-safe is enabled, the breaker is tripped when the relay control power fails. Normally, non-fail-safe position is selected.

4.17 Global Settings

As stated earlier, the global settings are common in all the three groups of settings. The global settings are as follows:

- Phase rotation (ABC or ACB)
- Rated frequency 50 or 60 Hz
- Date format
- Fault conditions

Event messenger points (up to 32 messages, text up to 148 characters): The relay can automatically send ASCII message on a communication port or an ASCII compatible device, when the trigger conditions are satisfied.

Time and date management settings: IRIG-B, synchrophasor measurement unit (PMU), SNTP applications, and time update from a DNP master. Also, automatic daylight saving time settings are available.

Breaker failure settings: The logic is shown in Figure 4.20. The assertion of relay word bit TRIP starts BFD timer if the sum of positive- and negative-sequence currents exceeds 0.02' INOM. The BFT output is used to trip appropriate backup breakers. Also, see the logic depicted in Chapter 11.

Arc-flash protection: This protection is discussed in Chapter 17. The relay supports both types of sensors—omni-directional point sensor and clear jacketed fiber loop sensor optimized for protection of long, distributed sources such as switchgear bus compartments.

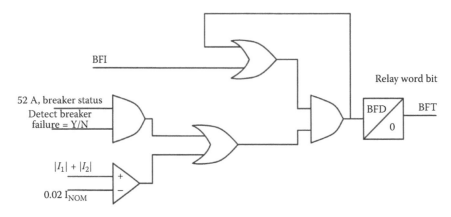

FIGURE 4.20
Breaker failure logic.

Analog inputs: The analog type, high and low input levels (say 4–20 mA), and engineering units are required to be set. Also set low and high warning values.

Analog outputs: Enter the same parameters as for the inputs.

Station battery monitor: This can alarm for under- or overvoltage DC battery conditions and also gives a view of how much the station battery voltage dips when tripping, closing, and other DC control functions take place.

Digital input debounce: To comply with different control voltages, AC and DC debounce modes are available. The debounce refers to the qualifying time delay before processing the change of state of a digital input. The debounce times are adjustable.

4.18 Port Settings

The settings allow us to configure the parameters of the communication ports. Port F (front panel) is an EIA-232 port, Port 1 is an Ethernet port, Port 2 is an optional fiber optic serial port, and Port 3 (rear) is an EIA-232 or EIA-485 port. Port 4 can be selected as either EIA-232 or EIA-485 port with COMMINF setting. Tables 4.10 through 4.13 show port settings. For substation automation and communication protocols, see Chapter 17.

4.19 Breaker Monitor

The breaker monitor can be set with the maintenance information supplied by the vendor. It lists the number of close/open operations that are permitted for a given current interruption level. An example of breaker maintenance information is shown in Table 4.14. Based on these data, a curve can be plotted as shown in Figure 4.21. Then the relay accepts three set points: (1) the maximum number of close/open operations with corresponding current interruption level, (2) the number of close/open operations that correspond to the midpoint current interruption level, and (3) the number of close/open operations that correspond to the maximum current interruption level. The relay generates 10%, 25%, 50%, and 100% wear curves. An output can be assigned for an alarm.

TABLE 4.10

Front Panel Serial Port Settings

Setting	Setting Range
PRPTOCOL	MOD, EVMSG, PMU, SEL
SPEED	300–38,400 bps
DATA BITS	7, 8 bits
PARITY	O, E, N
STOP BITS	1, 2 bits
PORT TIME-OUT	0–30 min
SEND AUTOMESSAGE	Y, N
HDWR HANDSHAKING	Y, N
MODBUS SLAVE ID	1–247

TABLE 4.11

Ethernet Port Settings

Setting	Setting Range
IP ADDRESS	zzz,www,xxx,www
SUBNET MASK	15 characters
DEFAULT ROUTER	15 characters
Enable TCP Keep-Alive	Y, N
TCP Keep-Alive Idle Range	1–20 s
TCP Keep-Alive Interval Range	1–20 s
TCP Keep-Alive Count Range	1–20
FAST OP MESSAGES	Y, N
OPERATING MODE	FIXED<FAILOVER,SWICHED
FAILOVER TIMEOUT	0.10–65.0 s
PRIMARY NETPORT	A, B, D
NETWORK PORTA SPD	AUTO, 10,100 Mbps
NETWORK PORTB SPD	AUTO, 10,100 Mbps
TELNET PORT	23,1025–65,534
TELNET TIMEOUT	1–30 min
FTP USER NAME	20 characters
Enable IEC 61850- Protocol	Y, N
Enable IEC 61850 GSE	Y, N
Enable Modbus Sessions	0–2
Modbus TCP Port 1	1–65,534
Modbus TCP Port 2	1–65,534
Enable DNP Session	0–3
Modbus Timeout 1	15–900 s
Modbus Timeout 2	15–900 s

TABLE 4.12

Fiber-Optic Serial Port Settings

Setting	Setting Range
PRPTOCOL	MOD,DNP, MBA, MBB,MB8A, MB8b, MBTA, MBTBEVMSG, PMU, SEL
SPEED	300–38,400 bps
DATA BITS	7, 8 bits
PARITY	O, E, N
STOP BITS	1, 2 bits
PORT TIME-OUT	0–30 min
SEND AUTOMESSAGE	Y, N
HDWR HANDSHAKING	Y, N
MODBUS SLAVE ID	1–247

TABLE 4.13
Rear Panel Serial Port (EIA-232/EIA-485 Settings)

Setting	Setting Range
PRPTOCOL	MOD, DNP, MBA ,MBB, MB8A, MB8b, MBTA, MBTBEVMSG, PMU, SEL
SPEED	300–38,400 bps
DATA BITS	7, 8 bits
PARITY	O, E, N
STOP BITS	1, 2 bits
PORT TIME-OUT	0–30 min
SEND AUTOMESSAGE	Y, N
HDWR HANDSHAKING	Y, N
FAST OP MESSAGES	Y, N
MODBUS SLAVE ID	1–247

TABLE 4.14
Breaker Maintenance Information Supplied by a Vendor

Current Interruption Level (kA)	Permissible Number of Close/Open Operations[a]
0.00–1.20	10,000
2.0	3,700
3.0	1,500
5.0	400
8.0	150
10.0	85
20.0	12

[a] The Close/Open is counted as one operation.

4.20 Front Panel Operations

The front panel has the following features:

- Rotating display on human-machine-interface (HMI)
- Programmable target LEDs
- Programmable pushbuttons, with two programmable LEDs associated with each pushbutton
- Configurable front panel labels to change identification of targets. The main layout of the front panel is shown in Figure. 4.22.

An example of the programmable six LEDs on the left-hand side of the relay is as follows:

LED 1: All instantaneous overcurrent trips, for example, phase, ground, residual, and neutral current

LED 2: Phase overcurrent time delay trips

LED 3: Ground/neutral overcurrent trips

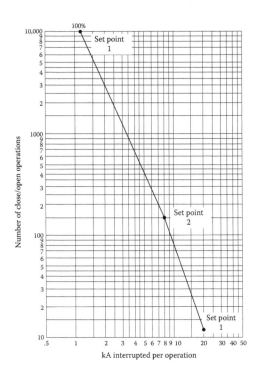

FIGURE 4.21
Breaker maintenance curve.

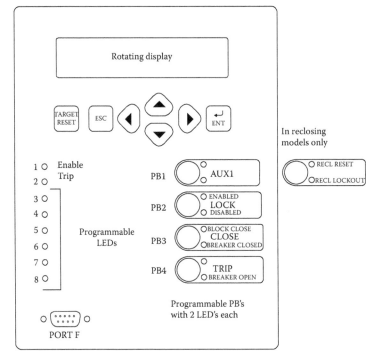

FIGURE 4.22
Layout of the front panel of the relay.

LED 4: Negative-sequence overcurrent trip
LED 5: Over/underfrequency and RTD
LED 6: Breaker failure

For a trip event, the relay latches the trip target LED. The target reset button can be pressed to reset target LEDs. When a new trip event occurs and previously latched trips have not reset, the relay clears the latched targets and displays the new trip targets.

The front pushbuttons and their LEDs can be set to the desired functions. For example, see Figure 4.23 for the reclosing operation and settings on LEDs associated with pushbuttons.

The front panel menu accesses most of the information that the relay measures and stores. The front panel controls can be used to view or modify the relay settings. The front panel control details are as follows:

- Up and down arrows: move up or down within a menu or data list
- Left and right arrows: move the cursor to the right or left
- ESC: escape from the present menu or display
- ENT: move from the rotating display to the main menu

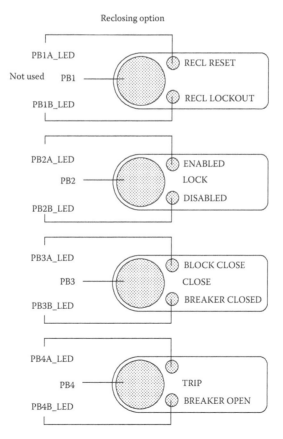

FIGURE 4.23
Pushbuttons and their programmable LED settings.

The main menu screen shows the following:

- Main meter
- Events
- Targets
- Control
- Set/show
- Status
- Breaker

These can be selected through the pushbuttons described.

4.20.1 Rotating Display

As many as 32 display points are available. The contrast of the front panel display can be modified, depending on the ambient light. An example of the rotating display points is as follows:

- Breaker closed
- Breaker open
- Breaker in draw-out position
- Remote trip
- Ground current magnitude
- Phase currents and voltages
- Active and reactive power and power factor
- DC monitor alarm

In fact, any function can be programmed into the rotating display, depending on the users choice and inputs assigned to the display points.

4.21 Analyzing Events

The relay provides comprehensive tools to analyze the relay operations.

Event reporting: Event reporting includes event summary reports, event history reports, and event reports. Event reports are stored in nonvolatile memory for later retrieval and detailed analysis.

Sequential event recording report has a resolution of 1 ms and an accuracy of ±1/4 cycle.

Event summaries: The relay can be enabled to send automatic messaging of event summaries to a serial port. A summary report provides a quick overview of the event.

Event history: The relay keeps an index of the stored nonvolatile event reports. The HISTORY command can be used to obtain this index to retrieve the latest or the earliest report.

Each time an event occurs, a new history and summary reports are created. Event report information includes the following:

- Date and time of event
- Individual sample analog inputs (currents and voltages)

TABLE 4.15

An Example of an SER Report

#	Date	Time	Element	State
8	11/28/2016	13.54.09.602	Three-phase OC time delay	Asserted
7		13.54.09.602	Phase overcurrent time delay	Asserted
6		1.54.10.003	Three-phase OC time delay timeout	Asserted
5		13.54.10.003	Trip	Asserted
4		13.54.10.219	Three-phase OC time delay	Deasserted
3		13.54.10.219	Phase overcurrent time delay	Deasserted
2		13.54.10.236	Three-phase OC time delay timeout	Deasserted
1		13.54.10.511	Trip	Deasserted

- Digital states of selected relay word bits
- Event summary, including front panel target states at the time of tripping and fault type
- Group, logic, global, and report settings (that were in service when the event was retrieved)
- Compressed ASCII event reports to facilitate event report storage and display

4.21.1 Sequential Event Recorder

The SER captures digital element state changes over time. In addition to the automatically generated triggers from relay power up, setting changes, and active setting group changes, settings allow as many as 96 relay word bits to be monitored. State changes are tagged to the nearest millisecond. As an example, see Table 4.15 for the SER. These events occurred before the beginning of event summary report.

4.21.2 Triggering

The relay will generate a report when any of the following occurs:

- Relay word bit trip asserts.
- Programmable event report settings assert.
- Trigger event reports serial port command is executed.

An example of 15-cycle event report (only for one cycle) is shown in Table 4.16. The data are captured every 1/4 of a cycle. The waveforms can be plotted. Figure 4.24a shows the phasors and sequence components for a system ground fault and Figure 4.24b is the captured waveform of the event.

4.21.3 Aliases

Aliases can be used to rename 20 trigger conditions. This changes relay word bit to a convenient understandable description. For example:

Alias1: = PB2 "FP_LOCK PICKUP DROP OUT."

Note that description is arbitrary as chosen by the user. It could be:

TABLE 4.16

One-Cycle Reproduction of a 15-Cycle Event Report

IA	IB	IC	IN	IG	VA	VB	VC	VS	VDC
−1,742	456	1,278	−0.0	1.8	−7,418	−3,332	10,679	−7,171	110
454	−1,737	1,258	−0.0	−25.2	7,999	−10,397	2,245	8,084	110
1,738	−65	−1,283	0.0	−10.2	7,412	3,334	−10,695	7,162	110
−456	1,736	−1,259	0.0	20.4	−8,006	10,391	−2,243	−8,097	110

Alias1: = PB2 "LOCKPICKUP DROP OUT-FRONT PANEL."

4.22 Setting the Relay

The manufacturers of the MMPRs provide free software for entering the relay settings. The program can check some errors, e.g., out of range entries, but not all. The accuracy of

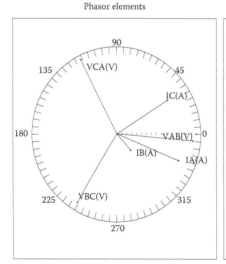
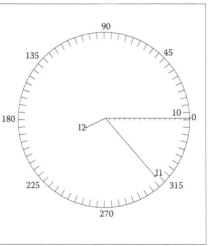

FIGURE 4.24
(a) Phasors for a two-phase-to-ground fault and (b) oscillogram captured for a two-phase-to-ground fault.

(*Continued*)

Microprocessor-Based Multifunction Relays 133

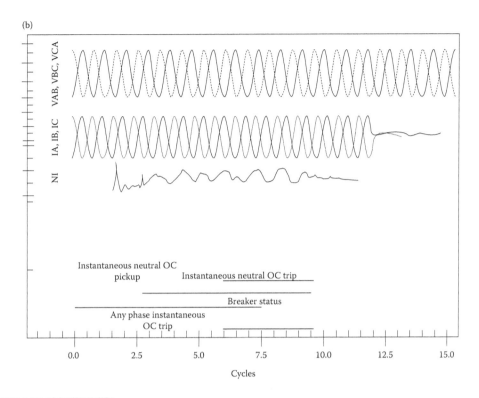

FIGURE 4.24 (CONTINUED)
(a) Phasors for a two-phase-to-ground fault and (b) oscillogram captured for a two-phase-to-ground fault.

the settings depends on the expertise and knowledge of the protection engineer. The settings may run into 30–40 pages. Once the settings are finalized, these can be uploaded into the relay through a port. It is possible to alter the settings at any later time; the new settings are overwritten online without an interruption.

The relay manufacturers have a complete family of relays for the power system protection. Usually, the programming logic is Boolean algebra. Each vendor has dedicated software for setting the relay and invariable settings are loaded through a port. If a relay of one manufacturer is selected for an application and its settings worked out, these cannot be translated or transformed into the settings of an equivalent relay of the second manufacturer.

It can be said that the functionality of the relay is limited by the expertise of the programmer.

The description in this chapter is based on Reference [1]. Other quoted references provide further reading.

Reference

1. Instruction Manual. SEL 751A Feeder Protection Relay. No. 20150206, 2015.

Further Readings

1. A Sahai, DJ Pandya. A novel design and development of microprocessor based current protection relay, in conference record, *International Conference on Recent Advancements in Electrical, Electronics and Control Engineering*, pp. 309–313, 2011.
2. MA Zamai, TS Sidhu, A Yazdani. A protection strategy and microprocessor based relay for low-voltage microgrids. *IEEE Trans Power Delivery*, 26(3), 1873–1863, 2011.
3. B Osorno. Application of microprocessor based protective relays in power systems. *IEEE Industry Application Society Annual Meeting*, pp. 1–8, 2009.
4. RD Kirby, RA Schwartz. Microprocessor based protective relays. *IEEE Ind Appl Mag*, 15(5), 43–50, 2009.
5. MM Mansour. A multi-microprocessor based traveling wave relay-theory and realization. *IEEE Trans Power Delivery*, 1(1), 272–279, 1986.
6. MA Al-Nema, SM Bashi, AA Ubaid. Microprocessor based overcurrent relays. *IEEE Trans Ind Electron*, IE-33(1), 49–51, 1986.
7. Y Akimoto, T Matsuda, K Matsuzawa, M Yamaura, R Kondow, T Matsushima. Microprocessor based digital relays application in Tepco. *IEEE Trans Power Appar Syst*, PAS-100(5), 2390–2398, 1981.
8. G Benmouyai. A log-table based algorithm for implementing microprocessor time overcurrent relays. *IEEE Trans Power Appar Syst*, PAS-101(9), 3563–3567, 1982.
9. JS Thorp, AG Phadke. A microprocessor based three-phase transformer differential relay. *IEEE Trans Power Appar Syst*, PAS-101(2), 426–432, 1982.
10. B Jeyasurya, WJ Smolinski. Design and testing of a microprocessor based distance relay. *IEEE Trans Power Appar Syst*, PAS-1013(5), 1104–1110, 1984.

5
Current Interruption Devices and Battery Systems

5.1 High-Voltage Circuit Breakers

The rating structure of the high-voltage circuit breakers according to ANSI and IEC standards, the current interruption theories, operating mechanisms, stresses in the circuit breakers on current interruption, and the failure modes are discussed in Volume 1. This insight into the high-voltage circuit breakers is desirable but not necessary from protective relaying point of view. It is, however, understood that all switching devices in a power system are selected and applied after rigorous short-circuit studies in the worst condition. The short-circuit calculations and selection of duties for the interrupting devices is covered in Volume 1. With respect to protective relaying, the following are of interest.

1. Circuit breaker operating time or interrupting time is the summation of relaying time, opening time, and arcing time. The circuit breaker timing diagram is shown in Figure 8.6 of Volume 1, which is reproduced again as Figure 5.1 for this chapter.

2. Reclosing is an important parameter for relaying. There is a certain time delay which has to be applied for reclosing. Before a circuit is reenergized, there has to be some dead time for the arc path to become deionized. A dead time of 135 ms is normally applied for circuit breakers of 115–138 kV. These data must be obtained from the manufacturer. As discussed in chapter 15, even a half cycle delay in clearing a fault may result in instability. Thus, the interrupting time of the breaker is of importance.

3. ANSI/IEEE standards on the circuit breakers specify the following interrupting times:
   - Indoor oilless circuit breakers to 36.0 kV: 5 cycles also 3 cycles
   - Outdoor circuit breakers 121–245 kV: 3 cycles
   - Outdoor circuit breakers 362–800 kV: 2 cycles, also 1 cycle (see the below explanation)

   As discussed in Volume 1, 8 cycle breakers, rated on the total current basis, are no longer being manufactured. Recent revisions to IEEE standards also specify an interrupting time of 3 cycles for indoor oilless circuit breakers (see bullet point 1 above). These are commercially available. Also, one cycle breakers using laser technology for detecting current zero crossing are available and have been applied at higher voltages, for example, 500 kV.

4. The high-voltage circuit breakers are interrupting devices only. These are provided with shunt trip coils, which are actuated by the protective relays. For reliability, the breakers can be fitted with duplicate trip coils.

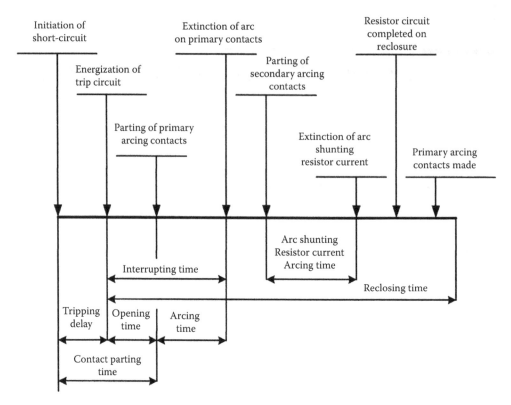

FIGURE 5.1
Characteristics of a current-limiting fuse.

5. The close and trip circuits of the circuit breakers are electrically operated. A reliable source of DC power is normally used at a voltage of 125 V DC, though control circuit voltages of 48 and 250 V DC are also applicable. The requirements of reliable DC source and its protection are discussed in Section 5.2. Catastrophic failures are on record due to loss of DC tripping power. The reliability of the closing and trip supplies is of importance, NERC classifies battery systems as an important part of the relaying systems.

6. Where DC power source is not available or the breakers are located remotely, capacitive trip devices discussed in Section 5.4 can be provided. This is applicable for medium-voltage circuit breakers in an industrial environment. However, these devices cannot be used for control power supply to microprocessor-based multi-function relays (MMPRs).

7. The failure rate of circuit breakers all over the world is reducing; however, failure to trip is one factor which should be considered. Table 5.1 is from the IEEE Standard 493 [1] and shows failure modes of the circuit breakers. If a breaker fails to trip, proper breaker failure protection, discussed in Chapter 4, will direct the trip to an upstream breaker. For all important breakers like generator breakers, this protection must be considered (see also Chapter 11). Yet, this protection is not always provided. Consider, for example, a feeder breaker at 2.4 kV, which is provided with phase overcurrent and ground fault relays only. If the breaker fails to trip, an upstream breaker will trip with additional time-delay, back-up protection.

TABLE 5.1
Failure Modes of Circuit Breakers (Percentage of Total Failures in Each Failure Mode)

Percentage of Totals Failures, All Voltages	Failure Characteristics
9	Failed while opening—backup protective equipment is required
Other Breaker Failures	
7	Damaged while successfully opening
32	Failed in service (not while opening or closing)
5	Failed to close when it should
2	Damaged while closing
42	Opened when it should not
1	Failed during testing and maintenance
1	Damage discovered during testing or maintenance
1	Other

Note: Total 100%.

8. The transient recovery voltage of the HV breakers is of interest for transient analysis, which is not a subject of this volume. Table 5.1 also shows that major failures occurred due to nuisance trips—the breaker tripped when it should not have.

5.1.1 DC Control Schematics

Figure 5.2 shows a DC control schematic of the close and trip circuit of a generator breaker, see Chapter 11. The control schematic can only be designed after the protective systems and the relays to be applied are finalized. Figure 5.2 is based on a generator MMPR relay, showing interlocks in the close circuit and the trip philosophy with simultaneous and sequential trips. With respect to control schematics, some points of importance are as follows:

1. On any trip the close circuit is locked out by the contacts of the lockout relay. Generally, hand reset lockout relays are used so that a fault can be investigated before the breaker is reclosed.
2. All HV breakers are "trip-free." All manufacturers design the control circuits so that if a trip signal is present and an attempt is made to close the breaker, it will not "pump." This means that closing and tripping will not occur repeatedly.
3. When remote trips are required, the DC trip supply is extended to the remote relays. In this case, the remote power circuit should be separately fused. Generally, a separate lockout relay is provided to indicate remote trips. Also, the trip and close circuits are separately fused.
4. The lockout relays, discussed in Section 5.5, are monitored with a white lamp in series with the relay coil. This takes a small amount of current through the relay coil. Discrete relays are available to monitor continuously the trip and close coils. Input contacts of MMPRs can be applied (Chapter 4).
5. Figure 5.2 shows that the DC power supply is monitored through a DC undervoltage relay. This is a desirable addition to all trip and close circuits.
6. An operation counter is a desirable device to be provided.

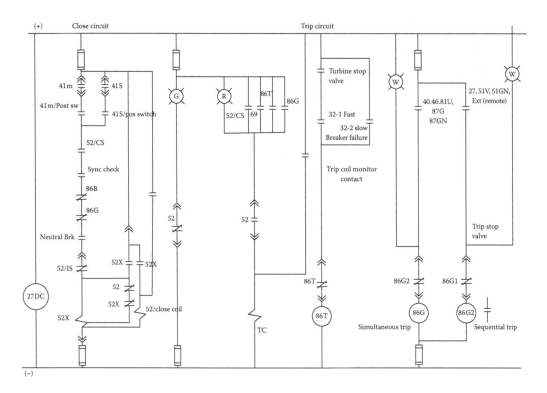

FIGURE 5.2
DC schematic of close and trip circuit of a generator breaker.

5.2 Battery Systems

The battery systems can be studied in the following categories:

- Type of batteries
- Sizing the battery
- Short-circuit calculations of the DC battery systems
- Coordinated short-circuit protection
- Battery and charger connections and redundancy

5.2.1 Battery Types

There are two types of batteries for use in substations and for switchgear service.

- Lead acid batteries, which are more popular in the United States
- Nickel–cadmium batteries, which are more popular in Europe

In 1796, Allesandro Volta's research into electromechanical power sources produced the first battery known as Volta Pile. Since then, the research has been rapid. At present, the lead acid batteries can be divided into the following main categories:

- Plante type
- Faure or pasted plate designs
- Tubular or gauntlet plate designs
- Valve-regulated lead acid

The positive plate consists of PbO_2 (lead peroxide), the negative plate consists of Pb (spongy lead), the electrolyte is sulfuric acid, and the specific gravity is 1.2–1.230. With all lead acid batteries, the chemical reaction is given by the following equations:

At the positive plate:

$$PbO_2 + SO_4^{2-} + 4H^+ + 2e^- \leftrightarrow PbSO_4 + 2H_2O. \tag{5.1}$$

At the negative plate:

$$Pb + SO_4^{2-} \leftrightarrow PbSO_4 + 2e^-. \tag{5.2}$$

The overall cell reaction is

$$PbO_2 + 2H_2SO_4 + Pb \leftrightarrow PbSO_4 + 2H_2O + PbSO_4, \tag{5.3}$$

where

Pb = lead
SO_4^{2-} = sulfate ion
$4H^+$ = hydrogen ion
$2e^-$ = electrons
$PbSO_4$ = lead sulfate
$2H_2O$ = water

At the positive plate, lead dioxide active material on discharge is converted into lead sulfate with the consumption of electrons; while at the negative plate, the spongy lead active material is also converted into lead sulfate with production of ions. This results in flow of electrons in the external circuit from negative to positive plates.

5.2.2 Plante Batteries

Developed in the nineteenth century, this is the oldest design of any rechargeable battery. The plate looks like a car radiator with fine lamellas giving the plate an extremely large surface, approximately 12 times the geometric area. The cells demonstrate a typical life of 25 years. In a temperature-controlled environment, the major drawback is the high initial cost. The advantages and limitations are summarized in Table 5.2. Many utilities still prefer this design.

TABLE 5.2

Lead Acid Batteries: Plante Type

Advantages	Limitations
High short-rate performance, so very suitable for switchgear service	High initial cost
Stable float charge characteristics throughout life	Larger floor area than other designs
Easy maintenance	Life is shortened at high temperature. Must be provided in a temperature-controlled environment
No loss of capacity—no need to size up for aging (normally a 20% increase is recommended in IEEE standards)	
No alloy related problems	
Long, proven, trouble-free life	

5.2.3 Pasted Plate Batteries

These utilize a lead alloy grid with a mixture of active material paste. The grid is designed for mechanical strength and to conduct electricity to terminal posts. Modern batteries are grouped into three categories:

- Lead antimony
- Lead calcium
- Lead selenium

The alloys are introduced to stiffen the grid.

Lead antimony: The alloy typically contains 4%–6% antimony. It will give approximately 800 cycles during its life. Its life expectancy is about 15 years in a temperature-controlled environment. The drawbak is antimonial poising of negative plate. See Table 5.3 for details.

Lead calcium: The development of these batteries took place in Bell Telephone Laboratories in 1951. By adding a small amount of calcium to grid material, it was found that the float charge becomes stable. Grid growth is caused by intergranular corrosion. As oxides form, the grid corrodes. See Table 5.3.

Lead selenium: By reducing the antimony percentage to about 2% and adding selenium in small amounts 0.02% desirable results are obtained (see Table 5.3).

5.2.4 Tubular Plate Batteries

The tubular plate designs utilize an alloy grid that resembles along toothed comb, which is placed inside a polyester pan pipe. The space between the lead spines and the polyester pipe is filled with active materials. These types of batteries are most tolerant of the cycle service, which is approximately 1500 cycles. The batteries have poor high discharge rate performance.
See Table 5.4.

5.2.5 Sealed (Valve-Regulated) Lead Acid Batteries

These batteries are claimed to be maintenance-free, but not really. The electrolyte may not need replenishment, but all other maintenance and periodic tests are required. These have

TABLE 5.3

Lead Acid Batteries: Pasted Plate Type

Lead Antimony		Lead Calcium		Lead Selenium	
Advantages	Limitations	Advantages	Limitations	Advantages	Limitations
Good deep charge capacity	Antimonial poisoning	Low water consumption	Unpredictable failure due to plate growth	Stable float charge capability	Shorter life at high temperatures
Cycle life ~800 cycles	Increased open circuit losses	Stable float charge characteristics throughout	Poor deep discharge capability	No positive plate growth	
	Reduced charging efficiency		Poor cycle life (950–100 cycles)	No antimonial poisoning	
	Increased water consumption		Passivation of positive plate requires capacity testing	Good deep charge capabilities	
	Shorter life at high temperatures			Excellent cycle life (up to 1000 cycles)	
				Low water consumption	

TABLE 5.4

Tubular Lead Acid Batteries

Advantages	Limitations
Rugged design	Relatively poor high rate discharge capability
Low water consumption	Larger floor area than other designs
Enhanced cycle life (1500 cycles)	Life is shortened at high temperature

immobilized electrolyte inside the cells. The acid is absorbed into fiberglass mat separators, which wrap around the plates or it is gelled to immobilize it. The cells will have a pressure relief valve to prevent dangerous pressure building inside the batteries. The cells can be stacked vertically and horizontally in rows, placed above each other, resulting in space savings. Very little hydrogen emission is claimed and unlike other flooded cell batteries the hydrogen monitoring, ventilation, and control may not be required. The hydrogen emission from the flooded type lead acid batteries can be estimated. The manufacturers provide estimating data. *It is very important that the hydrogen emission must be accurately calculated and hydrogen monitoring and ventilation must be provided to avoid fire hazards.*

The useful service life of valve-regulated batteries is only 7–8 years. The operation below 40°F is prohibited.

5.2.6 Battery Monitoring System

An automatic battery checking system is a valuable addition to any battery installation, especially for critical switchgear, substation, and generating station services:

- Scan and record all battery parameters such as overall voltage, each cell voltage, current, and temperature
- Perform scheduled internal resistance tests for all cells and interiors and store data for trending analysis

- Provide an indication of ampere-hour capacity remaining in the battery in real time
- Auto-detect discharges based on overall voltages or discharge currents and store data for real-time or accelerated playback
- Provide form C alarm contacts if the parameter is outside user programmed limits
- Communicate via RS 232, RS 485, and other communication protocols

This requires each cell of the battery system to be wired to a controller.

5.2.7 Nickel–Cadmium Batteries

These batteries have been widely used and have advantages over lead acid batteries. The reliability has been demonstrated from the frozen climate of Siberia to tropical heat in Zaire. The initial cost is a consideration. The positive plate consists of nickel hydrate Ni(OH) mixed with other ingredients, for example, specially treated graphite. The negative plate consists of cadmium and iron oxide. The electrolyte is KOH in distilled water of specific gravity 1–1.90. The nickel–cadmium battery equations are as follows:

At the positive plate:

$$Ni(OH)_3 + e^- \leftrightarrow Ni(OH)_2 + OH^-. \tag{5.4}$$

At the negative plate:

$$Cd + 2OH^- \leftrightarrow Cd(OH)_2 + 2e^-. \tag{5.5}$$

The plate reactions when added together yield overall cell reaction:

$$2Ni(OH)_3 + Cd \leftrightarrow 2Ni(OH)_2 + Cd(OH)_2, \tag{5.6}$$

where
 $Ni(OH)_3$ = nickel hydrate
 Cd = cadmium
 $2OH^-$ = hydroxide ions
 $Cd(OH)_2$ = cadmium hydrate

The important point is that although the potassium hydroxide electrolyte features in plate equations, with hydroxide ions liberated at the positive plate and incorporated at the negative plate during discharging, the net effect is zero in relation to the cell equation.

5.2.8 Pocket Plate Nickel–Cadmium Batteries

These represent 90% of the nickel–cadmium batteries used today. The plate is formed of finely perforated nickel-plated steel strip. Each strip is formed into a channel section to accept the powdered active material. As many as 20 strips or more are linked together.

TABLE 5.5

A Comparison of Maintenance Requirements

Maintenance Procedure	IEEE 450 Lead Acid	IEEE 1106 Nickel Cadmium	IEEE P1168 VRLA (Valve-Regulated Lead Acid)
Visual inspection	Monthly	Quarterly	Monthly
Pilot cell reading	Monthly	Quarterly	Monthly
Float voltage—battery	Monthly	Quarterly	Monthly
Float voltage—cells	Quarterly	Semi-annually	Semi-annually
Specific gravity	Monthly—pilot Quarterly—10% Annually—100%	N/A	N/A
Temperature	Monthly—pilot Quarterly—10%	Quarterly—pilot	Quarterly—100%
Connection resistance	Annually	Re-torque only	Quarterly—25% Annually—100%
Ohmic measurement	N/A	N/A	Quarterly—100%
Discharge tests	5 years/1 year	5 years/1 year	1 year/6 months

The plates are then cut to size, spot welded to a collector lug, and finally pressed under 400 tons to consolidate the plate. Other type of nickel–cadmium battery construction is valve regulated, which is less popular. The main advantages are as follows:

- Wide tolerance to temperature extremes.
- The electrical reactions of battery yield no change in the concentration of KOH.
- The mechanical strength does not degrade with age as that of a lead acid battery.
- Very long life—25 years or more.
- Storage in any state of charge—they can be held in discharged condition for several years and in charged condition up to 2 years.
- Low maintenance.
- Resistant to electrical abuse (overcharge and overdischarge).
- Excellent charge retention.
- Excellent resistance to shock and vibrations.
- Accepts high rates of charge.

See Table 5.5 for a comparison of maintenance. This amply shows that the nickel–cadmium batteries require even less maintenance compared to valve-regulated lead acid batteries.

5.3 Sizing the Batteries

Selecting an appropriate battery technology is only the first step in assuring system reliability. To correctly size a battery load profile is required. All current requirements must be included. The battery profile should consider the following points:

- The standing load on the battery indicates the number of lights or other devices which are connected to the battery and draw continuous current.
- The number of breakers that will be simultaneously tripped and reclosed after a small delay. The worst condition of impact load occurs when a number of breakers on the bus are tripped due to operation of differential protection and then subsequently closed. On loss of power source, the circuit breakers, if provided with undervoltage trips, will trip. Generally, the undervoltage trips are provided to prevent large inrush currents, say, due to energization of a number of transformers simultaneously, which may not be acceptable.
- The trip and close currents of the circuit breakers should be obtained from the manufacturer.
- Battery voltage window should be carefully determined. This defines the normal voltage range over which the battery is allowed to operate. Normally, alarms are provided for low battery voltage. The normal operating range of close and trip coil voltages is specified by the manufacturer.
- The temperature, especially the minimum temperature, plays an important role in determining the adequate capacity of the battery. It is recommended that the battery systems are located in a temperature-controlled environment.
- At the end of the life, for most battery technologies, only 80% of the initial performance is available. In general, an aging factor of 25% is used in sizing the batteries.
- The substation batteries should be selected for a life expectancy of 20 years. This will limit the choice to certain types as discussed before.

5.3.1 Standards for Sizing the Batteries

The following standards are applicable for sizing the batteries:

- ANSI/IEEE 485: For sizing lead acid batteries for generating stations and substations
- IEEE 1115: For sizing nickel–cadmium batteries for stationary applications
- IEEE 1184 (draft): For sizing batteries for uninterruptible power supply systems
- IEEE P1189: For selection and sizing of valve-regulated lead acid batteries.

The guidelines in these standards must be followed. An example of sizing the batteries is not included here. See appropriate references; however, a battery load profile for a certain application is shown in Figure 5.3.

5.3.2 System Configurations for Batteries

A simple system configuration without redundancy is shown in Figure 5.4. This shows a single battery and a charger in parallel connected to a DC bus, which serves the DC loads. This is, generally, not recommended.

Figure 5.5 shows a system configuration with a complete standby redundant system. In case of failure of system 1, the undervoltage is sensed on the source side of the automatic transfer switch (ATS) and the entire load is transferred to the standby system.

A further refinement is shown in Figure 5.6. In Figure 5.5, redundancy is achieved with respect to main system failure. In Figure 5.6, by adding another transfer switch, redundancy can be achieved with respect to each load circuit.

Current Interruption Devices and Battery Systems

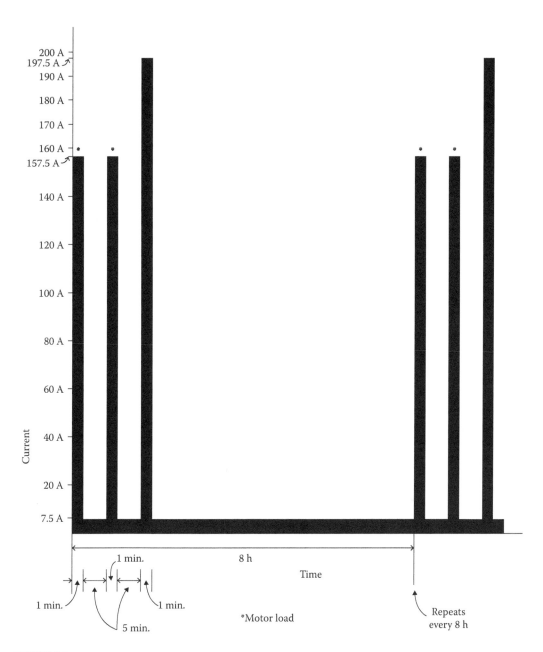

FIGURE 5.3
DC load profile for sizing a battery.

5.3.3 Automatic Transfer Switches

The automatic transfer switches (ATSs) can be made to close on differential sensing of both the normal and standby sources. The following features are normally provided:

- Switch position LED indicators
- A selector switch for manual/automatic operation

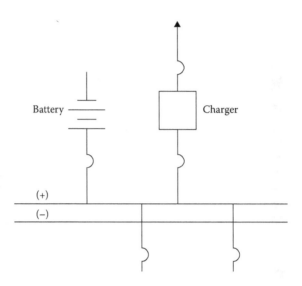

FIGURE 5.4
Single charger and battery in parallel, no redundancy.

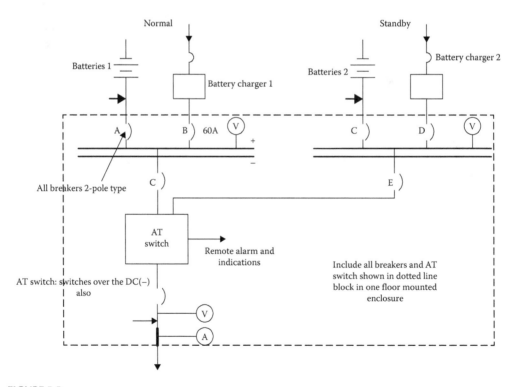

FIGURE 5.5
Redundant battery system with ATS.

Current Interruption Devices and Battery Systems

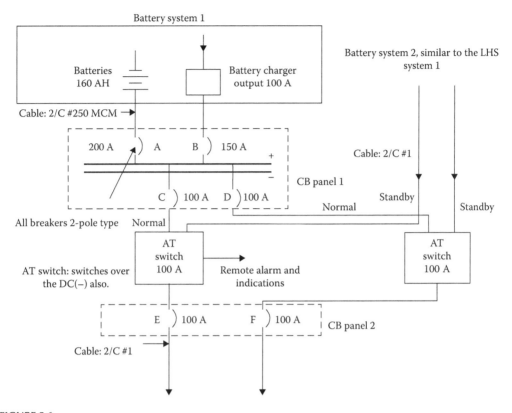

FIGURE 5.6
Redundant battery system with load redundancy added through another ATS.

- A test switch to simulate normal source failure
- User adjustable time delay to transfer to the standby
- Remote alarms and indications, hardwired or through communication links.

The switch is blocked from transferring on failure of standby source and these meet the requirements of UL 1008. The ATS may not have required short-circuit interrupting rating and must be protected by an upstream adequately rated device.

5.3.4 Battery Chargers

The battery can be float-charged with voltages ranging from 2.23 to 2.25 V/cell (lead acid batteries). The voltage for the rapid charge is 1.35–2.40 V/cell. Charging may be carried out under either a floating or switch operating mode.

5.3.4.1 Floating Operation

The battery and the critical load circuit are continuously connected in parallel. The charger must be designed for the following objectives:

- Charging the battery from the discharge condition while supplying the load.

- Providing required constant float charge. The charger voltage should be 2.23 V ±1% multiplied by the number of cells in the battery.
- Providing voltage for equalizing the battery.

5.3.4.2 Equalizing Charge

An equalizing charge is given when

- The temperature-corrected specific gravity has fallen by more than 10 points (0.01).
- More than one cell falls below 2.15 on float (corrected for temperature).

5.3.4.3 Switch Mode Operation

In this mode, the battery is separated from the load. Toward the end of charging, the cell voltage is 2.6–2.7 V. The charger system has automatic transfer mode; that is, when the battery is fully charged, it transfers to float charge mode. As a standard, the chargers are provided with

- AC and DC side circuit breakers
- AC and DC voltmeters and ammeters
- Float charge potentiometer for adjustments and adjustable equalizing voltage potentiometer
- Indicating lights
- Loss of AC input power alarm
- Low- and high-voltage DC alarms
- Surge voltage suppressors

5.3.5 Battery Charger as a Battery Eliminator

The battery chargers can be supplied as battery eliminators, which imply that in the event the battery is disconnected, the charger can supply the required substation DC load profile including tripping and closing of the circuit breakers. This means that

- The charger should be sized for the required loads.
- The transient performance should guarantee that the voltage dip on sudden impact load is within acceptable limits.
- When the charger is operating in parallel with the battery, the battery acts like a sink to spikes and harmonics. When the charger is designed as a battery eliminator, additional passive filtering of the output is required.

5.3.6 Short-Circuit and Coordination Considerations

Chapter 12 of Volume 1 discusses the short-circuit calculations of batteries and rectifiers. Also, the chapter provides a specimen example of the coordination and selection of the short-circuit interrupting devices, DC breakers and fuses. Often, these calculations are ignored and the applied DC circuit breakers may be underrated. Table 12.1 in Volume 1 shows the preferred ratings of general-purpose DC power circuit breakers with or without instantaneous direct acting trips.

5.4 Capacitive Trip Devices

An energy storage device can be used to trip the breaker. It uses an auxiliary long life dry battery and a capacitor is charged from the source voltage of 120 V single-phase AC or other source voltages. The charge can be held for 72 h and the trip signal can be maintained for 10 s. Enough charge can be stored on the capacitor for more than one trip operation on failure of source power. There is a small drain of 2 W continuously, representing losses in the system. The limitations are as follows:

- Indicating lamps are not permissible to conserve the capacitor charge.
- The capacitor is charged to much higher voltage than the nominal 120 V DC—may be up to 330 V. While this is acceptable for operation of a lockout relay, *this is not suitable for powering the control circuit of the digital relays or MMPRs.*

5.5 Lockout Relays

The lockout relays are of high speed operation, their operating time is as low as 8 ms, and they are primarily used for tripping and lockout of the circuit breakers. It is a two-position device having manual operation to the "reset" position and electrical trip, spring operated to the "trip" position. The escutcheon is marked "trip" and "rest." The trip coil is factory wired to a coil cutout contact. It has a low continuous current rating of about 50 mA so that a white indicating light can be provided in series with the coils for continuous monitoring (separate more elaborate trip coil monitoring relays are available; MMPRs have the functionality of continuously monitoring the close and trip coils). There are three types of lockout relays:

- Hand reset
- Electrically reset
- Self-reset instantaneous rest or time delay reset

Normally, the hand reset relays are applied. The fault needs to be investigated before the breaker is reclosed. The unit will stay in the lockout position till manually reset. Figure 5.7a shows the circuit of a hand reset lockout relay. The operation of the protective relay contact causes the device to snap to rip position, and locks into this position. Simultaneously, the relay contacts in series with the coil are opened, removing it from the service.

The circuit of an electrically reset lockout relay is shown in Figure 5.7b. It operates from the control bus voltage like a hand reset relay. The LOR/R coil represents the rotary solenoid that is used to reset the LOR/T electrically. K1 controls the rotary solenoid. The contact

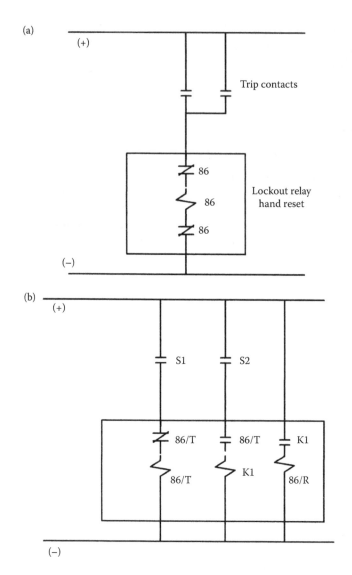

FIGURE 5.7
(a) Hand-reset lockout relay and (b) electrically reset lockout relay, see text.

S2 controls only the K1 relay coil. Closing of contact Se energizes the LOR/R, which resets LOR/T. S1 and S2 are momentary contacts and should not stay closed at the same time. A "pumping" action will result with LOR/R indexing back and forth.

The self-reset relays are not discussed here.

5.6 Remote Trips

Often, it is required to trip a breaker located remotely with the operation of a relay or intertripping of the breakers may be required. With hard-wired trips, there is a limit of the distance to which the breakers can be controlled or intertripped due to voltage drop in the control cables. This voltage drop is dependent on the distance involved as well as on the inrush currents drawn by the LOR which can vary over large limits, depending on the manufacture. Sometimes, it is possible to select coils of different burdens on the same lockout relay. However, the coils with higher resistance may slow the operating time of the LOR. See Figure 5.8. Considering #10 AWG control cable and limiting the voltage dip in the circuit to 10 V, with 2.5 A inrush current trip coil, ~0.50 km is the maximum distance for hard wire trip.

When tripping over a longer distance, the following are needed:

- Communication protocols that may be specific to a vendor over multimode or single-mode fiber
- Radio link with Ethernet interface

Figure 5.9a shows hard wire trips. Note that by running an extra conductor and providing an indicating lamp IL2, the trip control circuit can be continuously monitored for a break in the control conductors. Figure 5.9b shows the remote trip by using transreceivers and fibers. Figure 5.9c shows the remote trips through radio link.

5.7 CT and PT Test Switches

The CT and PT test switches are provided. The CT test switches will permit testing of the relays connected in CT circuit and will short out the CTS terminals and isolate the relay circuit (Figure 5.10). In this figure, relays 1 and 2 cannot be taken out *individually.*

Figure 5.11 shows CT test switches and their contact numbers. These are utilized in Figure 5.12. Note that here are two relays, each with phase- and ground-fault protection and also a power meter. By using three test switches of Figure 5.11, each device can be taken out of service individually, while the others remain in service. This is a desirable CT test circuit. A total of three test switches as shown in Figure 5.11 are required. Note that each test switch is shown in two parts, for example, TB1 and TB1A, for illustration of wiring connections.

5.8 Fuses

We discussed the characteristics of following fuse types in Section 8.15 of Volume 1:

- Current-limiting fuses
- Low-voltage fuses
- High-voltage fuses
- Interrupting ratings

DC-rated fuses are discussed in Section 12.9 of Volume 1, and an introduction to semiconductor fuses is provided in Section 12.10 of Volume 1. These data are not repeated

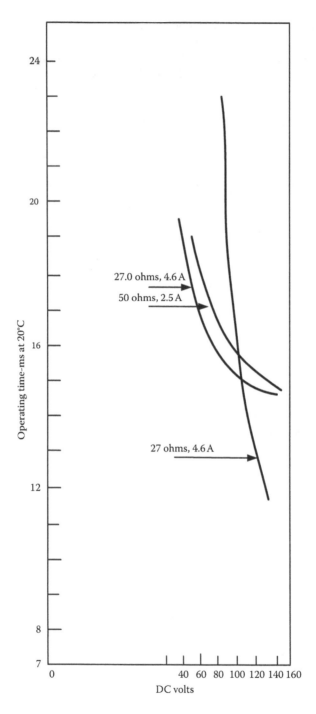

FIGURE 5.8
Operating times and voltage window of various lockout relay types, 125 V DC service.

Current Interruption Devices and Battery Systems

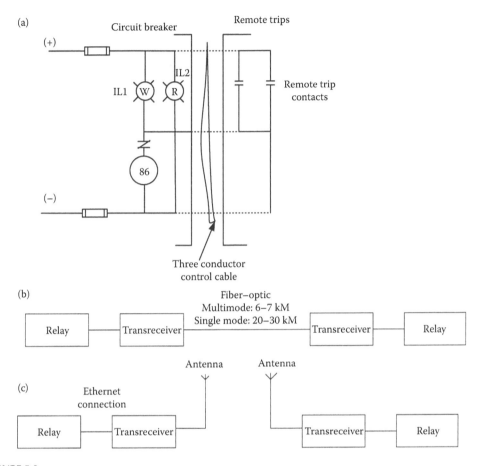

FIGURE 5.9
(a) Hard-wired remote trips, (b) remote trips through transreceivers and fibers, and (c) remote trips through radio link.

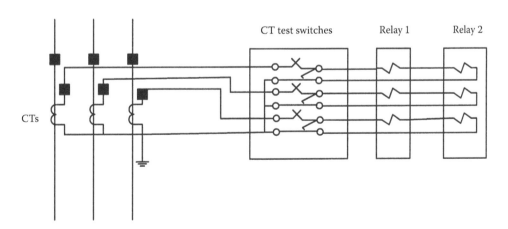

FIGURE 5.10
CT test switches introduced in relay 1 and relay 2 circuit. Both relays will be out of service on operation of test switches.

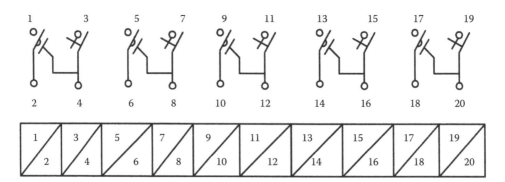

FIGURE 5.11
Circuit profile of composite CT test switches (the number of switches can be added or subtracted).

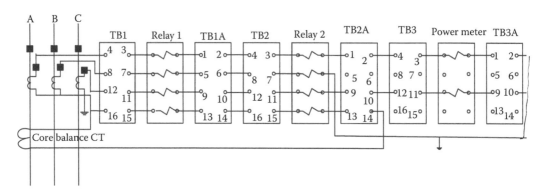

FIGURE 5.12
CT test switches wired so that any relay or meter can be taken out of service, while the others remain in service.

here. Figure 5.13 from ANSI/IEEE 242 shows the low-voltage fuse types (0–600 V). Table 5.6 shows popular low-voltage AC and DC fuses, and their ratings and applications. The following points may be noted:

- The time–current characteristics are published as a band, minimum melting time and the total operation time. For the coordination, the maximum of total operating time must be considered for an upstream device and the minimum melting time for the downstream device with a safe margin, generally 10%.
- The time–current characteristics of the fuses will vary with the type of fuse, fast acting or time delay type.
- There are variations in the time–current characteristics between various manufacturers. For coordination, the actual fuse time–current characteristics of the specific manufacturer should be plotted. As an example, the time–current characteristics of RK1 fuses and the melting time for fuses rated from 15 to 600 A are shown in Figure 5.14.
- The let-through characteristics of the various types of fuses vary. A typical let-through characteristic for RK1 fuses is shown in Figure 5.15.
- A current-limiting fuse limits the perspective fault current and opens in half cycle or less. Thus, the let-through energy to the fault is much reduced. See Figure 8.24

Current Interruption Devices and Battery Systems

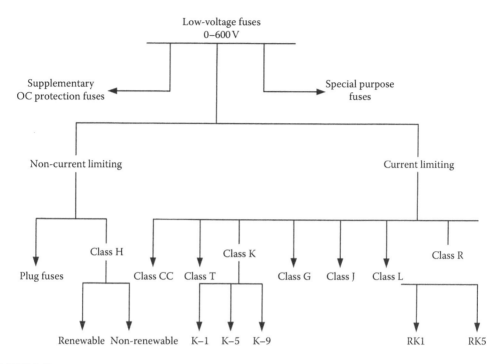

FIGURE 5.13
Skeleton diagram of types of low-voltage fuses.

in Volume 1 for the operation of the current-limiting fuses. Also see Chapter 6 for the coordination between the fuses and coordination on an instantaneous current basis.

- Table 5.6 shows that for a certain application, there is an overlapping choice of type of fuse that can be applied. For low-voltage motor starters, the application of fuses is going out of vogue, because of (1) the replacement cost of fuses on operation and (2) the possibility of single phasing, which can be damaging to the motor. Currently, molded case circuit breakers (MCCBs) or motor circuit protectors (MCPs) (see Section 5.9) are being applied to low-voltage motor starters.
- The series-connected devices and their ratings are discussed in Chapter 8 of Volume 1. See Figure 5.16 for motor-connected loads, when the primary fuse is used to protect lower-rated feeders on a bus. There is a limit to the motor loads that can be connected downstream. Series rating is not permissible if the motor full-load current exceeds 1% of the lower-rated circuit breaker.

5.8.1 Medium-Voltage Fuses

See Section 8.15.3 for the high-voltage fuses and also Table 8.15 in Volume 1 for the interrupting ratings of the high-voltage fuses. The standards governing the HV fuses are ANSI/IEEE Standards C37.40, C37.41, C37.42, C37.46, C37.47, and C37.48 [2–7]. NEMA and UL are provided in References [8–14]. The high-voltage fuses are classified as follows:

- Distribution cutouts
- Power fuses

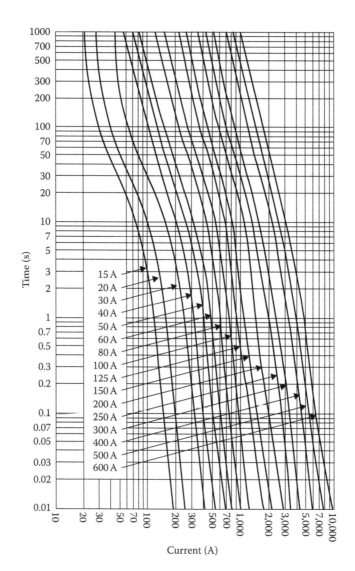

FIGURE 5.14
Time–current characteristics of RK1 fuses rated from 15 A to 600 A.

According to ANSI/IEEE C37.100, the distribution cutout has the following characteristics:

- BIL at distribution levels
- Application primarily on distribution feeders and circuits
- Mechanical construction adopted to pole or cross-arm mounting

Operating voltage limits corresponding to distribution system voltages

- Figure 5.17 shows a distribution cutout of 100 A. The fuse holder is lined with an organic material. Interruption of an overcurrent takes place within the fuse holder by action of deionizing gases liberated when the liner is exposed to arc heat.

TABLE 5.6

Popular Low-Voltage Fuses—Data

Fuse Type	Ratings AC	Ratings DC	Standards	Applications
Class J, time delay	1–600 A, 200 kA IR, 600 VAC, 300 kA IR	1–600 A, 500 V DC, 100 kA IR	UL: 248-8, 198L CSA C22.2, 248.8 IEC 269-2-1	Motor circuits, mains, branch circuits Lighting, heating, and general loads Transformers Control panels, circuit breaker backup Bus duct, load centers
Class L, time delay	100–6000 A, 200 kA IR, 600 V AC, 300 kA IR, 4 s delay at 500% rating	601–3000 A, 500 V DC, 100 kA IR	UL: 248-8, 198L CSA C22.2, 248.10 IEC 269-2-1	Mains, feeders Large motors Lighting, heating, and general loads Circuit breaker backup UPS DC links, battery disconnects, and other DC applications
RK1, time delay	1/10 to 600 A, 200 kA IR, 600 V AC, 300 kA IR		UL: 248-12, CSA C22.2, 248.12	Motors Safety switches Transformers and branch circuit protection Disconnects, control panels, and general-purpose circuits
CC, time delay	1/4 to 30 A, 200 kA IR, 699 V AC	1/4 to 30 A, 300 V DC, 100 kA IR	UL: 248-4, 198L CSA C22.2, 248.4	Small motors Contactors Lighting, heating, and general loads
RK5, time delay	1/10 to 600 A, 200 kA IR, 600 V AC,	1/10 to 400 A, 250 V DC, 20 kA IR 70–600 A, 600 V DC, 100 kA IR	UL: 248-12, 198L CSA C22.2, 248.12	Motor circuits, mains, branch circuits Transformers Service entrance equipment, general-purpose protection
Class J, fast acting	1–600 A, 200 kA IR, 600 V AC	1–600 A, 300 V DC, 20 kA IR	UL: 248-8, 198L CSA C22.2, 248.8 IEC 269-2-1	Excellent protection as current limiters Capacitors, feeder circuits. Lighting, heating, and general loads Switchboards, panels, circuit breakers Bus duct, load centers
Class T, fast acting	1–1200 A, 200,000 kA IR, 300 V AC,	1–1200 A, 160 V DC, 50 kA IR	UL: 248-15, 198L CSA C22.2, 248.15	Extremely current limiting Load centers Panel boards Switchboards Circuit breakers Metering centers
Class G, time delay	1/2 to 60 A, 100 kA IR, 480 V AC		UL: 248-5, CSA C22.2, 248.5	Lighting Heating Appliances Branch circuits
K-5	1–600 A, 50 kA IR, 250 V AC		UL: 248-9, CSA C22.2, 248.9	Feeders and branch circuits Resistive heating Residential and small commercial installations
Class H, renewable	1–600 A, 10 kA IR, 250 V AC		UL: 248-7, CSA C22.2, 248.7	General-purpose renewable links, where the IR is no more than 10 kA
Midget fuses	1/10 to 30 A, 10 kA IR, 500 VAC		UL: 248-14, 198L CSA C22.2, 248.14	These are of small dimensions 1-1/2" × 13/32" for supplemental protection of circuits up to 30 A
PC mount fuses	1–30 A, 100 kA IR, 600 V AC	1–30 A, 500 V DC, 100 kA IR	UL recognized components	Fast acting directly mountable for PC boards

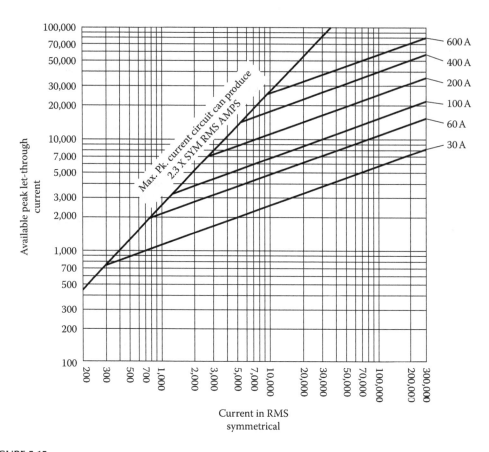

FIGURE 5.15
Let-through characteristics of RK1 fuses in Figure 5.14.

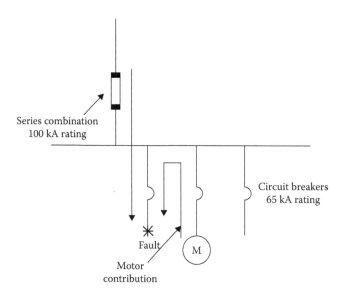

FIGURE 5.16
Limitations of motor loads for series-connected devices, as per NEC, see text.

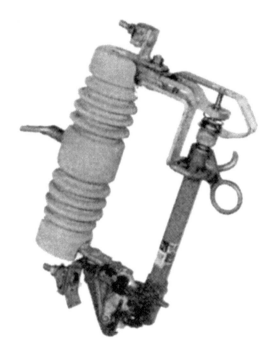

FIGURE 5.17
A distribution system cutout of 100 A.

Class E fuses are suitable for protection of voltage transformers, power transformers, and capacitor banks, while class R fuses are applied for medium-voltage motor starters. *All class E-rated fuses are not current limiting*; E rating merely signifies that class E-rated power fuses in ratings of 100E or less will open in 300 s at currents between 200% and 240% of their E rating. Fuses rated above 100E open in 600 s at currents between 220% and 264% of their E ratings. The application of class E fuses is discussed in Chapter 12 and that of class R fuses in Chapter 10. Figure 5.18 shows the application of class E current-limiting power fuses for transformer protection. For the medium-voltage motors, such general tables are not supplied, as the fuse size must coordinate with the motor locked rotor and starting time curve, which may vary for the similar rated motors. See Chapter 10 for coordination examples.

5.8.1.1 Variations in the Fuse Time–Current Characteristics

For the same rating and type of fuse, the time–current characteristics vary between various manufacturers. This is so because IEEE standards only define a point for the operation time at a certain current, see Chapter 6. Furthermore, there can be much difference in the time–current characteristics of current-limiting type and expulsion type of fuses. Figure 5.19 shows four characteristics of 150-A class E fuses: two fuses are current-limiting type and two are expulsion type.

5.8.2 Selection of Fuse Types and Ratings

The selection of fuses for medium-voltage applications requires careful consideration and study. The fundamental considerations are as follows.

Suggested minimum current limiting fuse current ratings for self-cooled 2.4–15.5 kV transformer applications—E–rated fuses

System Nominal kV	2.4		4.16		4.8		7.2		12.0		13.2		13.8		14.4	
Fuses Maximum kV	2.75		5.5		5.5		8.3		15.5		15.5		15.5		15.5	
Transformer kVA Rating Self-Cooled	Full Load Current Amps	Fuse Rating Amps E	Full Load Current Amps	Fuse Rating Amps E	Full Load Current Amps	Fuse Rating Amps E	Full Load Current Amps	Fuse Rating Amps E	Full Load Current Amps	Fuse Rating Amps E	Full Load Current Amps	Fuse Rating Amps E	Full Load Current Amps	Fuse Rating Amps E	Full Load Current Amps	Fuse Rating Amps E
112.5	27.1	50E	15.6	25E	13.5	20E	9.0	15E	5.4	10E	4.9	10E	4.7	10E	4.5	10E
150	36.1	65E	20.8	30E	18.0	25E	12.0	20E	7.2	15E	6.6	10E	6.3	10E	6.0	10E
225	54.1	80E	31.2	50E	27.1	50E	18.0	25E	10.8	15E	9.8	15E	9.4	15E	9.0	15E
300	72.2	125E	41.6	80E	36.1	65E	24.1	40E	14.4	20E	13.1	20E	12.6	20E	12.0	20E
500	120.3	200E	69.4	125E	60.1	100E	40.1	65E	24.1	50E	21.9	30E	20.9	30E	20.0	30E
750	180.4	300E	104.1	150E	90.2	150E	60.1	100E	36.1	65E	32.8	65E	31.4	65E	30.1	65E
1,000	240.6	350E	138.8	200E	120.3	175E	80.2	125E	48.1	80E	43.7	80E	41.8	80E	40.1	80E
1,500	360.8	600E	208.2	300E	180.4	250E	120.3	175E	72.2	100E	65.6	100E	62.8	100E	60.1	100E
2,000	481.1	750E	277.6	400E	240.6	350E	160.4	250E	96.2	150E	87.5	125E	83.7	150E	80.2	125E
2,500	601.4	1100E	347.0	600E	300.7	450E	200.5	300E	120.3	200E	109.3	175E	104.6	175E	100.2	175E
3,000	721.7	1100E	416.4	600E	360.8	600E	240.6	350E	144.3	250E	131.2	200E	125.5	200E	120.3	200E
3,750	902.1	1350E	520.4	750E	451.1	750E	300.7	—	180.4	250E	164.0	250E	156.9	250E	150.4	250E
5,000	1202.8	—	693.9	1100E	601.4	1100E	400.9	—	240.6	—	218.7	300E	209.2	300E	200.5	300E
7,500	1804.2	—	1040.9	—	902.1	1350E	601.4	—	360.8	—	328.0	—	313.8	—	300.7	—
10,000	2405.6	—	1387.9	—	1202.8	—	801.9	—	481.1	—	437.4	—	418.4	—	400.9	—

Note: Fuse ratings represent the fuse that will withstand transformer inrush (12 × FLC for 0.1 s and 25 × FLC for 0.01 s) and be able to handle temporary overloads (133% of FLC, 150% for 15.5 kV).

FIGURE 5.18
Recommended class E fuses for transformers primary protection.

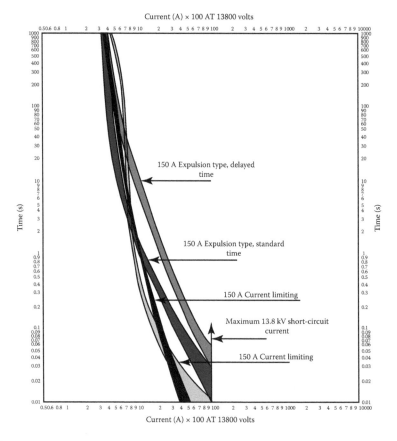

FIGURE 5.19
Variations in the time–current characteristics of 150 A fuses: two fuses are of current-limiting type and two are of expulsion type.

Voltage rating: The voltage rating must be selected as next higher standard voltage of the fuse above the maximum operating voltage level of the system. The boric acid power fuses are not so voltage critical in that they can be safely applied at voltage less than their rated voltage with no detrimental effects. These fuses have a constant current interrupting ability or at best an increased current interrupting ability when applied to systems operating at one or more voltage levels below fuse rating. The current-limiting power fuse is voltage critical.

Current rating: The continuous current rating becomes derated at higher ambient temperatures and in outdoor locations. These will give rated continuous rating with an ambient temperature of 40°C. In outdoor locations, the temperature inside the fuse enclosure can be much higher. The transient inrush current of transformers and the coordination with respect to these are discussed in Chapter 12. This may require a fuse rating which *is two times higher* than the transformer full load current. Here, a compromise is made that small overloads will not operate the fuse. The coordination with respect to motor starting is discussed in Chapter 10. Normal repetitive overloads may be higher than the normal loads. Boric acid fuses have inherent overload capability; these data can be obtained from the manufacturers.

Interrupting ratings: System short-circuit calculations are required as discussed in Volume 1. The fuse should be selected for the maximum rms symmetrical current and also for the maximum rms asymmetrical three-phase fault currents.

Selectivity: This important aspect is discussed in Chapters 6, 10, and 13.

5.8.3 Semiconductor Fuses

The semiconductor fuses are designed for projection of diodes, power electronic components, and applications in UPS and ASD systems. These are available in various current ratings for both AC and DC circuit applications, and voltages up to 700 V.

Unlike transformers, generators, or motors, the solid-state devices do not have high let-through energy capabilities and can be damaged on operation of a conventional fuse. The semiconductor fuses have the following characteristics:

- Low I^2T let through
- Fast acting
- Current limiting

As an example, the let-through curves of semiconductor fuses of a manufacturer are shown in Figure 5.20. Compare this with RK1 fuses in Figure 5.15. The let-through current is much lower. Table 5.7 shows the I^2T let through in tabulated form. The I^2T of the semiconductor device to be protected can be obtained from the manufacturer and the fuse I^2T coordinated. Note that with two similar fuses in series the I^2T is much reduced.

A modern trend in the protection of semiconductor devices is grid control or gate control. In the event of short circuit, the short-circuit current can be limited fast to save damage to the semiconducting device. See Figure 12.16 in Volume 1.

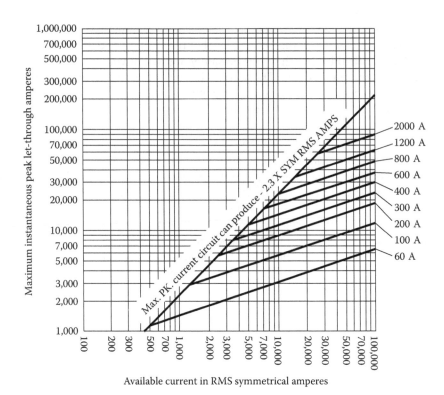

FIGURE 5.20
Let-through characteristics of semiconductor fuses, 700 V rating.

5.9 Low-Voltage Circuit Breakers

We discussed low-voltage circuit breakers in Section 8.14 of Volume 1. The description provided here may contain some repetitions. The three classifications of low-voltage circuit breakers are (1) MCCBs, (2) insulated case circuit breakers (ICCBs), and (3) low-voltage power circuit breakers [15–20].

5.9.1 Molded Case Circuit Breakers

In MCCBs, the current carrying parts, mechanism, and trip devices are completely contained in a molded case insulating material and these breakers are not maintainable. Available frame sizes range from 15 to 6000 A, interrupting ratings from 10 to 100 kA symmetrical without integral current-limiting fuses and to 200 kA symmetrical with current-limiting fuses. These can be provided with electronic- and microprocessor-based trip units, and have limited short-time delay and ground fault sensing capability. When provided with thermal magnetic trips, the trips may be adjustable or nonadjustable, and are instantaneous in nature. MCPs may be classed as a special category of MCCBs and are provided with instantaneous trips only. MCPs do not have an interrupting rating by themselves and are tested in conjunction with motor starters. An MCCB can be applied for a variety of applications, for example, in residential and industrial distribution panels, in main power feed panels, for controlling low-voltage motor starters, and for many other commercial applications.

TABLE 5.7

I²T Data, 600 V AC, 100 kA Interrupting Semiconductor Fuses

		I²T Data	
			Clearing at 600 V
Fuse Amps	Melting A²s	1 Fuse A²s	2 Fuses in Series A²s
15	35	100	70
20	60	170	130
25	95	275	210
30	140	400	290
35	270	1,800	1,200
40	350	2,400	1,500
45	450	3,000	2,000
50	550	3,600	2,400
60	800	5,400	3,600
70	4,000	13,000	9,800
80	5,300	17,000	13,000
90	6,700	22,000	16,000
100	8,300	27,000	20,000
125	13,000	42,000	31,000
150	19,000	60,000	45,000
175	25,000	80,000	61,000
200	33,000	110,000	80,000
225	42,000	140,000	100,000
250	52,000	170,000	125,000
300	75,000	240,000	180,000
350	100,000	340,000	240,000
400	130,000	490,000	320,000
450	170,000	620,000	400,000
500	210,000	770,000	500,000
600	300,000	1,100,000	720,000
700	430,000	1,700,000	1,000,000
800	560,000	2,250,000	1,400,000
1,000	875,000	3,500,000	2,200,000
12,00	1,250,000	5,000,000	3,100,000
1,500	2,000,000	7,900,000	4,900,000
1,600	2,200,000	9,000,000	5,600,000
1,800	2,800,000	11,000,000	7,100.000
2,000	3,500,000	14,000,000	8,900,000

The predominant standard is UL 489. For UL listing, the MCCB must undergo well-defined rigorous series of test sequences. These include temperature rise measurements at MCCB terminals while carrying rated current after the MCCB has interrupted an overload current of 600% of the full-load current 50 times.

Figure 5.21 shows the thermal overload trip action. The bimetallic design parameters controlling the thermal action are type of material, its resistance, thermal capacity, and overall length of the element. Figure 5.22 shows the magnetic trip action, and adjustable characteristics are obtained by varying air-gap length or adjustment of spring force.

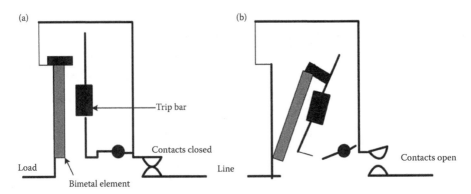

FIGURE 5.21
A bimetallic thermal trip device: (a) normal state and (b) tripped state.

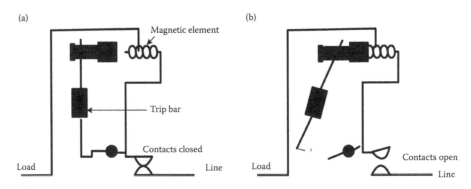

FIGURE 5.22
A magnetic instantaneous trip device: (a) normal state and (b) tripped state.

The electronic trip units permit different rating plugs in the same frame size, and they may use a flux transfer device requiring very small energy of 3 milli joules to shift the flux and trip. For high continuous currents of 1000 A or more, there are parallel main and arcing contacts; the main contacts make last and break first. The current commutation from mains to arcing is of the order of 300–600 μs. MCCBs use de-ion plates to form the arc chute. Much effort has gone into the optimization of arc runners for rapid arc motion.

MCCBs should not be applied at any current rating >80% of their nameplate rating, NEC. However, 100% rated breakers are available. These are tested for a minimum size enclosure to UL 489 for applications at 100% rating. These are equipped with electronic trip units and applied with 90°C cable rated at 75°C ampacity.

As per NEC:

Standard 80% rated designs
Noncontinuous load + 125% of the continuous load = total minimum load

Special 100% rated designs
Noncontinuous load + continuous load = total minimum load

5.9.2 Current-Limiting MCCBs

The 1970s saw the development of current-limiting MCCBs. The arc in an MCCB serves the additional function of suddenly injecting a resistive element into the circuit to limit the fault current. The current-limiting MCCBs have rapid contact motion after fault initiation and rapid arc voltage development which is achieved by arc runners or blow-out effect of de-ion plates, and fast gap recovery voltage. The effectiveness is given by both peak let-through current and $\int i^2 \, dt$ values. Figure 5.23 shows a representative arc voltage and current waveform for a current-limiting MCCB.

The operation of MCCBs not *classified as current limiting* is illustrated in Figure 5.24. This figure clearly depicts that to *some extent* the MCCBs are current limiting. Figure 5.25 illustrates the typical characteristics of a current-limiting MCCB. The current-limiting MCCBs can be thermal magnetic or provided with electronic trips. These trip units, much alike the trip units of low-voltage power circuit breakers (LVPCBs) (Figure 5.26), can have adjustable current pickups, adjustable LT delays, short time pickup and delays, and must be provided with instantaneous trips. Generally, all the adjustments shown in Figure 5.26 are not available in the electronic trip programmers for MCCBs. The MCCBs can be true rms sensing (for harmonic loads) and also be provided with shunt trip coils for operation from separate relaying devices. Undervoltage trips, zone interlocking, and a host of other features are possible.

For arc-flash reduction, the development of MCCBs, faster in operation with much higher short-circuit ratings, has attracted the manufacturers. The current-limiting action has three distinct benefits.

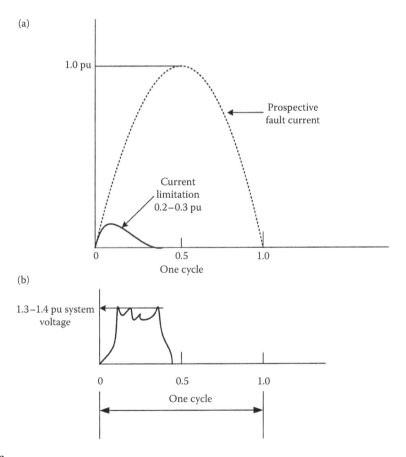

FIGURE 5.23
(a) Current-limiting operation of a current-limiting MCCB and (b) arc voltage generated during operation.

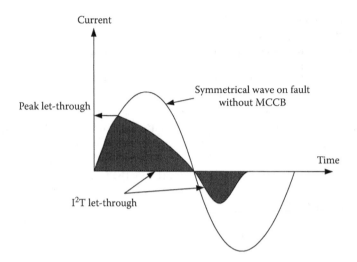

FIGURE 5.24
Operation of an MCCB, not specifically designed as current limiting.

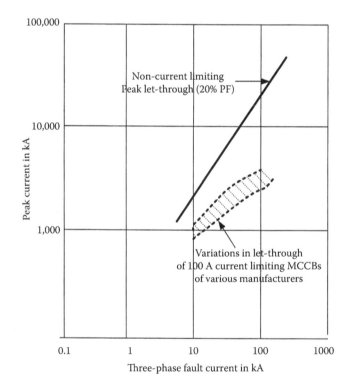

FIGURE 5.25
Let-through characteristics of a current-limiting MCCB.

- It provides lesser let-through energy and reduction in arc-flash hazard.
- Current limitation can help series ratings.
- Better coordination can be obtained with coordination on instantaneous basis (Chapter 6).

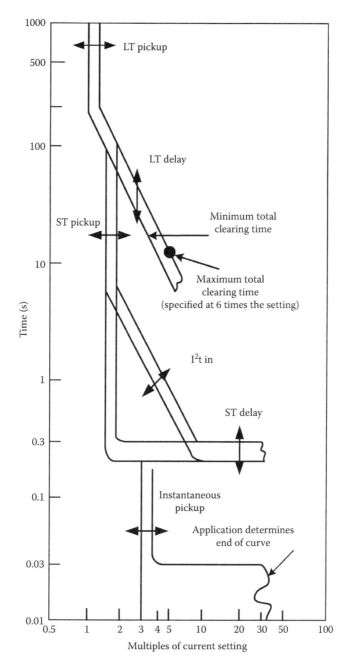

FIGURE 5.26
Time–current characteristics of a modern low-voltage trip programmer.

A common design feature in current-limiting MCCBs is the reverse current loop. The current is routed through parallel contact arms so that opposing magnetic forces are formed. During high fault currents, the magnetic repulsion forces force the contacts to overcome spring forces holding them together, so that these part quickly. These magnetic forces may give rise to current popping, where the contacts part temporarily.

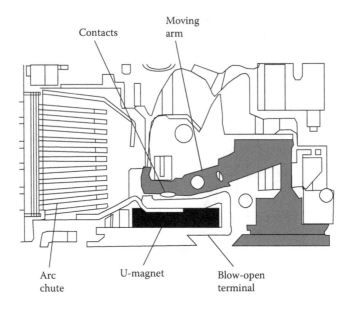

FIGURE 5.27
Low arc-flash circuit breaker design.

The MCCBs are sensitive to the peak current and peak energy delivered over the first few microseconds of a fault and then limit the energy they allow to flow on complete interruption.

For arc-flash considerations, many efforts are concentrated on the development of lighter mechanisms and faster operation. Figure 5.27 shows the low arc-flash circuit breaker design [21].

5.9.3 Insulated Case Circuit Breakers

ICCBs utilize characteristics of design from the power circuit breakers and MCCBs. These are not fast enough to qualify as current-limiting type and are partially field maintainable. These can be provided with electronic trip units and have short-time ratings and ground fault sensing capabilities. These are available in ratings up to 5000 A and 85 kA. These utilize stored energy mechanisms similar to LVPCBs.

MCCBs and ICCBs are rated and tested according to UL 489 standard. Both MCCBs and ICCBs are tested in open air without enclosure and are designed to carry 100% of their current rating in open air. When housed in an enclosure there is 20% derating, though some models and frame sizes may be listed for application at 100% of their continuous current rating in an enclosure. MCCBs are fixed mounted in switchboards and bolted to bus bars. ICCBs can be fixed mounted or provided in draw-out designs.

5.9.4 Low-Voltage Power Circuit Breakers

LVPCBs are rated and tested according to ANSI/IEEE C37.13 [15] and are used primarily in draw-out switchgear. These are the largest in physical size and are field maintainable. Electronic trip units are now almost standard with these circuit breakers, and these are available in frame sizes from 800 to 6000 A, interrupting ratings from 40 to 100 kA symmetrical without integral current-limiting fuses.

Current Interruption Devices and Battery Systems

Figure 5.26 shows a typical phase overcurrent time–current characteristics of an LVPCB. It is provided with an electronic trip programmer, designated as LSIG. Here, the letter L stands for the long time, S for the short time, I for the instantaneous, and G for the ground fault.

- Consider that the breaker is 1600 A frame and is provided with sensors 1600 A, call the sensor current rating "s."
- Plug ratings of 600, 800, 1000, 1100, 1200, and 1600 A can be provided; call this setting "x." This virtually changes the current rating of the circuit breaker.
- The long-time pickup can be adjusted at 0.5, 0.6, 0.7, 0.8, 0.9, 0.95, and 1.1 times the plug setting x. Call this setting "c." Say for a 1600 A plug, $x = 0.6$, $c = 960$ A.
- Long-time delay band can be selected at 2, 3, 4, 5, 6, 8, 12, 20, 24, and 32 s.
- Short-time pickup is adjustable from 1.5 to 9 times the c setting in increments of 0.5.
- Short-time delay band can be adjusted; there can be three to seven time delay bands.
- I^2T function of short time can be set in or set out. When the I^2T function is switched in, the shape of the curve slopes as shown in Figure 7.13. This slope is particularly helpful for coordination with the fuse characteristics.
- Instantaneous pickup is adjustable from, say, 1.5–12 times or more of the setting x.

These setting ranges vary with manufacturers and their various trip programmer types. A recent advancement is that the instantaneous or short-time pickup settings can be switched off when required, affording selective coordination.

Some trip programmers provide two distinct settings: one in the normal mode and the other brought in through maintenance mode switch. This switch is mounted directly on the trip programmer itself with indicating lights; however, it can also be remotely mounted and hard wired to the trip programmer.

There are a host of other functions such as zone interlocking, metering, energy management, front panel displays, and communications protocols that are provided in the trip programmers.

Figure 5.28 shows the control circuit of an electrically operated LVPCB. Motor is energized through LS contact and charges control spring. When the spring is fully charged, the LS contacts change state. Close contacts energize SR coil. In this figure, the contacts are shown for breaker not fully charged.

5.9.5 Short-Time Bands of LVPCBs' Trip Programmers

LVPCBs have short-circuit withstand capability of 30 cycles. MCCBs do not have any short-circuit withstand capability and must be provided with instantaneous protection. ICCBs may have short-circuit withstand capability of 15 cycles, yet these are provided with high-set instantaneous override. Thus, selection of appropriate low-voltage breaker types can be important criteria in selective coordination.

There has been an attempt to split the 500 ms (30 cycles on 60 Hz basis) time withstand of LVPCBs into much smaller short-time delay bands. Figure 5.29 shows three short-time time delay bands. Coordination is obtained between the bands, though the gap between the bands is small and sometimes the bands may be even overlapping. The device can trip

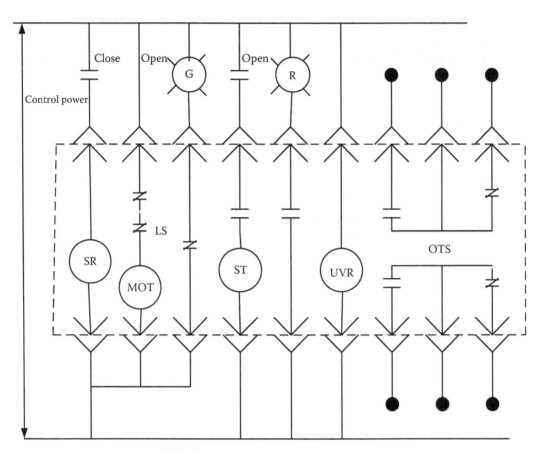

LS = Limit switch for closing spring
MOT = Motor for spring charging
ST = Shunt trip coil
SR = Spring release
UVR = Undervoltage release
OTS = Overcurrent trip switch

FIGURE 5.28
Control schematic of an LVPCB.

anywhere between the time zone of the band, but for conservatism, it is the maximum operating time that is considered for coordination. This figure shows only three ST delay bands: minimum, intermediate, and maximum. The time associated with these bands is 0.2, 0.3, and 0.5 s, respectively, which is shown with bold dots.

Table 5.8 shows a modern LVPCB, provided with an electronic trip device having seven short-time delay bands.

The MCCBs in smaller sizes, say up to 200 A, may have only thermal magnetic trips. The characteristics of an MCCB provided with an electronic rip programmer are illustrated in Figure 5.30. The MCCB is of 1220 A, which is set at 600 A pickup, with a time delay band of 1 (four bands are available), ST pickup of 4 (1.5–9 pickup range available), ST delay of band 1 (three bands are available), and instantaneous setting of 6 (range 2.5–10). Many a time, one or the other setting may be fixed; for example, LT delay and instantaneous may be made a function of ST pickup. In any case for MCCBs the instantaneous settings are a must, as these do not have any short-time short-circuit withstand ratings.

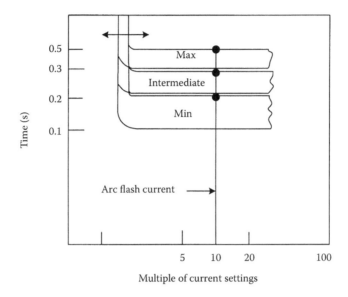

FIGURE 5.29
Short-time bands of a low-voltage trip programmer, see text.

TABLE 5.8

Short-Time Bands of Old versus New LVPCB Trip Programmers

Band	Breaker A Max Trip Time (ms)	Breaker B Max Trip Time (ms)
1	200	92
2	300	158
3	500	200
4		267
5		317
6		383
7		383
		500

5.9.6 Motor Circuit Protectors

MCPs have only magnetic pickups and no thermal pickups. These are applied for low-voltage motor protection. According to NEC 430.52, these can be set up to 1300% of the motor full-load current, except for NEMA design B high-efficiency motors where the MCP can be set up to 1700%. Compared to MCCBs, MCPs require higher magnetic settings to override the inrush currents of the motors (see NEC).

Figure 5.31 shows the time–current characteristics of a 400 A MCP, protecting a 200 HP, 480 V motor (starting time 10 s). Note that the curve of MCP above the crossing of the motor thermal protection curve is not valid. As we have motor protection curve whether the motor is protected with MCP or MCCB, the magnetic part of the curve for MCCB is really not required.

Unlike MCCBs, the MCPs are tested for short-circuit interrupting with the motor magnetic contactor in place.

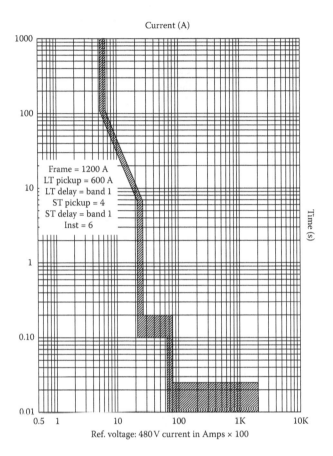

FIGURE 5.30
Time–current characteristics of an MCCB provided with electronic trip programmer.

Table 5.9 shows the application of MCCBs and MCPs for the protection of 480 V motors. These selections are based on the following:

- Ambient no more than 40°C
- Infrequent motor starting
- Locked rotor current a maximum of six times that of motor full-load amperes

Also the fuse, circuit breaker, and MCP selections in this table comply with NEC rules given in NEC 430.52 and Table 430.25 [22]. The current ratings are no more than the maximum limits set by NEC rules for motors with code letters F to V or without code letters. Motors with lower code letters require further considerations.

5.9.7 Other Pertinent Data of Low-Voltage Circuit Breakers

Refer to Chapter 8 of Volume 1. The following pertinent points are discussed but not repeated here.

- Short-circuit ratings, see Section 8.14.3
- Single-pole interrupting capability, see Section 8.14.3.1

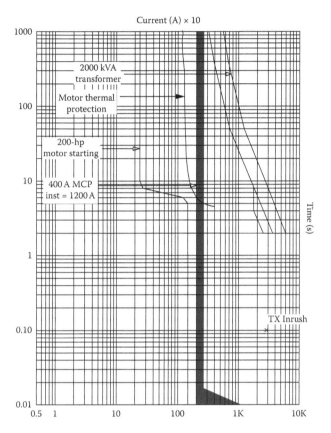

FIGURE 5.31
Time–current characteristics of an MCP of 400 A.

- Short-time ratings, see Section 8.14.3.2
- Series-connected ratings, see Section 8.14.3.3

5.10 Selective Zone Interlocking

The digital low-voltage trip programmers may be provided with zone interlocking facilities. Using LVPCBs, consider a simple distribution as shown in Figure 5.32. With proper settings on the three breakers in series, selective tripping will be obtained. However, fault downstream of breaker 2 will be cleared with a time delay of 0.3 s, and downstream of breaker 1 in 0.5 s. With zone interlocking, all faults can be cleared instantaneously. This requires additional zone interlocking wiring.

For a fault F1, no interlocking signals are required. The trip will be initiated instantaneously.

For a fault F2, the hardwired trip units send a restraining signal to the upstream trip unit, allowing the breaker closest to the fault to act instantaneously. Here, zone 2 will send a signal to zone 1 trip unit. The zone 1 trip unit will start timing out and in the event feeder

TABLE 5.9

Application of Fuses, MCCBs, and MCPs for Low-Voltage Motor Starters

		Fuse Size Max				
Motor HP	NEC FLA 460 V, Three-Phase	Time Delay	Nontime Delay	MCCB (A)	MCP (A)	Adjustment Range
1	1.8	6	6	15	7	21–70
1.5	2.6	6	10	15	7	21–70
2	3.4	6	15	15	7	21–70
3	4.8	10	15	15	15	45–150
5	7.6	15	25	15	15	45–150
7.5	11	20	35	25	30	890–300
10	14	25	45	35	30	90–300
15	21	40	70	45	50	150–500
20	27	50	90	50	50	150–500
25	34	60	110	70	70	210–700
30	40	70	125	70	100	300–1000
40	52	100	175	100	100	300–1000
50	66	125	200	110	150	450–1500
60	77	150	250	125	150	750–2500
75	96	175	300	150	150	750–2500
100	123	225	400	175	150	750–2500
125	156	300	500	225	250	1250–2500
150	180	350	600	250	400	2000–4000
200	240	400	800	350	400	2000–4000

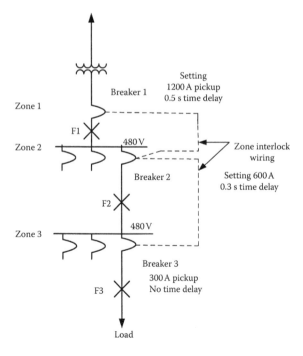

FIGURE 5.32
A system configuration to illustrate zone interlocking.

breaker in zone 2 is not tripped instantaneously, the main breaker in zone 1 will clear the fault in 0.5 s.

For a fault F3, zone 3 will send an interlock signal to zone 2 trip unit and zone 2 will send an interlocking signal to zone 1. Zone 1 and zone 2 trip units will start timing out and in the event the zone 3 would not clear the fault, the feeder breaker in zone 2 will clear the fault in 0.3 s.

Zone interlocking can be applied to phase- and ground-fault settings. When downstream motor loads are present, it adds complications (see Chapter 6).

Thus, the faults are cleared instantaneously and this reduces the equipment burnouts. The discussions are continued in Chapter 6.

5.11 Electronic Power Fuses

Electronic power fuses are a more recent innovation [23]. These are available for indoor distribution from 4.16 to 25 kV. These integrate a high power fuse with electronics, CT, and current sensing capabilities to provide TCC characteristics that are electronically derived.

A superior coordination can be provided through the application of these characteristics. Examples of protection and coordination are illustrated in the following figures.

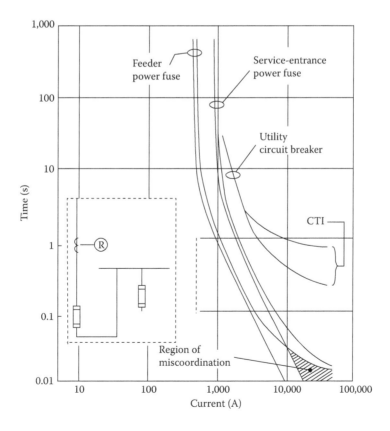

FIGURE 5.33
Time–current plot showing lack of coordination between main and feeder fuse characteristics, using conventional fuses.

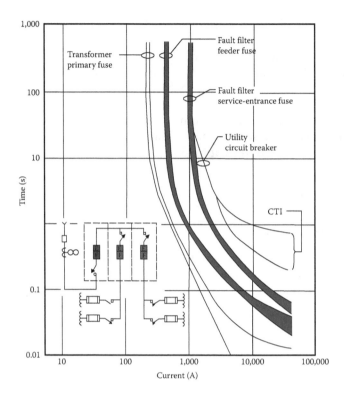

FIGURE 5.34
Time–current plot, but the lack of coordination is avoided with the use of an electronic fuse.

- Figure 5.33 shows lack of coordination between the service entrance conventional fuse and the feeder fuse. This is avoided in Figure 5.34 with application of electronic power fuse, Fault Filter.
- Figure 5.35 depicts a compound curve of the electronic fuse applied for transformer protection.
- Figure 5.36 shows the backup protection of elbow connectors and weak link transformers.

An overall assembly is shown in Figure 5.37. The control module provides the current sensing and time–current characteristics of the fuse as well as energy to initiate operation of interrupting module in the event of a fault.

The three types of curves shown in Figures 5.34 through 5.36 are not available in the same electronic fuse. These must be ordered for the required characteristics. The characteristics are fixed for a certain current rating and are not adjustable.

5.12 Low- and Medium-Voltage Contactors

The low- and medium-voltage contactors are used for motor starting, capacitor switching, and other services.

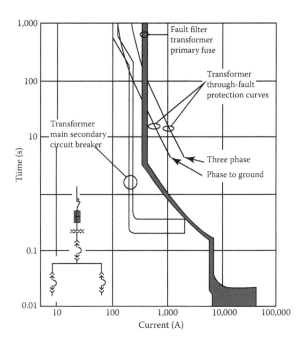

FIGURE 5.35
An electronic fuse coordinated with main secondary breaker, applied for primary protection of a transformer.

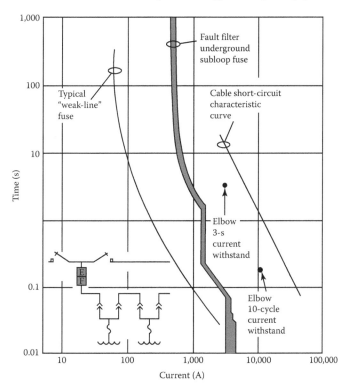

FIGURE 5.36
Electronic fuse applied for back-up protection of elbow connectors and weak link transformers.

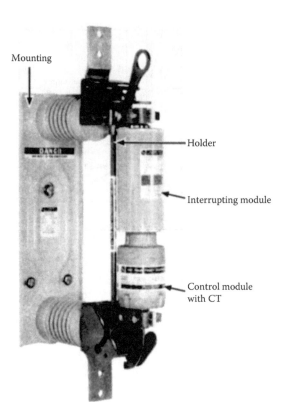

FIGURE 5.37
Assembly of an electronic fuse.

TABLE 5.10

Ratings of Three-Phase Single-Speed Full-Voltage Magnetic Controllers for Nonplugging and Nonjogging Duty

Size of Controller	Continuous Current Rating (A)	Horsepower at				Service Current Rating (A)
		60 Hz		50 Hz	60 Hz	
		200 V	230 V	380 V	460 or 575 V	
00	9	1.5	1.5	1.5	2	11
0	18	3	3	5	5	21
1	27	7.5	7.5	10	10	32
2	45	10	15	25	25	52
3	90	25	30	50	50	104
4	135	40	50	75	100	156
5	270	75	100	150	200	311
6	540	150	200	300	400	621
7	810		300		600	932
8	1215		450		900	1400
9	2250		800		1600	2590

Note: See ICS 2-321A.20; The HP ratings are based on locked rotor currents given in NEMA; motors with higher locked rotor currents may require a larger controller.

Table 5.10 shows the NEMA ratings of single-speed full-voltage magnetic controllers for nonplugging and nonjogging duty. Table 5.10 also shows similar ratings for plug-stop, plug-reverse, or jogging duty, and NEMA Part ICS 2-321B.

See NEMA standards for similar tables for other starting methods, i.e., reduced voltage starters. The electronic reduced voltage starters are not discussed here; see Chapter 7 of Volume 2.

The medium-voltage class E1 and E2 controllers are discussed in Chapter 10. Also see [24].

References

1. ANSI 493. IEEE Recommended Practice for Design of Reliable Industrial and Commercial Power Systems, 1980.
2. IEEE Standard C37.40. IEEE Standard Service Conditions and Definitions for High-Voltage Fuses, Distribution Enclosed Single-Pole Air Switches, Fuse Disconnecting Switches and Accessories, 1981.
3. IEEE Standard C37.41. IEEE Standard Design Tests for High-Voltage (>1000 V) Fuses, Fuse and Disconnecting Cutouts, Distribution Enclosed Single-Pole Air Switches, Fuse Disconnecting Switches, and Fuse Links and Accessories Used with These Devices, 2008.
4. IEEE Standard C37.42. IEEE Standard Specifications for High-Voltage (>1000 V) Expulsion-Type Distribution-Class Fuses, Fuse and Disconnecting Cutouts, Fuse Disconnecting Switches, and Fuse Links, and Accessories Used with These Devices, 2008.
5. ANSI Standard C37.46. American National Standard for High-Voltage Expulsion and Current Limiting Type Power Class Fuses and Fuse Disconnecting Switches, 2000.
6. ANSI Standard C37.47. American National Standard for High-Voltage Current-Limiting Type Distribution Class Fuses and Fuse Disconnecting Switches, 2000.
7. IEEE C37.48. IEEE Guide for Application, Operation and Maintenance of High-Voltage Fuses, Distribution Enclosed Single Pole Air Switches, Fuse Disconnect Switches, and Accessories, 1997.
8. ANSI/UL 198D. Safety Standards for Class K fuses, 1982.
9. ANSI/UL 198E. Safety Standards for class R Fuses, 1982.
10. ANSI/UL 198F. Safety Standards for Plug Fuses, 1982.
11. ANSI/UL 198G. Fuses for Supplementary Overcurrent Protection, 1981.
12. ANSI/UL 198H. Safety Standards for Class T fuses.
13. UL 198L. DC fuses for Industrial use, 1984.
14. NEMA Standard SG2 High-Voltage Fuses, 1981.
15. ANSI/IEEE. Standard C37.13. Standard for Low-Voltage AC Power Circuit Breakers used in Enclosures, 2008.
16. IEEE Standard C37.13.1. IEEE Standard for Definite – Purpose Switching Devices for use in Metal-Enclosed Low-Voltage Power Circuit Breaker Switchgear, 2006.
17. NEMA. Molded Case Circuit Breakers and Molded Case Switches, 1993, Standard AB-1.
18. UL Standard 489. Molded Case Circuit Breakers and Circuit-Breaker Enclosures, 1991.
19. IEC Standard 60947-2. Low-voltage Switchgear and Control Gear-Part 2: Circuit Breakers, 2009.
20. IEEE Standard 1015. Applying Low-Voltage Circuit Breakers Used in Industrial and Commercial Power Systems, 1997.
21. WA Brown, R Shapiro. Incident energy reduction techniques. *IEEE Ind Appl Mag*, 15(3), 53–61, 2009.
22. NEC, National Electric Code, NFPA 70, 2017.
23. S&C Fault Filter. Bulletin 441-30, 2007.
24. NEMA ICS2-234. AC General Purpose HV Contactor and Class E Controllers, 50 and 60 Hz.

6

Overcurrent Protection: Ideal and Practical

The phase overcurrent protection can be stand-alone for small systems, and is always provided as a backup protection for large systems too, which have unit protection. Primary and secondary distribution buses, main utility interconnection transformers, generators, and large substation transformers and large motors are, generally, covered in differential zones, yet the distribution system downstream like medium- and low-voltage motor control centers, panels, lighting circuits, and smaller distribution systems has time–current coordinated overcurrent protections. Extending the differential protection to low-voltage switchgear buses is a new trend. The objectives of the coordination in a radial system are to achieve selectivity without unduly sacrificing sensitivity and fast clearance times. An ideal time–current coordination is, however, rarely achieved. Though the setting ranges and characteristics of the protective devices may be as flexible as practical, compromises may be required. In case these compromises are not acceptable, additional protective devices, change in the type, and characteristics of the protective devices, and sometimes reorientation of the system being protected are required. Coordination is, generally, done in the last stages of the project, but it is necessary to consider it in the design stage itself.

6.1 Fundamental Considerations

The elements in a protective system include relays, direct-acting solid-state trip devices, trip programmers for the low-voltage circuit breakers, and electromagnetic trip devices. These are discussed in Chapter 5. A time–current coordination of these devices must ensure selectivity, so that only the faulted section is isolated. A downstream device must operate faster than the upstream device with a certain coordination time interval (CTI).

It should be ensured that all the system components like transformers, cables, and generators are protected with respect to constraints of overloads, fault withstand capabilities, and thermal damage curves according to standards.

The protection should meet the requirements of NEC. The OSHA has made strict compliance with NEC mandatory [1].

The coordination should meet the specific requirements of the operating processes. For a fire pump, it is possible to sacrifice some coordination. A nuisance trip may result in much loss of production.

Coordination is not a substitute for inadequate system planning and design.

6.2 Data for the Coordination Study

Prior to starting the study, an extensive data collection effort is involved:

1. A single-line diagram of the distribution system is required. As a minimum this should show all protective devices that are to be coordinated on a time–current basis. All the switching conditions and plant operations must be charted out; for example, loss of a source, alternate route of power, which breakers will be tripped, which breakers will be closed, how the system will be operated, which protective devices will go out of service, and which new protective devices will be brought into service should be known.

2. The equipment rated kVA, short-circuit ratings, and load flow currents under normal and contingency conditions should be known. Sometimes short-time overloads are allowed in the systems for continuity of operations. The equipment does have a short-time overload capability that can be calculated according to the guidelines established in various standards. For example, IEEE standard [2] details the guidelines for loading of the liquid immersed transformers, with and without derating in the life expectancy of the transformers.

3. CT ratios, their burden, secondary resistances, and relaying accuracies data are required (see Chapter 3). In this chapter, it is assumed that CTs are selected properly for the overcurrent relaying applications in accordance with the requirements discussed in Chapter 3.

4. Specific time–current characteristics of all protective devices to be coordinated should be available.

5. Full-load current, starting (locked rotor) current, thermal withstand curves of the motor (alternatively at least hot and cold safe locked-rotor times), motor protection fuse size for NEMA 2 [3], and medium-voltage starters should be available for plotting and coordination. For low-voltage motors, the ratings and characteristics of fused starters, thermal magnetic breakers, or motor circuit protectors (MCPs that have magnetic protection only) should be available.

6. For transformers, their percentage impedances, primary and secondary winding connections, and system grounding should be known. A computer program will generate a transformer frequent or infrequent fault withstand curve without manual calculations, see Chapter 12.

7. For conductors, short-circuit withstand curve should be available. The modern computer programs for coordination will plot a short-circuit curve without manual intervention based upon the input data of the conductors, their type, and installation methods.

8. The following short-circuit data should be available:
    a. The first-cycle asymmetrical current is required for instantaneous (also differential and distance) relays when the operation is fast and the asymmetry in the fault current should be accounted for. The asymmetry in the first cycle also impacts CT performance.

b. The maximum symmetrical current is required to establish CTI. The minimum interrupting current is required to ascertain whether the circuit protection sensitivity is adequate.

c. The 30-cycle short-circuit current is required for the application of time delay relays beyond six cycles. The induction and synchronous motor contribution to the fault currents decay to zero and generators are represented by transient or higher impedances related to their decaying short-circuit currents. See Volume 1 for detailed discussions on the decaying short-circuit currents.

It is evident that unlike other power system studies, lots of preparation and data collection are needed before starting a coordination study.

6.3 Computer-Based Coordination

Twenty years ago, the coordination of overcurrent devices was carried out manually using a light box; it had a translucent surface illuminated with lamps placed below in the box. The published curves by the manufacturer for the specific devices will be gathered, placed on the light box, and traced on a log–log paper. Appropriate coordination was ensured visually by adjusting the curves being plotted and analytically by calculating the time and current margins and tolerances from the characteristic plots. This was a rather tedious routine, yet academically instructive. *This methodology is no longer in use.*

Currently, the coordination is done using one or the other commercially available programs. These programs will run the short-circuit currents in a given system configuration, and input the results to the relay coordination program; if there are system changes, the short-circuit currents are automatically updated in the coordination. The time–current plots will be correctly terminated at the short-circuit currents seen by the protective devices in the system.

The protective device data and characteristics (of relays, low- and medium-voltage fuses, and low-voltage circuit breaker trip programmers) are built into the computer database. The software vendors have kept pace with the introduction of new devices and multifunction overcurrent relays with programmable characteristics and keep expanding these databases. Also the computer software has the capability that a user can program the time–current characteristics of a protective device not included in the computer database and add it to the database library. Similarly, circuit breaker types, low-voltage trip programmers, fuses, and other equipment data can be added to the library. Thus, it is much easier today to experiment with various characteristics and settings for coordination. When a time–current curve is plotted, it is possible to move it upward, downward, and sideways for the best coordination. The settings are automatically corrected and charted out with each movement and placement of the time–current curve.

Some programs are available which will also suggest the coordinating settings or provide a complete *auto-coordination* of the protective devices in a certain setup. Though these auto-set programs can be tried, a better coordination is often achieved with manual intervention. This is a field where experience of a coordination engineer comes to the forefront, and cannot be substituted with auto-coordination programs, especially in situations where compromises have to be made.

6.4 Initial Analysis

An initial analysis of the protection should reveal that none of the system components are exposed to damaging overloads or short-circuit currents which will go undetected under various operating conditions. Each equipment should be applied within its assigned continuous and short-circuit rating and thermal withstand capabilities, and meets the requirements of ANSI/IEEE, OSHA, NEC, NESC (National Electric Safety Code), and other national and international standards. While manipulating the settings, it is permissible to make an informed judgment with respect to lack of some selectivity, but not jeopardize the equipment protection.

6.5 Coordinating Time Interval

The sequential operation of series-connected protective devices depends upon maintaining a minimum CTI throughout the operating range. A graphical representation of the time–current characteristics is an accepted method, though it is possible to determine selectivity by comparing at the most three critical values of the fault currents and ascertaining the relay operating times.

The CTI takes into account the following:

- Circuit breaker interrupting time.
- Relay overtravel (also called impulse margin time), for electromechanical relays, explained next.
- An arbitrary safety factor to account for CT saturation and setting errors.

6.5.1 Relay Overtravel

Consider the coordination between two relays R1 and R2 in series. Relay R1 is upstream of relay R2, with little intervening impedance and a fault occurs downstream of relay R2. Both the relays see approximately the same magnitude of fault current. The operation of relay R1 should be delayed by CTI, so that relay R2 clears the fault and rely R1 does not operate. The impulse margin time is defined as

$$T_{IM} = T_{OP} - T_I \qquad (6.1)$$

where T_{IM} is the impulse margin time, T_{OP} is the time as read from the time–current curves of the relay for the specific current pickup and time dial setting, and T_I is the maximum impulse time during which sufficient inertia is supplied to the disc to eventually cause the disc to close its contacts, even when the fault seen by the relay is removed, in this case by operation of relay R2. This time is calculated by testing and manufacturers can supply data for the exact coordination. This should be taken into consideration for the CTI. Table 6.1 provides the guidelines for selecting CTI.

TABLE 6.1

Coordinating Time Intervals

Switching Device	Coordinating Time Interval

Relayed medium-voltage circuit breakers	*Very inverse and extremely inverse electromagnetic relays*
	  0.45 s: eight-cycle breakers (pre-1964 basis of rating, see Chapter 5)
	  0.40 s: five-cycle breakers
	  0.36 s: three-cycle breakers
	Inverse time characteristics electromagnetic relays
	  0.48 s: eight-cycle breakers (pre-1964 basis of rating)
	  0.43 s: five-cycle breakers
	  0.39 s: three-cycle breakers
	(The relay impulse travel time is longer for inverse relays compared to very inverse or extremely inverse relays.)
	Modern MMPR or solid-state relays
	  0.35 s: eight-cycle breakers (pre-1964 basis of rating)
	  0.30 s: five-cycle breakers
	  0.26 s: three-cycle breakers
	Properly calibrated and field tested relays
	Reduce the above CTIs by another 0.05 s. Much less calibration and testing of the timing is required for modern MMPRs.

Relayed medium-voltage circuit breakers and fuses	*Electromagnetic relays*
	  0.20 s
	MMPRs or solid-state relays
	  0.1 s
	Relay overtravel and breaker interrupting time is eliminated.

Low-voltage relayed power circuit breakers	The interrupting time may safely be considered as three cycles. Thus coordination with 0.1 s CTI possible.
Low-voltage power circuit breakers with solid-state trip programmers, see Figure 7.15	The slight time margin provided by the manufacturer between STD bands is adequate for coordination. Sometimes, these bands overlap when plotted graphically, yet coordination is achieved. Referring to Figure 7.15, which shows three ST bands, the coordination is achieved within the small gap shown between the bands.
Fuses, medium voltage and low voltage	Manufacturers provide minimum melting and maximum operating time curves. There should be a margin of at least 10% between the maximum operating characteristics of the downstream fuse and the minimum melting characteristics of the upstream fuse. For coordination below 0.05 s on a time–current basis should not be attempted.
Instantaneous functions, electromagnetic relays	The settings must recognize the asymmetrical nature of the fault, as the relays can operate equally well for ac and dc currents. Coordination without intervening impedance is not attempted.
Coordination instantaneous basis, current-limiting devices	See the text.

6.6 Fundamental Considerations for Coordination

- When protecting a delta–wye transformer, a 16% margin should be used between primary and secondary device characteristic curves; this is so because per unit primary current in one phase for a secondary phase-to-phase fault is 16% greater than the secondary current, see Chapter 12.

- In addition to short circuit, a load flow study is required to determine normal and emergency load currents that the system should carry. The continuous current ratings of the equipment, for example, cables, circuit breakers, and transformers, should not be exceeded. It is understood that each equipment is selected and designed to serve the continuous and emergency load currents, as a first protection requirement.
- All switching devices should be rated to meet the available short-circuit current requirements. The cables should be designed to withstand the required short-circuit currents in case of fault, for a duration dependent upon the protective device.
- Inrush currents of motors and transformers should not result in nuisance trips. The pickup settings should override these inrush currents.
- The individual load or branch circuit protection should meet the requirements of NEC, ANSI, IEEE, or similar codes. Sometimes to meet the requirements of NEC, higher cable sizes may be required.

Example 6.1

Consider a 2000 kVA, 13.8–0.48 kV transformer, of percentage impedance of 5.75% on 2000 kVA rating base, which is provided with primary overcurrent protection through a 150E fuse and has a secondary main circuit breaker of 3000 A provided with low-voltage trip programmer. According to Table 450.3(A) of NEC, the secondary circuit breaker should not have an overcurrent pickup setting of more than 125% of the transformer full-load rated current of 2405.6 A, that is, a setting of 3007 A. Though the actual load current of the transformer may be lower than its rated current, say only 80%, the secondary cables between the transformer and 3000 A circuit breaker must be rated for a minimum ampacity of 3000 A to meet the requirements of NEC.

6.6.1 Settings on Bends of Coordination Curves

This situation is illustrated in Figure 6.1. The intended settings are A and B, but these can translate into settings A' and B' providing much larger time delays.

6.7 Some Examples of Coordination

6.7.1 Low-Voltage Distribution System

Figure 6.2 shows a radial distribution system, typical of industrial distribution systems. The short-circuit and load flow calculation must be carried out and all equipment sized for the short-circuit currents involved. The load flow may require adjustment of off-load transformer taps to maintain acceptable voltages. The starting voltage dip on the largest motor must be calculated. These aspects are not discussed here, see Volumes 1 and 2.

Starting from downstream, the low-voltage MCC has motors and other loads; the largest motor rating is considered for the coordination. Generally, the industrial practice is to limit the largest rating of low-voltage motor to 200 hp; the higher rated motors are served from 2.4 kV or 4.16 kV systems. The motor is a squirrel-cage induction motor—the wound-rotor motors may be used sometimes but, generally, are extinct in the industry at low voltages.

The motor is protected from the short-circuit conditions by a class J fuse. The alternative to fuses is molded case circuit breakers (MCCBs) or MCPs, see Chapter 5. Each of these devices must clear the starting characteristics of the motor. For high-efficiency motors, the starting inrush currents are high and nuisance trips have occurred. Table 6.2 from the NEC provides the maximum settings of motor branch circuit short-circuit and ground fault protective devices.

Figure 6.3 shows the first attempt of overall coordination plot. Again starting with 200 hp motor, its starting curve is first plotted. This shows a starting time of 10 s. The starting time varies with the load type and its inertia. For accuracy, a dynamic starting study of the motor is required to ascertain the starting characteristics (see Volume 2 for details,

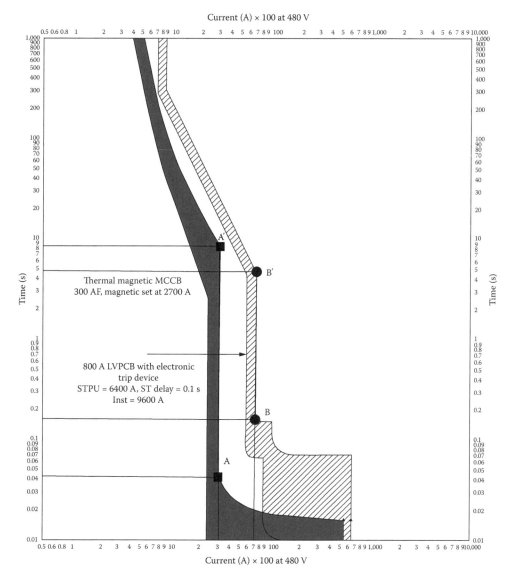

FIGURE 6.1
Settings on curve flexures, which can give increased fault clearance times.

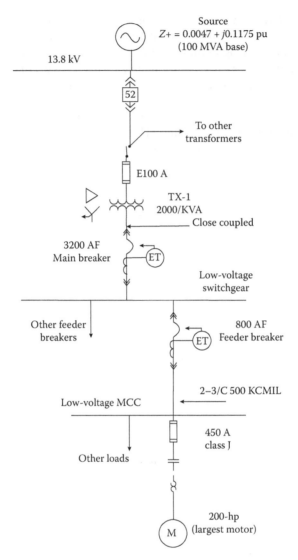

FIGURE 6.2
A low-voltage radial distribution system for coordination.

not repeated here). The locked-rotor current of the motor, the motor and load inertia, the motor torque-speed characteristics, and the load characteristics must be known to accurately calculate the starting time. The locked-rotor current of motors is designated by a code letter in the NEC (see Table 6.3). Ninety percent of the low-voltage motors in the industry are designed with a locked-rotor current equal to approximately six times the full-load current.

Figure 6.3 shows that a 450 A class J fuse admirably clears the starting curve of the motor and also the anticipated inrush current transient during the first few cycles of starting. The motor thermal element curve is then plotted. See Chapter 10 on motor protection. The thermal elements have given way to better digital relays for low-voltage motor protection.

TABLE 6.2

Maximum Rating or Setting of Motor Branch Circuit and Ground Fault Protective Devices

	Percentage of Full-Load Current			
Type of Motor	Nontime Delay Fuse	Dual-Element Time Delay Fuse	Instantaneous Trip Breaker	Inverse Time Breaker
Single–phase motors	300	175	800	250
AC polyphase motors other than wound-rotor motors	300	175	800	250
Squirrel cage other than design B, energy efficient	300	175	800	250
Design B energy efficient	300	175	1100	250
Synchronous	300	175	800	250
Wound rotor	150	150	800	150
Direct Current	150	150	250	150

Source: NEC Table 430.52.
Note: Certain exceptions as detailed in the NEC are applicable.

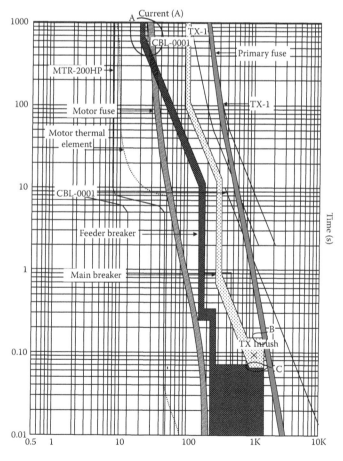

FIGURE 6.3
Coordination plot of the distribution system in Figure 6.2, showing the areas of lack of coordination.

TABLE 6.3

Locked Rotor Indicating Code Letters (Not Applicable to Wound-Rotor Motors)

Code Letter	kVA per hp with Locked Rotor
A	0–3.14
B	3.15–3.54
C	3.55–3.99
D	4.0–4.49
E	4.5–4.99
F	5.0–5.59
G	5.6–6.29
H	6.3–7.09
J	7.1–7.99
K	8.0–8.99
L	9.0–9.99
M	10.0–11.19
N	11.2–12.49
P	12.5–13.99
R	14.0–15.99
S	16.0–17.99
T	18.0–19.99
U	20.0–22.39
V	22.4 and above

Note: From NEC Table 430.7(B).

According to NEC guidelines, the selected device to trip shall be rated at no more than the following percent of the motor full-load amperes:

- Motors with a marked service factor of 1.15 or greater = 125%
- Motors with a marked temperature rise of 40°C or less = 125%
- All other motors = 115%

Next, the setting curve of the feeder breaker electronic trip programmer is plotted, see Chapter 5. The considerations are as follows.

The setting should carry the MCC standing load plus the starting impact of the largest motor. It is assumed that only one motor is started at a time. There can be exceptions to this, which must be considered.

- It should clear the motor short-circuit protection device, which is a 450 A class J fuse here.
- The pickup setting should protect the conductors from the low-voltage switchgear to the MCC according to the NEC. Here, we have 2–3/C 350 KCMIL conductors, also see Chapter 9.
- The pickup setting should not exceed the continuous current rating of the low-voltage MCC bus.
- It should coordinate with the main secondary circuit breaker settings, failing which all loads connected to the low-voltage switchgear will be lost.

Based on the above considerations, the low-voltage trip programmer, with 600 A sensors and 600 A plug (the sensor and plug rating must be based on the required settings), is set as:

- Long-time pickup (LTPU) = 1 (600 A)
- Long-time delay (LTD) = 24 s
- Short-time pickup (STPU) = 8 (4800 A)
- Short-time delay (STD) = 0.3 s
- $I^2T = In$
- Instantaneous = 12 (7200 A)

The termination of the instantaneous shows the maximum short circuit seen by the feeder breaker.

The following observations are of interest:

Around 700–100 s, there is a clash with the motor fuse characteristics. This is not of concern as the motor thermal element has the priority of fault clearance in this region. The instantaneous function at 0.01 s seems to coordinate with the motor fuse. This is not true and there is a possibility that on high short-circuit currents the feeder breaker and motor fuse may operate together. The coordination on instantaneous basis is further discussed in the sections to follow. An alternative will be not to use instantaneous setting at all. However, purposely, this coordination is sometimes sacrificed to reduce arc-flash damage to an MCC.

Settings on the main secondary breaker should coordinate with the feeder breaker. This coordination is important to prevent the possibility of complete shutdown of the loads. Also it should coordinate with the transformer fuse.

We cannot manipulate the fuse characteristics. The selection of fuses for transformer protection is discussed in Chapters 5 and 12. The pros and cons of the fuse versus relay protection with examples of coordination and protection of through-fault withstand curve are also discussed.

The pickup setting should ensure that all the loads connected to a 2000 kVA transformer are served. Also the setting should not exceed the low-voltage switchgear bus rating. Generally, the settings are provided considering future loads may develop, and the transformer may be provided with cooling fans to raise its rating. This is a prudent approach, because with each incremental load addition, we do not go about changing the settings. The settings provided are as follows:

LTPU = 0.9 (2880 A)

LTD = 4 s

STPU = 3 (8640 A)

STD = 0.1 s

$I^2T = In$

No instantaneous

A slight clash is shown at areas B and C, which is practically acceptable.

Thus, in the coordination so far, the major concern is lack of coordination between the feeder breaker and the motor fuse. By eliminating the instantaneous setting on the feeder breaker and adjusting the STD setting coordination is achieved as illustrated in Figure 6.4.

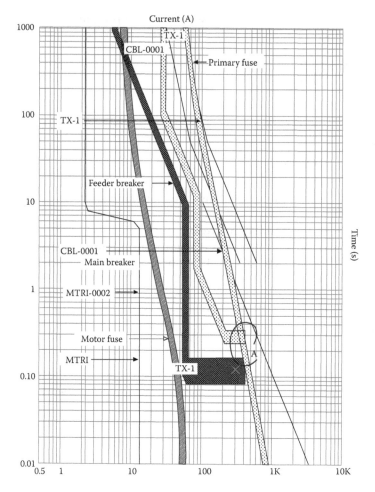

FIGURE 6.4
The lack of coordination with motor fuse in Figure 6.3 is avoided by selecting appropriate short-time bands.

However, note that, now there is some lack of coordination between the settings of the main breaker and the fuse, as the STD band of setting on the main breaker is raised. Generally, this small miscoordination is acceptable.

The primary fuse can be replaced with a relay. This coordination is depicted in Figure 6.5. Note that the characteristics of the primary relay plotted in this figure are obtainable with application of multifunction microprocessor-based relay which allow independent phase instantaneous overcurrent pickups and time delays in multiple steps.

Though the industry has used primary fuse protection for the transformers, the relay protection with primary fused switch replaced with a circuit breaker is the modern trend.

6.7.2 2.4 kV Distribution

Figure 6.6 shows a 2.4 kV distribution for the coordination.

The motor protection is discussed in Chapter 10. It is necessary to obtain the motor thermal withstand characteristics from the manufacturer. Second, the motor starting time

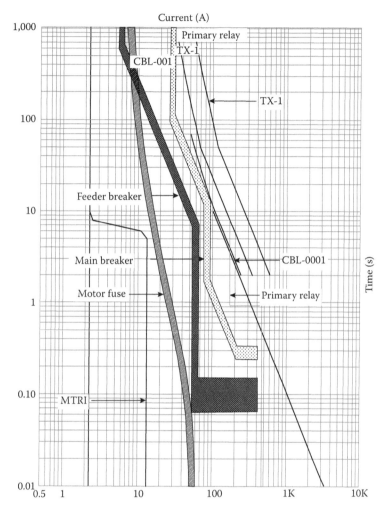

FIGURE 6.5
Further improvement in coordination by replacing transformer primary fuse with a multifunction relay.

should be accurately calculated, see Volume 2. Note that the starting time can vary over large limits; for example, starting of ID fans may take a minute.

Figure 6.6 shows that the largest motor is of 2500 hp. The motors to a certain rating are started with NEMA E2 starters, with contactors and fuses. Table 6.4 shows the maximum rating of induction motors that can be started at full voltage with vacuum contactors and Table 6.5 shows the rated currents and interrupting currents of the unfused vacuum contactors. Similar data are available for reduced voltage starters and starters for the synchronous motors. For larger motors, circuit breakers are used for starting. The current ratings may differ when installed in the enclosures.

The fundamentals of the medium-voltage motor starting coordination are shown in Figure 6.7.

The fuse is selected so that it clears the motor starting curve. It should also not operate on the motor locked-rotor current multiplied by a factor of 1.1.

The motor protection relay curve is sandwiched between the fuse curve and the motor starting curve.

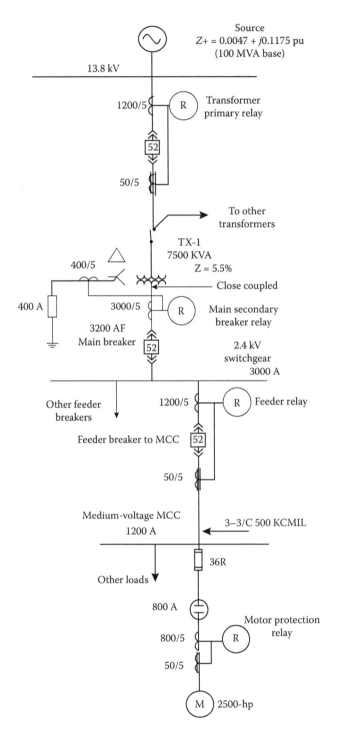

FIGURE 6.6
A medium-voltage radial distribution system for coordination.

TABLE 6.4

Medium-Voltage NEMA E2 Starters—Full Voltage, Squirrel-Cage Induction Motors

Motor Voltage (V)	Maximum hp Reversing	Maximum hp Nonreversing	Starter Interrupting Rating (kVA)
2,300	3,000	3,000	200,000
4,600	5,500	5,500	400,000
6,600	8,000	8,000	570,000
12,400–13,800	5,000	5,000	1,190 MVA at 13.8 kV

TABLE 6.5

Current and Short-Circuit Ratings of the Medium-Voltage Motor Starting Contactors

Contactor Rating (A)	Utilization Voltage NEMA E-1 Ratings, kA			
	2200–2500 V	3000–3600 V	3600–4800 V	6000–7200 V
400	8.5	8.5	8.5	8.5
800	12.5	12.5	12.5	12.5

- It should protect the thermal withstand of the motor.
- It should clear the motor locked-rotor current.
- It should not operate on the motor starting current inrush. In the modern digital relay, the pickup can be set close to the motor full-load current, say a factor of 1.05 times the motor full-load current.
- The instantaneous should be so set that the contactor does not interrupt a short-circuit current greater than its interrupting capability, see Table 6.5. The fuse should open for faults greater than the contactor's interrupting rating. Referring to Figure 6.8, the drop out time of the contactor may vary depending on the trapped residual magnetism in the coil. It may be of the order of two to three cycles. This figure shows improper coordination of the fuse and the contactor.

Referring to Figure 6.9, it is seen that all the above constraints are complied when setting the motor protection relay. The instantaneous is set so that the motor fuse will always clear the fault exceeding the contactor rating, irrespective of the drop out time of the contactor.

Next, the overcurrent settings of the feeder breaker to MCC, the main secondary breaker of a 7500 kVA transformer, and the primary circuit breaker of a 7500 kVA transformer are plotted. These curves illustrate that these coordinates well. Also, the settings on the primary circuit breaker of 7500 kVA transformer admirably protect the shifted thermal withstand curve of the transformer. These settings are tabulated in Table 6.6. Note that the settings shown require MMPR; an integrated curve of the inverse time characteristics and instantaneous overcurrent setting with required time delay cannot be obtained with electromagnetic devices.

Note that the transformer is protected by 13.8 kV main breaker, which serves other transformers, too. This is acceptable so long as the other transformers in radial connection are small and protected on the primary side with appropriate fuses and the settings allow total load current to be carried.

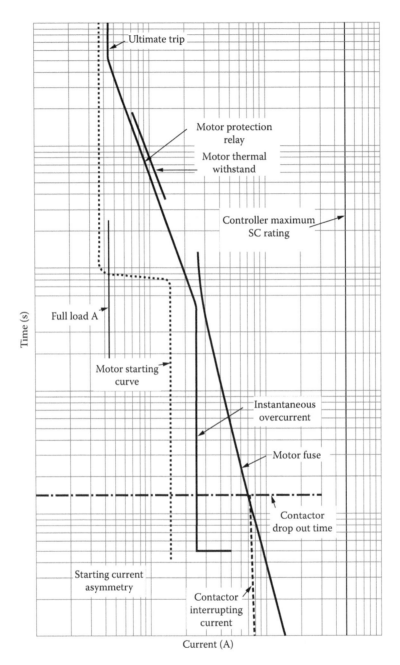

FIGURE 6.7
Fundamentals of coordination of motor starting curve with motor fuse.

6.7.3 Ground Fault Protection

The ground fault coordination and protection is discussed in Chapter 8. Here, the ground fault coordination is shown in Figure 6.10. Note the following.

The motor ground fault protection is set at 5 A pickup and six-cycle time delay. Experience shows that this setting is appropriate even for large motors, and, with an available ground

Overcurrent Protection: Ideal and Practical

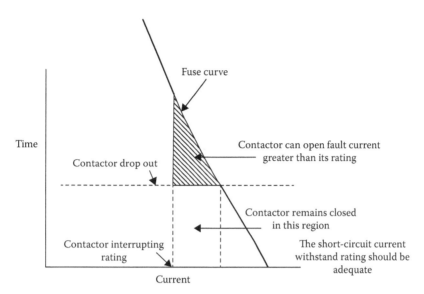

FIGURE 6.8
The motor contactor should not interrupt any current greater than its interrupting capability. Coordination with motor fuse is required.

fault current of 400 A, protects 98.75% of the motor windings from the line end, assuming a linear reduction in the fault current as the neutral end is approached.

The settings on the feeder breaker and transformer neutral relay are shown in Figure 6.10 and also tabulated in Table 6.7. Note that the main secondary breaker is not provided with a zero-sequence CT, as the transformer is close coupled and installing a core-balance CT on the bus bars is not so practical. The transformer neutral-connected CT (400/5) has two settings. The lower setting 1 is at a pickup of 16 A and at a time delay of 0.7s. If the line-to-ground fault happens to be in the 2.4 kV switchgear bus or in the connections from the transformer to the switchgear bus, this setting should operate. If the fault resides in the transformer windings, setting 2 should operate.

6.7.4 Coordination in a Cogeneration System

Figure 6.11 shows a cogeneration facility; the 35 MVA generator operates in synchronism with the utility source. Consider that each section of the bus, the generator, and 25 MVA transformer are all protected in the overlapping differential zone of protection. Yet, for a fault location shown at F1, outside the differential zone, the fault will be fed from two sources. The utility relays will isolate the fault by tripping an upstream breaker, not shown in this figure, but the fault will continue to be fed from the 35 MVA generator. There are two options: (1) intertripping; that is, the utility should trip the customer's 13.8 kV breaker in such a situation, and (2) a directional overcurrent relay can be added as shown. The three-phase fault current contributions from the generator are shown.

Though each bus section is covered in a differential zone, backup selective tripping is provided with overcurrent relays on the bus section circuit breaker. For a fault in one or the other section of the bus, the breaker should be opened. Also, the settings should coordinate with the generator breaker settings and the main secondary breaker of transformer TX2 settings.

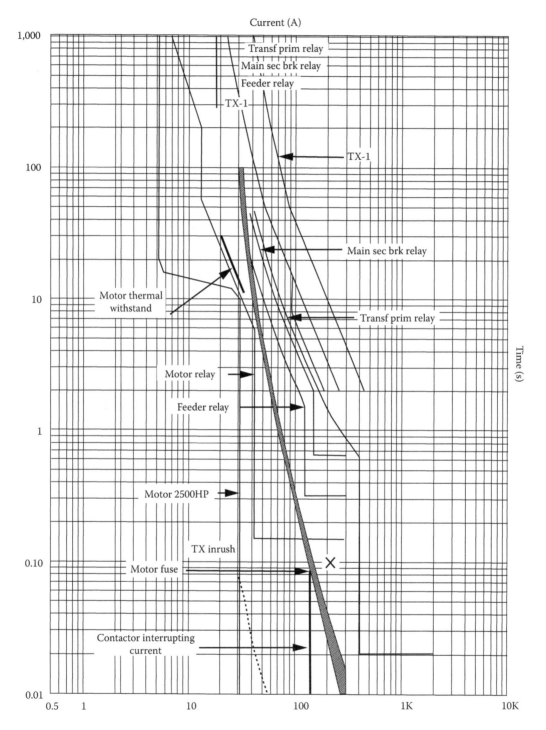

FIGURE 6.9
Coordination plot of a system configuration shown in Figure 6.6.

TABLE 6.6

Phase Overcurrent Settings, 2.4 kV Distribution (Figure 6.6)

Breaker	CT Ratio	Characteristics	Pickup	Time Dial	Instantaneous	Instantaneous TD
Feeder to MCC, 2.4 kV	1200:5	Extremely inverse	10 A (=2400 A)	6	52 A (=12480 A)	0.32
Main secondary, 2.4 kV	3000:5	Extremely inverse	4.2 A (=2520 A)	10	25 A (=15000 A)	0.65
Transformer primary, 13.8 kV	1200:5	Extremely inverse	2 A (=480 A)	10.5	30 A (=7200 A)	0.02

The settings on the directional overcurrent relay should allow the maximum power that can be supplied into the utility system by the generator. Considering that the minimum load on the system is 20 MVA, approximately 15 MVA maximum can be supplied into the utility system. Thus, the pickup setting of directional overcurrent is adjusted at 800 A.

With these considerations, the coordination on the utility-side protection is shown in Figure 6.12 and Table 6.8 depicts the details of the settings.

The generator-side coordination is shown in Figure 6.13 and Table 6.8 tabulates the settings. The generator fault decrement curve is plotted as illustrated in Chapter 11. The application of voltage-restraint for backup generator protection is discussed in Chapter 11.

6.8 Coordination on Instantaneous Basis

This is a new frontier of much importance. The selectivity is accessed from the time–current coordination curves (TCCs) of devices plotted to the same scale and noting down the overlaps. The TCCs are not extended below 0.01 s and when a current-limiting fuse or circuit breaker operates faster in the subcycle range, the TCC plots cannot determine the selectivity with any accuracy. When there is device interaction, the TCC plots will not indicate how two fast operating devices will behave together when connected in series.

The current-limiting MCCBs are available, and even the normal MCCBs not really marked as current limiting according to UL 489 may be current limiting to an extent. The circuit breaker instantaneous trip is sensitive to peak amperes, and the faults of the same rms value but of a different X/R ratio will be sensed differently by the circuit breaker trip systems.

The characteristics of current-limiting fuses are also discussed in Volume 1 and not repeated here. Above approximately one cycle, the selectivity between the two fuses is determined by maintaining a separation between the total clearing time of the downstream fuse and the minimum melting curve of the upstream fuse.

The current-limiting behavior has the following advantages:

- Reduced let-through energy.
- The downstream faults can be cleared faster.
- Selectivity can be obtained with proper application considerations.

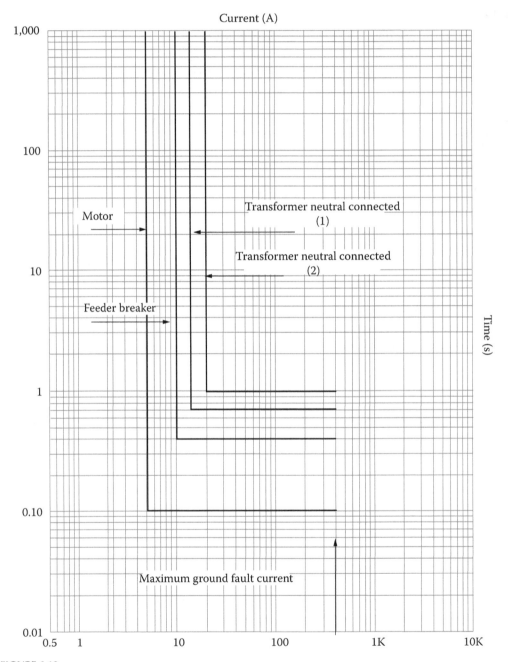

FIGURE 6.10
Ground fault coordination of the distribution system shown in Figure 6.6.

6.8.1 Selectivity between Two Series-Connected Current-Limiting Fuses

Fuse melting and clearing I^2T values can be compared to access selectivity in their current-limiting range. The total I^2T of the downstream fuse must be less than the melting I^2T of the upstream fuse for events lasting less than 0.01 s. This is shown in Figure 6.14. The

TABLE 6.7

Ground Fault Coordination

Protective Device	CT Ratio	Characteristics	Pickup	Time Delay (s)
Motor protection relay	Core balance CT 50/5	Definite time	0.5 A (=5 A)	0.1
Feeder protection relay	Core balance CT 50/5	Definite time	1.0 A (=10 A)	0.4
Transformer neutral-connected relay setting 1	400/5 neutral CT	Definite time	0.2A (=16 A)	0.7
Transformer neutral-connected Relay setting 2	400/5 neutral CT	Definite time	0.25A (=20 A)	1.0

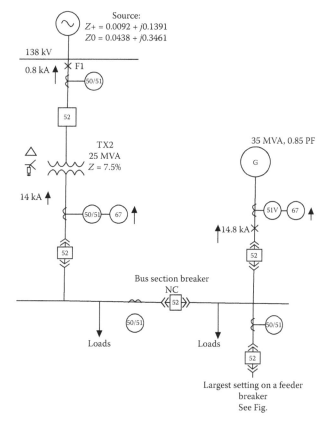

FIGURE 6.11
A cogeneration facility for the time–current coordination.

curves must be plotted on the same test power factor. The fuse of rating A will coordinate with fuse of rating C, but not with fuse of rating B. Another factor to be considered is the number of fuses in parallel clearing the fault. When two fuses in parallel clear the fault, these share line-to-line voltage and yield a lower arcing I^2T than the single fuse. Figure 6.15 shows the let-through characteristics of medium-voltage power fuses of 4–40 A of a manufacturer. The curves are based on the fuse being at an ambient temperature of 25°C

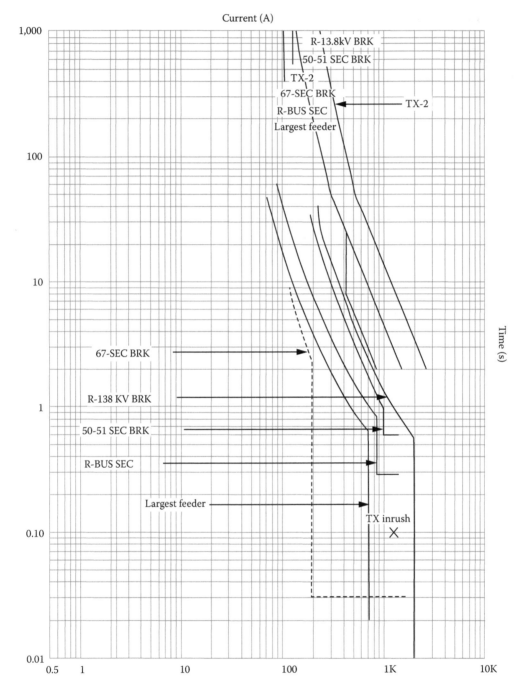

FIGURE 6.12
Coordination of the distribution shown in Figure 6.11 for the utility source.

TABLE 6.8
Coordination Settings for Selective Overcurrent Protection (Figure 6.11)

Breaker	CT Ratio	Characteristics	Pickup	TD	Instantaneous	Instantaneous TD
Largest feeder breaker	1200:5	Extremely inverse	2 A (=480 A)	10.5	30 A (=7200 A)	2 A (=480 A)
Bus section breaker	2000:5	Extremely inverse	1.5 A (=600 A)	13.3	22 A (=8800 A)	0.30
Main secondary breaker TX2 50/51	2000:5	Extremely inverse	3.2 (=1280 A)	7.5	25 A (=10,000 A)	0.7
67		Extremely inverse	2 (=800 A)	2	5 A (=2000 A)	0.02
138 kV breaker	200:5	Extremely inverse	3.7 (=148 A)	8.6	50 A (=2000 A)	0.00
Generator breaker 51 V	2000:5	Very inverse	6 A—100% 1.4 A—25%	15	—	—

and with no-load; all variations should be negative. Figure 6.16 illustrates the let-through characteristics of RK5 time delay fuses of a manufacturer.

6.8.2 Selectivity of a Current-Limiting Fuse Downstream of Noncurrent-Limiting Circuit Breaker

In this case, traditional TCC is suitable for determining selectivity. If the fuse curve crosses the instantaneous foot of the circuit breaker curve, it is likely that the combination may not coordinate and a proper gap between the maximum operating time of the fuse and the instantaneous setting of the circuit breaker is required. Sometimes, it may be misleading to visually examine and place the fuse and circuit breaker curves with some separation gap at 0.01 s and conclude that the system is selective.

Example 6.2

Figure 6.17 depicts an upstream 800 AF LVPCB provided with a programmable trip device having LSI functions and a sensor and plug rating of 800 A, and its coordination with a 100 A RK5 fuse on the downstream side. The upstream circuit breaker is set as follows:

LTPU = 800 A (=1)
LTD = 4 s
STPU = 3200 A (=4)
I^2T = In
STPU delay = 0.1 s
Instantaneous pickup = 9600 A (=12)

This figure shows that apparently there is coordination between the fuse and the circuit breaker, which is true for time duration of 0.01 s; however, it will be demonstrated that on an instantaneous basis, these devices do not coordinate.

Circuit breaker instantaneous trip units are typically of magnetic or electronic types. Both these types' of trips respond to instantaneous current values. Time is also influential as the current needs to last long enough to generate sufficient force in a magnetic

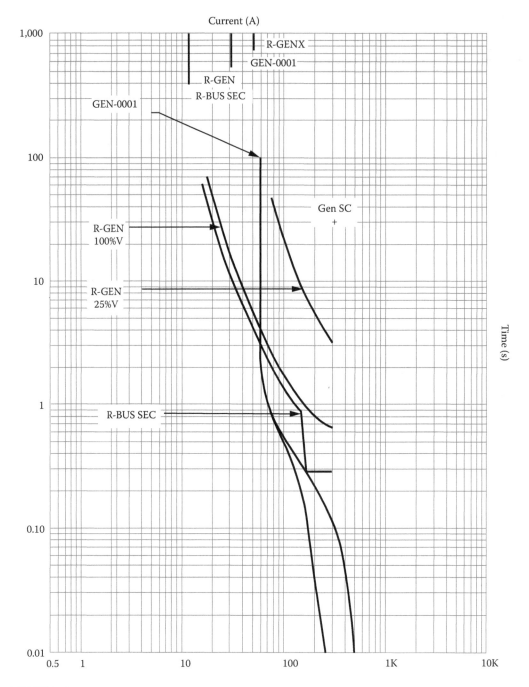

FIGURE 6.13
Coordination of the distribution shown in Figure 6.11 for generator relaying.

trip to overcome friction forces and inertia. Similarly, electronic trip units may require several samples of currents and may have filtering to minimize risk of nuisance trip. When the circuit breaker is set, a tripping threshold is defined and if one or more samples of measured current exceed this set value, then the circuit breaker is tripped.

Overcurrent Protection: Ideal and Practical

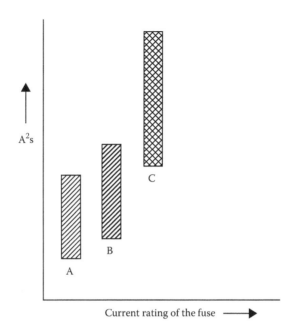

FIGURE 6.14
Selectivity criteria based upon I²T let-through of current-limiting devices.

This may be somewhat confusing because the circuit breaker time–current characteristics are published by the manufacturers in terms of rms currents. For coordinating settings, it is customary to use asymmetrical currents to account for dc component, which decays fast in low-voltage systems.

For the purpose of coordination, if a downstream current-limiting device is considered with an upstream electronic trip programmer, then the following coordination procedure can be adopted based upon Reference [4].

- Obtain the let-through curve of the downstream current-limiting device from the manufacturer. For the coordination shown in Figure 6.17, the let-through characteristics of the RK5 fuses are shown in Figure 6.16.
- Calculate the bolted three-phase symmetrical rms fault current through the devices.
- Using the let-through curve, calculate the peak let-through current at the calculated short-circuit current.
- Calculate the corresponding rms current.
- Multiply the rms current thus calculated by a factor of √2. This gives the peak at unity power factor, also see Figure 6.18. This shows a peak curve (dotted) drawn at 1.414 times the rms current and it runs parallel to the peak let-through curve of the manufacturer at 2.3 times the rms current.

Applying above criteria the following calculations results:
Peak let-through of 100 A fuse at 50 kA rms short-circuit current = 18.3 kA.

$$\text{Peak at unity power factor} = \frac{18.3}{\sqrt{2}} = 13.5 \text{ kA}.$$

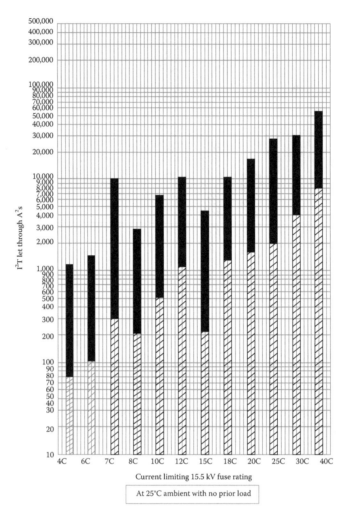

FIGURE 6.15
I^2T let-through curves of medium-voltage power fuses, 4–40 A, of a manufacturer.

Set the instantaneous of upstream circuit breaker at a minimum of 13.5 kA to coordinate with downstream current-limiting fuse. The coordination in Figure 6.16 shows the instantaneous set at 9600 A. As this is <13,500 A, coordination will not be obtained in the instantaneous zone. The instantaneous setting of 800 A breaker should be raised.

The instantaneous settings of electronic trip programmers are generally specified in terms of sensor rating. (Sometimes these can also be in terms of plug rating.) Depending upon the circuit breaker type, the instantaneous setting range is up to 12 times the sensor rating and in some cases it is up to 15 times the sensor rating. To provide the required instantaneous setting of 13.5 kA, we must have a sensor rating of 1200 A. As a sensor rating of 1200 A cannot be accommodated in an 800 AF breaker, the breaker frame rating is changed to 1600 A and it is provided with 1200 A plug. For the low-voltage systems, the LTPU must protect the conductors within their rated ampacities, see Chapter 9. The new settings are as follows:

LTPU = 840 A (=0.7)
LTD = 4 s

STPU = 3200 A (=4)
I²T = In
STPU delay = 0.1 s
Instantaneous pickup = 14,400 A (=12)

This shows that the 800 A circuit breaker has to be replaced with a 1600 A breaker and required sensor and plug to accommodate new instantaneous settings. The coordination is shown in Figure 6.19: the coordination in the instantaneous zone will be achieved.

6.8.3 Selectivity of Current-Limiting Devices in Series

The following observations can be made:

1. In a series circuit, any current and energy limiting by either device will impact both the devices in series.
2. The device with the lowest current-limiting threshold and fastest response will affect the current magnitude to operate the less sensitive and slower device.
3. The faster device limits the let-through energy as well as let-through current.

The operating responses of the circuit breakers and fuses respond to different parameters. Circuit breaker response is primarily a function of current while that of a current-limiting fuse is based on energy. Evaluating device coordination requires a technique that includes both variables and their interaction [5].

On a simplified basis, if it is ensured that the two current-limiting devices, say an upstream fuse and a downstream circuit breaker, are so sized that for the maximum fault

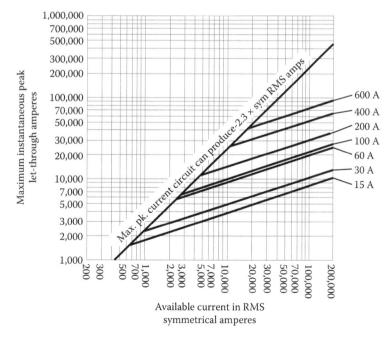

FIGURE 6.16
Peak let-through current of time delay RK5 fuses of a manufacturer.

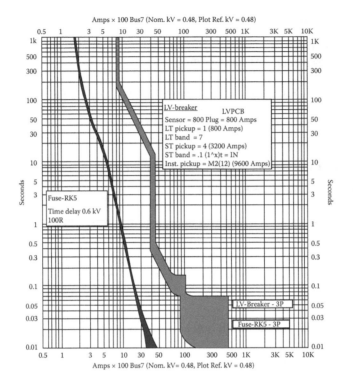

FIGURE 6.17
Coordination on instantaneous basis between an upstream LVPCB and a downstream 100 A, RK5 fuse; the system is not selective on instantaneous basis (see the text).

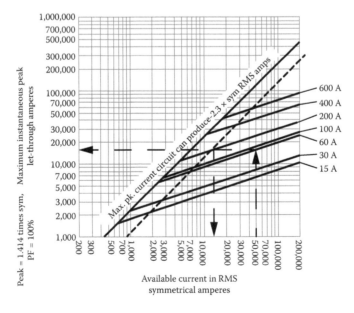

FIGURE 6.18
Graphical construction for determining the peak current setting of an upstream non-current-limiting device.

current at which the selectivity is desired, the upstream device does not see at all the peak let-through current or follow current of the downstream device, a coordination will be obtained. This method may give much larger rating of the upstream device compared to the rigorous evaluation of the let-through current and energy.

An electronic trip implementing waveform recognition (WFR) can consider a combination of peak current and time to determine if the fault current shows a characteristic wave shape of an energy and current-limiting interruption. A trip which is able to distinguish that waveform is energy limited can be set more sensitive than the one that considers peak current only, sometimes resulting in settings that are 50% of the peak current method alone. WFR is specific to one manufacturer only [5].

Figure 6.20 shows a comparison of different selective circuit breaker settings. The minimum TCC setting can be identified as the setting determined by the prospective fault current peak, as determined from the TCC considering dc offset. The minimum peak setting is based upon the downstream current-limiting overcurrent device. The WFR setting can be still lower if the trip employs advanced WFR or energy sensing algorithm.

For current-limiting circuit breakers, the manufacturer's data should be applied. Figure 6.21 shows the curve for a current-limiting MCCB. The threshold current is around 9 kA rms symmetrical. At 50 kA rms short-circuit current, the peak current let-through is 20 kA. Also the test power factor is 20%, while that of the RK5 fuses is 15%. The 20% power factor corresponds to an X/R of 4.899, which gives a peak of 2.16, while the 15% power factor

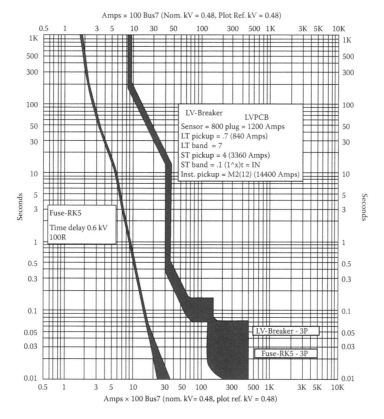

FIGURE 6.19
Coordination on instantaneous basis between an upstream LVPCB and a downstream 100 A, RK5 fuse; the system is selective on instantaneous basis (see the text).

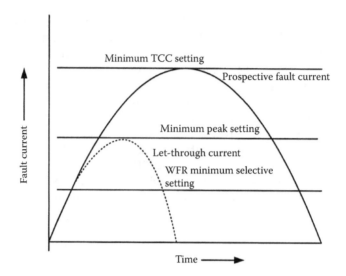

FIGURE 6.20
To illustrate selective settings with various sensing methods, TCC method, peak let-through method, and WFR.

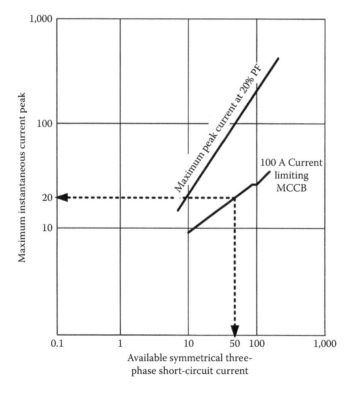

FIGURE 6.21
Peak let-through current of a current-limiting 100 A MCCB.

TABLE 6.9

Typical Minimum Instantaneous Thresholds for Upstream Feeder Current-Limiting Overcurrent Protective Devices (Settings for Circuit Breakers Rated 480 V and below, Settings for Fuses Rated 600 V and below)

Current-Limiting OCPD Type	Current-Limiting OCPD Rating	WFR TRIP[a,b] (A)	Peak Sensing Trip for 55 kA Selectivity (A)
MCP with current limiters	50 A	3800	9740
	150 A	10,750	19,200
MCCB	150 AF	9600	27,610
	600 AF	20,350	33,810
Class J time delay fuse	100 A	3810	7730
	600 A	26,730	28,330
RK1 time delay fuse	100 A	4300	8370
	600 A	32,380	32,190
RK5 time delay fuse	100 A	8820	13,520
	600 A	57,120	47,000

Source: Valdes, M. et al., Method of determining selective capability of current limiting overcurrent devices using peak let-through current, in *Conference Record, IEEE Pulp and Paper Industry Technical Conference*, Birmingham, AL, pp. 145–153, June 2009.

[a] Selectivity will range up to the short-circuit rating of the lowest rated device in pair, or withstand of the line-side device whichever is lower.

[b] Minimum setting assumes 10% tolerance.

corresponds to an X/R of 6.59. Most conservatively, a factor of 1.414 is considered for an entirely resistive load. This gives a peak practically independent of the test circuit power factor and actual X/R of the short-circuit current.

Table 6.9 shows the minimum instantaneous thresholds for upstream feeder overcurrent protective device selectivity, WFR recognition, and peak settings (see Figure 6.20).

6.9 NEC Requirements of Selectivity

The following articles from NEC, 2017 [1] relating to overcurrent selectivity are of interest:
Article 240.12: Electrical System Coordination.

Where an orderly shutdown is required to minimize the hazards to personnel and equipment, a system of coordination based on the following conditions shall be permitted:
Coordinated short-circuit protection

Here, the intention is that the faulty circuit is isolated by operation of the immediate overcurrent device, without impacting the overcurrent devices in the rest of the system.

Overload indication based on monitoring systems or devices

The monitoring system may cause the system to go to alarm, allowing corrective action for an orderly shutdown.

Article 240.100 deals with overcurrent protection of feeders and branch circuits (over 600 V, nominal).

Article 517.17(B) and (C) specify the ground fault coordination.

Article 700.27 and Article 701.27 specify the coordination for emergency and legally required standby systems. Here, it is stated that emergency systems overcurrent devices shall be selectively coordinated with all supply side overcurrent protective devices. An exception is that selective coordination shall not be required between two overcurrent devices located in series if no-loads are connected in parallel with the downstream device. An example of such a case will be transformer through-fault selectivity. If the selectivity between the primary and secondary overcurrent protections is sacrificed, it meets the intent of NEC.

Continuity of operation of illumination for occupant evacuation or maintaining continuity of essential safety equipment such as smoke detectors is necessary for occupant safety during a fire or other emergency. It implies that a coordination study should be performed.

Some states in the United States will accept a coordination study up to 0.01 s and the others up to 0.1 s to comply with NEC requirements. However, as we have examined, such systems cannot be said to be fully selective.

6.9.1 Fully Selective Systems

Systems designed for higher level of selectivity may lead to higher arc-flash energy. It becomes a challenge to have a fully selective system and also reduce arc-flash energy simultaneously. This can only be achieved by selecting suitable devices which can coordinate on instantaneous basis.

The systems for complete selectivity may require the following:

- Higher sizes of cables and trip devices.
- Larger size (kVA rating) of equipment, more than what is required to serve the loads.
- The systems may be more expensive to implement.
- Selectivity between three low-voltage circuit breakers in series is generally the maximum that can be attempted on instantaneous basis.
- If all the three circuit breakers see the same amount of short-circuit currents, that is, the interconnecting cables between these breakers are of short length, it will be all the more difficult to design for 100% *selectivity and simultaneously reduce the arc-flash hazards.*

Example 6.3

Figure 6.22 illustrates an essential service distribution, where a selective coordination is required. A 400 A automatic changeover switch is provided between the emergency diesel generator and the normal source of power. The essential loads are served from the emergency panel and only a 75 kVA lighting transformer is shown.

The normal power source is provided from a 1500 kVA, delta–wye, 13.8–0.48 kV transformer, 480 V system high-resistance grounded. The transformer primary is protected with a current-limiting fuse of 150E and there is a main secondary power circuit breaker of 2000 A, provided with electronic trip programmer with LSI functions. The feeder power circuit breakers of 800 A are also provided with electronic trip programmers. One such feeder circuit breaker serves a 480 V MCC, which in turn serves the emergency distribution panel through a breaker "BRK-PNL" and a changeover switch of 400 A. The 75 kVA lighting transformer is provided with primary and secondary breakers and serves a lighting distribution board at 240 V.

Overcurrent Protection: Ideal and Practical

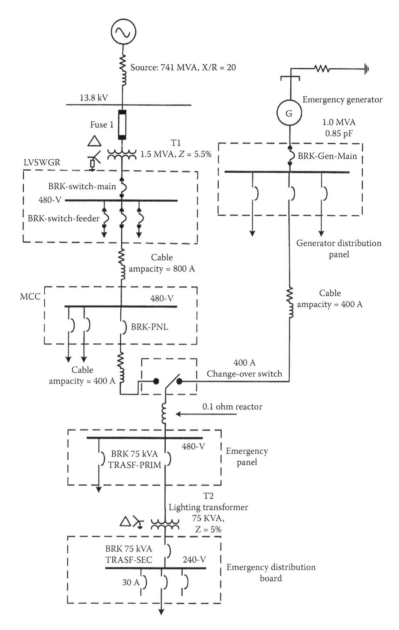

FIGURE 6.22
An essential service system configuration for full selectivity and arc-flash hazard reduction.

The emergency generator is rated at 1.0 MVA, 0.85 PF, and has a main generator LVPCB, with electronic trip programmer. A feeder circuit breaker from the generator panel connects to the 400 A changeover switch as an alternate source of power.

When the system is served from normal power source:

- The lighting board feeder circuit breaker of 30 A should fully coordinate with 75 kVA lighting transformer secondary circuit breaker of 250 A.

- Coordination between 75 kVA lighting transformers primary and secondary breakers is not necessary, according to the NEC.
- 75 kVA lighting transformer primary circuit breaker should coordinate with MCC feeder circuit breaker, BRK-PNL.
- MCC feeder circuit breaker BRK-PNL should coordinate with low-voltage feeder circuit breaker, BRK-SWITCH-FEEDER.
- The low-voltage feeder circuit breaker BRK-SWITCH-FEEDER should coordinate with main 2000 A secondary circuit breaker of 1.5 MVA transformer.
- The main secondary circuit breaker should coordinate with 1.5 MVA transformer fuse.

In total, seven devices in series are required to be coordinated.

Similarly when the system is served from the emergency generator, five devices in series must be coordinated.

Not only full coordination is required, but also arc-flash hazard should remain low.

It is obvious that current-limiting MCCBs are required for selectivity. A manufacturer's published data on its low-voltage circuit breakers showing coordination on instantaneous basis are used to develop the settings. These manufacturer's data are not reproduced [6].

6.9.2 Selection of Equipment Ratings and Trip Devices

It is necessary to reiterate between the selection of breakers, their trip devices, and system design. Note that the 0.1 Ω current-limiting reactor is provided on the feeder circuit to the lighting transformer T2 (see Figure 6.22). The calculated short-circuit level at the emergency panel is 14.9 kA and at the MCC it is 20.64 kA rms symmetrical with changeover switch connected to normal power source. The 400 A current-limiting MCCB, breaker BRK-PNL, to the emergency panel will coordinate with the feeder LVPCB on the low-voltage switchgear, instantaneous set at maximum up to 65 kA short-circuit current level, but the T2 transformer primary current-limiting breaker, BRK 75 KVA-TRANSF-PRIM, of 150 A cannot be made to coordinate with MCC breaker for any currents above 3.212 kA on the instantaneous basis. The 0.1 Ω reactor reduces the short-circuit current on the source side of the T2 primary breaker to 2.37 kA. A perusal with the manufacturer's data shows that there are no combinations of tested breakers which can achieve coordination for up to 15 kA short-circuit currents and be of required size to protect the small T2 transformer of 75 kVA.

Similarly to achieve coordination with 250 A secondary breaker of T2, BRK 75 KVA-TRASF-SEC, with downstream breaker of 30 A, the short-circuit current at 240 V should be reduced to 2 kA. This necessitates that 75 kVA transformer must have an impedance of 5%. This is too large, and commercially a 75 kVA transformer of 5% impedance may not be available. A larger transformer will require reselection and design of the trip devices.

It is also obvious that it may be difficult to match the circuit breakers which will coordinate with the actual load requirements and larger circuit breakers may be needed. This will necessitate that the cables be upgraded for conductor protection according to the NEC.

Figures 6.23 through 6.25 illustrate the coordination. Though the devices instantaneous functions apparently overlap, selectivity is achieved as per manufacturer's published data. Table 6.10 gives a summary of the protective device types and their settings. Table 6.11 shows the arc-flash hazard calculations, see Chapter 17.

Overcurrent Protection: Ideal and Practical

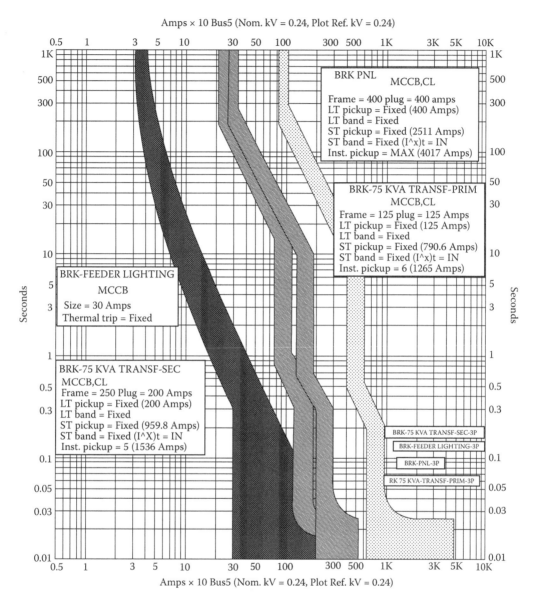

FIGURE 6.23
Time–current coordination of the OCPDs in Figure 6.22.

6.10 The Art of Compromise

Practically, an ideal coordination is rarely achieved and it becomes necessary to make compromises [7,8]. Two experienced protection engineers may take different approaches to a particular coordination situation. When compromises are necessary, these need to be made carefully, still observing the basis coordination principles. As an example:

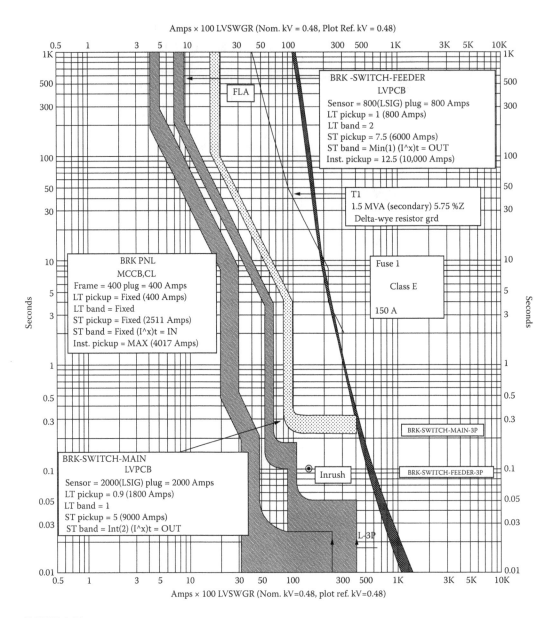

FIGURE 6.24
Time–current coordination of the OCPDs in Figure 6.22.

1. Coordination can be sacrificed when the area of shutdown is not much impacted and the faults can be limited to the faulty section.

2. It is desirable not to accept lack of coordination with fuses, as the replacement costs of fuses can be high. Operation of a fuse in a solidly grounded system in one phase can give rise to single phasing.

3. For a tie interconnecting circuits with two way flow of power, the overcurrent coordination between the breakers on either end of the tie circuit can normally be sacrificed.

Overcurrent Protection: Ideal and Practical

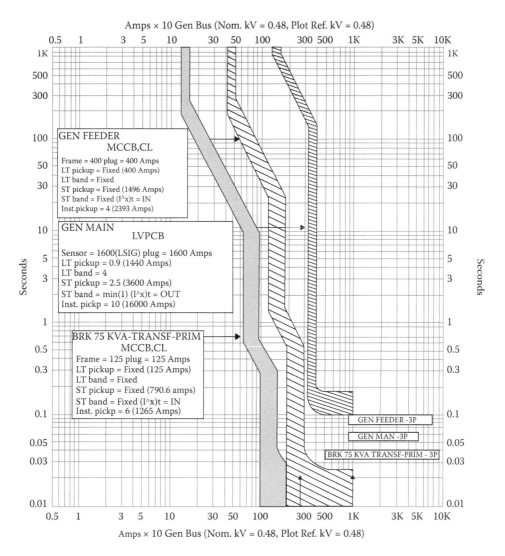

FIGURE 6.25
Time–current coordination of the OCPDs in Figure 6.22.

4. For the protection of a transformer, where both the secondary and primary protections are provided, the primary protection can cover, to an extent, the secondary faults, and is more important than the secondary protection. The NEC allows only primary protection of the transformers, with no secondary protection, provided the requirements of NEC overcurrent settings are met.

5. For large motors, some lack of coordination at the motor starter level is acceptable rather than perpetuating it upstream.

6. Coordination should be obtained for the maximum fault currents in the system. However, bolted three-phase faults are rare. Some lack of coordination at the maximum fault currents can be acceptable.

TABLE 6.10

Circuit Breaker Selection and Trip Settings for Selective Coordination (Figure 6.22)

Circuit Breaker Designation	Circuit Breaker Frame	Sensor/Plug (A)	Circuit Breaker Type	Trip Device	Coordinating Settings
BRK-SWITCH-MAIN	2000 AF	2000/2000	LVPCB	Digital LS	LTPU = 0.9, LTD band = 1, STPU = 2, STD band = 2, I^2T = out, no instantaneous
BRK-SWITCH-FEEDER	800 AF	800/800	LVPCB	Digital LSI	LTPU = 0.9, LTD band = 2, STPU = 7.5, STD band = 1, I^2T = out, instantaneous = 10,000 A
BRK-PNL	400 AF	400/400	CL MCCB	Digital LSI	LTPU = fixed, LTD band = fixed, STPU = fixed, STD band = fixed, instantaneous = 4020 A
BRK 75 KVA-TRANSF-PRIM	125 A	125/125	CL MCCB	Digital LSI	LTPU = fixed, LTD band = fixed, STPU = fixed, STD band = fixed, instantaneous = 1265 A
BRK 75 KVA-TRANSF-SEC	250 A (240 V)	250/200	CL MCCB	Digital LSI	LTPU = fixed, LTD band = fixed, STPU = fixed, STD band = fixed, instantaneous = 1535 A
BRK-FEEDER-LIGHTING	30 A	—	MCCB	Thermal magnetic	Fixed
BRK-GEN MAIN	1600 AF	1600/1600	LVPCB	Digital LS	LTPU = 0.9, LTD band = 4, STPU = 2.5, STD band = 1, I^2T = out, no instantaneous
BRK-GEN FEEDER	400 AF	400/400	CL MCCB	Digital LSI	LTPU = fixed, LTD band = fixed, STPU = fixed, ST delay band = fixed, instantaneous = 3165 A

TABLE 6.11

Arc-Flash Hazard Calculations (Example 6.3), ATS Connected to Normal Power Supply

Equipment ID	Bolted Fault Current (kA)	Arcing Current kA (85% Ia)	Arcing Time (s)	Working Distance (In.)	Ground	Arc-Flash Boundary (ft)	Incident Energy (cal/cm^2)	PPE
LV switchgear	30.32	15.86	0.33	24	No	11.03	14.86	3
MCC	20.64	10.20	0.18	18	No	4.50	7.9	2
Transfer switch	14.88	9.21	0.025	18	No	1.33	0.986	0
Emergency panel	14.88	9.21	0.025	18	No	1.33	0.986	0
Emergency distribution board, 240 V	2.07	1.25	0.909	18	Yes	2.72	3.19	1

Note: See Chapter 17.

Figure 6.26 shows a medium-voltage distribution system for coordination. This identifies the largest motor and the largest 4.16 and 13.8 kV feeder breakers to be coordinated. If the coordination is achieved with respect to the largest load, say a motor, served from a bus, it can be concluded that for smaller loads connected to the same bus the coordination will be

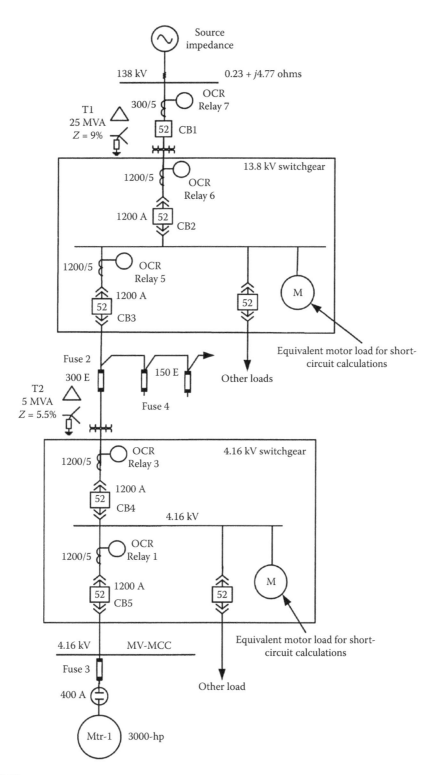

FIGURE 6.26
A medium-voltage distribution system for selective overcurrent coordination.

achieved; some care is required in exercising this generality—for example, a smaller motor may have a much longer starting time.

4.16 kV circuit breaker CB5 feeds a medium-voltage MCC, bus rating of 1200 A, which has a large motor starter for 3000 hp motor. Transformer T2 is rated 5.0 MVA, and is protected by 300E current-limiting fuse.

Figure 6.27 shows the motor starting curve, the thermal withstand curves, and the curves for motor fuse of 24R and transformer 300E primary fuse. As seen from this coordination plot, the motor fuse is properly selected for the application, and so also 300E transformer primary fuse. However, the feeder breaker CB5 and the main 4.16 kV secondary

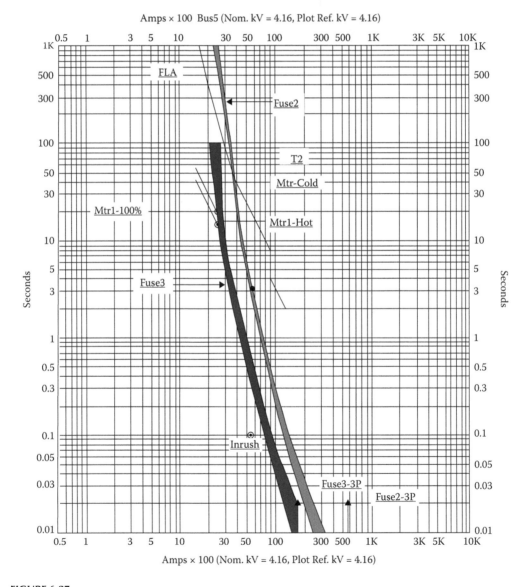

FIGURE 6.27
Time–current coordination in Figure 6.26, showing close spacing between the motor and transformer protection fuses (see the text).

breaker CB4 overcurrent settings, howsoever manipulated, cannot fit between these two fuse curves.

It will not be desirable to omit altogether feeder overcurrent, CB5, and main secondary circuit breaker CB4 overcurrent protections and let the 300E fuse protect the 4.16 kV switchgear bus, medium-voltage MCC bus, and feeder connections to the medium-voltage MCC bus.

One possibility will be to replace the 300E fuse with an overcurrent relay for 5.0 MVA transformer primary protection, which can be set closer to the transformer T2 thermal damage curve. This can open up the space toward the right side of motor fuse and the transformer thermal withstand to accommodate the overcurrent settings on breakers CB4 and CB5.

It is also possible to eliminate the 300E fuse as well as any overcurrent relay for the primary protection of 5 MVA transformer, and rely upon the protection provided by 13.8 kV breaker CB3 overcurrent relay settings. Let us apply the latter approach.

Breaker CB3 serves other transformers too, daisy chained, of no more than 1750 kVA rating, which are protected with 150E current-limiting primary fuses. Consider that three such transformers, each protected with a 150E fuse, are in service.

Then, the settings on breaker CB3 should consider the following:

- The primary protection settings are not exceeded more than six times the 5 MVA transformer full-load current as per NEC.
- The maximum overcurrent settings on the secondary breaker CB4 should not exceed six times the 5 MVA transformer full-load current as per NEC.
- The instantaneous overcurrent should be set to account for all the feeder inrush currents; that is, all the transformers connected to breaker CB3 can be energized together. This inrush current is calculated as 6000 A.

Figure 6.28 depicts the revised coordination. Some coordination is sacrificed between the overcurrent relays 3 (breaker CB4) and 1 (breaker CB5) at the maximum short-circuit levels, but no coordination is sacrificed between the overcurrent relays 3 (breaker CB4) and 5 (breaker CB3). This is desirable as breaker CB3 feeds an entire chain of transformers. The settings on breaker CB3 relay adequately protect the thermal withstand of 5 MVA transformer. The instantaneous is delayed by 0.05 s to coordinate with motor circuit and other fuses. The instantaneous is set at 12 kA, which clears relay 1 and 3 curves and allows for the feeder inrush current.

Proceeding further, the coordination with 25 MVA transformer T1 withstand and its primary and secondary relays is obtained adequately with appropriate CTI, as shown in Figure 6.29. A summary of the protective device settings is provided in Table 6.12.

This is illustrative of the compromises that can be made to the protective relaying and their settings. The strategy of accepting lack of coordination will vary from system to system, selected protected devices, loads to be served, and equipment to be protected.

6.11 Zone Selective Interlocking

We discussed zone selective interlocking for a simple low-voltage radial system in Chapter 5. It can also be applied to medium-voltage systems and preserves the selective coordination

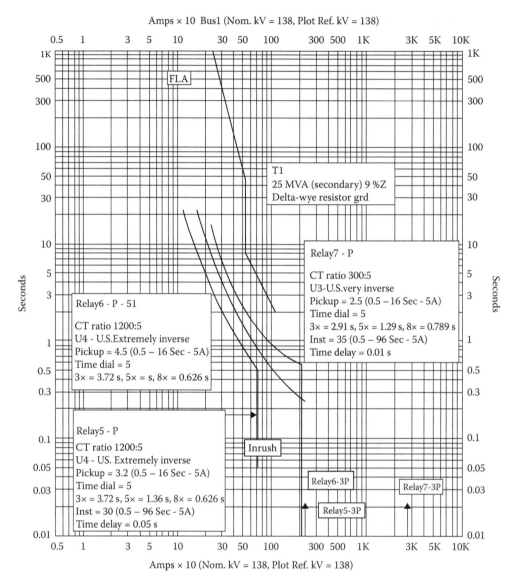

FIGURE 6.28
Time–current coordination achieved by omitting the 5 MVA transformer primary protection (see the text).

between main, tie, and feeder circuit breakers allowing fast tripping between device desired zones. This is done through wired connections between trip units and relays. If a feeder detects a fault, it sends a restraint signal to the main circuit breaker, but for a fault on the bus, the main circuit breaker does not get a downstream restraint signal and trips without delay. The restraint logic is not instantaneous and there is some time delay associated with it, so that there is no unrestrained tripping of the main. For conservatism, a delay of 20 ms can be added, though it varies from manufacturer to manufacturer. Also, care has to be exercised with motor loads. A motor load will contribute to the bus short-circuit current and the feeder circuit breaker should not send a restraint signal upstream when the motor contribution fault current flows through it. There can be more than one source

Overcurrent Protection: Ideal and Practical

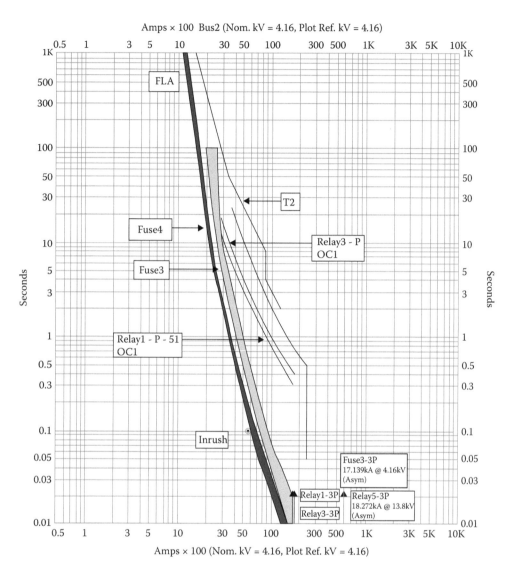

FIGURE 6.29
Time–current coordination for a 25 MVA main transformer (Figure 6.26).

TABLE 6.12

Final Settings on OCPDs of Figure 6.26

Breaker No.	Relay	CT Ratio	Final Settings
CB5	1	1200:5	Extremely inverse, pickup = 8 (=1920 A), time dial = 3, no instantaneous
CB4	3	1200:5	Extremely inverse, pickup = 8 (=1920 A), time dial = 4, no instantaneous
CB3	5	1200:5	Extremely inverse, pickup = 3.2 (=768 A), time dial = 5, instantaneous = 30 (12 kA), instantaneous time delay = 0.05 s
CB2	6	1200:5	Extremely inverse, pickup = 4.5 (=1080 A), time dial = 4, no instantaneous
CB1	7	300:5	Very inverse, pickup = 2.5 (=150 A), time dial = 5, instantaneous = 35 (2.1 kA), instantaneous time delay = 0.01 s

Example 6.4

A real-world zone interlocking illustrated here with reference to Figure 6.30 shows a section of a distribution from a 1500 kVA substation transformer. The main circuit breaker BK1 and the feeder circuit breaker BK2 are zone interlocked. Figure 6.31 shows the three-step coordination. The 400 A circuit breaker feeding the panel is a current-limiting circuit breaker. Circuit breakers BK1 and BK2 are LVPCBs with electronic trip programmers.

The coordination shown in Figure 6.31 illustrates the zone interlocking features. For a bus fault, no signal to block the trip is received from the feeder circuit breaker BK2, and the circuit breaker BK1 clears the fault with its STD band moved down as demonstrated with dotted lines, which shows a delay of 20 ms. In coordination, a conservative approach is taken and the maximum clearing time of the short-time band is considered (in this case 0.07 ms). This means that instead of clearing the arcing current in time A in 0.34 s, it is cleared in time B, 0.07 s.

For a fault on the downstream of circuit breaker BK2, the main circuit breaker BK1 does not receive any restraint signal and the normal coordination applies. That means the STD band stays where it is, as shown with the solid lines. Thus, the feeder circuit breaker clears the fault selectively in 0.05 s (point C).

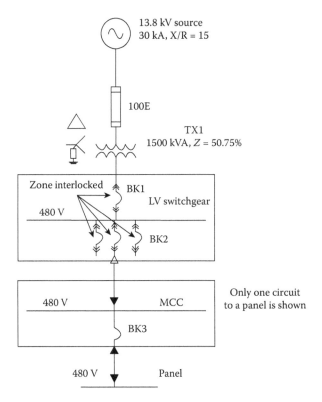

FIGURE 6.30
A low-voltage radial distribution for zone interlocking.

Overcurrent Protection: Ideal and Practical

In Figure 6.31, it seems that there is no coordination between the instantaneous settings on circuit breakers BK2 and BK3. However, circuit breaker BK3 is a current-limiting circuit breaker and the coordination on instantaneous basis is discussed above. Though these two circuit breakers do not seem to coordinate in the TCC plot, but on a current let-through basis, these do.

Example 6.5

This example illustrates the problem that can occur with zone interlocking when large motor loads are present. Consider the system configuration shown in Figure 6.32. The coordination is shown in Figure 6.33. The 200 hp motor starting curve is plotted. Also,

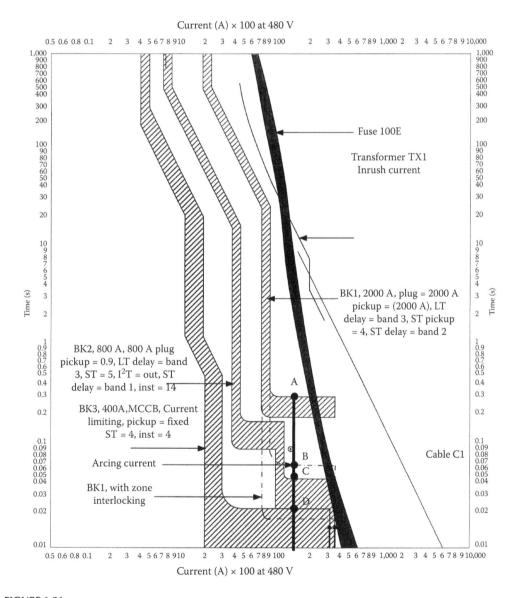

FIGURE 6.31
Time–current plot of Figure 6.30 (see the text).

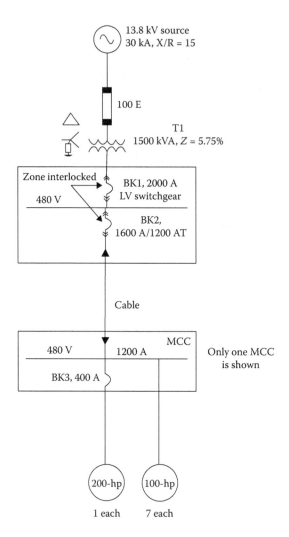

FIGURE 6.32
Zone interlocking with motor load contributions (system configuration).

a curve illustrating the starting load plus the running load of 7100 hp motor is plotted. The short-circuit current profile of the motor loads crosses the feeder short-time setting band 1. This could result in tripping of the feeder circuit breaker BK2. There are two solutions to this problem:

1. The short-time band of the feeder circuit breaker relay can be raised, so that it clears the motor fault decrement curve. This implies that the incident energy release and arc-flash hazard will increase.
2. The second method utilizes the microprocessor-based technology to sense the direction of the fault current.

Zone interlocking can be extended to medium-voltage systems, though it is uncommon.

Overcurrent Protection: Ideal and Practical

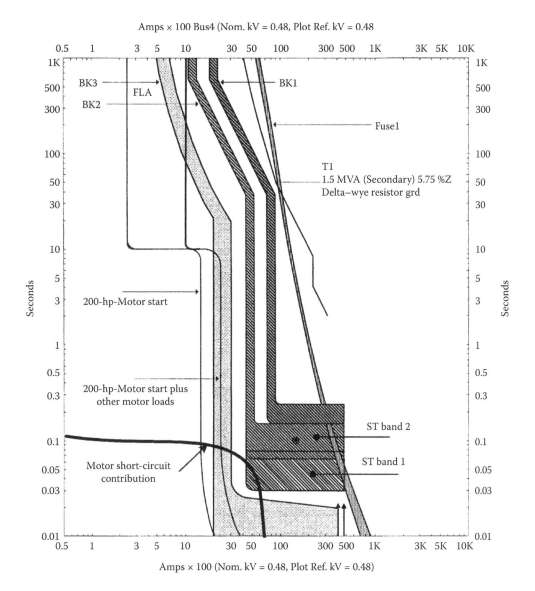

FIGURE 6.33
Coordination plot of Figure 6.32 (see the text).

6.12 Protection and Coordination of UPS Systems

Figure 6.34 shows a schematic arrangement of a UPS (uninterruptible power supply) system with a static bypass. The short-circuit current on the output of the UPS is limited to a low value for protection of the semiconductors by gate control (see Volume 1, Chapter 13). A manufacturer will generally specify the magnitude and duration of this current. It is

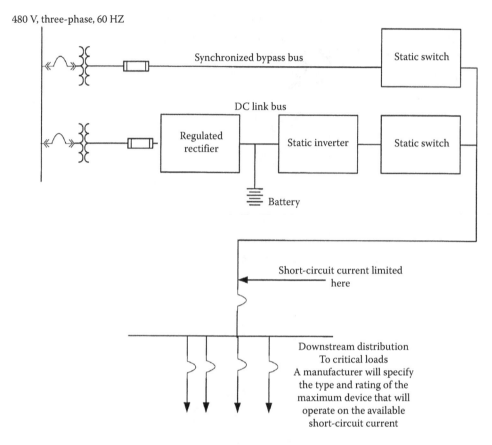

FIGURE 6.34
Schematic of a UPS system. The short-circuit current is limited at the output (see the text).

imperative that the downstream protection devices must be selected and coordinated so that these trip selectively with the limited available short-circuit current.

Figure 6.35 shows the profile of the short-circuit current supplied by a manufacturer for 4 MVA static frequency converter, 50–60 Hz and a part of the downstream 60 Hz distribution system. It is a challenge to remove the fault selectively with the short-circuit current profile supplied by the manufacturer. All 4.16 kV systems can be wrapped in the differential zone of protection, retaining overcurrent devices as backup. Some lack of coordination can be allowed. The 4.16 kV system behaves like an ungrounded system and no ground fault current from the SFC (Static Frequency Convertor) is contributed.

Overcurrent Protection: Ideal and Practical

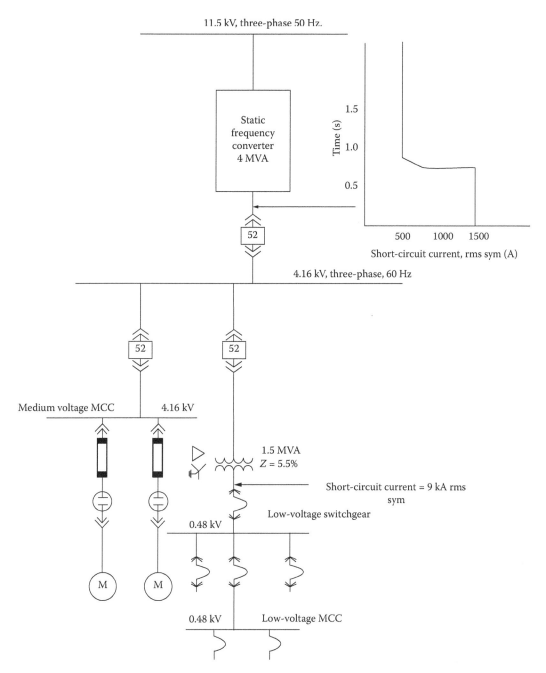

FIGURE 6.35
A static converter of 4 MVA, with distribution and short-circuit current profile supplied by the manufacturer.

References

1. NEC. National Electric Code NFPA 70, 2017.
2. IEEE Standard. C.57.91 (Also Cor.1-2002). IEEE Guide for Loading Mineral-Oil Immersed Transformers, 1995.
3. NEMA. Industrial Control and Systems, Part ICS 2-324. AC General Purpose High Voltage Contactors and Class E Controllers, 50 Hz and 60 Hz, 1974.
4. V Marcelo, H Steve, P Sutherland. Improving selectivity and arc-flash protection through optimized instantaneous protection settings. *Conference Record, IEEE Pulp and Paper Industry Technical Conference*, Nashville, TN, pp. 34–41, June 2011.
5. M Valdes, T Papallo, A Crabtree. Method of determining selective capability of current limiting overcurrent devices using peak let-through current. *Conference Record, IEEE Pulp and Paper Industry Technical Conference*, Birmingham, AL, pp. 145–153, June 2009.
6. GE. GE Overcurrent Device Instantaneous Selectivity Tables, Publication DET 537.
7. ANSI/IEEE Standard 242. IEEE Recommended Practice for Protection and Coordination of Industrial and Commercial Power Systems, Chapter 14: Overcurrent Coordination, 1986.
8. JC Das. Protective Relay Coordination—Ideal and Practical, *IEEE IAS Annual Meeting*, San Diego, CA, pp. 1861–1874, 1989.

7
System Grounding

7.1 Study of Grounding Systems

The grounding systems can be studied under three classifications:

1. System grounding
2. Equipment grounding
3. Transients in grounding systems

System grounding refers to the electrical connection between the phase conductors and the ground and dictates the manner in which the neutral points of wye-connected transformers and generators or artificially derived neutral systems through delta–wye or zig-zag transformers are grounded.

Equipment grounding refers to the grounding of the exposed metallic parts of the electrical equipment, which can become energized and create a potential to ground—say due to breakdown of insulation or fault—and can be a potential safety hazard. In the grounding grids for substations, the criteria of safety are limiting the step, touch, and transfer potentials to safe limits, see Reference [1].

Transients in grounding systems occur under lightning impulse currents, and the impulse response of the grounding grids must be studied.

In this volume, we will confine to system grounding.

Further aspects of grounding are as follows:

- Grounding in mine distribution systems
- Grounding of portable electrical equipment
- Hybrid grounding
- Grounding of adjustable speed drive (ASD) systems
- Lightning protection of buildings, chimneys and structures, and grounding
- Grounding of surge arresters
- Lightning protection of substations
- Grounding of electronic equipment, data cables, and computers.

All of these aspects are not discussed. The grounding systems are also discussed in Volume 1. *There is some repletion of the concepts presented in Volume 1, which is carried over in this chapter.*

7.2 Solidly Grounded Systems

Figure 7.1 from IEEE Standard 142 [2] illustrates various methods of system grounding. In a solidly grounded system, there is no intentional impedance between the system neutral and the ground. A power system is solidly grounded when the generator, power transformer, or grounding transformer neutral is directly connected to the ground. *A solidly grounded system is not a zero impedance circuit due to the sequence impedances of the grounded equipment*, like a generator or transformer itself. These systems, in general, meet the requirements of an "effectively grounded" system in which the ratio X_0/X_1 is positive and less than 3.0 and the ratio R_0/X_0 is less than 1, where X_1, X_0, and R_0 are the positive-sequence reactance, zero-sequence reactance, and zero-sequence resistance, respectively.

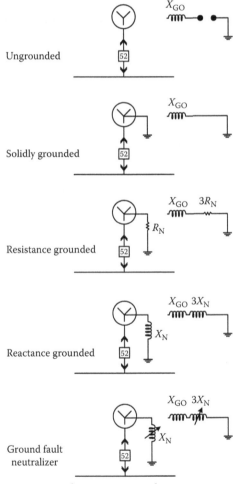

X_{GO} = Zero sequence reactance of generator or transformer
X_N = Reactance of grounding reactor
R_N = Resistance of grounding resistor

FIGURE 7.1
Various methods of system grounding.

The coefficient of grounding (COG) is the ratio of E_{Lg}/E_{LL} in percentage, where E_{Lg} is the highest rms voltage on an unfaulted phase, at a selected location, during a fault effecting, one or more phases to ground and E_{LL} is the rms phase-to-phase power frequency voltage obtained at that location with the fault removed. Solidly grounded systems are, generally, characterized by COG of 80%. Approximately, a surge arrester with its rated voltage calculated on the basis of the system voltage multiplied by 0.8 can be applied. See Volume 1 for further discussions and calculations of COG.

The utility systems at transmission, subtransmission, and distribution levels are solidly grounded. The main reason for this is that on occurrence of a ground fault, enough ground fault current should be available to selectively trip the faulty circuit.

The utility generators, connected in step-up configuration to a generator transformer, are invariably high-resistance grounded. If a generator neutral is left ungrounded, there is a possibility of generating high voltages through inductive–capacitive couplings. Ferroresonance can also occur due to the presence of generator PTs.

The utility substations serving large chunks of power at high voltages for industrial plants through delta–wye transformers have low-resistance grounded secondary wye windings. The most common voltages of distributions for the industrial plants are 13.8, 4.16, and 2.4 kV.

The low-voltage systems in industrial power distribution systems used to be solidly grounded. However, this trend is changing and high-resistance grounding (HRG) is being adopted.

The solidly grounded systems have an advantage of providing effective control of overvoltages, which become impressed on or are self-generated in the power system by insulation breakdowns and restriking faults. Yet, these give the highest arc fault current and consequent damage and require immediate isolation of the faulty section. Single-line-to-ground fault currents can be higher than the three-phase fault currents. These high magnitudes of fault currents have a twofold effect:

- Higher burning or equipment damage.
- Interruption of the processes, as the faulty section must be selectively isolated without escalation of the fault to unfaulted sections. This is a major consideration of transition from solidly grounded to high-resistance grounded systems for the low-voltage systems.

Yet, some systems according to National Electric Code (NEC) must be solidly grounded. These include systems below 120 V ac, control systems, lighting systems, and commercial distributions. Also solidly grounded systems are required where three-phase four-wire loads are required to be served.

The arc fault damage to the *equipment* for low-voltage 480 V systems has been investigated using laboratory models [3]. Stanback reported that for single-phase 277 V arcing fault tests using spacing of 1–4 in from bus bars to ground and for currents from 3000 to 26,000 A, the burning damage can be approximated by the following equation:

$$\text{Fault damage} \propto (I)^{1.5} t \tag{7.1}$$

where I is the arc fault current and t is the duration in seconds

$$V_D = K_s (I)^{1.5} t (\text{in.})^3 \tag{7.2}$$

where K_s is the burning rate of material in in.^{3}/As$^{1.5}$, V_D is the acceptable damage to material in in.^{3}, I is the arc fault current, t is the duration of flow of fault current, and K_s depends upon the type of material and is given by

$$K_s = 0.72 \times 10^{-6} \text{ for copper}$$
$$= 1.52 \times 10^{-6} \text{ for aluminum} \quad (7.3)$$
$$= 0.66 \times 10^{-6} \text{ for steel}$$

The NEMA [4] assumes a practical limit for the ground fault protective devices, so that

$$(I)^{1.5} t \triangleleft 250 I_r \quad (7.4)$$

where I_r is the rated current of the conductor, bus, disconnect, or circuit breaker to be protected.

Combining these equations, we can write

$$V_D = 250 K_s I_r \quad (7.5)$$

Example 7.1

As an example, consider a circuit of 4000 A. Then, the NEMA practical limit is 1.0×10^6 (A)$^{1.5}$ s and the permissible damage to copper, from (7.5), is 0.72 in.^{3}. To limit the arc fault damage to this value, the maximum fault clearing time can be calculated. Consider that the arc fault current is 20 kA. Then, the maximum fault clearing time including the relay operating time and breaker interrupting time is 0.35 s. It is obvious that vaporizing 0.72 in.^{3} of copper on a ground fault which is cleared according to established standards is still damaging to the operation of the equipment. A shutdown and repairs will be needed after the fault incidence.

Due to high arc fault damage and interruption of processes, the solidly grounded systems are not in much use in the industrial distribution systems. However, ac circuits of less than 50 V and circuits of 50–1000 V for supplying premises wiring systems and single-phase 120/240 V control circuits must be solidly grounded according to NEC [5].

Figure 7.2 shows a sustained arc fault current in a 3/16 in. gap in a 480 V three-phase system [3]. Experimentally, an arc is established between phase c of the bus and ground, and a current of 1100 A flows. After three-cycle phase a is involved and the arc current for two-line to enclosure fault is 18,000 A, the arc energy equals 7790 kW cycles.

Approximately 70% of the faults in the electrical systems are line-to-ground faults. Sometimes, these may be self-clearing and of transient nature (e.g., in over head (OH) line systems) or may evolve into three-phase faults over a period of time.

Thus, the probability of a worker being subject to arc flash due to ground faults is much higher. A footnote in NFPA 70E reads as follows:

> High resistance grounding of low-voltage and 5 kV (nominal) systems, current limitations, and specifications of covered bus within equipment are techniques available to reduce the hazard of the system.

Thus, high-resistance systems are recommended, though at medium voltages these should be carefully evaluated.

System Grounding

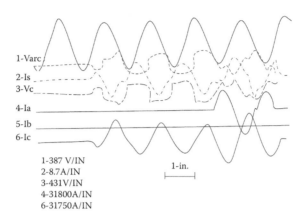

FIGURE 7.2
Arcing fault in a 3/16 in. gap, 480 V three-phase system.

Another consideration is selective protective relaying for various systems, see Chapter 8. It is documented that though the selective ground fault clearance is possible in a solidly grounded low-voltage system, practically, compromises are made (see Chapter 8).

7.2.1 Hazards in Solidly Grounded Systems

Let us revisit Figure 7.2, which shows that a single line-to-ground fault quickly involves the other unfaulted phases. Dunki-Jacobs [6] states that the escalation time of a single-phase fault to three-phase faults is rather small, of the order of—one to two cycles. Further, single-phase faults are more difficult to sustain than three-phase faults. Dunki-Jacobs illustrated the arcing line-to-ground faults as a discontinuous sine wave. It seems that the safest type of 480 V grounded systems will be the ones that are selectively coordinated to isolate the low-level faults as soon as these occur. However, there are limitations in designing selectively coordinated HRG systems and solidly grounded systems as illustrated in the examples in Chapter 8.

Lucks [7] contends that the arcs for ground fault can be sustained even at low values of ground fault current of the order of 800 A, which may not be any more than the load current and the arcing will continue unless this low-level ground fault current is cleared by the protective relaying. This is based on the work of Land [8], arcing ground faults on a low-voltage system with 3.81 cm phase-to-phase spacing and 1.9 cm phase-to-ground spacing.

A significant finding of Land is that the arc started as a line-to-ground fault does not escalate into a phase-to-phase arcing fault, but as a phase-to-ground-to phase fault. The magnetic fields, surrounding each bus, constrain the arc to jump phase to phase, which results in more slowly evolving phase-to-phase fault, and for a greater period of time, it remains as a phase-to-ground fault. A phase-to-ground fault may be initiated with a conductor of larger cross section, that is, tools or hardware inadvertently left inside the enclosures.

Ground faults are more likely to occur than phase faults. IEEE 493-2007 [9] states that the ground faults are:

- 2.3 times more than phase faults in bus ducts
- 7.3 times more than phase faults in cables
- 7.8 times more than phase faults in cable joints

The insulation of the bus system impacts the arc transfer to other phases. With insulated buses, the arcing fault can be expected to remain stationary at the point of fault and it can easily burn down 480/277 V systems.

Thus, it becomes imperative that low-level ground faults are cleared quickly to prevent conductor and equipment damage. As the current reduces, the time to clear a fault on inverse characteristics relays will increase. The problem will be to precisely calculate the ground fault current as the return path involves a number of factors. Lucks [7] lists the following factors:

- Metallurgy of enclosure
- Size and type of ground conductor
- Corrosion on pressure contacts
- Rectification from dissimilar materials
- Mutual inductance
- Skin effect
- Phase to neutral voltage
- Earth conductivity

7.3 Low-Resistance Grounded Systems

An impedance grounded system has a resistance or reactance connected in the neutral circuit to ground, see Figure 7.1. In a low-resistance grounded system, the resistance in the neutral circuit is so chosen that the ground fault is limited to approximately full-load current or even lower, typically 200–400 A. The arc fault damage is reduced, and these systems provide effective control of the overvoltages generated in the system by resonant capacitive–inductive couplings and restriking ground faults. Though the ground fault current is much reduced, it cannot be allowed to be sustained and selective tripping must be provided to isolate the faulty section. For a ground fault current limited to 400 A, the pick-up sensitivity of modern ground fault devices can be even lower than 5 A. Considering an available fault current of 400 A and the relay pickup of 5 A, approximately 98.75% of the transformer or generator windings from the line terminal to neutral are protected. This assumes a linear distribution of voltage across the winding. (Practically, the pickup will be higher than the low set point of 5 A.) The incidence of ground fault occurrence toward the neutral decreases as a square of the winding turns.

It has been a general practice to limit the ground fault current to 400 A for the grounded sources such as generators and wye-connected windings of the transformers. With modern MMPRs that provide greater sensitivity, the trend is to reduce it to 100 A or sometimes even less.

The low-resistance grounded systems are adopted at medium voltages, 13.8, 4.16, and 2.4 kV, for industrial distribution systems. Also industrial bus-connected generators are commonly low-resistance grounded. Hybrid grounding systems for industrial bus-connected generators are a recent trend in industrial bus-connected medium-voltage generator grounding.

7.4 High-Resistance Grounded Systems

High-resistance grounded systems limit the ground fault current to a low value, so that an immediate disconnection on occurrence of a ground fault is not required. It is well documented that to control over voltages in the high-resistance grounded systems, the grounding resistor should be so chosen that

$$R_n = \frac{V_{ln}}{3I_c} \tag{7.6}$$

where V_{ln} is the line-to-neutral voltage and I_c is the stray capacitance current of each phase conductor. Figure 7.3 depicts transient voltage in percent of normal line-to-ground crest voltage versus the resistor kW/charging capacitive kVA [10]. The transients are a minimum when this ratio is unity. This leads to the requirement of accurately calculating the stray capacitance currents in the system. These calculations for a high-resistance system are not documented here. Cables, motors, transformers, OH lines, surge arresters, and generator windings have distributed stray capacitances to ground—all contribute to the stray capacitance current. For the purpose of HRG, we can consider all these distributed stray capacitances lumped together. Surge capacitors connected line-to-ground must be considered in the calculations. The authors of [11–13] provide charging current (stray capacitance current) data of electrical system components. Once the system stray capacitance is determined, then the charging current per phase, I_c, is given by

$$I_c = \frac{V_{ln}}{X_{co}} \tag{7.7}$$

where X_{co} is the capacitive reactance of each phase, stray capacitance considered lumped together.

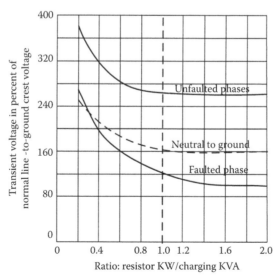

FIGURE 7.3
Overvoltage versus ratio of resistor kW/charging kVA. (From Westinghouse, *Transmission and Distribution Handbook*, Westinghouse, East Pittsburg, PA, 1964.)

Example 7.2

An HRG system for a wye-connected neutral of a 13.8 kV-0.48 transformer is shown in Figure 7.4a. This shows that the stray capacitance current per phase of all the distribution system connected to the secondary of the transformer is 0.21 A per phase. In a three-phase system, the three phases are symmetrical (though not perfectly) with respect to each other; we can assume that the charging currents of all three phases are equal. Figure 7.4b shows that under no-fault condition, the vector sum of three capacitance currents is zero, as these are 90° displaced with respect to each voltage vector and therefore 120° displaced with respect to each other:

$$\vec{I}_a + \vec{I}_b + \vec{I}_c = 0 \tag{7.8}$$

Thus, the grounded neutral does not carry any current and the neutral of the system is held at the ground potential; *no capacitance current flows into the ground* or in the neutral-connected grounding resistor.

On occurrence of a ground fault, say in phase *a*, the situation is depicted in Figure 7.4c and d. The capacitance of faulted *a* phase is short circuited to ground, and this phase does not contribute to any capacitance current. The faulted phase, assuming zero fault resistance, is at the ground potential (see Figure 7.4d) and the other two phases have line-to-line voltages with respect to ground. Therefore, the capacitance current of the

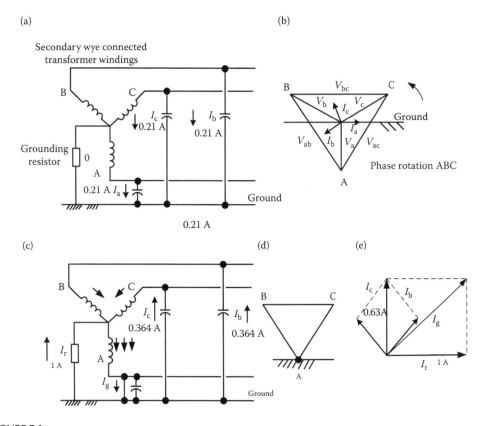

FIGURE 7.4
(a and b) The stray capacitance currents and voltages in a wye-connected HRG system under no-fault conditions, (c) the flow of capacitance and ground currents, phase *a* faulted to ground, (d) voltages to ground, phase *a* grounded, and (e) phasor diagram of summation of capacitive and resistance components of current.

System Grounding

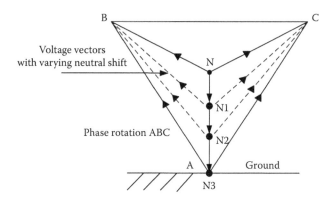

FIGURE 7.5
The neutral shift in high-resistance grounded systems, depending on the fault resistance.

unfaulted phases *b* and *c* increases proportional to the voltage, that is, $\sqrt{3} \times 0.21 = 0.365$ A. Moreover, this current in phase *b* and *c* reverses, flows through the transformer windings, and sums up in the transformer winding of phase *a*. Figure 7.4e shows that this vector sum of the capacitance currents in phases *b* and *c* is equal to 0.63 A. This is conceptually important. *Note that no capacitance current returns to the faulted point through the grounding resistor, also see Chapter 8.*

Now consider that the ground current through the grounding resistor is limited to 1 A only. This is acceptable according to Equation 7.6 as the total stray capacitance current is 0.63 A. This resistor ground current also flows through transformer phase winding *a* to the fault (Figure 7.4e) and the total ground fault current is $I_g = \sqrt{1^2 + 0.63^2} = 1.182$ A.

The above analysis assumes a full neutral shift, ignores the fault impedance, and assumes that the ground grid resistance and the system zero-sequence impedances are zero. Practically, the neutral shift will vary, see Figure 7.5.

7.5 Ungrounded Systems

In an ungrounded system, there is no intentional connection to ground except through potential transformers or metering devices of high impedance. In reality, an ungrounded system is coupled to ground through distributed phase capacitances. It is difficult to assign X_0/X_1 and R_0/X_0 values for ungrounded systems. The ratio X_0/X_1 is negative and may vary from low to high values and COG may approach 120%. These systems provide no effective control of transient and steady-state voltages above ground. A possibility of resonance with high voltage generation, approaching five times or more of the system voltage, exists for values of X_0/X_1 between 0 and −40. For the first phase-to-ground fault, the continuity of operations can be sustained, though unfaulted phases have $\sqrt{3}$ times the normal line-to-ground voltage. All unremoved faults, thus, put greater than normal voltage on system insulation, and an increased level of conductor and motor insulation may be required. The grounding practices in the industry are withdrawing from this method of grounding.

If an inductor of certain size gets connected to ground, then the possibility of high voltages exist, as the overvoltage is given by:

$$V_{ov} = \frac{X_L}{(X_{co}/3) - X_L} V_t \qquad (7.9)$$

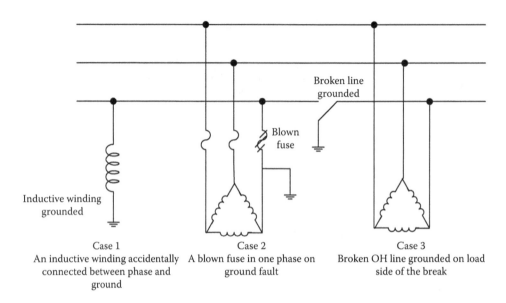

FIGURE 7.6
Connections of an inductance to ground for faults in ungrounded systems.

where V_t is the applied terminal voltage and X_L is the inductance of the grounded inductor. Figure 7.6 shows three possible cases of a high inductance that may get connected to a phase, and give rise to resonant voltages.

1. The coil of a motor starter may be inadvertently connected between phase and ground due to a ground fault.
2. A fuse can operate in one phase due to a ground fault, which connects reactances of the other two phases in parallel between phases and ground.
3. A broken conductor on the load side of the transformer connects reactances of two phases in parallel between phase and ground.

The phenomena of arcing grounds and resulting overvoltages which may escalate to five to six times the normal rated voltage can be explained with reference to Figure 7.7. It is somewhat similar to restrikes in circuit breakers discussed in Volume 1. Intermittent ground faults can give rise to these phenomena.

Figure 7.7a shows normal voltage vectors rotating counterclockwise. Consider that phase *a* is grounded, see Figure 7.7b. The current in a capacitor is zero when the voltage is at its peak and therefore at the instant shown in Figure 7.7b, the capacitance is charged to the line voltage and the current is zero; thus, the arc tends to extinguish. During the next half cycle, as the voltage vectors rotate, the phase *a* charges from zero (at the neutral point) to twice the line voltage, which is denoted by dotted lines. This value of line-to-ground potential of phase *a* may be sufficient to break down the gap in the ground fault circuit, which got extinguished a half cycle before. Thus, a pulse current flows and the phase *a* voltage may swing between plus and minus two times the rated voltage at a frequency of 20–100 times the fundamental, due to the presence of reactance in the circuit. If there were a solid metallic connection between the phase *a* and the ground, it would leave the phase

System Grounding

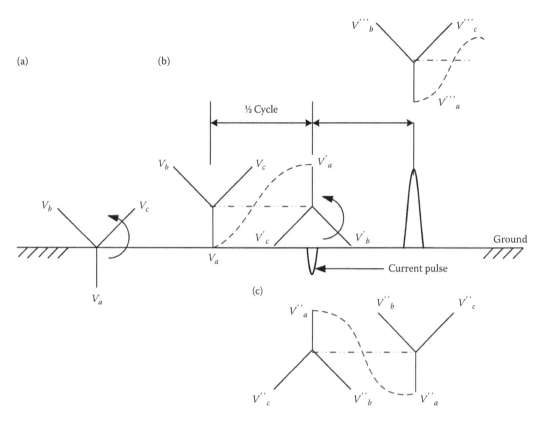

FIGURE 7.7
(a–c) Illustration of arcing faults in ungrounded systems and consequent escalation of voltages.

a conductor at ground potential. Associated with the transitory oscillation of the voltage, there will be a corresponding oscillatory charge current. This transient charging current or restrike current will again reach zero when the system voltage angle is at its maximum excursion in the negative direction, see the lower part of Figure 7.7c. In the next half cycle as the voltage vectors rotate further, the phase *a* voltage will escalate from –2 to –4, as indicated in the lower part of Figure 7.7c. This increased voltage across the gap, which may again result in a restrike.

7.6 Reactance Grounding

In reactance grounding, a reactor is connected between the system neutral and the ground; the magnitude of the ground fault current that will flow depends upon the size of the reactor. The ground fault current should be at least 25% and preferably 60% of the three-phase fault current to prevent serious transient overvoltages ($X_0 < 10X_1$). This current is considerably higher than that in a resistance grounded system and the reactance grounding is not an alternative to resistance grounding. The system is generally used for grounding of small generators, so that the generator ground fault current does not exceed three-phase fault current and three-phase four-wire loads could be served. Reactance grounded systems are not common in the United States.

Up to around the 1940s, the utility left their generators ungrounded to limit the internal ground fault currents in the generator to very low values. However, when these generators were connected to transmission lines many insulation failures occurred, due to high-voltage transients as discussed in ungrounded systems.

The next step was to ground the generator neutral through primary of a potential transformer, the secondary connected to a voltage relay—the actuation of voltage relay will trip the generator. Practically, it increased the generator failures due to arcing grounds.

Examine circuit of a generator shown in Figure 7.1; the generator grounded through a high reactance X_n, with a line-to-ground fault near one terminal. The small arcing current through X_n tries to extinguish and reignite with increasingly high voltages. The high-reactance grounded systems are no longer in use.

7.7 Resonant Grounding

Figure 7.1 also shows a resonant grounding system, with ground fault neutralizer. A reactor can be connected between the neutral of a system and the ground and tuned to the system charging current so that the resulting current is resistive and of low magnitude. This current is in phase with the line-to-neutral voltage, so that the current and voltage zeros occur simultaneously. The system is used for voltages above 15 kV, consisting of overhead transmission or distribution lines. The system is rarely used for industrial and commercial establishments. A disadvantage of the system is that the resonant tuning can change due to switching conditions, that is, when a part of the system may be out of service or when the system expansion takes place. This grounding method is not common in the United States, though sometimes used in Europe and Russia.

7.8 Corner of Delta Grounded Systems

Low-voltage systems which in the past were supplied from delta-connected secondary were inherently ungrounded. These were grounded by connecting one corner of the secondary delta windings to ground, through an adequate connection and there was no additional impedance inserted in this connection to obtain a grounded system (see Figure 7.8a). These are no longer in use. It is necessary to positively identify the grounded phase throughout the system to avoid connecting meters, instruments, and fuses in the grounded phase. A higher line-to-neutral voltage will occur on two phases than in a neutral grounded system. There is a possibility of exceeding the interrupting ratings of marginally applied circuit breakers, because for a ground fault, the interrupting duty on the affected circuit breaker pole may exceed the three-phase fault duty. There is a high probability of sustaining arcing for 480 V or higher phase-to-phase, single-phase circuit extension without escalating to a three-phase fault.

System Grounding

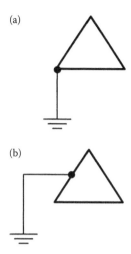

FIGURE 7.8
(a) Corner delta grounded system and (b) midpoint grounded delta grounded system.

The NEC Table 430–37 requires that for a three-phase system three overloads are used, one in each phase for motor protection. The control circuit for motor starting has to be carefully designed, so that a ground fault in it does not automatically start or stop a motor.

Some advantages of the system are low cost of establishing a grounded system, the effective control of transient overvoltages, though a maximum of 1.73 times the normal phase-to-ground voltage can exist between two conductors and ground, and a fault from phase-to-ground can be easily detected.

One phase of a delta system can be grounded at the midpoint, see Figure 7.8b. The system is not recommended for voltages above 240 V. Serious arc flash hazard from a phase-to-ground fault can exist because of high fault levels [2]. Grounding of one phase of a delta system at midpoint of that phase for three-phase systems with phase-to-phase voltages over 240 V has little application. The shock hazard of the high phase leg to ground is 1.73 times the voltage from the other two phases. These systems are no longer in use.

Note that mostly molded case circuit breakers (MCCBs) are suitable for high-resistance grounded systems, but all MCCBs may not be suitable for corner-grounded systems.

7.9 Artificially Derived Neutrals

Many times, it is required to ground delta-connected transformer windings and other ungrounded separately derived systems. A neutral can be artificially derived in a delta-connected system with zigzag transformer or a wye–delta connected grounding transformer. Figure 7.9a shows a zigzag transformer. Windings a_1 and a_2 are on the same limb and have the same number of turns, but are wound in opposite directions. The zero-sequence currents in the two windings on the same limb have, therefore, canceling ampere-turn effect. The impedance to the zero-sequence currents is due to leakage flux of the windings. For the positive- and negative-sequence currents, neglecting magnetizing currents, the connection has infinite impedance. Figure 7.9b shows the distribution of zero-sequence currents on a ground fault ahead of a zigzag transformer. A wye–delta grounding transformer

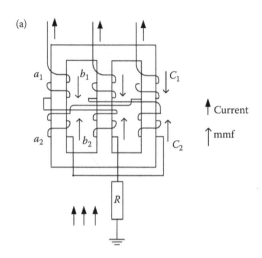

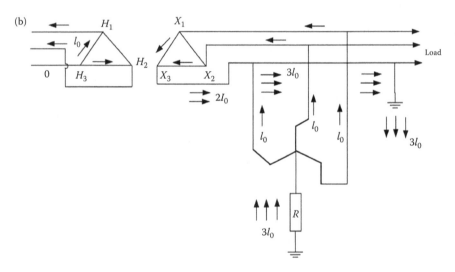

FIGURE 7.9
(a) Flow of ground fault current in the windings of a zigzag transformer. (b) Application of a zigzag transformer to derive an artificial neutral and current flows for a downstream fault.

circuit can be similarly drawn. The neutrals of large utility generators directly connected to the step-up transformers generator step up (GSUs) are not directly connected to the ground through a high resistance but through a distribution transformer which is loaded on the secondary windings with a grounding resistor. An example will clarify the calculations.

Example 7.3

See Example 5.3 in Volume 1. This conveys important concepts for sizing of the grounding resistor in HRG or low-resistance grounded (LRG) systems.

This example showed that all the sequence impedances, including that of the grounding transformer, can be ignored without an appreciable error in sizing the grounding resistor. The reason is that all system sequence impedances are much lower than the grounding resistor zero-sequence impedance.

7.10 Multiple Grounded Systems

Figure 7.10 shows typical grounding practice for wye service entrance served by a wye multiple grounded medium-voltage system in North America. Note the multiple grounds of the neutral conductor (PEN—protected neutral). The practice of grounding of commercial and residential facilities in the United States requires that the neutral conductor is bonded to the ground conductor at the service entrance, and both are bonded to the building ground. There cannot be N-G surge at the service entrance. However, L-N surges within the building can produce N-G surges at the end of a branch circuit.

Further implications of multiple grounded distribution systems are shown in Figure 7.11. The National Electric Safety Code (NESC) [14] requires that the neutral on multiple grounded-wye distribution systems have a minimum of four earth connections per mile.

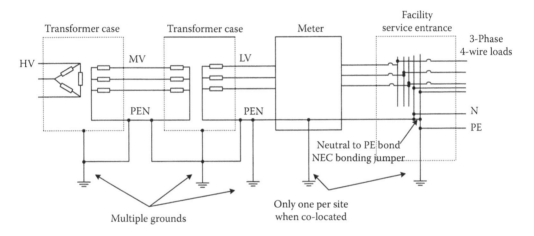

FIGURE 7.10
Typical grounding practice for a wye-service entrance, served by a multiple grounded system in North America.

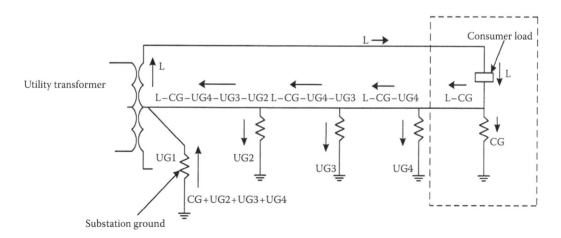

FIGURE 7.11
Distribution of load currents in phase and neutral/ground conductor in multiple grounded system; the ground/neutral conductor develops differential voltages.

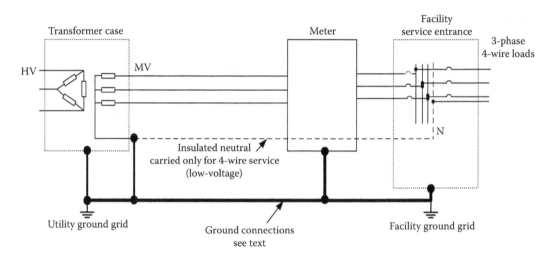

FIGURE 7.12
Typical grounding practice in industrial distribution systems; the transformer neutral is grounded only at the source in North American systems.

This also applies to direct buried underground cables. The voltage between neutral and earth can originate from variety of sources. A 60 Hz voltage can exist between objects connected to neutral and earth. A short-duration transient can exist, when the lightning current is dissipated into the earth. A differential voltage between the neutral and the ground is more likely to occur when the same service transformer feeds two or more consumers.

Figure 7.12 illustrates the grounding practice for industrial establishments. Here, the neutral is grounded only at one point, which is at the source. This figure shows that the neutral from the utility transformer is not required to be run for industrial plant medium-voltage three-phase loads. In case the industrial plant needs some loads like lighting and controls to be served from low-voltage grounded systems, these lower voltages are served from a separate transformer with artificially derived neutral. In case the service is at low voltage, a neutral may be run to supply phase-to-neutral loads, *but it is* not grounded anywhere in the plant except at the service transformer. There is no bonding of neutral conductor with the ground at the service entrance, a practice which is invariably followed for industrial medium- or high-voltage grounded systems or separately derived industrial systems.

7.10.1 Equivalent Circuit of Multiple Grounded Systems

Figure 7.11 of a multiple grounded system shows that the grounds at various points cannot be at the same potential. This figure shows the current flow in the multiple grounded neutral under normal operation. The load current flows through line-to-neutral, but as the neutral is grounded at the consumer premises (ground CG) and also at multiple points the neutral current returns to the utility transformer through multiple paths, and the sharing of current depends upon the relative impedances on the grounding circuit. An equivalent impedance diagram is shown in Figure 7.13. The system may be analyzed using symmetrical components or equivalent circuit concepts. The line may pass through a region of high soil resistivity; each grounding point can be modeled individually along with the section of the feeder separating it from the adjacent grounds.

System Grounding

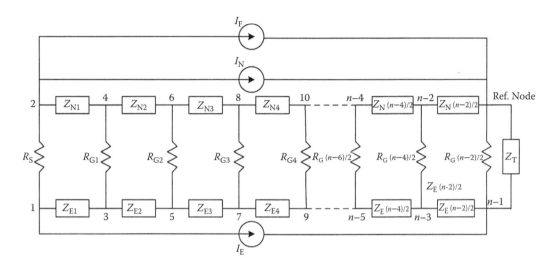

FIGURE 7.13
An equivalent circuit diagram of a multiple grounded system for a line-to-ground fault at remote node.

7.11 NEC and NESC Requirements

The NEC [5] article 250.106 requires that lightning protection system ground terminals shall be bonded to the building or structure electrode system (so that the potential differential between these is minimized). Further NEC articles mandate that electrical equipment is bonded and connected in a manner to establish a path of sufficiently low impedance. Where the ground resistance of a single ground electrode exceeds 25 Ω, additional ground rods are mandated to lower the resistance to 25 Ω. This value of 25 Ω in the NEC is too high; an ideal ground should provide a near zero resistance between bonded components and the ground electrode of the facility to limit ground potential rise on a surge current. With the grounding resistance of 25 Ω allowed by the NEC, high surge currents will produce high ground potential rise, which can damage the surge protection devices.

Consider the dissipation of a lightning surge near the distribution system. If the lightning current is not effectively dissipated through the arrester, the result can be flashover of the insulation and impinging the surge on the consumer apparatus. A surge voltage will appear on the neutral conductor and the consumer premises as the neutral is bonded to the consumer ground. This surge voltage will depend upon a number of factors—the downward lead length, the resistance of the grounding electrodes, and the surge impedances of the various paths. Generally, it may not be detrimental to the premises, especially when the recommendations of surge protection of low-voltage systems and the category of installations are followed; however, a possibility of flashover cannot be ruled out. The grounding of underground cable distribution is more important, in the sense that lightning surges and wave fronts may double on cables. The voltage is reduced by close connection of the surge arrester to the cable terminations.

It may be necessary to state here the grounding and bonding requirements laid down in the NESC and the NEC.

- The NESC requires grounded items on joint poles (e.g., for power and communication) to be bonded together either using single grounding conductor or bonding the supply grounding conductor to the communication grounding conductor, except where a certain separation is maintained (Rule 97A), in which case there should be insulation between the grounding conductors. A hazardous potential difference can exist between the two conductors. Rule 215C3 requires bonding between messengers at typically four times per mile.
- It is required that a common ground electrode system should be created by the two utilities for the communication and power supply systems. If separate electrode systems are used, these should be bonded together with a minimum #6 AWG conductor. This is to ensure that dangerous potentials do not exist between the two grounding systems. A user of computer modem, fax machine, answering machines, and other communication equipment could be exposed to an electrical shock apart from damage to the equipment.
- The NEC (250.24(A)(5)) prohibits bonding of the equipment grounding conductor and neutral inside the premises. Note that the neutral will carry a current and a difference of potential exists between the neutral and the ground conductor. Again, bonding is required to metal water pipes, which limits the possibility of a potential difference between the water system and other noncurrent carrying parts within the building. (A coupling can occur through soil resistivity.) It is prohibited using *interior metal water piping system* located more than 5 in. from the entrance to the building from being used to interconnect grounding electrodes within the building.

7.12 Hybrid Grounding System for Industrial Bus-Connected Generators

Louie Powell [15] investigated a number of industrial bus-connected generator failures. He considered the basic configuration as shown in Figure 7.14. The generator operates in synchronism with the utility source through a transformer. Consider a fault in the stator windings. The energy released into the fault can be written as

$$E = \int i^k dt \tag{7.10}$$

The value of k varies from 1 to 2, and for a purely resistive circuit, this value is 2.

Note that the source and the generator are both grounded through 400 A resistors which have been a common industrial practice.

Considering fast relaying, the source current of 400 A is interrupted in six cycles, which is the sum of relay operating time of one cycle (assumption) and interrupting time of breaker which is equal to five cycles.

The curves in Figure 7.15 show the damage indices. Figure 7.15a shows the energy with respect to the magnitude of the fault current and Figure 7.15b shows the energy accumulation. These figures show the energy associated with the source current.

In six cycles, the generator breaker is tripped; however, the energy to the fault is not interrupted. It continues to be fed for a relatively higher time as the generator current decays slowly in about 0.8–1 s, depending on the generator single-line-to-ground fault constant.

System Grounding

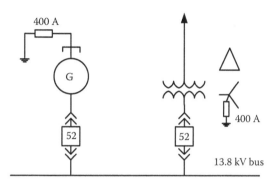

FIGURE 7.14
A typical grounding system, the transformer and generator are each grounded through a 400 A resistor.

Figure 7.15c shows this energy accumulation. Therefore, in 1 s, the energy into the fault is approximately six times the energy released by the source fault current.

In this scenario, no neutral breaker is considered. A neutral breaker is often provided and tripped simultaneously with the line breaker for faults such as differential phase and ground and also ground faults. If the neutral breaker is present, it will limit the energy into the fault to a much lower value. This has been studied by other authors.

The IEEE/IAS working group published four papers in *IEEE Transactions on Industry Applications*, which are referenced in [16].

The concept of hybrid grounding is illustrated in Figure 7.16. For industrial bus-connected generators, selective ground fault protection is required, and the generators cannot be simply grounded through a 10 or 8 A resistor. The generators are generally the last to trip on an external ground fault and the other system protection must operate faster to clear the fault. This cannot be achieved if the industrial generators are high-resistance grounded like the utility unit-connected generators.

In Figure 7.16, the low-resistance section of 400 A is connected through a neutral breaker. The neutral breaker is tripped simultaneously with the main line breaker. After the neutral breaker is tripped, the high-resistance ground section through 8 A remains connected and limits the fault energy. The low-resistance grounded section provides ground fault coordination with the rest of the system. See Chapter 8 for an example of ground fault relaying.

7.13 Grounding of ASDs

A brief reference can be made to the grounding of variable speed drives (VSD) systems, where special considerations apply. Consider a three-phase six-pulse bridge rectifier circuit in Figure 7.17. Only two phases conduct at a time, and the dc plus and negative voltages to midpoint are shown. These voltages do not add to zero and the midpoint oscillates at thrice the ac supply system frequency. The dc positive and negative buses have common-mode voltages and its magnitude changes with the firing angle. The peak of the voltage is approximately 0.5 V_{ln}, where V_{ln} is the peak line-to-neutral point input voltage.

Consider common-mode voltages on the motor isolated neutral to ground of a 4160 V drive system gate turn-off (GTO) inverter. The peak line-to-ground V_{l-g} voltage is 4100 V and the

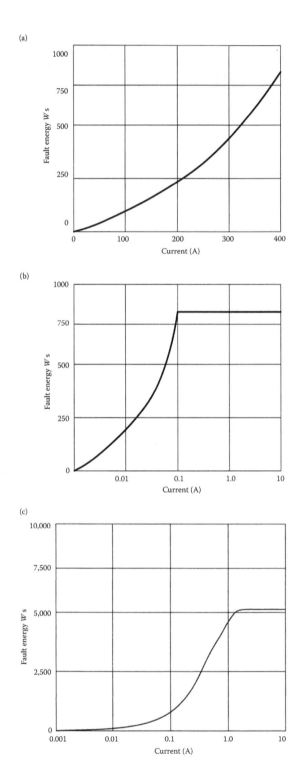

FIGURE 7.15
(a and b) Fault energy for the system source fault and (c) fault energy for the generator fault through the neutral connection.

System Grounding

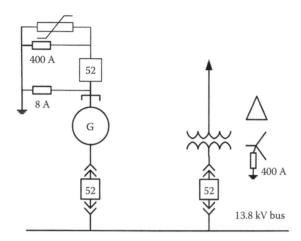

FIGURE 7.16
Hybrid grounding system for the generator, see the text.

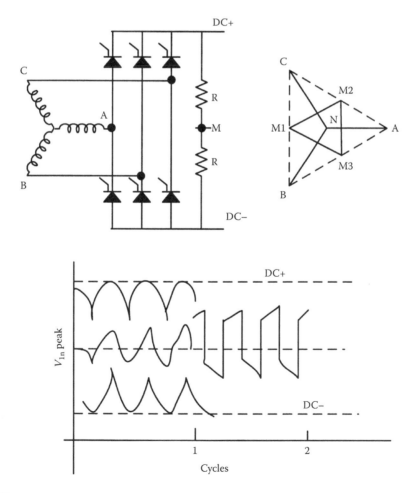

FIGURE 7.17
Generation of common-mode voltages in a six-pulse converter.

waveform has a frequency of 60 Hz. The peak neutral-to-ground voltage is 2500 V and has a frequency of 180 Hz. The operation of the output bridge creates a common-mode voltage by exactly the same mechanism as the input bridge does, where the back EMF of the motor is analogous to the line voltage.

Thus, the worst case condition for the common-mode voltage is no-load, full-speed operation, as the phase-back angle is 90° for both the converters, and the motor voltage is essentially equal to the line voltage. The sum of both common-mode voltages is approximately V_{ln} at six times the input frequency. Since the input and output frequencies are generally different, the motor experiences a waveform with beat frequencies of both input and output frequencies and there will be instances when twice the rated voltage is experienced, see Reference [17].

The grounding must assure that the motor insulation system is not stressed beyond its design level. Figure 7.18 shows three possible methods. Figure 7.18a shows an output drive isolation transformer with its secondary grounded. The transformer primary winding insulation can tolerate the common-mode voltage swings better than the motor insulation. This option is tricky as the drive isolation transformer is subjected to residual dc offset in the drive and high harmonic content passing through the transformer windings. A better location for the drive isolation transformer will be at the input line end (Figure 7.18b), and the secondary winding is left ungrounded. The neutral of the filter capacitors on the output of the inverter provides a convenient place to ground the load side of the inverter and a small resistor rated 1 A or so is enough to do it. The line-to-neutral voltages must, then, be taken into account for the insulation of the secondary of the transformer. Also the cables forming the transformer secondary to the input rectifier bridge must be rated for higher voltage to ground, for example, for a 4.16 kV drive system, cables with 173% insulation level will be required; alternatively, 8 kV cables can be used.

An input or output drive transformer increases the cost and reduces the efficiency of the drive system. In a transformerless GTO drive system, a line reactor may be used on the input side to affect commutation notches, but the motor insulation must withstand twice the normal voltage stress to ground, see Figure 7.18c.

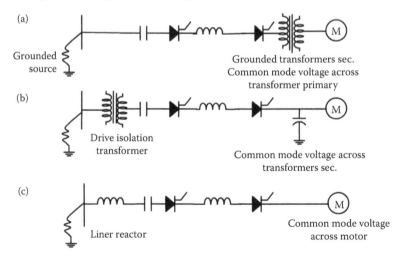

FIGURE 7.18
(a) Isolation transformer with grounded secondary to withstand common-mode voltage. (b) Drive transformer at input, common-mode voltage across secondary of the transformer. (c) Input line reactor, common-mode voltage across drive motor windings.

7.14 Grounding in Mine Installations

Figure 7.19 is extracted from IEEE Standard 141, and illustrated the general features of grounding in mine installations. Adequate grounding has been a difficult problem [2] in the mining industries. Hazards associated with ground faults are amplified by the portable and mobile nature of the equipment, and system and equipment grounding are interrelated.

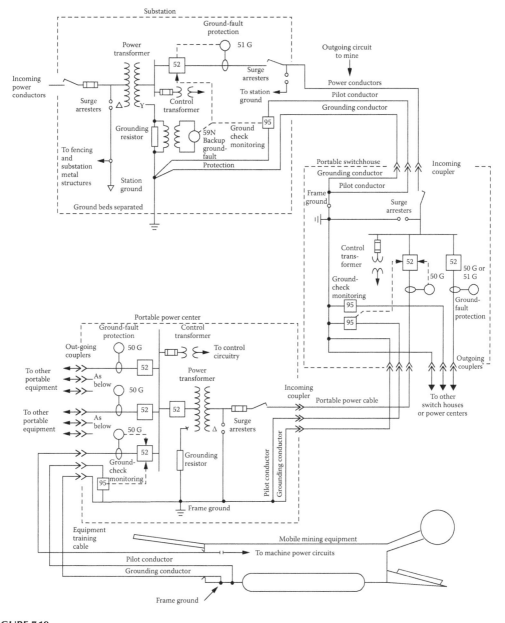

FIGURE 7.19
A simplified mine power distribution system with a safety grounded system, see the text. (IEEE Standard 142, IEEE Recommended Practice for Grounding of Industrial and Commercial Power Systems, 1991.)

A surface mining can have a substantial power demand of 18,000 hp or more, at potentials up to 25 kV or greater. The power demand of an underground mining machine can exceed 1100 hp at potentials up to 4160 V. In Figure 7.19, substations transform the incoming utility power to distribution level. In surface mines, large utilization equipment, such as continuous miners, longwalls, load hauling-dump units, etc., are powered at distribution voltage, and a trailing cable completes the power circuit from switch house to the machine. Mine distribution is always radial and OH lies or cables are used to supply power houses (portable switchgear).

The recommended grounding techniques for these portable or mobile equipment are a safety ground system that employs resistance grounding. The substation contains two separate beds maintained some distance apart. Substation surge arresters, fencing, and equipment frames are tied to the system ground bed under the substation area. The substation transformer is either delta–wye, wye–delta, or delta–delta connected, wye–wye connection is not recommended. The secondary neutral direct or derived is connected to safety ground bed through neutral grounding resistor.

Separation between safety and system ground is needed to isolate high ground voltage rise (a temporary rise of 5 kV or more is not unusual). A resistance of 5 Ω or less is recommended. A much larger separation is required to provide the required isolation, the design of these ground beds is complex, and many variables must be examined to obtain an acceptable configuration.

Note that the two ground systems will be coupled together through the soil, depending upon soil resistivity and the separation between the beds. IEEE Standard 367 [18] contains important information about ground bed separation.

At each transformation step, within the distribution system, such as in a portable power center, an additional neutral is established at the transformer secondary, and tied through a grounding resistor to the equipment frame and, thus, via the grounding conductors to the safety ground bed.

Because of considerable cable lengths, the stray capacitance current to ground is higher than in HR grounded systems. Most substations serving mines use a 25 A ground current limit, but typical practice is 15 A.

Correct selection and coordination of the protective circuitry are essential to the safety ground system. In Figure 7.19 all phase overcurrent devices are omitted for clarity. In order that the safety ground system is effective, grounding conductors must be continuous and ground-check monitors (relays) are used to verify continuity. Pilot conductors are shown with monitoring, but these are not needed when pilotless relays are employed. All the sensors operate to trip the associated circuit breaker and remove power on the effected segment.

References

1. ANSI/IEEE Std. 80. IEEE Guide for Safety in Substation Grounding, 2000.
2. IEEE Standard 142. IEEE Recommended Practice for Grounding of Industrial and Commercial Power Systems, 1991.
3. HI Stanback. Predicting damage from 277 volt single phase to ground arcing faults. *IEEE Transactions on Industry Applications* vol. IA-13, no. 4, pp. 307–314, July/August, 1977.
4. NEMA PB1-2. Application Guide for Ground Fault Protective Devices for Equipment, 1977.
5. NEC National Electrical Code, NFPA 70, 2017.
6. JR Dunki-Jacobs. The escalating arcing ground fault phenomenon. *IEEE Transaction on Industry Applications* vol. 22, no. 6, pp. 1156–1161, November 1986.

7. DG Lucks. Calculating incident energy released with varying ground fault magnitudes on solidly grounded systems. *IEEE Transactions on Industry Applications* vol. 46, no. 2, pp. 761–769, March/April 2010.
8. HB Land, III. The behavior of arcing faults in low-voltage switchboards. *IEEE Transactions on Industry Applications* vol. 44, no. 2, pp. 437–444, March/April, 2008.
9. IEEE Std. 493 (Gold Book). Recommended Practice for Design of Reliable Industrial and Commercial Power System, 2007.
10. Westinghouse. *Transmission and Distribution Handbook*. Westinghouse, East Pittsburg, PA, 1964.
11. DS Baker. Charging current data for guess-free design of high-resistance grounded systems. *IEEE Transactions on Industry Applications* vol. 15, no. 2, pp. 136–140, March/April 1979.
12. *GE Industrial Power Systems Data Book*, GE, Schenectady, New York.
13. JR Dunki-Jacobs. The reality of high-resistance grounding. *IEEE Transactions on Industry Applications* vol. 13, pp. 469–475, September/October 1977.
14. National Electrical Safety Code, C-2.
15. LJ Powell. The impact of system grounding practices on generator fault damage. *IEEE Transactions on Industry Applications* vol. 34, no. 5, pp. 923–927, September/October 1998.
16. IEEE/IAS Working Group Report. Grounding and ground fault protection of multiple generator installations on medium voltage industrial and commercial power systems. *Parts I, II, III and IV, IEEE Transactions on Industry Applications* vol. 40, no. 1, pp. 11–28, January/February 2004.
17. JC Das, H Osman. Grounding of AC and DC low-voltage and medium-voltage drive systems. *IEEE Transactions on Industry Applications* vol. 34, no. 1. pp. 295–216, January/February 1998.
18. ANSI/IEEE Std. 367-1987. IEEE Recommended Practice for Determining the Electrical Power Station Ground Potential Rise and Induced Voltages form a Power Fault.

8
Ground Fault Protection

8.1 Protection and Coordination in Solidly Grounded Systems

8.1.1 NEC Requirements

NEC Article 230.95 [1] requires that for solidly grounded wye electrical services 150–600 V, the service disconnect rated 1000 A or more, ground fault protection should be provided. The maximum permissible settings are 1200 A pickup and 3000 A ground fault current, and the maximum trip time is 1 s. For separately derived systems through a transformer in industrial establishment, these provisions can be applied. These NEC recommendations are based on a number of incidents of ground fault damage, where the ground fault protection was not provided.

Figure 8.1 shows exhibits from NEC. Note that two methods of ground fault protection are shown: (1) through a ground-fault sensor (for example, a core balance CT) encircling all conductors including neutral and (2) through a ground sensor encircling only bonding jumper connection. The neutral is again connected to the ground at the service entrance. For a single-line-to-ground fault downstream, part of the ground fault current returns to transformer grounded neutral through ground. This is true of multiple grounded systems for commercial installations; see multiple grounded systems in Chapter 7. In industrial establishments, we ground the transformer neutral only at one point.

In solidly grounded systems, as the ground fault currents can be even greater than three-phase fault currents, a selective ground fault protection cannot be achieved unless dedicated ground fault protection is included on all the circuits and subcircuits served from a low-voltage transformer substation.

Delta-connected windings of three-phase transformers act as a sink to the zero-sequence currents, and these currents circulate in the delta winding, but do not appear in the line conductors feeding the transformer delta-connected windings.

- Though NEC requirements apply to service disconnects of 1000 A or more, much damage can take place even for services of lower ratings. The omitting of dedicated ground fault protection means that the ground faults will be cleared as phase faults.
- NEC is nonspecific about coordination with respect to ground faults downstream of the distributions. More sensitive ground fault protection, compared to phase fault protection is required. Thus, with no ground fault protection downstream, the phase overcurrent relays relied upon tripping for ground fault can result in serious lack of protection and process shutdowns.

The protection philosophy of solidly grounded systems is illustrated in the following examples.

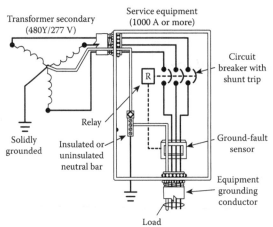

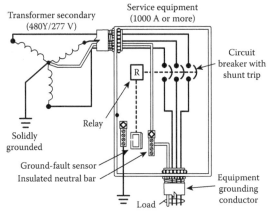

FIGURE 8.1
Exhibits from NEC, grounding, and ground fault protection of services.

Example 8.1

A simple low-voltage system derived from a 2500 kVA, 13.8–0.48 kV, primary windings delta connected and the secondary windings wye connected, solidly grounded, is shown in Figure 8.2. The secondary of the transformer is connected to a low-voltage switchgear assembly, which in turn serves a number of low-voltage motor control centers, only one shown. The main 4000 A low-voltage power circuit breaker (LVPCB) is provided with electronic trip programmer, having LSIG functions (long-time pickup, long-time delay, short-time pickup, short-time delay, and instantaneous pickup settings for phase faults; also adjustable pickup and short-time delay settings for ground faults. Both phase and ground fault short-time functions are provided with I^2T setup that can be switched in or switched out. Similarly, the 800 A feeder LVPCBs to MCCs are provided with electronic trip programmers with LSIG functions. A three-phase, three-wire distribution is shown. The transformer wye-connected winding neutral is grounded at one point only, close to the transformer. This means that any ground fault current must return through the single neutral connection. For solidly grounded neutrals, it is necessary to provide a neutral-connected ground relay, see Chapter 12 for further discussion.

Ground Fault Protection

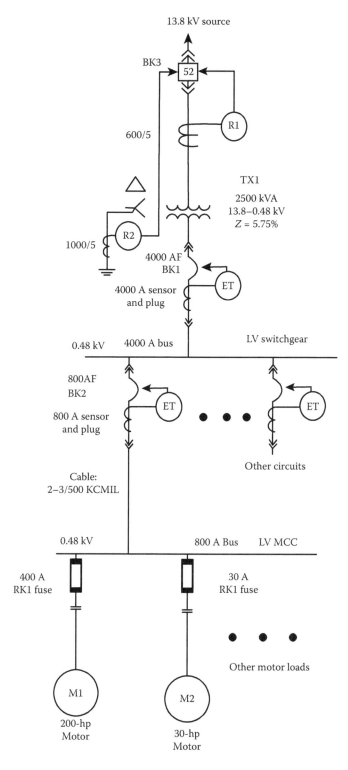

FIGURE 8.2
A solidly grounded low-voltage system for ground fault coordination.

Ground fault settings and coordination using *minimum* pickup settings on the electronic trip programmers of the low-voltage circuit breakers are shown in Figure 8.3. The low-voltage circuit breaker trip programmers have a minimum pickup ground fault setting of 0.25 times the sensor rating, and in this case, these settings are 200 A for the 800 A feeder breakers (BK2) and 1000 A for the main 4000 A breaker (BK1). Also the neutral-connected relay, R2, is set up to coordinate with these settings, through a definite time characteristic, with a pickup setting of 1200 A and a time delay setting of 0.6 s. The settings on R2 meet NEC requirements. Relay R2 must trip an upstream breaker as a ground fault may be in the transformer windings, or the cable connection between the transformer and main secondary breaker at the low-voltage switchgear. This fault cannot be cleared by tripping the secondary breaker.

Though the ground fault functions on trip programmers are commonly used in the industry, their settings have limitations (Figure 8.3). The 4000 A main circuit breaker pickup could easily be set at 400 A to coordinate with the feeder circuit breaker, but it is not possible to do so as the minimum pickup is prefixed at 0.25 of the sensor rating, which is very typical of the trip programmers in the industry.

The damage curves of 4000, 800, and 400 A circuits are superimposed in Figure 8.3, based on Equations 7.1 through 7.5. These are straight lines drawn for:

$$4000\text{ A}: \quad 1 \times 10^6 \, A^{1.5} \, s$$

$$800\text{ A}: \quad 0.2 \times 10^6 \, A^{1.5} \, s$$

$$400\text{ A}: \quad 0.1 \times 10^6 \, A^{1.5} \, s$$

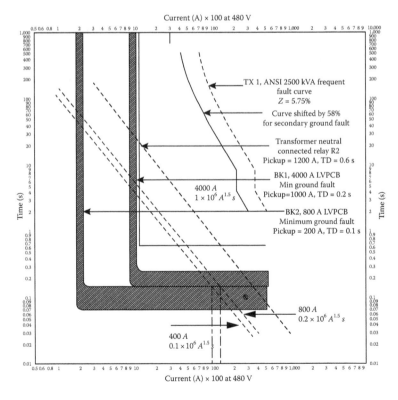

FIGURE 8.3
Three-step ground fault coordination in the radial system of Figure 8.2, with damage curves of 4000, 800, and 400 A devices.

Ground Fault Protection

With respect to the ground fault settings, these damage lines show to what extent the circuits of these ampacities are protected. Figure 8.3 shows that it is a three-step coordinated ground fault protection. For a ground fault on the load side of 800 A feeder, anywhere in the MCC feeders, cable to MCC, or the MCC bus, the ground fault feeder protection pickup of 200 A on 800 A feeder circuit breaker is operative. This shows that 800 A circuits are protected for ground faults from 200 to 12 kA (time delay approximately 0.15 s) while 400 A circuits from 200 to 10 kA to limit the damage to permissible limits, according to Equation 7.5, that is 0.144 in.³ of copper for 800 A circuit, and 0.072 in.³ of copper for 400 A circuit. Faster fault clearance times will be required to protect these for higher levels of ground fault current.

The phase fault coordination is shown in Figure 8.4, and the phase fault coordination superimposed on the ground fault coordination is shown in Figure 8.5. This shows that as the fault current increases, the phase fault protection takes over the ground fault

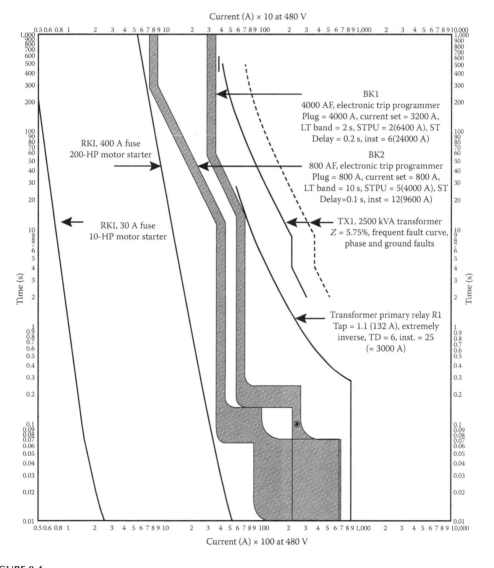

FIGURE 8.4
Phase fault coordination of Figure 8.2.

protection. Therefore, a ground fault will be cleared by the phase overcurrent protection. Referring to Figure 8.5 this happens when:

- Feeder circuit breaker instantaneous takes over at fault current >9 kA.
- For feeder faults >4 kA and <9 kA, either the ground fault or the phase fault ST pickup will operate.
- For a bus fault on the low-voltage switchgear the main circuit breaker overcurrent operates at >22 kA.

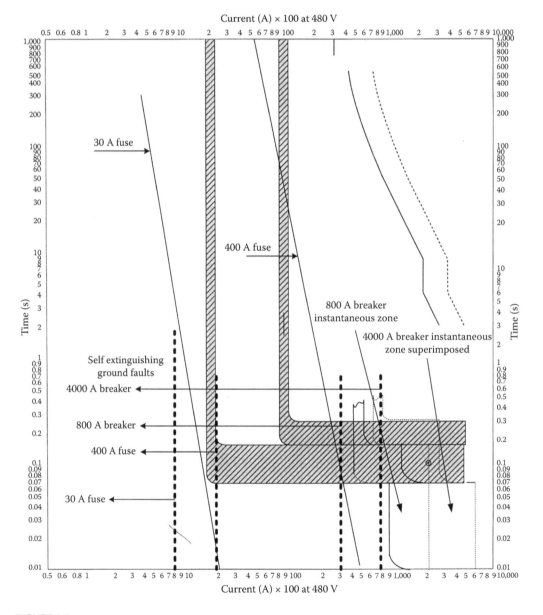

FIGURE 8.5
Ground fault coordination and protection with superimposed phase fault protection and self-clearing limits of ground faults in Figure 8.2.

8.1.2 Self-Extinguishing Ground Faults

Conservatively, faults below 38% of the bolted phase-to-ground fault will be self-extinguishing for a 480 V three-phase system, with 277 V to ground [2]. *This means that ground fault protection is not necessary where bolted fault current at every point on a branch or feeder circuit is at least 263% of the trip setting protecting the circuit;* see Figure 8.5.

This does not imply that no arc flash damage will occur, though it will be minimal. Figure 8.5 shows that a 200 hp motor starter protected with a 400 A fuse has a self-clearing zone below 2.1 kA.

As a safe stance, it is desirable that selective and sensitive ground fault protection is provided for the ground fault currents, from a low level to the highest calculated level. If a fuse in one phase operates due to a single-line-to-ground fault, this results in single phasing and a fully loaded induction motor may stall. Both stator windings and rotor will be seriously overloaded due to negative-sequence currents, which produce damaging overloads. The negative-sequence impedance of an induction motor with a locked rotor current of six times the full-load current is approximately 1/6 of the positive-sequence impedance and, therefore, even a 5% negative-sequence component can produce 30% negative-sequence current in the motor. Thermal relays generally used for low-voltage motor protection will be insensitive to operate fast and the motor may be damaged. Also an uncleared ground fault, even of low value, can be damaging to the motor windings and core damage can occur.

Though the advantage of current-limiting fuses is being extolled, this fact of single phasing should be borne in mind for grounded systems. The single phasing can be detected with negative-sequence current relays.

8.1.3 Improving Coordination in Solidly Grounded Low-Voltage Systems

Example 8.1 illustrates the following:

1. A 200 A ground fault in the motor windings will not be cleared by its 400 A fuse in the 200 hp motor starter, and it can be very damaging to the motor insulation.
2. For a ground fault in a feeder or motor circuit in the MCC, the main feeder breaker BK2 at low-voltage switchgear trips, resulting in complete shutdown of the MCC.

These limitations can be circumvented as illustrated in Example 8.2.

Example 8.2

This example explores how the sensitivity of the ground fault settings given in Example 8.1 can be improved. The ground fault protection is carried a step further and applied on every circuit from the source to the load. Each feeder or motor starter in the low-voltage MCC is provided with a core-balance CT and a dedicated ground fault relay. Now, each feeder circuit breaker must be electrically tripped through a shunt trip coil. If we provide electrically operated circuit breakers, with shunt trip coils, and dedicated ground fault relays, as shown in Figure 8.6, the coordination can be arranged as shown in Figure 8.7. The following ground fault settings are provided:

- Each feeder on low-voltage MCC, controlled through an MCCB: 20 A, time delay 0.03 s
- Each feeder from the 800 A LVPCB in the low-voltage switchgear: 50 A, time delay 0.1 s

- Main 4000 A LVPCB in the low-voltage switchgear: 100 A, time delay 0.18 s
- Transformer neutral-connected ground relay: 150 A, time delay 0.26 s

The neutral-connected ground relays must trip a transformer primary protective device, in this case breaker BK3 shown in Figure 8.2 or 8.6. This gives a well-coordinated and

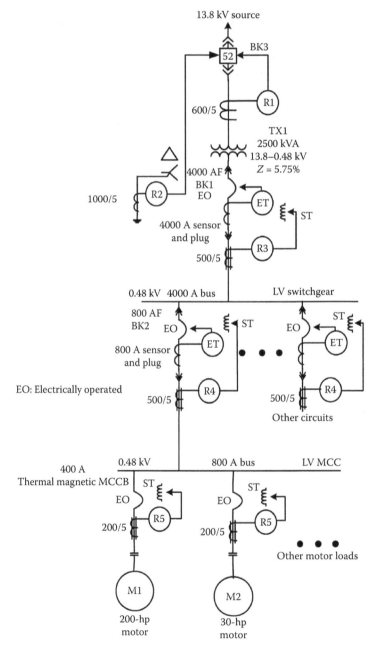

FIGURE 8.6
An improved ground fault protection over that shown in Figure 8.2 with added protective devices and electrically operated circuit breakers.

Ground Fault Protection

sensitive ground fault protection. Though this is an ideal situation for solidly grounded systems, many times, some lack of coordination is accepted and a three-step coordination, as shown in Figure 8.3, is provided, i.e., the provision of dedicated ground fault protection on all circuits from MCC is omitted. This means that the ground faults may be cleared as phase faults, and if these are of low magnitude, these have to escalate or result in three-phase faults. This can cause increased equipment damage and arc-flash hazard.

The saturation of zero-sequence CTs when applied in circuits with large short-circuit currents is a consideration, discussed in Chapter 3.

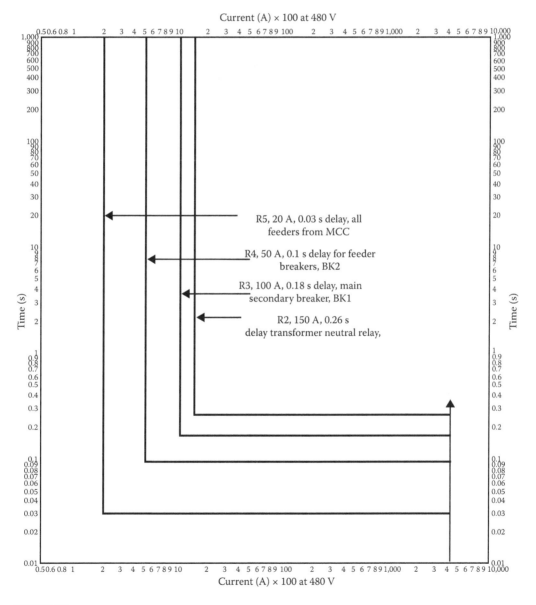

FIGURE 8.7
Ground fault protection and coordination with low-ground fault pickup settings in Figure 8.6.

8.2 Ground Fault Coordination in Low-Resistance Grounded Medium-Voltage Systems

We will illustrate the coordination in a 5000 kVA low-resistance grounded substation and a large 13.8 kV primary distribution system through the following examples:

Example 8.3

First we consider a simple medium-voltage substation, see Figure 8.8. A transformer of 5000 kVA, delta–wye connected, and the wye neutral grounded through a 400 A resistor serves a lineup of a 2.4 kV metal-clad switchgear, which, in turn, serves medium-voltage

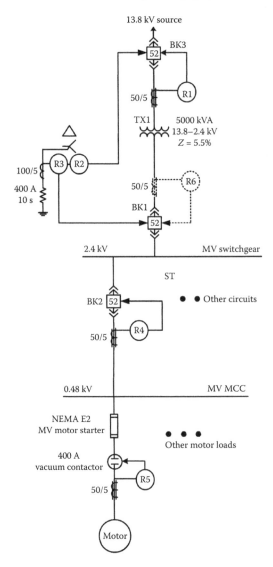

FIGURE 8.8
A medium-voltage low-resistance grounded system for ground fault protection and coordination.

Ground Fault Protection

motor control centers. The figure shows only ground fault protection. The four-step coordination is shown in Figure 8.9. The transformer neutral-connected relay must trip the primary 13.8 kV feeder breaker, as for the low-voltage substations. Microprocessor-based multifunction relays (MMPRs) are used, and with these relays a coordinating time interval (CTI, Chapter 6) of 0.3 s is adequate for selectivity. Figure 8.9 shows the coordination achieved with extremely overcurrent characteristics and also with definite time instantaneous protections. It is evident that the latter are preferable. The time delay increases in steps and the transformer neutral-connected relay R2 has a time delay of 1 s at the maximum ground current of 400 A.

Observe that there are two neutral-connected relays, relay R2 trips the upstream breaker BK3, while relay R3 trips the downstream breaker BK1. These two relays are coordinated with respect to pickup settings and the time delay. If the ground fault lies on the MV switchgear bus and on the load side of the main secondary breaker BK1, relay

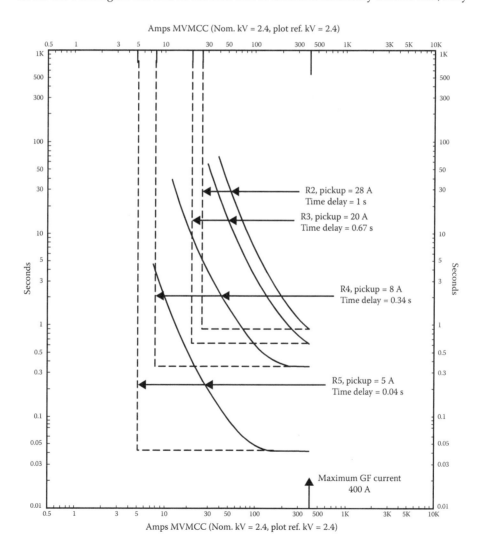

FIGURE 8.9
Ground fault coordination and settings in Figure 8.8, with definite time and inverse characteristic ground fault relays.

R3 will operate faster to clear the fault, and the primary breaker BK3 is not tripped. If the fault is in the transformer or upstream of breaker BK1, breaker BK3 is tripped with a further time delay.

The same operation could have been achieved by relay R6, connected to a core balance CT on the source side of the breaker BK1. In practical application, this has a problem that large number of incoming cables from transformer to BK1 has to pass through a core balance CT, and a CT of large window diameter may not be commercially available to accommodate all the required cables. Also in close-coupled substations connected to the transformer through a metal-clad bus, a zero-sequence CT will be hard to install.

8.3 Remote Tripping

We have discussed remote tripping in Section 5.6. In an industrial distribution system, a primary distribution circuit breaker, say at 13.8 kV, serves a number of substations in a daisy chain, sometimes as many as 10 unit substations over a considerable distance may be connected to one feeder breaker. The transformer primary protection is provided through current-limiting fuses, Chapter 12. For clearing of ground faults through transformer neutral-connected relays, a primary or source breaker is required to be tripped. The following options exist:

- Relayed circuit breakers can be provided on the primary of each substation for local trips. These circuit breakers are also helpful for better primary protection of the transformers and when secondary arc-flash protection is added, see Chapter 12.
- A capacitor trip device can be added for local trip circuit control of the local transformer primary breakers, see Chapter 5 for discussions and limitations.
- Redundant local tripping battery-charger systems or UPS can be provided.
- The possibility of adopting high-resistance grounding systems can be explored, so that no ground fault tripping is required. Even then, a high-resistance fault should be reliably alarmed for remedial actions. Practically, sometimes the alarms may be ignored causing extensive equipment damage.

8.4 Ground Fault Protection in Ungrounded Systems

Though ungrounded systems are mostly no longer in use, see Chapter 7, where they are still in use, protection, alarms, and ground fault detection can be provided. It is necessary that, even in ungrounded systems, the presence of a ground fault must be detected.

Figure 8.10 shows an ungrounded system and the corresponding sequence network connections for a single-line-to-ground fault, see Volume 1, for construction of sequence networks for fault conditions.

In this figure, X_{C1}, X_{C2}, and X_{C0} are the positive-, negative-, and zero-sequence distributed capacitance to the fault point. $X_{C1} = X_{C2}$ and both are much smaller than X_{C0}. Therefore, approximately the ground fault current I_a is

Ground Fault Protection

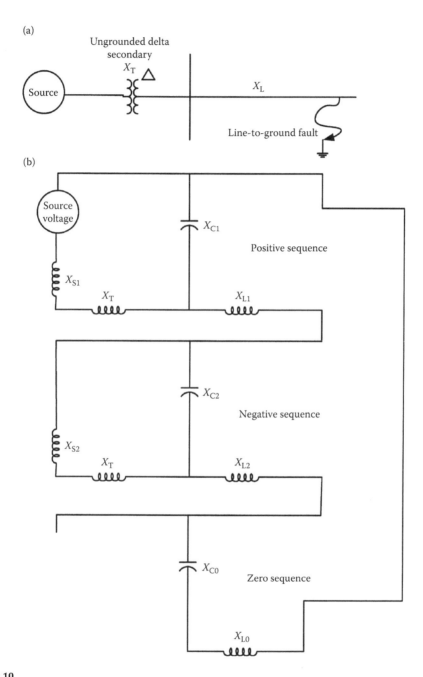

FIGURE 8.10
A sequence network connection for a ground fault in an ungrounded system.

$$I_a = 3I_0 = \frac{3V_N}{X_{C0}} \qquad (8.1)$$

This is good for steady-state conditions. Transient voltages can be produced as illustrated in Chapter 7.

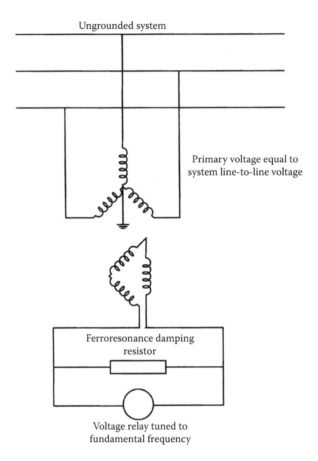

FIGURE 8.11
Protection/alarm in an ungrounded system through a voltage relay and ferroresonance suppressor resistor.

The preferred ground detection system is shown in Figure 8.11. The voltage transformer has a rated primary winding voltage equal to the line-to-line system voltage. (This is the voltage that can be impressed on two unfaulted phases during a line-to-ground fault.) With no ground fault, the voltage across the relay is zero. With a ground fault, the voltage is 3 times the secondary voltage of the PT, say for 69 V, it is 207 V. The relay burden is too small for the system voltage class, and the transformer can be subjected to ferroresonance. Without the ferroresonance suppressing shunt resistance, the grounding amounts to very high impedance grounding, practically ungrounded.

8.4.1 Nondiscriminatory Alarms and Trips

Nondiscriminatory alarms and inductions are a requirement. Figure 8.12a indicates that lamps with high series resistances connected to ground will burn equally bright under no fault conditions. On a single-line-to-ground fault two lamps will burn brighter. Figure 8.12b indicates that the voltmeters will read equal line-to-neutral voltages, with no single-line-to-ground fault and two voltmeters will read line-to-line voltages in case of a single-line-to-ground fault. The voltmeters can be provided with alarm contacts that can be wired in the external circuit.

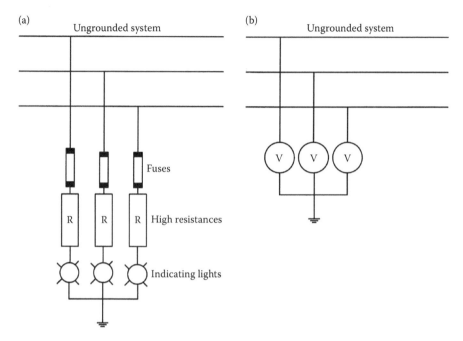

FIGURE 8.12
Ground fault indications in ungrounded systems: (a) indicating lamps and (b) voltmeters.

8.5 Ground Fault Protection in High-Resistance Grounded Systems

8.5.1 Nondiscriminatory Alarms and Trips

Figure 8.13 shows that a voltage relay can be connected across the grounding resistor, which will provide required alarm or trip on a line-to-ground fault in the system. A single-line-to-ground fault results in flow of resistive current through the grounding resistor, which will develop a voltage across it; this voltage can be detected through a voltage relay, tuned to fundamental frequency. The voltage relay can also be replaced with a CT-connected current relay, sensitive enough to detect the low-magnitude of ground current through the resistor for alarm/trip. As the phase-to-ground voltages change by a factor of $\sqrt{3}$ on unfaulted phases, these can be monitored with three indicating lamps connected through resistors or potential transformers. The indicating lamps can also be replaced with voltmeters, same as for ungrounded systems, see Figure 8.12.

8.5.2 Selective Ground Fault Clearance

As the ground fault currents are low, special means of fault detection and isolation are required; see Figure 18.14a and b. In the case provided in Figure 18.14a, alarm/trip can be provided through neutral-connected voltage relay, device 59, connected across a part of the neutral grounding resistor. This relay should preferably be a rectifier type of relay to sense harmonic currents that may flow through the ground resistor. Consider a feeder ground fault; the flows of capacitive and resistive components of the current are shown in Figure 8.14a. The sensitive ground fault sensor and relay on each unfaulted feeder see the capacitive current related to that feeder only, but the sensor and relay on the faulted circuit see

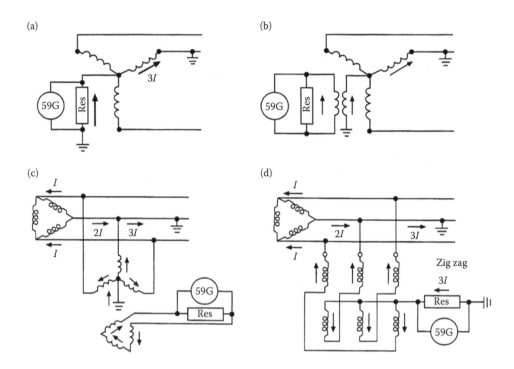

FIGURE 8.13
Nondiscriminatory alarms/trips through a voltage relay in high-resistance grounded systems.

total ground fault current less the feeder's own capacitive current. The sensitive ammeter can indicate the state of the system by monitoring the capacitance current in each phase. In some cases of unequal loads on the feeders, the charging current on a feeder may be much reduced. Suppose that only two unequally loaded feeders are connected as shown in Figure 8.14a, and a ground fault occurs on the heavily loaded feeder. The capacitance of the faulted phase is short-circuited by the ground fault and the charging current flowing through the feeder may be only a fraction of the total charging current.

Ground fault sensors of sensitivity of the order of 10–20 mA are commercially available. These have to be installed carefully and tested for the application. The load cables have to be centrally installed through the window of the sensor occupying not more than 40%–50% of the window diameter. For sensitive ground fault settings, effect of harmonic currents that may be returning through neutral ground must be accounted for.

8.5.3 Pulsing-Type Ground Fault Detection System

Figure 8.14b shows a ground fault localization scheme, popular in the industry. It is called a "pulsing-type high-resistance grounding system." The pulses are created by a current sensing relay and cyclic timer, 62, at a frequency of approximately 20/min by alternatively shorting and opening a part of the grounding resistor through a contactor. These can be traced to the faulty circuit with a clip-on ammeter. Figure 8.15 shows the control circuit of the pulsing-type equipment.

If a switchboard is having many feeders emanating from the bus, each circuit has to be tested with a portable clip-on ground detector, which must go around each cable conductor to identify the faulty phase. Due to the laying of cables, there may not be enough spacing

Ground Fault Protection

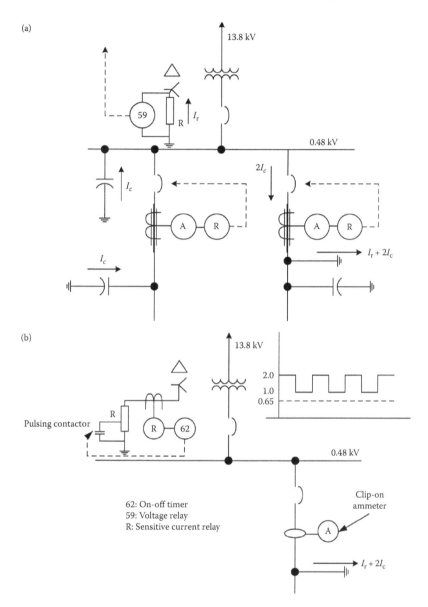

FIGURE 8.14
(a) Discriminatory alarms/trips in a high-resistance grounded system and (b) a pulsing-type high-resistance grounded system.

between them to easily thread the clip-on ground detector. In this regard the ground fault selective system shown in Figure 8.14a is better.

8.5.4 Protection of Motors

Low levels of fault currents, if sustained for a long time, may cause irreparable damage to the rotating machines. Though the burning rate is slow, the heat energy released over the course of time can damage cores and windings of rotating machines even for ground currents as low as 3–4 A [3].

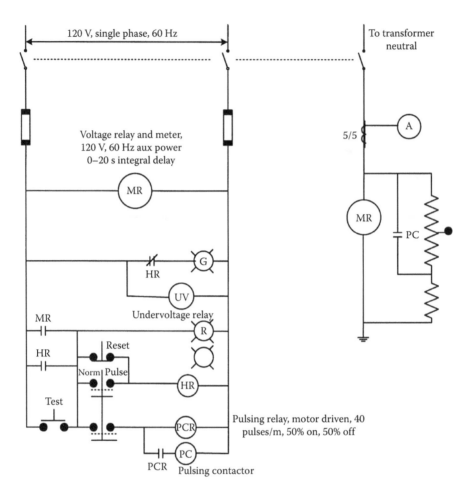

FIGURE 8.15
A control circuit diagram of the pulsing-type equipment for high-resistance fault detection.

This is of importance when a high-resistance grounding system is applied. Even if the charging current in the system is low, provisions must be made to selectively trip out a motor with a short time delay of a couple of cycles. Consider that the ground fault in an HR grounded system is limited to 8 A. The motor ground fault protection is arranged through a core balance CT to trip at 4 A current, time delay of 0.06 s or less. Experience shows that these low pickup and time delay settings even on motors of thousands of horse-power have not given rise to nuisance trips and provide effective ground fault protection. Similar low settings can be used for low-resistance grounded systems. Exceptions are where cable lengths or OH lines of more than 1000–1500 ft are involved. Even in these cases, a pickup setting of no more than 7–8 A and time delay of 0.08–0.1 s is adequate.

8.5.5 Protection against Second Ground Fault

The purpose of adopting high-resistance grounded system is that an immediate shutdown can be avoided. This does not mean that a first ground fault can be left unattended for long periods of time. A relatively dangerous condition may arise on occurrence of a second ground fault. The operation of the system with a single ground fault increases the possibility of second ground fault due to increased insulation stresses. If a second ground fault occurs before the first is cleared, the fault current is no longer controlled by the grounding resistor. It is limited by the supply system impedance and the zero-sequence impedance between the two faults. A second ground fault on the same feeder circuit that has the first ground fault results in a two-phase fault and the fault current is high. This may be cleared by the phase overcurrent relays. However, if the fault is on a different feeder some distance away or embedded in the apparatus windings, the fault current may remain below the pickup settings of the phase overcurrent relays and cause potential damage.

Protection for the second ground fault can be provided. Similar relays are provided and these are set a little higher than the first ground fault current settings and shunt trip the respective circuit breakers. Another effective protection for single phasing is device 46, negative-sequence current relays. In an induction motor:

$$\frac{Z_1}{Z_2} = \frac{I_s}{I_f} \tag{8.2}$$

where Z_1 and Z_2 are the positive and negative-sequence impedances, I_s is the motor starting current or the locked rotor current and I_f is the full-load current. For an induction motor with typical starting current of six times the full-load current, the negative-sequence impedance is 1/6 of the positive-sequence impedance. A 5% negative-sequence component in the voltage will give rise to 30% negative-sequence currents, and practically even higher.

8.5.6 Insulation Stresses and Cable Selection for HR Grounded Systems

As stated above, a ground fault gives rise to voltage stresses equal to $\sqrt{3}$ times the normal. The following review is of interest for cable systems:

Cables: ICEA publication S-61-401 and NEMA WC5 [4] specify the following:

- 100% insulation level: Ground fault is cleared as early as possible, but in any case within 1 min.
- 133% insulation level: This corresponds to that formerly designated for underground systems. Ground fault is cleared within 1 h.
- 173% insulation level: The time required to de-energize is indefinite.

Thus, a 173% level of conductor insulation is required when operation with a single phase to ground exceeds 1 h. But the insulation thicknesses specified in NEC for 600 V grade cables, Table 310.13(A) of NEC [1], for various insulation types, do not specify insulation levels.

Some manufacturers are of the opinion that intrinsic strength of thin insulation section of insulation used for cable insulations is 500 V/mil. Consequently, thickness of insulations specified in NEC for 600 V grade cables is more than adequate to withstand any voltage encountered even in 600 V high-resistance grounded systems. Thus, the mechanical considerations overweigh the electrical concerns. Other cable manufacturers take a conservative approach and recommend 1000 or 2000 V grade cables. Considerations should also be given to the higher system operating voltages. NEMA standard WC5 specifies that operating voltages on cables should not exceed the rated voltage by more than 5% during continuous operation or 10% during emergencies lasting for not more than 15 min. This is of importance for three-phase 600 V systems. Further the DC loads served through six-pulse rectifier systems, connected to 480 or 600 V three-phase systems, will have a no-load voltage of 648 and 810 V, respectively. For 2.4 kV HR grounded systems, 5 kV grade cables, 100% insulation level normally applied, will be acceptable.

For all other electrical equipment such as motors, controllers, switchgear, and transformers, NEMA and ANSI standards do not specify specific tests for grounded or ungrounded systems, and the dielectric and impulse test voltages should be acceptable for these systems. The higher voltages to ground may have minimal effect on the life of equipment, which is difficult to predict. *Generally*, IEEE standards do not distinguish BIL tests for grounded and ungrounded equipment.

Some observations pertaining to high-resistance grounded systems are as follows:

- The resistance limits the ground fault current and, therefore, reduces burning and arcing effects in switchgear, transformers, cables, and rotating equipment.
- It reduces mechanical stresses in circuits and apparatus carrying fault current.
- It reduces arc-blast or -flash hazard to personnel who happen to be in close proximity of ground fault.
- It reduces line-to-line voltage dips due to ground fault and three-phase loads can be served, even with a single-line-to-ground fault on the systems.
- Control of transient over voltages is secured by proper selection of the resistor.

The limitation of the system is that the capacitance current should not exceed approximately 10 A to prevent immediate shutdowns. As the system voltage increases, so does the capacitance currents.

Though immediate shutdown is prevented, the fault situation should not be prolonged; the fault should be localized and removed to avoid insulation stresses and increase the possibility of a second ground fault in the system. See References [5–12] for further studies.

8.6 Ground Fault Protection in Resonant Grounded Systems

As discussed in Chapter 7, the resonant grounding is reactance grounding, and if the reactor perfectly matches the system capacitance, a line-to-ground fault will produce zero current; the transient arc path will be deionized. While perfect neutralization of the entire system may not be possible, the reactor will minimize a large portion of the line-to-ground fault currents. Approximately 75% of the line-to-ground fault currents may be self-extinguishing. The rest must be cleared. Figure 8.16 shows a partial protection scheme. On a ground fault, the relay 50 picks up and energizes operating coil of latch or toggle relay 50X/0. Normally closed contacts 50X open the operating coil, and setup reset relay 50X/R and start timer 2. If after a preset time 0–15 s, fault is not cleared, timer 2 contacts close breaker 52N, solidly grounding the neutral so that conventional ground relays can clear the fault. Breaker 52N also energizes 50X/R to reset the toggle relay and sound an alarm. If fault clears before timer 2 contacts actuate, 50X/R is energized to reset toggle relay.

Ground Fault Protection

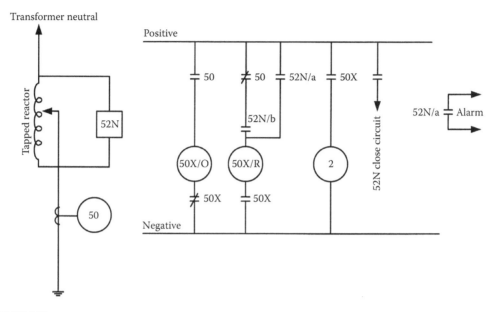

FIGURE 8.16
A ground fault protection system for resonant grounded systems.

Some limitations are:

- A section of the distribution system may be out of service or new distribution may be added, and each time this occurs the reactor value must be adjusted.
- All transformer and equipment must be designed for full line-to-line voltages and insulation.

8.7 Studies of Protection and Coordination in Practical Systems

8.7.1 Ground Fault Protection of Industrial Bus-Connected Generators

The ground fault protection of industrial bus-connected generators requires special considerations and ground fault protective relaying. This is illustrated through the following example:

Example 8.4

This example is based on References [13,14]. Figure 8.17 shows the system configuration of a large industrial distribution system with three plant generators and two utility interconnections and ground fault protection of the 13.8 kV distributions. Each generator is hybrid grounded and has a neutral breaker. The blocks in dotted lines show phase differential protection zones, which indicate that the entire system is covered in overlapping differential zones for phase faults (except feeder circuits from the 13.8 kV buses). The phase differential protections on generators are not shown. *Phase differential protection systems may not provide sensitive ground fault protection;* the minimum pickup current

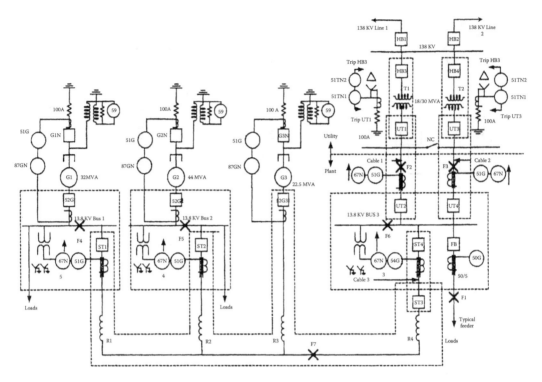

FIGURE 8.17
A system configuration for selective ground fault protection in a large distribution system with cogeneration facilities.

of a phase differential protection zone can be calculated, using the manufacturer's data, CT accuracies, number of protected circuits [15]. It may be too high for the low-level ground fault currents. In Figure 8.17 each generator is grounded through 100 A resistor and also two utility transformers T1 and T2 are grounded through 100 A resistor, limiting the ground fault current anywhere in the 13.8 kV distributions to 500 A, which reduces the ground fault damage.

See Chapter 7 for the fundamental concepts of hybrid grounding. On opening of the neutral breaker in response to operation of generator differential ground fault relay, device 87GN or standby ground fault relay 51G, the grounding system reverts to high-resistance grounding, and throttles the ground fault current flow because high resistance is introduced in the neutral circuit.

It is imperative that a bus-connected industrial generator, even of a small rating, is provided with ground fault differential protection, device 87GN, because 51G device has to coordinate with other system ground relays and there will be a time delay associated with its coordination, resulting in some ground fault damage to the generator windings. The device 87GN will operate instantaneously, trip out generator and neutral breakers, and high-resistance grounding branch is active to limit the ground fault damage. Also see generator protection, Chapter 11.

Table 8.1 shows the tripping matrix for selective ground fault clearance and Table 8.2 shows the area of shutdown for the respective fault locations in Figure 8.17. It can be seen that without the provision of directional ground fault relays, selective tripping cannot be obtained.

Figure 8.18 shows that coordination is achieved throughout the distribution system including the utility ground fault relays, which are electromechanical type, though

Ground Fault Protection

these could be replaced with modern relays and provide similar characteristics as other ground fault relays. The low pickups of the protective devices are of interest:

- 50G, all 13.8 kV feeder breakers: 5 A, instantaneous. (All transformers in the downstream distribution system have delta-connected 13.8 kV windings, permitting these settings.)
- 67N, breakers UT2, UT4: 10 A
- 67N, breakers ST1, ST2, ST4: 16 A
- 51G, generator 3, breakers ST1, ST2 and ST4: 20 A
- 51G, generators 1 and 2, UT2, UT4: 25 A

These low levels of ground fault pickups in a complex distribution system as shown in Figure 8.18 are admirable from personal safety, reduced equipment damage. The

TABLE 8.1

Relay Operation for Selective Ground Fault Tripping

Fault at → Relay ↓	F1	F2	F3	F4	F5	F6	F7
Feeder 50G	X	NO	NO	NO	NO	NO	NO
Gen1, 51G	O	O	O	X	O	O	O
Gen2, 51G	O	O	O	O	X	O	O
Gen3, 51G	O	O	O	O	O	O	X
ST1, 51G	O	O	O	O	O	O	X
ST1, 67N	NO	NO	NO	X	NO	NO	NO
ST2, 51G	O	O	O	O	O	O	X
ST2, 67N	NO	NO	NO	NO	X	NO	NO
ST4, 51G	O	O	O	O	O	O	X
ST4, 67N	O	O	O	NO	NO	X	NO
UT2, 51G	O	O	O	O	O	O	O
UT2, 67N	NO	X	NO	NO	NO	NO	NO
UT4, 51G	O	O	X	O	O	O	O
UT4, 67N	NO	NO	O	NO	NO	NO	NO
T1, 51TN1	O	X	O	O	O	O	O
T2, 51TN2	O	O	X	O	O	O	O

Note: X, relays that operate to clear the fault; O, relays that are operative, and coordinated; NO, relays that are not operative for the direction of fault current flow.

TABLE 8.2

Area of Shutdown for Fault Locations Shown in Figure 8.17

Fault Location	Area Isolated
F1	Only feeder breaker tripped
F2	Breakers UT2 and UT1 tripped
F3	Breakers UT4 and UT3 tripped
F4	Breakers 52G1 and ST1 tripped
F5	Breakers 52G2 and ST2 tripped
F6	Breakers ST4, UT1, UT2, UT3, and UT4 tripped
F7	Breakers ST1, ST2, ST4, and 52G3 tripped

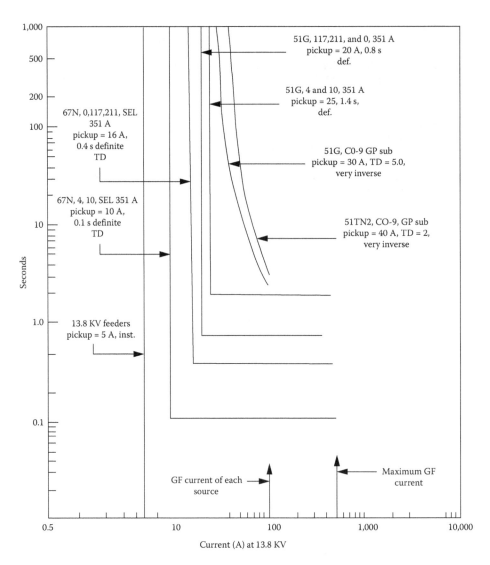

FIGURE 8.18
Ground fault settings and coordination in Figure 8.17.

system shown in this figure is designed utilizing the modern relaying technology and grounding concepts for industrial generators. These low pickup settings require a reliable directional ground fault relay.

8.7.2 Directional Ground Fault Relays

Selection and application of the directional ground protection are, therefore, the key issues in coordinated ground fault protection.

Single-phase-to-ground and double-phase-to-ground faults are unsymmetrical faults. All the sequence component currents and sequence voltages i.e., positive, negative, and zero sequence, are produced. Figure 8.19 shows the interconnections of sequence networks and sequence current flows for a single-line-to-ground fault:

$$I_0 = I_1 = I_2 = \frac{1}{3}Ia \tag{8.3}$$

$$I_a = 3I_0 = \frac{3V_a}{(Z_1 + Z_2 + Z_0) + 3Z_f + 3R_n} \tag{8.4}$$

where I_0, I_1, I_2, I_a are the zero-sequence, positive-sequence, negative-sequence, and phase-to-ground fault currents, Z_1, Z_2, Z_0 are sequence impedances and Z_f is the fault impedance. V_a is the prefault voltage to neutral at the fault location, I_a is the single-line-to-ground fault current and R_n is neutral resistor. See Volume 1 for further details.

Figure 8.20 shows the zero-sequence network of Figure 8.17 for a single-line-to-ground fault at F1. When the neutrals are grounded through resistances, these impedances predominate and all other sequence impedances can be ignored in the calculations. Thus the ground fault current anywhere in the 13.8 kV system is 500 A and a rigorous computer calculation with all the sequence impedances modeled shows only a minor difference, i.e., a fault at bus 1 gives 499 A ground fault current and the sequence impedances in per unit at 100 MVA base are as follows:

$$Z_0 = 2.5116 + j0.0290$$
$$Z_1 = 0.0005 + j0.0233 \tag{8.5}$$
$$Z_2 = 0.0008 + j0.0221$$

The same concept was illustrated in Chapter 7, Example 7.3.

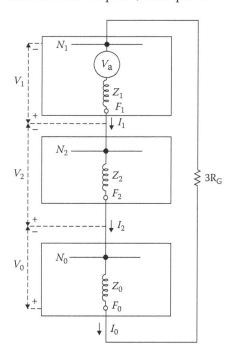

FIGURE 8.19
Sequence impedance connections and flow of currents for a single-line-to-ground fault.

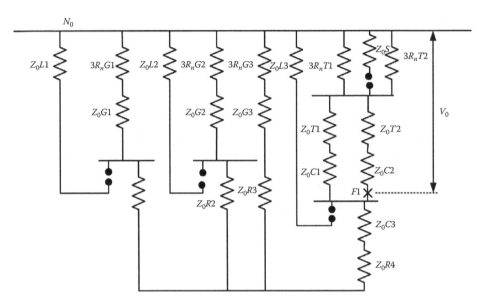

FIGURE 8.20
A zero-sequence circuit for fault at F1 in Figure 8.17.

Directional elements use a reference against which the quantities are compared, this reference is also known as polarizing quantity. These references are either zero-sequence or negative-sequence quantities. With mutual inductance problems between transmission lines use of negative-sequence quantities provides further security.

Classical relays used current or voltage polarization or both, derived from external sources. Figure 3.22 shows an open delta-connected potential transformer for zero-sequence voltage polarization of the directional element.

A proper source for the external current polarization is required. While a CT in the neutral circuit of a delta–wye transformer provides a suitable polarization source, a CT in the wye–wye connected, grounded or ungrounded transformers and in zig-zag transformers is not a proper current polarization source [16,17].

This is obvious when the zero-sequence impedance connections of these transformers are examined. With these classical relays the sequence quantity for each application must be selected for the application.

Modern ground directional relays (GDRs) consist of combination of three directional elements:

1. Zero-sequence current polarized
2. Negative-sequence voltage polarized
3. Zero-sequence voltage polarized

The selection logic makes it possible to select one or two or the best choice logic for the application.

The zero-sequence voltage polarization does not require external PTs in modern GDRs. This voltage can be internally generated from wye–wye PT inputs (but not from open delta-connected PT inputs).

External zero-sequence current source is not necessarily required. The relay algorithm calculates the sequence components from the line current inputs.

Ground Fault Protection

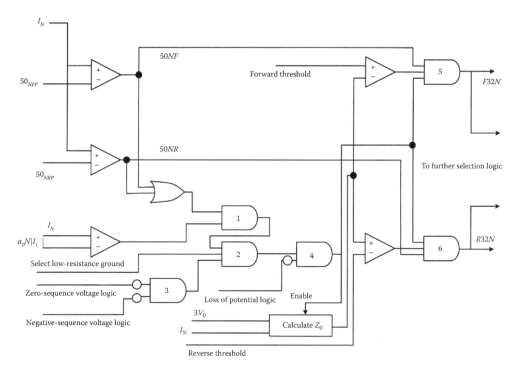

FIGURE 8.21
A logic circuit diagram for directional ground fault relay.

Figure 8.21 shows a much simplified logic diagram for the ground directional element of the relay for low-resistance grounded generators. Inputs from zero- and negative-sequence voltages are shown at NAND gate 3. The low-resistance logic operating from neutral channel current is also shown. The logic for the negative-sequence voltage polarized and current polarized elements are similar, and not shown in Figure 8.21. Note that though the positive-sequence, negative-sequence, and zero-sequence currents are equal in magnitude on a single-line-to-ground fault, the sequence voltages will be much different, as the sequence impedances have large variations, Equation 8.5.

In Figure 8.21, Z_0 is measured by the relay, based upon the equation:

$$Z_0 = \frac{\text{Re}\left[3V_0 \left(I_N < \theta_0\right)^*\right]}{|I_N|^2} \tag{8.6}$$

Similarly in the negative-sequence logic, Z_2 is calculated by the equation:

$$Z_2 = \frac{\text{Re}\left[V_2 (I_2 < \theta_1)^*\right]}{|I_2|^2} \tag{8.7}$$

where θ_0 and θ_1 are the zero-sequence and positive-sequence line angles, respectively, i.e., from (5), $\theta_0 \approx 1^0$ (the resistance predominates). The asterisk shows conjugate of the current.

Figure 8.21 shows that I_N should be greater than positive-sequence current multiplied by a settable factor a_{0N} and its settings 50_{NFP}, or 50_{NRP}, the set currents for the directional logic to operate. The setting of $a_{0N} = I_N/I_1$ increases the security of directional element and prevents nuisance tripping for zero-sequence currents of system unbalance, CT saturation etc. For industrial systems, the unbalance currents will be small and the setting should be lower than the intended pickup of zero-sequence current settings.

For an output F32N or R32N to assert, the measured Z_0 in the forward or reverse direction must overcome the set forward or reverse thresholds, Figure 8.22. These are controlled by the equations:

Z_0F setting ≤ 0, forward threshold:

$$0.75.Z_0F - 0.25\left|\frac{3V_0}{I_N}\right| \tag{8.8}$$

Z_0F setting >0, forward threshold:

$$1.25.Z_0F - 0.25\left|\frac{3V_0}{I_N}\right| \tag{8.9}$$

Z_0R setting ≥ 0, reverse threshold:

$$0.75.Z_0R + 0.25\left|\frac{3V_0}{I_N}\right| \tag{8.10}$$

Z_0R setting <0, Reverse threshold:

$$1.25.Z_0F + 0.25\left|\frac{3V_0}{I_N}\right| \tag{8.11}$$

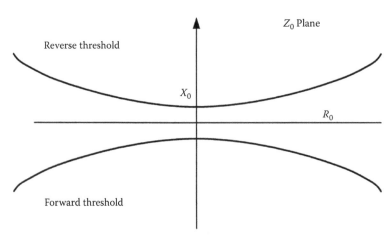

FIGURE 8.22
Characteristics of forward/reverse setting of a ground fault directional relay.

TABLE 8.3
Sequence Currents Seen by the Directional Ground Relays for Single-Line-to-Ground Fault in Figure 8.17 (in the Direction of the Operation of the Relay)

67N Relay Location	I_1 (A)	I_2 (A)	I_0 (A)
UT2	119 < −2.48°	119 < −2.48°	133 < −1.48°
UT4	119 < −2.48°	119 < −2.48°	133 < −1.48°
ST1	93.7 < −1.81°	93.7 < −1.81°	133 < −1.81°
ST2	73.2 < −1.52°	73.2 < −1.52°	133 < −1.65°
ST4	71.9 < −2.48°	71.9 < −2.48°	100 < −1.48°

Note: $I_1 + I_2 + I_0$ do not give 499 A as the sequence currents are partial currents.

8.7.3 Operating Logic Selection for Directional Elements

8.7.3.1 Single-Line-to-Ground Fault

Table 8.3 shows the sequence currents seen by each relay at its location. For a zero resistance fault in the resistance grounded system, i.e., $Z_f = 0$, the zero- and positive-sequence voltages are approximately equal to line voltage (full displacement of neutral-to-ground potential) and the negative-sequence voltage is approximately zero. With some resistance to the fault, the positive-sequence and zero-sequence voltage will be reduced and the negative-sequence voltage will be 180° out of phase with the positive-sequence voltage. The sequence currents are used for setting zero-sequence voltage logic.

As the relay is totally blocked in one direction of operation only one threshold setting, forward or reverse need be made.

8.7.3.2 Double-Line-to-Ground Fault

A double-line-to-ground fault gives high positive- and negative-sequence currents, of the order of 5–8 kA. The positive- and negative-sequence currents are equal and at a phase difference of 180°. The zero-sequence current is small, 66 A. The voltages of faulted phases are zero and of the unfaulted phase rise to 20.7 kV. Thus, the positive-, negative-, and zero-sequence voltages are all equal in magnitude, approximately 7 kV. The zero-sequence voltage polarized algorithm of the relay will not operate and the negative-sequence polarization will operate positively.

8.8 Selective High-Resistance Grounding Systems

The advantages of high-resistance grounding systems are well documented. At medium-voltage level, however, the low-resistance grounded systems are commonly applied, the main criteria being that the ground fault current should be of sufficient magnitude, so that the faults are selectively cleared. Figure 8.23 shows the hybrid grounding system, see Chapter 7.

If we can clear low-level ground faults selectively, then hybrid grounding systems are not required—these can be high-resistance grounding systems, with selective tripping as in a low-resistance grounded system. With this concept, the items shown by asterisks in this Figure 8.23 are not required.

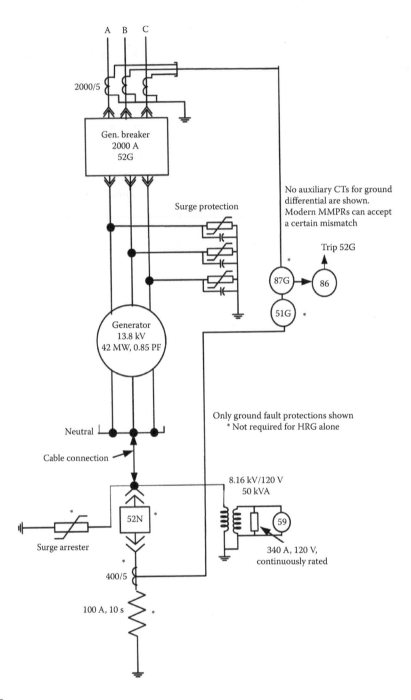

FIGURE 8.23
A hybrid grounding system for a generator. The items shown with * are not required for selective HRG system, low-resistance grounding section, and differential ground fault relay eliminated.

According to Dunki-Jacobs [3], the selective high-resistance grounding systems are nonexistent. This concept needs to be revised in the light of modern sensitive ground fault relaying and protection, as demonstrated in the example to follow.

Ground Fault Protection

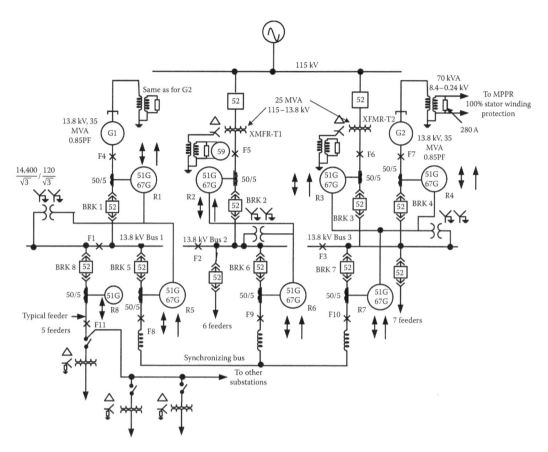

FIGURE 8.24
A large system configuration, with two utility interconnections and two generators, all high-resistance grounded for selective ground fault protection.

Example 8.5

Figure 8.24 shows a system configuration for implementation of HRG. It shows a fairly large size industrial system, consisting of two utility tie transformers of 115–13.8 kV, 25 MVA connected to 13.8 kV buses 2 and 3, two 13.8 kV, 35 MVA generators connected to buses 1 and 3. All three buses are connected through synchronizing tie reactors. The discussions that follow are based on Reference [18].

Only the ground fault protection is shown for selective ground fault tripping, the operation of ground fault directional element is as shown. A typical feeder serving downstream substations is shown connected to Bus 1.

8.8.1 EMTP Simulation of a HRG

In order to determine the minimum settings that will not result in a nuisance trip, some EMTP simulations of HRG are carried out [19]. Consider that a 10 MVA, three-phase 138–13.8 kV transformer, primary windings delta connected and the secondary windings wye connected, is high-resistance grounded, primary ground current limited to 5 A. The stray capacitance current per phase is 1.50 A (stray capacitance nearly = 0.5 μFD per phase). The transformer is connected to a 138 kV source of 25 kA, $X/R = 15$. A single-line-to-ground fault occurs on one of the phases, which is removed in about six cycles at 100 ms. The resultant transients, using EMTP simulations are shown in Figures 8.25 through 8.27.

Figure 8.25 shows that the voltage of the faulted phase is zero and rises to almost normal value after the fault is cleared. Soon after the occurrence of fault, transients occur in unfaulted phase voltages that reach a peak of approximately 2.22 times the phase-to-ground voltages.

Figure 8.26 shows the current through the grounding resistor and is indicative that no capacitance current is returning through it.

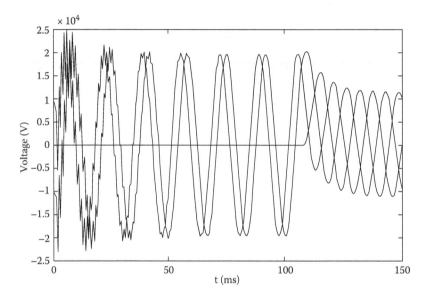

FIGURE 8.25
An EMTP simulation for a ground fault in a HRG system, voltages with ground fault removed in six cycles.

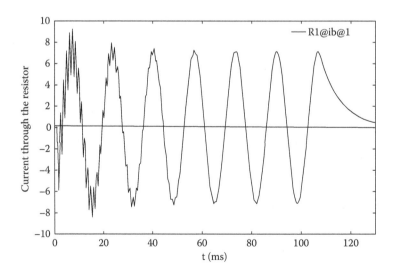

FIGURE 8.26
An EMTP simulation for a ground fault in a HRG system, current through the grounding resistor, with ground fault removed in six cycles.

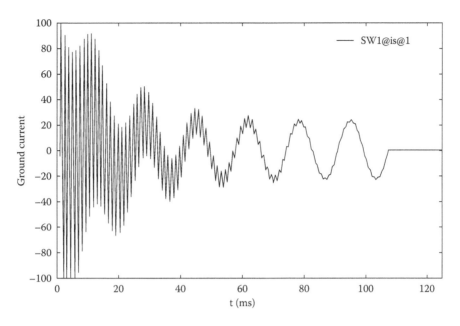

FIGURE 8.27
An EMTP simulation for a ground fault in a HRG system, total ground fault current transient with ground fault removed in six cycles.

Figure 8.27 illustrates the total current to ground. Note the transients before it settles down.

From these simulations, it can be concluded that in a HRG system, *do not set the ground fault protections to instantaneous values. Nuisance trips can occur. A time delay of approximately six cycles is desirable in any setting.*

The first step in designing the HRG system is to calculate accurately the capacitance charging current of the system. Note that the delta-connected primary windings of the transformers act as a sink to the zero-sequence currents. Any distribution served from the secondary of the delta-connected transformers does not contribute to the stray capacitance currents. Similarly, the delta windings of utility transformers block the currents from the utility systems. Fifty percent capacitance of transformer windings plus 13.8 kV bushing capacitance must be considered.

The generator winding capacitance data are generally provided by the vendors; if not, the capacitance can be estimated from the curves provided in Reference [20].

The generator surge capacitors contribute considerable amount of capacitance currents, similarly, the surge capacitors anywhere in the system should not be omitted. The details of these calculations are not further discussed. The following stray capacitance currents are calculated:

- Distribution from feeder breakers connected to 13.8 kV buses 1, 2, and 3: 0.65, 0.73, and 0.82 A, respectively, per phase
- Generators G1 and G2, cable connections, surge capacitors of 0.25 μFD, and generator windings = 1.2 A per phase, each
- Transformers T1 and T2; 0.01 A per phase each.

Thus, total stray capacitance current = 4.62 A per phase, capacitance current on single-line-to-ground fault = 13.86 A. References [21,22] may be referred to for charging current data.

There are four sources to be high-resistance grounded, two generators and two transformers. *Before the resistive components of the ground fault currents are decided, it is necessary that outage of sources should be considered.*

The plant operating loads can be served fully, even when any two sources are out of service. Therefore, ground all sources through 8 A resistors. All sources are grounded through a single-phase distribution transformer, 70 kVA, *continuously rated, with secondary loaded through 280 A continuously rated resistor.* (Considering transformer turns ratio this gives 8 A on the primary of the transformer.) The grounding system of the generator is connected to an MMPR—which can be set for 100% stator winding protection. This will trip the respective generator breaker, while the transformer grounding resistor is connected to an overvoltage device 59, time delayed, which will trip the associated primary transformer breaker. These settings are calculated further in the paper.

Current Flow and Selective Tripping Matrix

Figure 8.24 shows 11 fault points, F1 through F11. Before any selective tripping is attempted, it is necessary to generate a current matrix that shows the capacitive and resistive components of the currents, in upward (away from the bus) and downward (toward the bus) for each operating condition. Without first generating such a matrix it is not possible to decide the pickup sensitivity of the nondirectional and directional ground fault elements. Table 8.4 shows that for selective tripping in any of the fault locations, which breakers should be tripped.

Table 8.5 shows that for fault locations F1 through F11, which ground fault functions, which are in operating mode (denoted by symbol "O"), and which are blocked due to reverse flow of current (denoted by X).

1. All sources in service

    With all sources in service, the fault current at any point in the system is 32 A resistive and 13.86 A capacitive, total fault current = 34.87 A.

TABLE 8.4

Selective Tripping Requirements for Ground Fault

Fault at	Requirements of Selective Tripping	Relay Operation
F1	Breakers 1 and 5 trip	R1 and R5
F2	Breakers 2 and 6 trip	R2 and R6
F3	Breakers 3, 4, and 7 trip	R3, R4, and R7
F4	Breaker 1 trips	R1
F5	Breaker 2 trips	R2
F6	Breaker 3 trips	R3
F7	Breaker 4 trips	R4
F8	Breakers 5, 6, and 7 trip	R5, R6, R7
F9	Breakers 5, 6, and 7 trip	R5, R6, R7
F10	Breakers 5, 6, and 7 trip	R5, R6, R7
F11	Breaker 8 trips	R8

TABLE 8.5

Tripping Matrix for Selective Fault Clearance

Fault at	Relay R1 Trips Brk 1		Relay R2 Trips Brk 2		Relay R3 Trips Brk 3		Relay R4 Trips Brk 4		Relay R5 Trips Brk 5		Relay R6 Trips Brk 6		Relay R7 Trips Brk 7		Relay R8 Trips Brk 8
	51G	67G	51G	67G	51G	67G	51G	67G	51G	67G	51G	67G	51G	67G	51G
F1	O	X	O	X	O	X	O	X	O	O	O	X	O	X	O
F2	O	X	O	X	O	X	O	X	O	X	O	O	O	X	O
F3	O	X	O	X	O	X	O	X	O	X	O	X	O	O	O
F4	O	O	O	X	O	X	O	X	O	O	O	X	O	X	O
F5	O	X	O	O	O	X	O	X	O	X	O	X	O	X	O
F6	O	X	O	X	O	O	O	X	O	X	O	X	O	O	O
F7	O	X	O	X	O	X	O	O	O	X	O	X	O	O	O
F8	O	X	O	X	O	X	O	X	O	O	O	X	O	X	O
F9	O	X	O	X	O	X	O	X	O	X	O	O	O	X	O
F10	O	X	O	X	O	X	O	X	O	X	O	X	O	O	O
F11	O	X	O	X	O	X	O	X	O	O	O	O	O	O	O

The resistive component of the current from all sources is arithmetically summed. The flow of currents through protective elements is shown in Table 8.6.

2. Generator G2 and transformer T2 out of service

   When generator G2 and transformer T2 are out of service, the fault current at any point is 10.26 A capacitive and 16 A resistive, which gives a total current of 19.01 A. Table 8.7 shows a matrix of current flows in the protective ground fault elements. *Note that the total current at the fault point does not change.* Table 8.7 gives the minimum current flows for which the ground fault functions should respond.

An interesting observation is that for the faults at locations F8, F9, or F10, 51G elements of relays R5, R6, and R7 should operate to isolate the fault. This islands the system, all reactor ties to the synchronizing bus are lost and each bus will be served by the sources connected to it.

When we consider an operating condition, where all the sources connected to a bus are lost, the loads connected to that bus will be served through the synchronizing bus reactor ties. Consider the operating condition when both sources on bus 3, that is, generator G2 and transformer T2 are out of service and a single-line-to-ground fault occurs at F10. Breakers 5 and 6 will be tripped by 51G function on relays R5 and R6. *After this tripping*, 51G function of relay R7 carries only the capacitive current and no resistive component, as generator G2 and transformer T2 are out of service. No loads can be served as the reactor ties are lost. The fault locations on the reactor ties must be investigated before putting back the system on line.

TABLE 8.6
Fault Current Flows, All Sources in Service

Fault at	R1 51G	R1 67G	R2 51G	R2 67G	R3 51G	R3 67G	R4 51G	R4 67G	R5 51G	R5 67G	R6 51G	R6 67G	R7 51G	R7 67G	R8 51G
F1	3.6/8														
F2			0.03/8									11.64/24			
F3					0.03/8		3.6/8							7.77/16	
F4, up		10.26/24													
F5, up				13.83/24											
F6, up						13.83/24									
F7, up								10.26/24							
F8									5.55/8		5.82/8		6.09/16		
F9									5.55/8		5.82/8		6.09/16		
F10									5.55/8		5.82/8		6.09/16		
F11										8.31/24					13.86[a]/32

[a] Fault current at any point is 13.86 A capacitive and 32 A resistive, except that for a fault at F11, the capacitance current is reduced by a small amount contributed by the distribution connected to that feeder.

TABLE 8.7
Fault Current Flows Generator G2 ABD Transformer T2 Out of Service

Fault at	R1 51G	R1 67G	R2 51G	R2 67G	R3 51G	R3 67G	R4 51G	R4 67G	R5 51G	R5 67G	R6 51G	R6 67G	R7 51G	R7 67G	R8 51G
F1	3.6/8														
F2		6.63/8	0.03/8												
F3				10.2/8	0.03/8		3.6/8								
F4, up										4.68/8					
F5, up												8.01/8			
F6, up						6.63/8								7.77/16	
F7, up								10.2/8							
F8									5.55/8		5.82/8		1.95/0		
F9									5.55/8		5.82/8		1.95/0		
F10									5.55/8		5.82/8		1.95/0		
F11															10.23[a]/16

[a] Fault current at any point is 10.23 A capacitive, except that for a fault at F11, the capacitive current is reduced by the distribution connected to that feeder.

Coordinating Ground Fault Settings
Based on the calculations, the following settings will provide coordinated protection for the HRG system:

- Set all feeder relays (R8) 51G function, pickup 2.5 A, definite time characteristics, time delay = 0.07 s.
- Set 67G function of relays R1, R2, R3, and R4 at 2.5 A, definite time characteristics, time delay = 0.07 s.
- Set 67G function of relays R5, R6, and R7, pickup at 4.5 A, time delay 0.12 s. All 13.8 kV breakers are three cycle breakers, rated on symmetrical current basis.
- Set 51G function of relays R5, R6, R7, pickup at 6 A, definite time characteristics, time delay = 0.18 s.
- Set 51G function of relays R1, R2, R3, R4 pickup at 6 A, time delay = 0.25 s.

The time–current coordination is shown in Figure 8.28.

8.8.2 Generator 100% Stator Winding Protection

Tables 8.6 and 8.7 show that the ground fault current for a line-to-ground fault at F4 and F6, as seen by relays R1 and R3 is 26.1 A with all sources in service, and will reduce to approximately 10.39 A when two sources are out of service. The 67G elements of relays R1 and R4 are set to pickup at 2.5 A, time delay 0.07 s. The fundamental frequency voltage distribution across the generator windings can be assumed to be linear from line-to-neutral voltage at the line end to zero at the neutral terminal. If the fault is embedded in the generator windings toward the neutral, the generator breaker will not be tripped and the source fault current plus the neutral fault current contributions can be damaging to the generator stator.

The 100% stator winding ground fault protection using third harmonic voltage distributions across the generator stator windings from line-to-neutral terminals is discussed in IEEE Guide for AC generator protection [23,24]. The third harmonic voltages act as zero-sequence components. Therefore, the ground fault, which generates zero-sequence currents, will alter the third harmonic voltage distributions across the generator windings. This philosophy is used to detect a ground fault toward the neutral terminals of the windings. The third harmonic voltage distribution under no-fault conditions varies considerably depending upon the following:

- Rating of the generator.
- Even for the same generator rating, it can vary from manufacturer to manufacturer.
- It varies with active and reactive power load on the generator.

Due to these variations, a third harmonic voltage setting is not made till these voltages are measured during the commissioning stage of the generator. Sometimes, the manufacturers can suggest the settings based on their experience of installation of similar machines.

Figure 8.29 shows the third harmonic voltage distribution under no-fault and under a single-line-to-ground fault condition close to the neutral.

The third harmonic voltage increases at the line terminals and decreases at the neutral terminal. This figure is of general nature and not specific for any machine size.

Ground Fault Protection

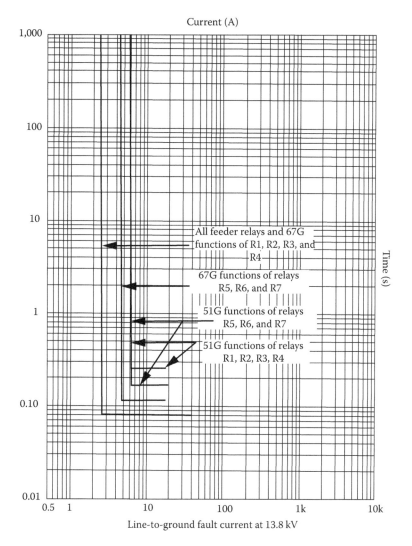

FIGURE 8.28
Ground fault settings and coordination in a system of Figure 8.24.

A circuit diagram of the 100% stator winding protection from IEEE Guide [23] is shown in Figure 8.30a and its tripping logic in Figure 8.30b. This guide also describes third harmonic differential protection and subharmonic injection methods.

Settings
The pickup setting of the fundamental frequency overvoltage element is 7 V. This will protect generator stator windings for a fault within 2%–5% of the windings from the neutral end, a trip time delay of 1 s is adequate.

Consider that the actual measurements at the time of commissioning a third harmonic voltage = 400 V. Then the setting can be

$$(0.5) \times (400) \times (14,400/240) = 12.0 \text{ V}$$

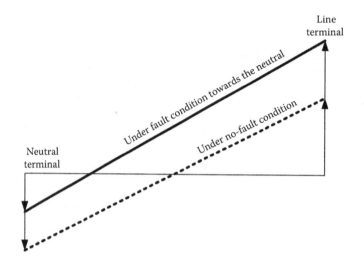

FIGURE 8.29
Third harmonic voltage distribution across the stator windings of a generator under normal operating and fault conditions.

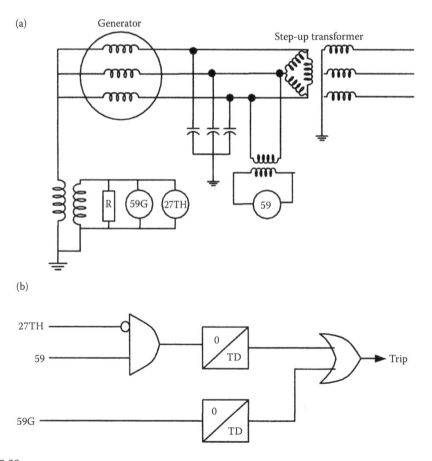

FIGURE 8.30
(a) A scheme for 100% stator winding protection of a generator, with HRG and (b) tripping logic.

Ground Fault Protection

Again a time delay of 1s is adequate, 59 is instantaneous overvoltage relay, 27TH is instantaneous UV relay tuned to third harmonic frequency, and this is supervised by device 59. The relay 59G is the instantaneous overvoltage relay tuned to the fundamental frequency.

Transformer Ground Fault Protection

A line-to-ground fault at F5 or F7 can be embedded in the transformer secondary windings or transformer secondary terminal chamber; and will not be removed by tripping of associated secondary breakers 2 or 3. It will continue to be fed from the transformer primary source, till it is tripped.

Details of this tripping and monitoring scheme are shown in Figure 8.31. Under no fault conditions, no current is expected through the transformer neutral grounding resistor. But, practically, some small current does return, as the three-phase systems are not *perfectly symmetrical*. For example, consider the cable connections of number of parallel conductors, which will not be of exactly the same lengths.

This small retuning current can be measured on the ammeter and the voltage thus developed can be indicated on the voltmeter as shown in Figure 8.31. This return current is not expected to be more than approximately 100–200 mA even for large distribution systems.

On a ground fault, embedded in the transformer secondary windings, close to the neutral, the fault point driving voltage is reduced and so the ground fault current, proportionally. A pickup setting of 15 A on 51G relay will detect ground faults within approximately 95% of the wye connected–transformer windings. The corresponding voltage settings on

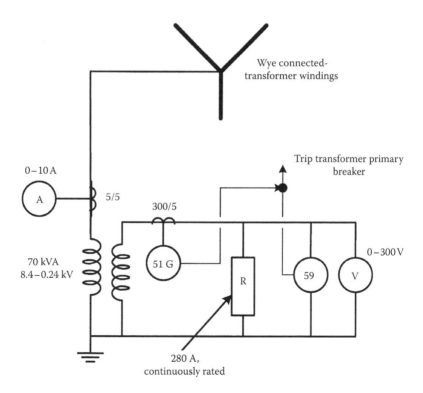

FIGURE 8.31
Details of a high-resistance grounded transformer ground fault protection.

59 will be approximately 12 V. The 51G and 59 devices should trip a transformer primary circuit breaker.

An MMPR will provide all the protection functions and continuous monitoring of current and voltage with recording capability.

Core-Balance CT

The core-balance CTs for the ground fault protection, ratio 50:5 should have a large window diameter. The outline of a CT is shown in Figure 8.32, which has a window of 14″ × 5″, and ANSI relaying class accuracy of C10, CT secondary winding resistance = 0.02 Ω.

The CT saturation is addressed in Chapter 3. Both steady-state and transient performance must be considered. A rigorous equation ensuring the integrity of CT, with DC offset is given in Equation 3.24.

In this application, the maximum fault current is limited to approximately 47 A, the burden of the MMPR relays is small, and the lead lengths are small. Consider a total burden of 0.03 Ω.

The zero-sequence resistance predominates. Table 8.8 shows the calculated values of positive, negative, and zero-sequence currents, with solidly grounded sources and also when the ground current at each source is limited to 8 A. The impedances are shown in per unit values on a 100 MVA base.

The fault point X/R ratio from this table, say at bus 1 is

$$\frac{(0.119+0.119+0.158)}{(0.003+0.003+392.225)} = 1.0096 \times 10^{-3}$$

Thus, CT saturation is not a problem even with C10 accuracy CT.

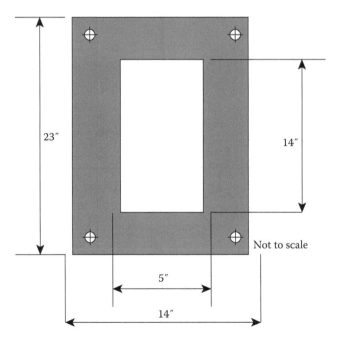

FIGURE 8.32
A special large-dimension core balance CT, ratio 50% required for sensitive ground fault protection.

TABLE 8.8
Calculated Sequence Impedance Values

Fault at	Solidly Grounded Sources		Resistance Grounded Sources	
	Z_+, Z_-	Z_0	Z_+, Z_-	Z_0
Bus 1	0.004 + j0.164	0.004 + j0.198	0.004 + j0.164	392.225 + j0.237
Bus 2	0.005 + j0.158	0.005 + j0.184	0.005 + j0.158	392.225 + j0.237
Bus 3	0.003 + j0.119	0.004 + j0.135	0.003 + j0.119	392.225 + j0.158

8.8.3 Accuracy of Low Pickup Settings in MMPR

The accuracy of low current pickup settings specified by a manufacturer is ±5% of the setting or ±0.025 A in terms of secondary current. Thus, the recommended settings are conservative in this respect.

8.9 Monitoring of Grounding Resistors

Neutral grounding resistors should meet the requirement of IEEE 32 [25,26]. These are rated line-to-neutral voltage, though for HR grounding of generators it is not unusual to see full line-to-line rated systems, grounded through distribution transformers. Generally edge-wound design is used for neutral grounding systems. These resistors are constructed of flat stainless steel alloy, wound on its edge to form a helical resistive element. The element is then mounted on solid porcelain core and welded terminals at both ends providing the means of connections and terminations. The IEEE temperature rise limits are 760°C for 10 s or 1 min, 610°C for 10 min or extended time, and 385°C for steady-state continuous rating. The stainless steel alloy used should have low temperature coefficient of resistivity. A variation not more than ± 10% in the resistance values is commonly specified.

The other types of resistor constructions are as follows:

- Continuous flat one piece SS elements grid resistors
- Wire wound, nichrome wire, helically wound
- Smooth wound, nickel chromium wire wound
- Cast iron grid

Whatever resistor type is used for grounding, it should be monitored for continuity and short-circuit continuously, under no-fault and under fault conditions. An open circuit of a grounding resistor in low-resistance or high-resistance grounding system will result in the system to become ungrounded, with all its perils. This is not acceptable. A short circuit in the grounding resistor, though not so common, will result in solidly grounding the system. Without proper monitoring, these faults in the resistor may remain undetected for a long time, till next scheduled routine maintenance.

Yu [26] describes how a combination of measurements of voltage across the resistor, current through the resistor and ohms measurement of the resistor to ground can achieve an effective monitoring system, both under fault conditions and under normal operation. Many commercial products are available.

References

1. National Electric Code, NFPA 70, 2017.
2. HI Stanback. Predicting damage from 277 volt single-phase-to-ground arcing faults. *IEEE Trans Ind Appl*, IA-13(4), 307–314, 1977.
3. JR Dunki-Jacobs. The reality of high-resistance grounding. *Trans Ind Appl*, 13, 469–475, 1977.
4. ICEA Publication S-61-401, NEMA WC5. Thermoplastic Insulated Wire and Cable for the Transmission and Distribution of Electrical Energy (ICEA S-61.402, Third Edition).
5. AP Sakis Meliopoulos. *Power System Grounding and Transients*. Marcel Dekker, New York, 1988.
6. JC Das, H Osman. Grounding of AC and DC low-voltage and medium voltage drive systems. *Trans Ind Appl*, 34(1), 205–216, 1988.
7. JR Dunki-Jacobs, FJ Shields, CS Pierre. *Industrial Power System Grounding Design Handbook*. Thomson Shore, Dexter, MI, 2007.
8. IEEE Stand. 142. IEEE Recommended Practice for Grounding of Industrial and Commercial Power Systems, 2007.
9. AS Locker, MS Scarborough. Advancement in technology create safer and smarter high resistance grounded systems. *IEEE Pulp and Paper Industry Technical Conference*, 2009, pp. 102–103.
10. R Beltz, I Peacock, W Vilcheck. Application considerations of high-resistance grounding retrofits in pulp and paper mills. *IEEE Pulp and Paper Industry Conference*, 2000, pp. 33–40.
11. D Murry, J Dickens, RA Hanna. High resistance grounding: Avoiding unnecessary pitfalls. *IEEE Trans Ind Appl*, 45(3), 1146–1154, 2009.
12. D Paul, P Sutherland. High-resistance grounded power system equivalent circuit damage at the line-to-ground fault location. *IAS Annual Meeting*, 2013, pp. 1–9.
13. JC Das. 13.8-kV selective high-resistance grounding system for a geothermal generating plant—A case study. *IEEE Trans Ind Appl*, 49(3), 1234–1343, 2013.
14. JC Das. Ground fault protection of bus connected generators in an interconnected 13.8 kV system. *IEEE Trans Ind Appl*, 43(2), 453–461, 2007.
15. GE. PVD Voltage Differential Relay, Manual GEK-45405-1982.
16. Westinghouse. *Applied Protective Relaying*. Westinghouse Electric Corporation, Coral Springs, FL, 1982.
17. A Guzman, J Roberts, D Hue. New ground directional elements operate reliably for changing system conditions. *51st Annual Georgia Tech Protective Relaying Conference*, Atlanta, GA, 1997.
18. JC Das. Selective high-resistance grounding system for a cogeneration facility. *IEEE Trans Ind Appl*, 49(3), May/June 2013.
19. JC Das. *Transients in Electrical Systems, Recognition, Analysis and Mitigation*. McGraw Hill, New York, 2010.
20. Westinghouse. *Electrical Transmission and Distribution Reference Book*, 4th Edition. Westinghouse Corporation, East Pittsburg, PA, 1964.
21. DS Baker. Charging current data for guess free design of high-resistance grounded systems. *IEEE Trans Ind Appl*, IA-15, 136–140, 1979.
22. D Paul, S Panetta, P Sutherland. A novel method of measuring inherent power system charging current. *IEEE Trans Ind Appl*, 47(6), 2330–2342, 2011.
23. IEEE Standard C37.101. IEEE Guide for Generator Ground Protection, 1993.
24. IEEE Standard C37.102. IEEE Guide for AC Generator Protection, 2007.
25. ANSI/IEEE Std. 32. IEEE Standard Requirements, Terminology, and Test Procedure for Neutral Grounding Devices, 1972.
26. L Yu. Selection of neutral grounding resistor and ground fault protection for industrial power systems. *38th Petroleum and Chemical Industry Conference*, September 1991, pp. 147–153.

9
Bus-Bar Protection and Autotransfer of Loads

9.1 Bus Faults

A bus may interconnect multiple input sources and loads; the loads may be connected at different voltages, other than the bus voltage through step-up or step-down transformers. A bus fault can shut down a large area; furthermore, the fault currents at bus are high due to interconnection of many sources and loads. High-speed fault clearance is essential to limit damage to the buses, supply partial loads, if possible, and maintain stability. Without a bus protection, the remote terminals of lines connected to faulty bus must be tripped, resulting in loss of loads. Factors such as bus configuration relay input sources, operating time, sensitivity, and existing facilities are important for selection of bus protection schemes.

The most sensitive method of bus protection is differential relaying. The principles of differential protection are covered in Chapter 2.

9.2 Bus Differential Relays

Simple overcurrent relays can be used and partial differential schemes are discussed in Chapter 2. Fundamentally there are two types of differential protection schemes for bus bars: (1) high-impedance relays and (2) low-impedance relays.

9.2.1 Low-Voltage Bus Bars

Rarely, bus differential protection is applied to low-voltage bus systems. Generally, these are single bus systems, or sectionalized buses with bus tie breakers which can be open or closed. Though application of differential protection with or without zone interlocking is a modern trend (see Chapter 6 for more details).

9.3 High-Impedance Differential Relays

Electromechanical single-phase high-impedance relays have been popular in the industry for the last 50 years. Standard bushing-type CTs are used and the protection can be easily extended if additional circuits are added. The CTs are normally dedicated and of the same ratio and accuracy classification (see Chapter 3).

A basic circuit of the high-impedance differential relay is shown in Figure 9.1, and a general equivalent circuit is illustrated in Figure 9.2. Consider that there are n circuits to be protected in the differential zone; only three circuits are shown in Figure 9.1. The circulating current I_1 on one side of the differential relay 87 in Figure 9.2 is driven by all the nonfaulted CTs. There magnetizing circuit impedance is shown as $X_m/(n-1)$ and series impedance including lead resistances is shown as $R_{CT}/(n-1)$, assuming that all CTs are identical. The faulted CT resistance and magnetizing impedance are on the right side of the differential relay 87: R_{CTF} and X_{mF}, respectively. Thus, the currents I_1 and I_2

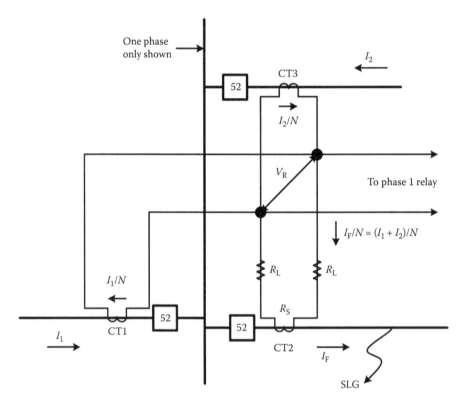

FIGURE 9.1
Principle of a high-impedance bus differential relay.

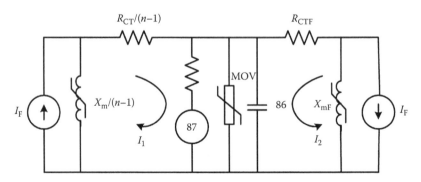

FIGURE 9.2
Equivalent circuit of a high-impedance bus differential relay for an external fault.

on the two sides of the relay cannot be equal. If no resistance is introduced in the relay coil circuit, the transient currents will be high and not much voltage will be developed across the relay coil.

A series resistance in the relay coil is introduced, which reduces the current through the coil to milliampere range. A nonlinear resistor (thyrite) is used to limit the overvoltage that can be developed across the relay during internal faults to safe values. The thyrite blocks will dissipate energy during an internal fault and to protect these from thermal damage, lockout relay (which trips and locks out the required circuit breakers to remove the fault) contact is connected as shown. As the relay picks up, it short-circuits the thyrite blocks. The relay also has a high-set element, 87H. Figure 9.3 shows a practical circuit of connections.

A manufacturer provides the following expression for calculating the 87L unit setting, considering CT performance and CT burden [1]:

$$V_R = 1.6 K \left(R_s + PR_L \right) \frac{I_F}{N} \qquad (9.1)$$

where V_R = voltage developed across the relay coil, K = CT performance factor, R_s = CT secondary winding and lead resistance, R_L = one way cable resistance from the junction point to CT, I_F = primary fault current, N = CT ratio, and P = 1 for three-phase faults and 2 for single-line-to-ground faults. If the calculated V_R remains below the knee point voltage of the CT, the application is acceptable. A value higher than the knee point voltage is acceptable if the calculated V_R is below 0.67 times the excitation voltage at 10 A excitation of the CT. A CT magnetization curve is required for the application (see Chapter 3).

The CT performance factor is given by a graph provided by the manufacturer, not reproduced here. To ascertain K factor, first calculate

$$\frac{\left(R_S + PR_L \right) I_F}{E_S N} \qquad (9.2)$$

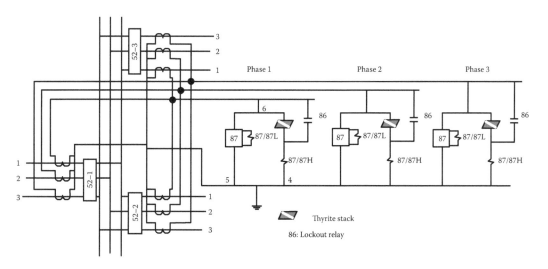

FIGURE 9.3
Connections of a three-phase (three single-phase units) of a high-impedance bus differential relay. (From GE. PVD Differential Voltage Relay, Manual: GEK-45405.)

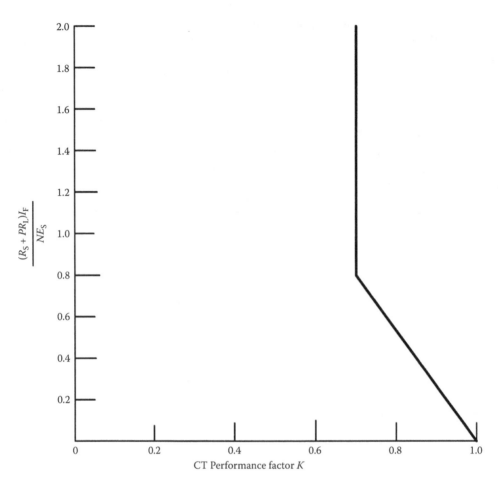

FIGURE 9.4
CT performance factor K. (From GE. PVD Differential Voltage Relay, Manual: GEK-45405.)

where E_s is the CT saturation voltage. For numerical values between 0.8 and 2.0 given by Equation 9.2, $K = 0.7$. Between the numerical values 0 and 0.8, it increases linearly to 1.0 (see Figure 9.4).

Example 9.1

Consider a system fault current of 40 kA, five-cycle symmetrical breakers, CT ratio = 1200:5, $R_s = 0.61\ \Omega$, and the CT lead resistance corrected for temperature at 50°C = 0.548 Ω. For this example, consider that the CT characteristics shown in Figure 3.1 for 1200/5 CT are applicable.

For each circuit breaker being protected, the V_R can be separately calculated and then the maximum value used. Alternatively V_R is calculated using $P = 2$, even for three-phase faults.

To apply Equation 9.1, first the factor K given by Equation 9.2 is calculated. The CT saturation voltage is 300 V from Figure 3.1. Then,

$$\frac{(0.61 + 2 \times 0.548) \times 40 \times 10^3}{300 \times 240} = 0.94$$

Therefore, $K = 0.7$. Calculate V_R from Equation 9.1:

$$V_R = \frac{0.7 \times 1.6 \times (0.61 + 2 \times 0.548) \times 40 \times 10^3}{240} = 318 \text{ V}$$

This exceeds the CT saturation voltage of 300 V. Apply the second criteria of application and calculate the CT voltage at 10 A excitation from Figure 3.1; it is 495 V. Then, 0.67 multiplied by 495 gives 331.7 V, which is higher than the calculated value of 318 V. Therefore, the application is acceptable.

From Equation 9.1 it is obvious that the V_R can be reduced by increasing the CT ratio. With higher CT ratios, higher relaying C class accuracy is obtained, but as the CT ratio rises so will its secondary resistance (see Figure 3.1). If the voltage setting is above the knee point voltage, a more accurate calculation of the excitation current is required. The settings above knee point voltage up to the C rating of the CT are generally acceptable.

9.3.1 Sensitivity for Internal Faults

The sensitivity of the relay for internal faults is of consideration. Again a manufacturer provides the following relation:

$$I_{\min} = \left[\sum_{X=1}^{n} (I)X + I_R + I_1 \right] N \qquad (9.3)$$

where I_R = current in the 87L unit at pickup voltage, I_1 = current in the thyrite unit at pickup voltage, $I_{\min}$ = minimum internal fault current to trip 87L, and I = secondary excitation current of the individual CT at the pickup of 87L.

The calculation will demonstrate that the sensitivity is dependent upon the voltage setting and the number of circuit breakers being protected. It can be of the order of 200–400 A. *Thus, for resistance-grounded system where the ground fault current can be limited to 200–400 A, separate ground fault differential protection is required.*

The high-set element settings in volts as a function of 87L settings are shown in Figure 9.5.

9.3.2 High-Impedance MMPRs

Microprocessor-based multifunction relays (MMPRs) duplicating the performance of electromechanical high-impedance relays are available. Apart from smaller dimensions and a single three-phase unit, panel, or rack mounted, these incorporate filters to account and correct to some extent the CT saturation. Nevertheless, varying CT ratios are not recommended. A certain manufacturer recommends that the CTs should have a minimum C200 relaying class accuracy. The relay can detect open circuit of a CT lead and has all the other functionalities of communication, fault capture, diagnostics, and oscillography as discussed in Chapter 4. The capabilities of a relay in one panel mounted enclosure are as follows:

- Three-phase high-impedance directional elements—each phase can be set separately and two independent settings per phase can be provided.
- Instantaneous overcurrent, phase, neutral, and negative sequence with or without definite time delay.
- Inverse time phase, neutral and negative sequence with different IEEE and IEC characteristics.

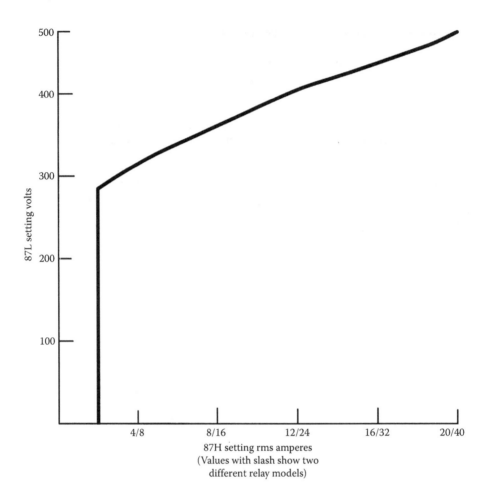

FIGURE 9.5
87H voltage settings based on 87L settings, high-impedance bus differential relay.

- Event reports.
- Sequential event recorder.
- Communication protocols.
- Remote and local control switches.
- Differential phase, negative sequence, and ground metering.
- Peak and maximum demand metering.
- Open CT connections alarm.
- Application to transformer protection.
- Metal oxide varistor (MOV) supervision.

Figure 9.6 shows the basic protection logic. Two separate filters optimize the input quantities to the elements. The half-cosine filter in high-impedance elements removes the DC component and also the odd harmonics. The full-cosine filter in the overcurrent elements filters all but the fundamental frequency. The differential elements compare the measured voltage with element setting, the element acts like a comparator.

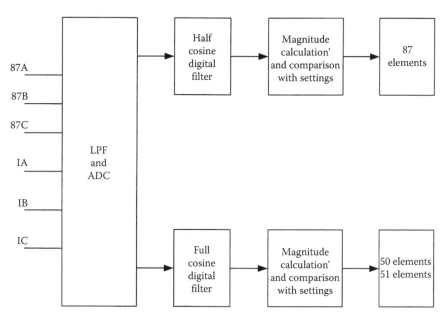

FIGURE 9.6
Block circuit diagram of a high-impedance bus differential MMPR.

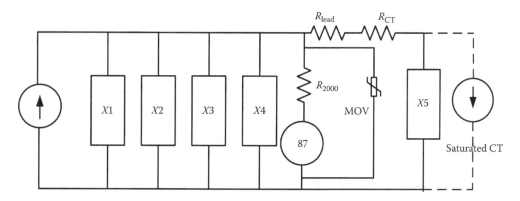

FIGURE 9.7
Equivalent circuit diagram of high-impedance MMPR, showing CT saturation.

The equivalent circuit for through-faults is shown in Figure 9.7. The voltage developed across the high-impedance element is

$$V_r = (R_{CT} + P \cdot R_{lead}) \cdot \frac{I_F}{N} \tag{9.4}$$

where
N = CT ratio
V_r = voltage across high-impedance element
I_F = maximum external fault current
R_{CT} = the CT secondary resistance and lead resistance to CT terminal
P = 1 for three-phase faults and 2 for single-phase-to-ground faults

If the CT ratio is 2000:5, knee point voltage 432 V, accuracy C800, $I_F = 50$ kA, $R_{CT} = 0.82\ \Omega$, and $R_{lead} = 0.6\ \Omega$, then $V_r = 178$ V. This should be below knee point voltage.

Then the relay is set:

$$V_s = K \cdot V_r \tag{9.5}$$

A manufacturer recommends a safety factor of 1.5; then the setting is 267 V.

The minimum primary current required to operate the relay is

$$I_{min} = (nI_e + I_r + I_m) \cdot N \tag{9.6}$$

where

N = number of CTs in parallel with the relay per single phase
I_e = CT exciting current for the relay setting voltage V_s
I_r = current through the relay at voltage setting V_s
I_m = current through MOV at relay setting V_s.

To read the exciting current value, the CT excitation characteristics are required (see Chapter 3). In this case, say, it is 80 mA.

I_r is given by

$$I_r = \frac{V_s}{R_{2000}} = 134\ \text{mA}$$

The current through MOV is read from a curve published by the manufacturer, not reproduced here. For $V_s = 267$ V, it is 0.

If there are eight breakers, then the minimum internal fault current to operate the relay is 342 A. Again for a low-resistance grounded system, commonly used in industrial distribution systems, the relay may not pickup.

The manufacturers offer choices with respect to MOV selection, based on the calculated knee point voltage and the secondary fault current. The recommended curves are published by the manufacturer, not reproduced here.

CTs of accuracy lower than C200 are not recommended irrespective of the fault currents involved. Also using a higher CT ratio will reduce the secondary current as well voltage setting V_s.

9.3.3 Open-Circuited CT

An open-circuited CT will produce a voltage which may exceed the relay settings, for a heavy load or external fault resulting in misoperations. One level of differential elements is set to a low level to detect this condition (two levels of settings are available for differential elements in each phase), and the second level based on the calculations as above.

9.4 Low-Impedance Current Differential Relays

Electromechanical low-impedance relays have been used in the past for protection of generators. These have percentage restraint characteristics, fixed or variable restraint (see

Chapter 2). Multirestraint differential systems were also used, though high-impedance electromechanical relays were the most popular and are still in use.

Figure 9.8 illustrates the CT connections for a low-impedance current differential relay for bus protection. Note that the CTs with different transformation ratios and secondary currents can be accommodated. The maximum allowable mismatch can be as high as 32:1 according to one manufacturer. The relay scales the secondary currents to the maximum primary current among the CTs, defining a given bus differential zone; I per unit corresponds to the highest rated primary current.

Figure 9.9 is the block circuit diagram of a digital bus protection relay [2]. A brief description of each block is as follows:

- Block1: Currents are digitally filtered to remove DC components and distortions.
- Block 2: A dynamic bus replica is created. The filtered input signals are brought to a common scale taking into account the transformation ratios of the connected CTs.

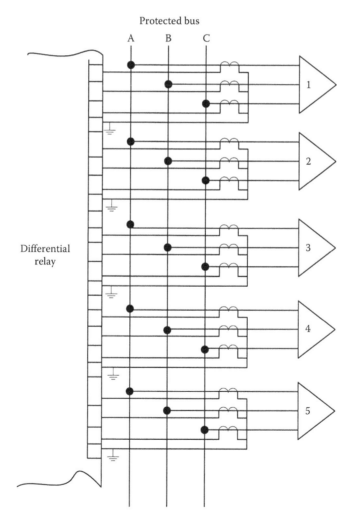

FIGURE 9.8
CT connections in a low-impedance bus differential relay.

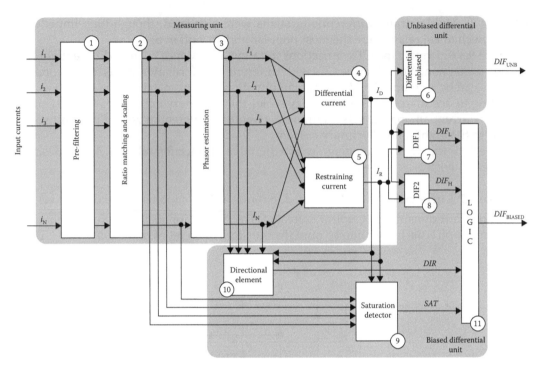

FIGURE 9.9
Block circuit functional diagram of a low-impedance MMPR. (GE. B30 Bus Differential Relay, Manual: GEK-113371.)

- Block 3: The phasors of differential zone currents are estimated digitally.
- Blocks 4 and 5: Restraining and differential signals are calculated.
- Block 6: The magnitudes of the differential signals are compared with a threshold and an appropriate flag indicating operation of the unbiased bus differential protection is produced.
- Blocks 7 and 8: The magnitudes of differential and restraining currents are compared and two auxiliary flags, which correspond to two separately shaped portions of differential characteristics, DIF1 and DIF2, are produced.
- Block 9: The saturation detector analyses the differential and restraining currents as well as samples of the input currents. This block sets up an output flag upon detecting CT saturation.
- Block 10: Directional element block 10 supervises the biased differential characteristics as necessary. The current directional principle is used.
- Block 11: The output block 11 combines the differential, directional, and saturation flags into biased differential operation flag.

A dual breakpoint operating characteristic is shown in Figure 9.10. It operates in conjunction with saturation detection and a directional comparison principle, as described above. The protected zone input current with the highest magnitude is used as the restraining signal. Stability during heavy faults is achieved by dynamic CT saturation detection and current flow direction supervision without affecting sensitivity for internal faults.

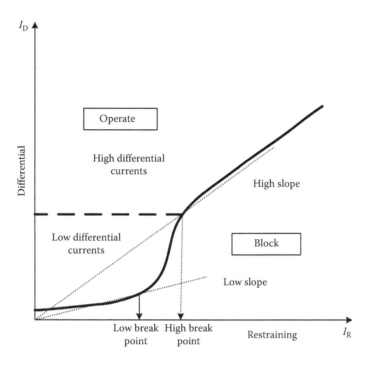

FIGURE 9.10
Two slope characteristics of the relay (see text for explanations and settings).

The differential operating characteristics are divided into two regions. A low pickup setting is provided to cope with the spurious signals when the bus carries a light load and there is no effective restraining signal. The first breakpoint (low breakpoint) specifies the linear operation of the CTs in the most unfavorable conditions of residual magnetism left in the cores or multiple autoclosure shots. The element operates on a 2-out-of-2 basis, applying both the differential and current directional tests. If the differential current and CT saturation are detected, both the directional and current and differential tests are applied. If the CT saturation is ruled out by the saturation detector, the differential principle alone is capable of operating the element. The saturation detractor is an integral part of the bus differential element. It has no settings and uses some of the directional characteristics parameters. Similarly, the directional principle is an integral part of the bus differential element and has no associated settings. It dynamically identifies what appears to be the faulted circuit and compares the current angle with that of the sum of remaining currents in the differential zone. The element declares a bus fault if the angle is less than 90°. The unbiased differential function checks the differential current against an adjustable threshold. Neither the bias nor the directional principles apply.

The second breakpoint is provided to specify the limits of operation of CTs without any substantial saturation. The higher slope acts as a percentage bias regardless of the value of restraining signal.

The directional comparison principle is as follows:

- If all the fault currents flow in one direction, the fault is internal.
- If at least one fault current flows in an opposite direction compared with the sum of remaining currents, the fault is external.

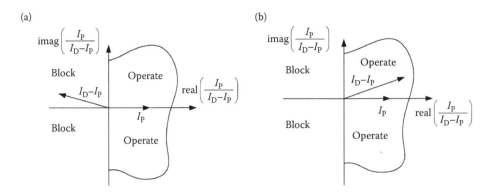

FIGURE 9.11
(a) Directional operating principle during external faults and (b) directional operating principle during internal faults.

At first, based upon the magnitude of fault current, it is determined whether it is fault current or load current; for example, if its magnitude is >2 times the CT rating or >K times the restraint current, where K varies with the number of feeders, it can be flagged as fault current.

Second, for the selected fault current, the phase angle between a given current and the sum of all the other currents is checked. The phase angle check is not initiated for the load current, as the direction will be out of the bus even during internal faults. Ideally, for external faults this angle is 180° and for internal fault it is 0°. The phase angle between a given current and the sum of all the remaining currents is checked. The sum of all the remaining currents is the differential current minus the current under consideration; say for a pth path the angle between I_P and I_D-I_P phasors is checked (see Figure 9.11a and b).

9.4.1 CT Saturation

A historical problem in bus protection systems has been unequal core saturation of the current transformers. This core saturation occurs due to large variations of current magnitudes and residual flux trapped in the individual current transformers. The criteria are that the necessary sensitivity should be provided for internal faults and restrain from operating on external faults. See Chapter 3 for bushing, window-type, wound-type, and auxiliary transformer characteristics. A CT has a certain finite time constant to saturate. Thus, even under high primary currents, it will take a finite time for the CT to saturate. As a result, for an external fault, the differential current will be low during initial period of the operation of the CT, while the restraint signal develops rapidly. Once the CT saturates, the differential current will increase, but restraint current does not change at least for a few milliseconds. For internal faults, both the differential and restraining currents develop simultaneously. The relay declares CT saturation if the restraining signal is higher than the second breakpoint; and at the same time differential current is below the low slope. In order to cope with fast saturation, another condition is checked which uses signals at the waveform level. The sample-based stage of the saturation detector uses a time derivative of the restraining signal di/dt to trace the saturation pattern and saturation detector is capable of detecting saturation occurring in approximately 2 ms into a fault.

Figure 9.12 shows the saturation detection operation and Figure 9.13 shows the output logic of biased differential protection. For low differential signals, both directional and differential principles are applied, DIF_L and DIR.

For high differential signals DIF_H, the directional principle is included only if determined by the saturation detector.

Figure 9.14 shows that even for a very severe CT saturation for the weakest CT, the relay remains stable and the external fault current is seen in the opposite direction. For an internal fault the relay operates in 10 ms [2].

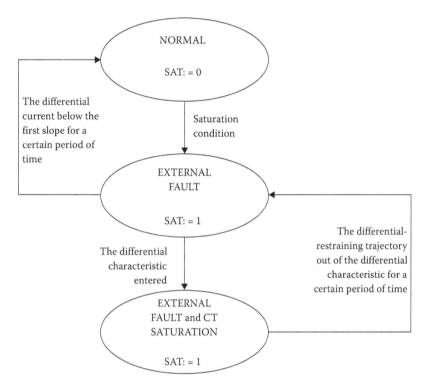

FIGURE 9.12
CT saturation detection in circuit of differential relay in Figure 9.9.

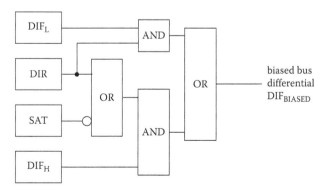

FIGURE 9.13
Output logic of biased differential protection.

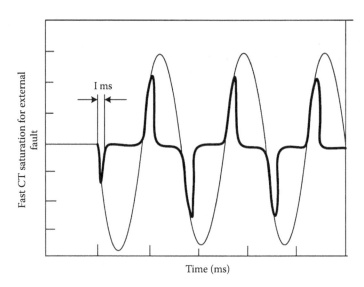

FIGURE 9.14
Illustrates severe CT saturation for an external fault.

9.4.2 Dynamic Bus Replica

Consider that the CTs installed are as follows:

- Circuit 1: 600:5
- Circuit 2: 500:1
- Circuit 3: 600:5
- Circuit 4: 1000:5
- Circuit 5: 500:1
- Circuit 6: 600:5

Note that the CTs have different rated secondaries. Scaling to a common base is performed internally by the relay; the maximum ratio mismatch is 32:1. The relay scales the secondary currents to the maximum primary current among the CTs. One pu corresponds to the highest rated primary current.

Thus, given the above CT ratios, 1000 A is selected as the base, which is 1 pu. Consider now that differential current is exclusively fed from circuits 1 and 2.

- Circuit 1 inputs 1 (one) kA primary current, and circuit 2 inputs 2 (two) kA primary current. These currents are in phase. Then pu current of source 1 is 1000 A: (600:5): 5 A/pu = 1.67 pu. The pu current of source 2 is 2000 A: (500:1): 1 A/pu = 4.0 pu. The pu differential current is (1000 A + 2000 A): 1000 A = 3 pu.
- In reverse, consider circuit 1 supplies 2 kA and circuit 2 1 kA. These currents are in phase. The pu current of circuit 1 is 2000 A: (600:5): 5 A/pu = 3.3 pu. The pu current of source 2 is 1000 A: (500:1) = 2.0 pu. The pu differential current is 3.0.

Continuing further, taking into account CT saturation characteristics, resistance of leads, and burden of CTs, the following currents are guaranteed without significant saturation:

- Circuit 1: 6.0 kA
- Circuit 2: 7.5 kA
- Circuit 3: 5.0 kA
- Circuit 4: 13.0 kA
- Circuit 5: 8.0 kA
- Circuit 6: 9.0 kA

Circuit 3 CT is most likely to saturate. During an external fault circuit 3 CT will carry all the fault currents contributed by the other circuits. Thus, the current of the circuit 3 CT will become the restraining signal for the biased differential characteristics for external faults. Consequently, the higher breakpoint of the differential characteristics should be set not higher than 5000 A:1000 A = 5 pu. The same approach applies to setting the lower breakpoint.

9.4.3 The Differential Settings

Differential pickup: The setting defines the minimum differential current required to operate the biased bus directional element The setting is chosen based on the maximum amount of differential current that may occur under no-load conditions. This setting prevents relay misoperations when the bus carries little power and the restraining signal is too low to provide enough bias in the first slope region of the differential characteristics. The setting may be set above maximum load level to ensure security during CT trouble conditions.

Typical setting = 0.1 pu (setting range 0.05–2.0 pu). Some nuisance trips have occurred if the setting is too low, that is, 0.05 pu).

Differential low slope: The setting defines percentage bias for the restraining currents from zero to lower breakpoint. This determines the sensitivity of the relay for low-current internal faults. The value chosen should be high enough to accommodate spurious differential current resulting from inaccuracies in the CTs operating in their linear mode, that is, under load conditions and distant external faults. The restraint current is created as the maximum of all input currents.

Typical setting: Low slope = 25% (setting range from 15% to 100%).

Differential low breakpoint: This defines the lower breakpoint of the dual-slope operating characteristic. The percentage bias applied for the restraint current from zero to the low breakpoint is given by the low-slope setting. This should be set above the maximum load current.

Typical setting: Low break point non-trip (BPNT) = 2.0 pu (setting range 1.0–30.0 pu).

Differential high slope: The setting defines the percentage bias for restraining currents above the high BPNT. This impacts the stability of the relay for heavy external faults. The setting should be high enough to ride through the spurious differential current due to CT saturation for high external faults.

Typical setting: Bus differential high slope = 60% (setting range from 50% to 100%).

High BPNT: The setting defines the higher BPNT of the dual-slope operating characteristics. The percentage bias applied for the restraining current above the value specified in high BPNT is given by the high slope setting. The high BPNT should be set below the minimum AC current that will saturate the weakest CT.

Typical setting: High BPNT = 8.0 pu (setting range 1.0–30.0 pu).

Differential high set: The setting defines the minimum differential current required for operation of the unbiased differential function. The setting is based on the maximum magnitude of the differential current that might be seen during heavy external faults causing deep CT saturation. The unbiased differential function uses full-cycle Fourier measuring algorithm and applies it to prefiltered samples of input current. It can be disabled by setting very high.

Typical setting: 15.0 pu (setting range 0.10–99.0 pu).

9.4.4 Comparison with High-Impedance Relays

Comparing briefly the high-impedance relays versus low-impedance relays, the following picture emerges:

- In low-resistance differential schemes, the CTs can be shared, that is, the same CTs can be used for, say, overcurrent protection; also multiple CT ratios can be accommodated. For the high-impedance differential, dedicated CTs of the same ratio are required.
- It is possible to detect a shorted CT in low-resistance differential, but not so in high-resistance differential. On an open-circuit CT, the high-impedance differential relay will trip, and a low-resistance differential relay can alarm the situation on unbalance.
- CT polarity compensation can be provided in low-resistance differential schemes but not in high-resistance differential relays.
- The low-resistance differential schemes are somewhat faster as compared to high-resistance differential schemes, 1 cycle typical versus approximately 1.5 cycle.
- A low-resistance differential scheme, generally, has selective circuit breaker failure protection, selective end-zone protection, direct circuit breaker tripping, and individual circuit metering facilities.

Also see References [3–6].

9.5 Direction Comparison Bus Protection

A direction comparison bus protection system compares the direction of current flow in each circuit connected to the bus. If the current in all circuits flows into the bus, a bus fault exists. If the current in one or more circuits is flowing away from the bus, an external fault exists. Fault detectors are used to initiate the system. The system can be used for phase and ground fault protection.

The system requires directional relays, fault detectors, and a timer. Directional relays are used in each circuit connected to the bus. Fault detectors are used to indicate a fault in the vicinity of the bus. Phase fault detectors are instantaneous overcurrent relays connected to bus tie breaker or to one or more of the circuit connections. Ground fault detectors are overcurrent relays connected to transformer neutrals or to one or more circuit connections. A timer is required to permit contact coordination of the directional relays. The contacts of the directional relays are connected in series with timer contact to initiate a trip signal. The system is rather complex and requires rigorous maintenance due to a number of relay contacts.

9.6 Bus Protection Using Linear Couplers

9.6.1 Linear Couplers

Linear couplers are air-cored mutual reactors wound on nonmagnetic toroidal cores. The adjacent circuits do not induce unwanted voltages. These can be designed to fit in the same space as conventional current transformers. Bushing or wound-type units are available for all voltage classes. By design, 5 V is induced per 1000 A of primary current; thus mutual impedance is 0.005 Ω.

The couplers have negligible DC response, and only steady-state conditions need to be considered. The linear coupler method of differential protection uses a series circuit and a voltage relay (see Figure 9.15).

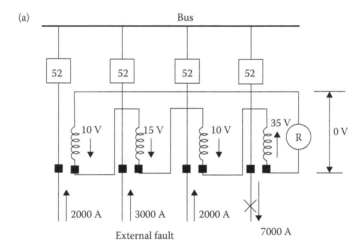

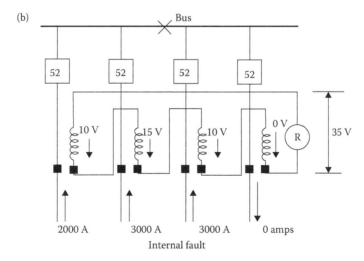

FIGURE 9.15
Differential protection using linear couplers: (a) external fault and (b) internal fault.

Define:
 I_R = current in the linear coupler secondary
 E_{sec} = voltage induced in linear coupler secondary
 I_{pri} = primary current in each circuit rms sym
 $M = 0.005\ \Omega$ = mutual reactance
 Z_c = self-impedance of linear coupler secondary
 Z_R = relay impedance.

Then

$$I_R = \frac{\sum E_{sec}}{Z_R + \sum Z_C} \qquad (9.7)$$
$$= \frac{\sum I_{pri} M}{Z_R + \sum Z_C}$$

Consider that the mutual impedance is accurate within ±1%. For the external fault shown in Figure 9.15a, the worst case occurs if linear couplers on all unfaulted sources were out by +1% and linear coupler on faulted phase is out by −1%. With this maximum error, if the relay is set to operate at X amperes for internal fault, it will also trip for 50X external fault. Apply the safety factor, the maximum range of external fault to internal fault is specified at 25:1.

Thus, calculate the internal and external faults accurately. The external fault current is divided by 25 to give minimum setting. Comparing this setting with the minimum fault current for a fault on the bus will indicate whether a ground relay will be required. When the ground fault currents are limited to 400 A or less, the protection for ground faults will be ineffective.

Due to much advancement in the microprocessor-based technology, the linear coupler differential systems are not much in use.

9.7 Differential Protection of Common Bus Configurations

9.7.1 Single Bus

This is the simplest system, the incoming sources and loads connected to the same bus. A fault on the bus will interrupt all the loads and sources. A coordinated overcurrent protection will clear the bus fault with some time delay, while a differential protection can act fast within one cycle and limit the fault damage. Figure 9.16 shows the differential protection on a single bus system.

9.7.2 Sectionalized Bus

A bus section breaker can be provided and the bus split into two sections. Figure 1.1 shows the differential protection applied to a sectionalized bus. With the overlapping differential zones as shown, only one section of the bus and its incoming and outgoing circuits will be

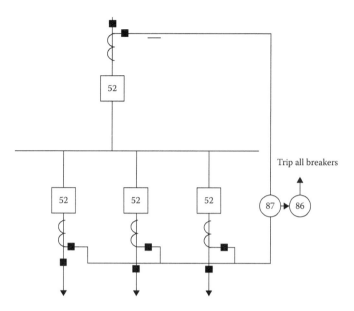

FIGURE 9.16
Differential protection of a single bus.

tripped. Single buses with a bus section breaker are common in medium-voltage industrial systems. The two sources may be operated in parallel, or the bus section can be left normally open, to be automatically closed, if one of the sources goes out of service.

9.7.3 Double Bus Double Breaker

The double bus double breaker scheme is shown in Figure 9.17. Each bus and its breaker are protected by a separate differential relay system. A fault on one bus clears that bus only, the other bus and all circuits remain in service.

9.7.4 Main and Transfer Bus

The main and transfer bus scheme shown in Figure 9.18 is a modification of the single bus scheme, with addition of a transfer bus, transfer breaker, and disconnects. The purpose of the transfer bus is to bypass a circuit breaker during maintenance without circuit interruption. Close the associated normally open disconnect switch connected to the transfer bus and take the circuit breaker out of service.

The transfer bus breaker is included in the differential zone. The transfer bus is a part of the main incoming line with bypass breaker and is not included in the differential zone.

9.7.5 Double Bus Single Breaker with Bus Tie

The double bus single breaker with bus tie, shown in Figure 9.19, provides the operating flexibility comparable to the double bus double breaker scheme. The use of only one circuit breaker results in some differential relaying problems. Both buses are main operating buses and either or both may be energized at a time. When both are energized they may or may not be electrically connected. Each circuit can be switched to either of the two buses. As shown in Figure 9.19, two bus differential zones can be provided, one for each bus with

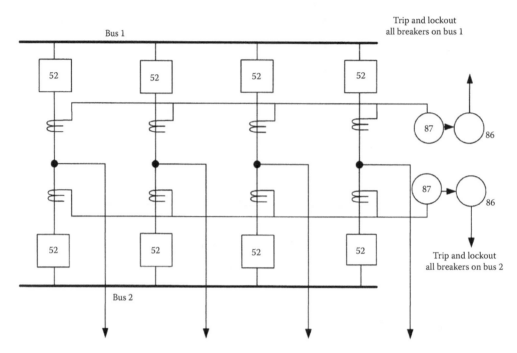

FIGURE 9.17
Differential protection of a double bus double breaker bus configuration.

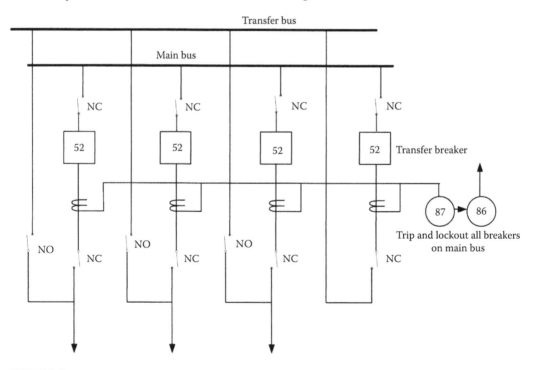

FIGURE 9.18
Differential protection of a main and transfer bus configuration.

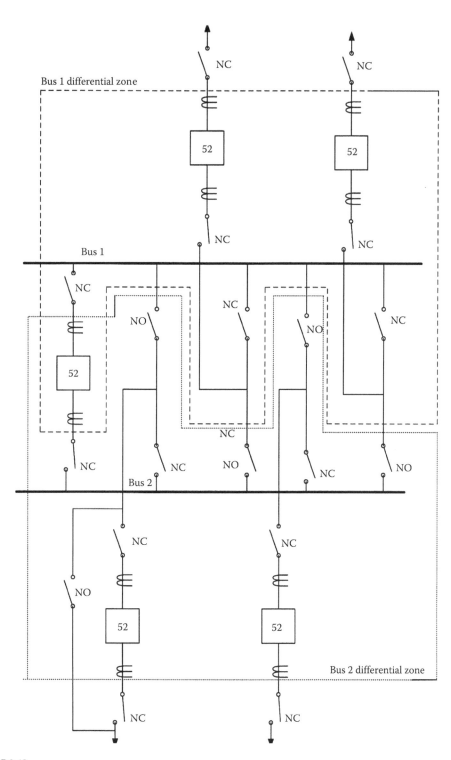

FIGURE 9.19
Differential protection of a double bus single breaker with bus tie circuit breaker configuration.

each zone overlapping the bus tie breaker. Because each primary circuit can be switched to either bus, each relay input circuit and breaker trip circuit must be capable of switching to the appropriate relay zone. Special precautions are required to prevent incorrect differential relay operations. Consider that a circuit is to be transferred without interruption; the buses are tied together through two sets of switches prior to opening the switches to the initial bus. With both the switches closed, currents may flow through this connection between buses, resulting in an inaccurate summation of currents in the differential circuits. One method is to deactivate both relay trip circuits before switching. Another method is to reconnect the relay input circuits to temporarily convert it into a single bus differential zone. Both operations are subject to human error. One may forget to activate the trip circuits after changeover.

Bypass switches can be installed on the main line breakers—one such switch is shown typically in Figure 9.19. During breaker maintenance, one of the buses serves as the main bus, while the other is used as a transfer bus supplied by the bus tie breaker. The bypassed breaker line is directly connected to the transfer bus without a breaker. All other lines are connected to the main bus through breakers. As the relay input devices to the bypassed breaker are also disabled, the differential scheme on the transfer bus now measures the current and may trip. Therefore, this bus section differential relay must be removed from service prior to bypassing the breaker.

9.7.6 Breaker and a Half Scheme

Figure 9.20 shows the breaker and a half scheme, so called because for four circuits six breakers are used; that is, one and a half breakers per circuit are used. It has similar advantages as bus-relaying as the double bus double breaker scheme. It is used extensively for large multi-circuit high-voltage systems. Two operating buses have separate differential protection. Each line section is supplied by the two buses through two circuit breakers. The center circuit breaker serves both lines. The CT interconnections for each line section are shown as dashed lines in this figure. The voltage for line relays must use line side voltage transformers. Line fault trips two breakers but do not cause interruption of service to other circuits.

Directional-type bus protection schemes are not recommended for this scheme because the direction of flow of fault current components in extremely low-impedance bus network is not predictable during bus faults.

9.7.7 Ring Bus

A ring bus scheme is shown in Figure 9.21. Bus differential protection is not required. Each breaker serves two lines and must be opened for faults on either of the lines. The bus section between the breakers becomes part of the line, and bus protection is not applied. The interconnection of CTs for each line is shown in dashed lines in this figure. If the ring is open due to any reason, a fault on the line may separate the other lines and the bus, resulting in significant interruption to the power system network. Line protection voltage is obtained from the VTs shown on each line.

9.7.8 Combined Bus Differential Zones

A combined bus differential zone is shown in Figure 9.22. Note that the bus and transformer are included in the differential zone. Generally, the transformer differential relays, rather than bus differential relays, are used in this application to accommodate transformation ratio, transformer inrush currents and harmonics.

Bus-Bar Protection and Autotransfer of Loads

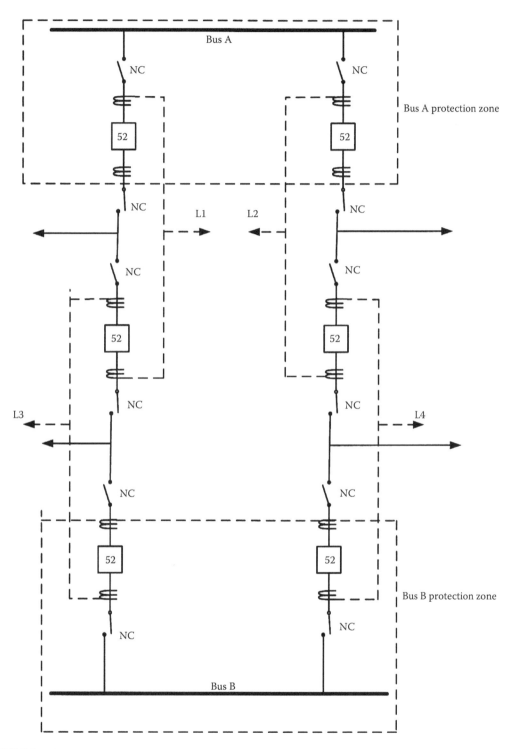

FIGURE 9.20
Differential protection of a breaker and half bus scheme.

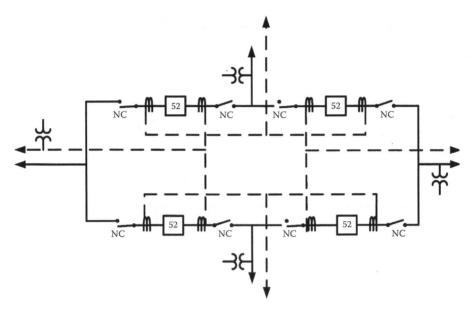

FIGURE 9.21
Typical four circuit ring bus. The differential protection is not applicable. Bus sections protected as part of line and connected equipment.

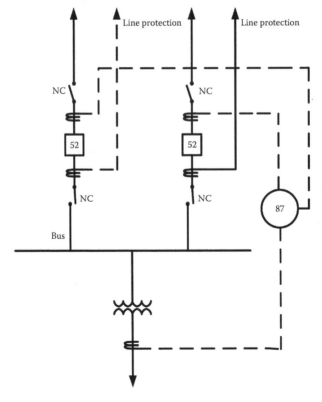

FIGURE 9.22
Combined differential protection of a transformer and bus.

9.7.9 Ground Fault Bus Differential Protection

We noted that the bus differential protection may not be sensitive enough to clear ground faults in the protected bus zones for line-to-ground faults, where the ground fault may be limited to 400 A or less in resistance-grounded systems in industrial distribution systems.

A ground fault bus differential scheme using core balance CTs is shown in Figure 9.23a. Depending on the current ratings involved, it may not be possible to use core balance CTs embracing large numbers of cables involved, even when the largest window CTs, commercially available, are applied. An alternative using residual connection of the CTs is shown in Figure 9.23b.

Simple overcurrent relays without any restraint characteristics have been applied in the industrial environment. Due to small magnitude of the ground fault currents available, the CT saturations are not a consideration.

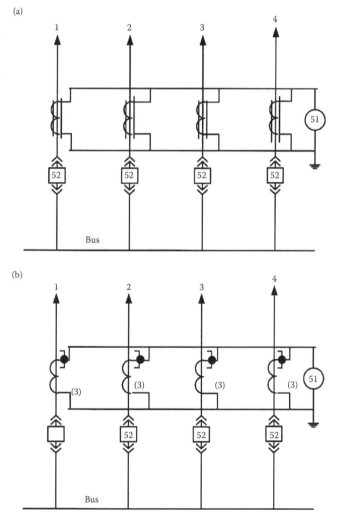

FIGURE 9.23
Ground fault differential protection of bus: (a) using zero sequence CTs and (b) with residual connections of phase CTs.

The scheme with core balance CTs will provide a more sensitive ground fault protection. Consider a ground fault of 400 A within the protected zone and core balance CTs of 50:5. With minimum pickup setting of 0.5 A on the overcurrent relay, low levels of ground fault current of 5 A primary can be theoretically picked up. On the other hand, consider 2000:5 ratio phase CTs. Then the minimum pickup for ground fault currents is 200 A. Sometimes auxiliary wound-type CTs are introduced in the residual circuit, but this should be avoided (see Chapter 3).

One solution to increase the sensitivity is to use an overcurrent relay rated for one ampere CT secondary, but still use 5 A secondary main CTs. Then the pickup setting can be reduced from 200 A to 50 A. The one ampere overcurrent relay coil, normally, can withstand the higher secondary current for a short duration. A time delay of six cycles is recommended.

9.8 Reclosing

We discussed reclosing in Chapter 4, with its logic and an example. For bus faults, reclosing the breaker after tripping may have a profound impact on system restoration and system stability.

The breakers can be closed manually, after repairs to the bus are made. This is the safest, but results in complete interruption of the loads and sources served from the bus.

Breakers can be closed through supervisory control, provided no lockout of breaker close circuit exists, or the lockout relays are set remotely.

Reclose the breakers with automatic reclosing relays, provided no lockout of breaker close circuits exists.

The choice depends on effects or system shutdown for a bus fault and consequences of closing the beakers on a bus fault. An air-insulated bus in the outdoor substations may be subjected to faults, such as flashovers, insulator pollution, and lightning. A rigid outdoor bus is an open strain bus and is an example of such systems. If these faults are promptly cleared, there is a good probability that reclosing without inspection may be successful. One of the breakers may be equipped with single-shot reclosing relay to close the breaker for a dead bus. However, the differential relay should be sensitive to trip on a reduced fault current. If the reclosing of this breaker is successful, the other breakers on the bus can be reclosed with automatic reclosing relays, programmed to close the breakers for a hot bus–hot line condition.

The system may be able to withstand the initial fault without exceeding the transient stability limit, but reclosing on to the fault again may exceed the transient stability limit. In this case delayed reclosing or delayed supervisory reclosing can be applied, if sufficient time is allowed for the system oscillations to decay.

Circuit breaker failure can be detected by providing each circuit breaker with a fault detector, which working with the bus relays, initiated a timer to trip other circuit breakers (see Chapter 4).

9.9 Bus Transfer Schemes

When a bus is provided with two sources of power, the bus transfer schemes are the strategies to quickly restore power to the bus, if one of the sources fails. It is obvious that the

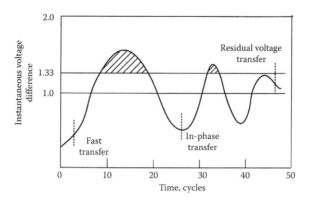

FIGURE 9.24
Fast bus transfer methods.

two sources must be in phase. Modern bus transfer relays will detect the synchronizing between the sources before allowing a transfer.

The transfer of rotating motor loads poses a problem, as discussed in Section 9.9.2. There are three schemes of bus transfer (Figure 9.24). These show the following:

- Fast bus transfer
- In-phase transfer, and
- Residual voltage transfer

Figure 9.24 shows the three modes of transfer with respect to the voltage *difference profile* across the bus section breaker—the fast transfer and in-phase transfer are better compared to residual voltage transfer with respect to motor stresses. If we consider the residual voltage of the motor to completely decay to zero, the voltage across the bus section breaker is 1 per unit, while for fast bus transfer and in-phase transfer it will be much lower and less stressful to the motor. The motor transients will be much reduced.

9.9.1 Fast Bus Transfer

Figure 9.24 suggests that large transients can be avoided by fast reclosing, so that the motor internal voltage is not much displaced, and the incoming supply voltage is almost in phase with the decaying motor EMF.

9.9.2 Residual Voltage Transfer

An alternative will be to let the motor speed fall, and the internal voltage decay to a low value, say 25% of the supply voltage to avoid transient torques. Tests made on motors of 100–1500 hp establish that the critical time for maximum inrush currents vary from 15 to 30 cycles. The voltage decay in a group of motors will be faster than a single large motor of an equivalent size. Figure 9.25 shows the decay of motor residual voltage and phase angle displacement and Figure 9.26 is a polar plot of the motor residual EMF on disconnection from the power supply. However, this mode of transfer defeats the purpose of maintaining continuity of operation of connected loads. The motors may decelerate to a level, where the reacceleration may not be possible on reconnection.

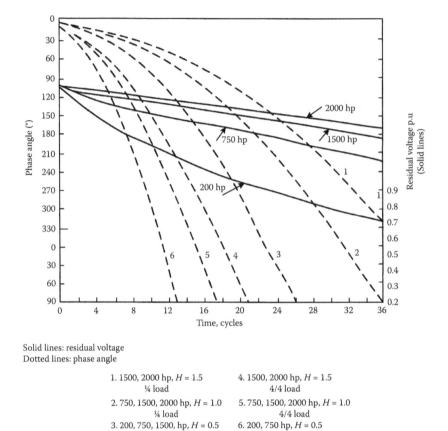

FIGURE 9.25
Decay of internal EMF of induction motors and phase angle separations on disconnection from power supply.

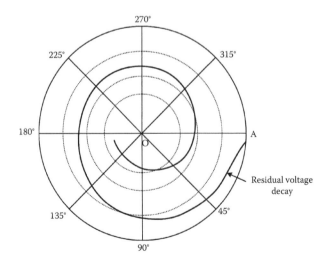

FIGURE 9.26
Polar circuit diagram of decay of induction motor internal EMF on disconnection from power supply.

9.9.3 In-Phase Transfer

When the motor is disconnected from the supply lines, its voltage and frequency fall, as amply illustrated above. The *supply* frequency, however, remains constant. The residual voltage falls in and out of phase with the supply voltage at increasing rate as the motors slow down. If the alternate supply breaker is closed exactly when the voltages are in phase, bumpless transfer can be achieved. Care has to be exercised, as the two sources may not be exactly in phase when the transfer is started and the alternate supply phase may change further when the normal supply breaker is opened, due to changes in the system interconnections.

The in-phase transfer must consider the circuit breaker closing time and close the circuit breaker ahead of zero phase angle by accounting for breaker closing time (see Figure 9.27). Assume that the breaker closing time is t_b seconds. Then the relay must initiate closing of the breaker t_b seconds, before the phase angle reaches 360°, which means at time t_a. In order to predict when the phase angle will be zero, Taylor's series can be used.

The phase time curve can be estimated by Taylor's series, considering first three terms:

$$\phi(t) = \phi(t_1) + \phi'(t_1)(t - t_1) + \frac{\phi''}{2}(t - t_1)^2 \tag{9.8}$$

where

$$\phi(t_1), \phi'(t_1), \phi''(t_1)$$

are the value of phase at t_1, the derivative at t_1, and second derivative at t_1, respectively.

For $\phi(t) = 0$,

$$\phi(t_1 + t_b) = \phi(t_1) + \phi'(t_1)t_b + \frac{\phi''(t_1)}{2}t_b^2 \tag{9.9}$$

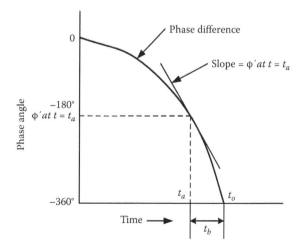

FIGURE 9.27
Illustration of phase bus transfer.

Equation 9.9 gives prospective phase value at t_b when $\phi(t_1 + t_b) = 0$, by predetermination of ϕ, ϕ' and knowing the value of t_b.

The in-phase transfer will be ineffective if at the time of transfer there is not enough bus voltage to support motor reacceleration.

A transfer logic controller is available that performs all the three modes of transfers, user selectable. Studies [7–10] provide further reading.

9.10 Momentary Paralleling

During planned load transfer, momentary paralleling of two sources is widely used, i.e., bus section breaker in Figure 9.28a is closed, before any of the two source breakers are opened. (Synchronism check features are incorporated.) A question arises that when the two sources can be paralleled in the shutdown or planned load transfer modes, why these cannot be continuously run in parallel? Appropriate protective relaying schemes are available for two continuously running parallel circuits, which will selectively trip the faulty circuit and the loads will experience lesser transients, as compared to switching on failure of a source. While continuous parallel operation is much desired from the operational point of view, paralleling increases the short-circuit duties on the circuit breakers. During momentary paralleling, a calculated risk is taken that for the time duration involved, the probability of a fault is low. If a fault does occur, it can destroy the equipment and can be hazardous to human life. This practice should be carefully reviewed.

9.10.1 Fault Conditions

The fault conditions should be distinguished from simple loss of source voltage, say, due to inadvertent switching. The fault voltage dips will be more severe, depending upon the location and nature of the fault. Obviously for any fault in the bus itself or in the transformers (Figure 9.28b), the autotransfer is blocked. A fault voltage dip will give rise to transients in the connected motor loads and these may not survive operation when the alternate source of power is restored.

Sometimes simultaneous opening and closing of the breakers is implemented. The control signals to open and close are applied at the same time. The dead time including sensing of the sources may be one to three cycles. There is some risk involved, in the sense that a breaker may fail to open. Alternatively, the normal breaker is first opened and interlocked so that the standby breaker cannot be closed, unless the normal breaker has opened. The dead times may be 10 cycles or more. The stability of motor loads will depend on how fast the power is restored.

9.10.2 Dropout of Motor Contactors

Another parameter of practical consideration is that magnetic contactors for motor starting will drop out quickly—for low-voltage motor starters, the contactors may drop out in the very first cycle of voltage dip exceeding 30%. NEMA E2 starters are used for starting and control of medium voltages (in a certain range), and these have DC contactors, which may take a little longer to drop out, depending upon the trapped residual magnetism in the

Bus-Bar Protection and Autotransfer of Loads

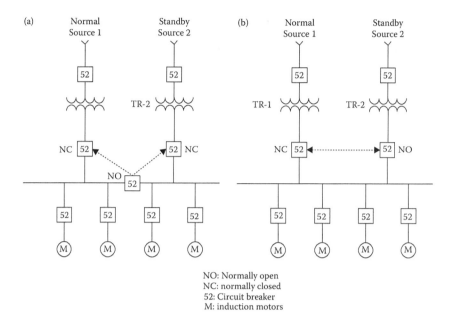

FIGURE 9.28
(a) Momentary paralleling for load transfer from one source to another and (b) a circuit for simulation of bus transfer with EMTP simulation.

coils. It is possible that the motors are tripped out from service because of dropout of contactors, while these may be still able to ride through the voltage dip.

Latched type of contactors and stabilizing the contactors for undervoltage dips for certain time duration are possible options. This has to be attempted carefully, as on voltage restoration all the connected motor loads, which lost speed during the voltage dip, will take high inrush currents to accelerate. This large current may cause further voltage dips in the system impedance, a cumulative effect, to precipitate a shutdown. Thus, a rigorous study is required.

9.10.3 Autotransfer of Synchronous Motors

Autotransfer of power on synchronous motors is not attempted. The phase angle between the motor-generated voltage and supply connection will vary from 0° to 360° per slip cycle. An autoclosing operation may produce short-circuit current equal to 2.5 times the normal short-circuit current. This will subject the windings to stresses approaching 6.5 times the normal short-circuit currents. The transient torques produced by the fault currents have oscillatory components and the peak torque may be as high as 30–40 times the normal full load torque.

On a voltage dip, the excitation to the motor may be removed and reapplied shortly thereafter if the system conditions are favorable.

Example 9.2

The transients on reconnection of induction motor loads are studied with EMTP modeling. Consider that five motors are connected to a 6.6 kV bus, served from two alternate sources. The normally closed breaker is opened and the standby breaker is closed, the total duration of loss of power (including relaying and closing time of breakers) is

500 ms (arbitrary here, for the example). Each of the two transformers in Figure 9.28a is rated at 80 MVA, and transformer impedance is 9% on 80 MVA base. Figure 9.28b shows that only one transformer is in service and on its failure the alternate transformer is brought in service.

Figure 9.29a shows that in approximately 500 ms, the bus voltage in phase *a* has decayed by 43%. The restoration of power at 500 ms does not give rise to much voltage escalation. The motors accelerate and remain stable, the slip which had increased to

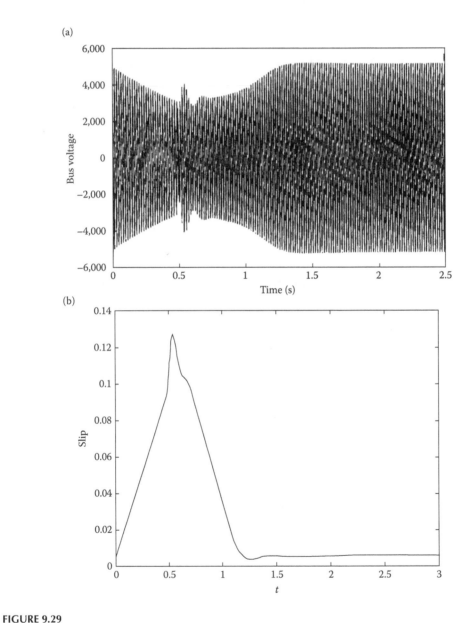

FIGURE 9.29
(a) Bus voltage decay in 500 ms after disconnecting the motor from power source, rebuilt of voltage on restoration of voltage source, (b) motor slip and its recovery, (c) phase *a* current of the motor on reconnection to power supply, and (d) transients in electromagnetic torque.

(*Continued*)

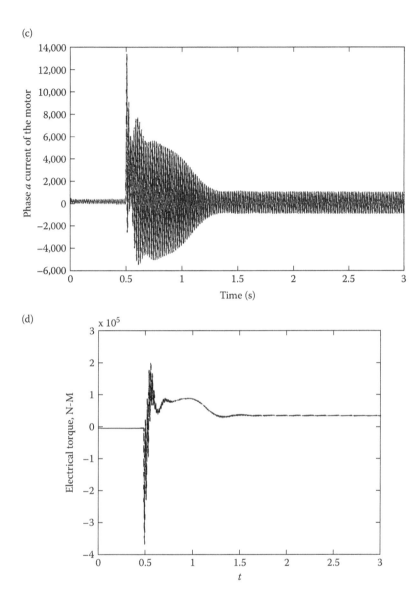

FIGURE 9.29 (CONTINUED)
(a) Bus voltage decay in 500 ms after disconnecting the motor from power source, rebuilt of voltage on restoration of voltage source, (b) motor slip and its recovery, (c) phase a current of the motor on reconnection to power supply, and (d) transients in electromagnetic torque.

approximately 13% rapidly decreases (see Figure 9.29b). The transient electromagnetic torque (Newton meters) of one of the motors are shown in Figure 9.29c and d, respectively. The current on reconnection jumps to 13,800 A, approximately 15.7 times the full load current of the motor. The 80 MVA transformer will therefore experience a sudden loading of 13.8×5 = 69 kA. The ANSI/IEEE short-time withstand capability of the transformer for through-fault is 77.76 kA for 2 s. The transient load for a couple of cycles can be safely allowed. The example illustrates a successful transfer. The rating of the transformers and the parameters of motors are important. (The parameters of motors are not shown.)

References

1. GE. PVD Differential Voltage Relay, Manual: GEK-45405.
2. GE. B30 Bus Differential Relay, Manual: GEK-113371.
3. Westinghouse Electric Corporation. *Applied Protective Relaying*, chapters 8, 9, and 14. Westinghouse Electric Corporation, Coral Springs, FL, 1982.
4. Westinghouse, ABB. Bus Protection Guide, Publication NO. PRSC-9, November 1989.
5. IEEE Standard C37.103. Guide for Differential and Polarizing Relay Circuit Testing.
6. IEEE Standard C37.97. Guide for Protective Relay Applications to Power System Buses, 1979.
7. CC Young, JD Jacobs. The concept of in-phase transfer applied to industrial systems serving essential service motors. *AIEEE Trans*, 79(6), 508–516, 1961.
8. KG Williams, A Khayer. Auto-transfer of induction motors in industrial systems. *Electr Rev*, 395–397, 1968.
9. JM Daly. Load transfer strategies for machine and other inrush loads. *IEEE Trans Ind Appl*, 34(6), 1404–1410, 1998.
10. DG Lewis, WD Marsh. Transfer of steam-electric generating station auxiliary bus. *AIEEE Trans*, 74(3), 322–331, 1955.

10
Motor Protection

10.1 Motor Characteristics

See Chapter 7 of Volume 2 for the following:

- Induction motor models
- Double-cage induction motor rotors
- NEMA starting characteristics
- Effects of voltage dips and frequency on operation of induction motors
- Calculations of voltage dips on impact loads and motor starting
- Motor starting methods—number of starts and load inertia, dynamic calculations of motor starting and starting time
- Starting of synchronous motors
- Stability considerations of induction and synchronous motor on starting
- Acceptable voltage dips and EMTP simulations
- Synchronous motors driving reciprocating compressors.

These aspects are not repeated here and knowledge of these is essential for motor protection.

10.2 Motor Protection

10.2.1 Medium-Voltage Induction Motors

Motors rated above 250 hp are normally connected for medium-voltage operation at 2.4 kV, 4.16 kV, or 6.6 kV distribution systems. Motors rated above 10,000 hp may be connected for 13.8 kV service.

Microprocessor-based multifunction relays (MMPRs) for protection of medium-voltage motors may have the following functionality:

- Motor thermal model, which will account for overload curves, unbalance biasing, hot/cold safe stall ratio, motor cooling time constants, start inhibit, emergency start, and resistance temperature detector (RTD) biasing
- Phase overcurrent protection as a backup or may replace thermal protection

- Short-circuit protection
- Locked rotor protection, including special distance relays
- Motor start supervision, signaling unsuccessful start
- Mechanical jam and acceleration times
- Negative-sequence current protection, time delay, alarm, and trip
- Sudden load rejection
- Stator and bearing temperature monitoring, polling, and trip
- Phase differential protection, full differential, or flux balance differential
- Ground fault protection, time delay, and instantaneous
- Voltage and frequency protection
- Phase reversal
- Protection system blocks, say from repeated starts or starting before the set time has elapsed
- Metering functions
- Zero-speed switch inputs
- RTD protection of stator windings and bearings, ambient temperature RTDs for temperature compensation
- In addition, it will have power elements, current and voltage inputs, digital and analog inputs and outputs, monitoring and metering, event recorder, advanced motor diagnostics and communications

The older bimetallic thermal elements are no longer in use for the medium-voltage motors. Figure 10.1 shows the functionality of a modern MMPR for motor protection.

10.2.2 Medium-Voltage Synchronous Motors

In addition to the protections described earlier for medium-voltage induction motors, the synchronous motors will have the following:

- Synchronizing protection, which will vary depending on whether the motor is brush type or brushless type; the brushless synchronous motors have practically replaced the brush-type motors
- Loss of excitation power, underexcitation, overload protection of the excitation system
- Pull-out (PO) protection—PO power factor can be calculated
- Automatic resynchronization

10.2.3 Low-Voltage Motors

The motors rated 250 hp or below are normally connected to low-voltage systems, 480 V or 600 V.

Bimetallic thermal elements of class 10 or 20 are still in use for the low-voltage motors, although digital relays providing some of the features of protection described for the medium-voltage motors are available. There is a recent study of the motor thermal overload relays and phase monitors to power quality events [1]. This paper describes the tests carried

Motor Protection

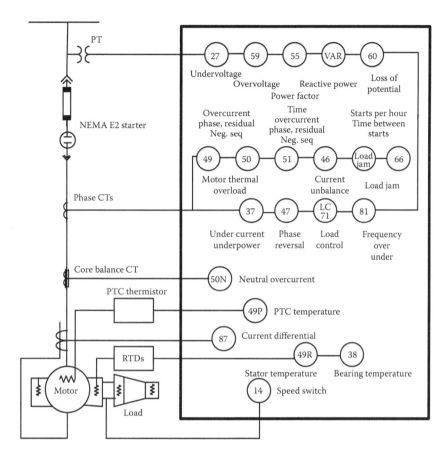

FIGURE 10.1
Functionality of a modern MMPR for motor protection.

out on nine thermal relays widely used for motor protection. The findings reveal that the devices did not follow industry definitions of voltage unbalance and did not meet their published specifications. Inconsistent behavior was observed for thermal overloads. The performance is sensitive to the phase that either caused the unbalance or was opened during single phasing. Nuisance trips may occur or failure to trip can occur when necessary.

Digital low-voltage motor protection relays are available. Like MMPRs for medium-voltage motor protection, the motor thermal characteristics can be coordinated with the thermal curves provided in the relay. Accurate thermal models for phase unbalance are available.

10.3 Motor Protection and Coordination Study

A motor coordination study is conducted as follows:

- It will start by first plotting the motor starting current time curve. Boiler ID (induced draft) fan motors driving large inertia may take 45 s or more to start. The calculation of the starting time of the motor for a given load requires computer simulation and dynamic motor starting studies, not discussed here (see Chapter 7 of Volume 2). These data should be accurate, based upon motor starting studies.

- Next, the motor thermal withstand curve is required. Special care is required when the locked rotor time of the motor is much less than the starting time. Though during starting some heat dissipation will take place due to rotation of the rotor, as compared to the full-voltage locked rotor withstand time, yet to be safe, the motor should be designed with locked rotor hot withstand time slightly greater than the starting time. If this cannot be achieved, additional protection features are required. One solution is to use impedance-type relay for protection, another method is to use zero-speed switch, which senses the rotation of the motor and its acceleration up to speed. This bypasses the protection for a predetermined time.

- The MMPRs have built-in thermal protection curves—maybe 10 or more, which can be coordinated with the motor thermal withstand curve. Also these relays provide the facility to generate a custom-made curve for the specific application. A proper standard curve built in the relay or custom designing a curve to protect the motor thermal withstand is required.

- Figure 6.9 shows the fundamental coordination principles with respect to motor thermal withstand curve, type R fuses, and motor starting characteristics. The short-circuit protection coordination, with type R fuses with contactors for NEMA E2 starters [2], is considered. Choosing an appropriate fuse size for the motor starting and coordinating with the contactor interrupting rating and motor starting curve has been discussed in Chapter 6.

- Low-voltage high-efficiency motors may draw a first cycle starting current exceeding 15 times the rated full-load current. NEC (National Electric Code) allows setting magnetic only [motor circuit protectors (MCPs)] up to 17 times the motor full-load current.

- Selection of appropriate motor starter fuses, MCPs or MCCBs (molded case circuit breakers) for low-voltage motors come into the picture. NEC Table 430.52 [3] specifies the maximum settings on nondelay fuse, dual element time delay fuse, instantaneous trip breaker (MCP), and inverse time breaker for various types of motors. Practically, much lower settings for coordination will be adequate, also see Chapter 6.

These aspects will be demonstrated with examples.

10.4 Coordination with Motor Thermal Damage Curve

IEEE Standard C37.96-2000, Guide for AC Motor Protection [4], recommends overcurrent relays for overload and locked rotor protection and IEEE standard 620-1996, Guide for Presentation of Thermal Withstand Curves for Squirrel Cage Induction Motors [5], lays down guidelines that a manufacturer must follow to supply the thermal damage curves. The allowable locked rotor thermal limit is given for rated locked rotor current. It can also be given as accelerating thermal limit curves both for cold and hot starts at various voltages, 100%, 90%, and 80%. Induction and synchronous motor starts are specified in NEMA MG-1 [6]. This provides for two starts in succession coasting to rest with the motor initially at ambient temperature and one start when the motor is at a temperature not exceeding its rated load operating temperature. The motors may be specifically designed for higher starts, inching, and jogging. Figure 7 of Standard C37.96 is reproduced as Figure 10.2. As thermal conditions are protected by overcurrent protection, a question of correlation between the two arises.

Figure 10.3 shows the thermal withstand curve supplied by a manufacturer for 400 hp, 2.4 kV motor, service factor (SF) = 1.15. The basic thermal protection model is given by

$$t_{\text{H-Curve}} = T_{\text{th}} \ln\left(\frac{I^2 - I_{\text{H}}^2}{I^2 - I_{\text{SF}}^2}\right)$$

$$t_{\text{C-Curve}} = T_{\text{th}} \ln\left(\frac{I^2 - I_{\text{C}}^2}{I^2 - I_{\text{SF}}^2}\right) \quad (10.1)$$

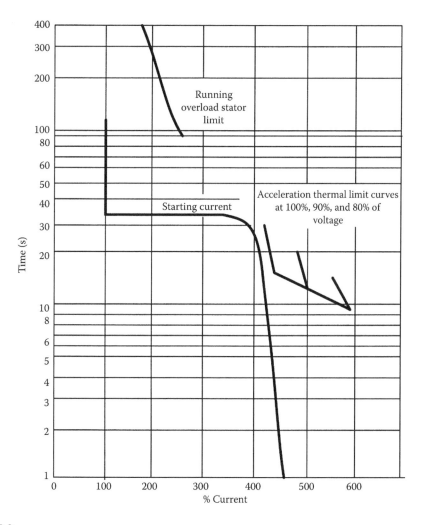

FIGURE 10.2
Thermal withstand curve of a motor. Reproduced from ANSI/IEEE C37.96, IEEE Guide for AC Motor Protection, 2012.

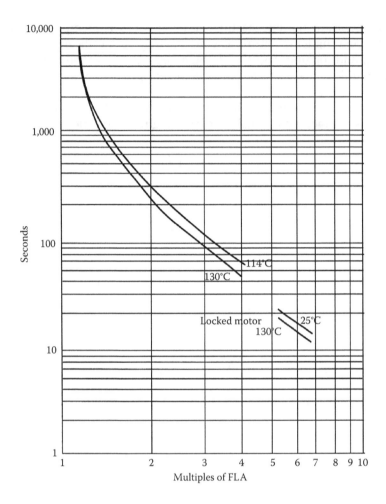

FIGURE 10.3
Thermal withstand curve of a 400 hp motor, hot and cold characteristics.

where T_{th} is thermal time constant, I is the motor current in pu of full-load current, and I_{SF} is the current at service factor. I_H is current that raised temperature to 130°C, and I_C is the current that raised the temperature to 114°C.

We can write the constraint that

$$\frac{I_H^2}{I_C^2} = \left(\frac{130-25}{114-25}\right) = 1.179 \qquad (10.2)$$

By solving the following equations in terms of SF a fit can be obtained:

$$I^2\left[1 - e^{-\frac{t_{H\text{-curve}}}{T_{th}}}\right] + I_H^2 e^{-\frac{t_{H\text{-curve}}}{T_{th}}} = 1.15^2$$

$$I^2\left[1 - e^{-\frac{t_{C\text{-Curve}}}{T_{th}}}\right] + \frac{I_H^2}{1.179} e^{-\frac{t_{C\text{-Curve}}}{T_{th}}} = 1.15^2 \qquad (10.3)$$

Consider a point on thermal limit curve, say at 2.0 per unit current, then from the hot curve in Figure 10.3, time is 223 s. Considering $T_{th} = 1370$ s gives $I^2_H = 0.846$ and $I^2_C = 0.717$. Then any point on the thermal curve can be calculated:

$$t_H = 1370 \ln\left(\frac{I^2 - 0.846}{I^2 - 1.15^2}\right)$$

$$t_C = 1370 \ln\left(\frac{I^2 - 0.717}{I^2 - 1.15^2}\right) \quad (10.4)$$

Equation 10.4 is the time–current characteristics. The overcurrent model is implemented by integrating the reciprocal of hot thermal limit curve as specified in IEEE Standard C37.112-1996 [7]. The incremental equations for this process are as follows:

For $I > 1.15$,

$$\theta_n = \theta_{n-1} + \frac{\Delta t}{t_H} \quad (10.5)$$

For $I < 1.15$,

$$\theta_n = \left(1 - \frac{\Delta t}{1370}\right)\theta_{n-1} \quad (10.6)$$

Equation 10.5 is used to calculate response of overcurrent relay above the pickup current. θ_n and θ_{n-1} are consecutive samples displaced by one time step. An overcurrent relay can trip; say on a cyclic load, even before the thermal limit is reached as the relay does not have a thermal memory. Exponential reset is used with thermal time constant to coordinate with cooling.

When selecting an overcurrent relay curve to protect a given thermal withstand characteristic of the motor, it is important that overcurrent characteristic is matched closely to the thermal characteristics. The use of microprocessor-based relays provides more accurate means of determining the coordination under starting conditions. As stated before, in modern MMPRs for motor protection, there are two options:

- A standard built-in curve can be selected to match the thermal curve.
- A user can create a curve to match the thermal curve.

The test of a thermal model is its ability to adequately protect the motor from overheating during cyclic overloads. To this end, see the two settings provided to protect the same motor, settings A and B, in Figure 10.4a and b. The response of the two settings to cyclic overloads is shown in Figure 10.5. Setting A prematurely trips the motor on cyclic overloads.

Though we talk of thermal time constant as a single number, practically, the thermal model of a machine is fairly complex. The slots embedded in the iron core—the overhangs in air, the frame, the end rings, the shaft, and the rotor structure—all have different materials, mass, and conductivity varying over large limits.

The overcurrent model and thermal model may not be in step and it is possible that the motor may be prematurely tripped or subjected to overloads, though attempts have been made that the two models correlate as much as possible.

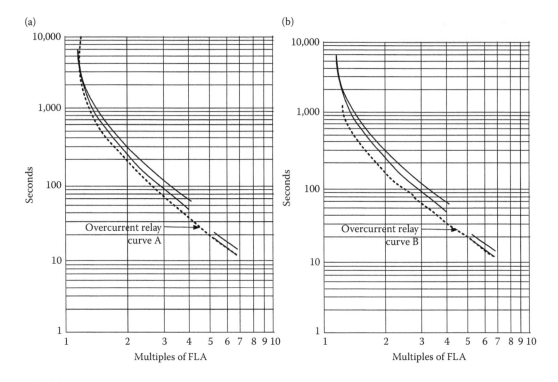

FIGURE 10.4
(a) Overcurrent setting A to protect the thermal withstand curve of Figure 10.3. (b) Overcurrent setting B to protect thermal withstand curve of Figure 10.3.

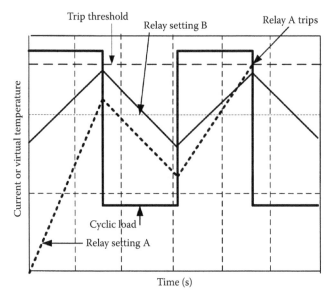

FIGURE 10.5
Nuisance trip due to setting A in Figure 10.4b on motor cyclic loads.

Motor Protection

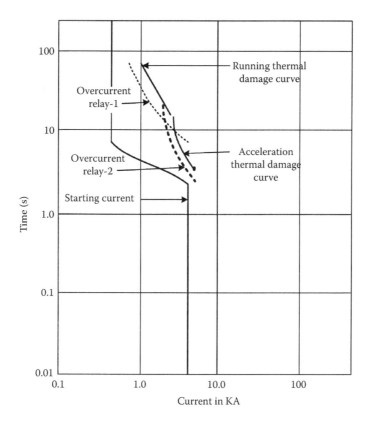

FIGURE 10.6
Two-step overcurrent characteristics for coordination with motor thermal withstand curve and starting curve.

Figure 10.6 shows a two-step overcurrent relay for coordinating with motor thermal withstand curve and also the starting characteristics. As stated before, a problem of starting can arise when the locked rotor withstand time is much lower than the motor starting time. With a properly set thermal protection system, it will not allow starting the motor. One solution is that the zero-speed switch (device 12 in Figure 10.7a) acts in conjunction with an overcurrent element. As soon as the motor starts to accelerate, the 51 start is disabled, leaving the overcurrent protection to longer time overcurrent relay 51. Also see Figure 107b.

Example 10.1

This is a practical coordination of protection of a 1000 hp, 2.3 kV motor, connected to a 2.5 MVA transformer. The selection of appropriate motor fuse size and motor protective relay settings is illustrated. Consider that the motor starting curve has been calculated and the thermal damage curve is supplied by the manufacturer. As a first step, these two curves can be plotted, see Figure 10.8. The motor is controlled through a NEMA E2 starter, with a 400 A vacuum contactor.

Increase the starting current (ignoring the increased inrush during the first cycle or so) by 10%, dotted line in this figure and select R-type fuse rating, so that it clears this dotted line as well as the locked rotor withstand of the motor. A fuse should not operate for the locked rotor condition of the motor, and the vacuum contactor should clear this condition. An R-type fuse of 18R is selected and it meets these criteria. The symmetrical

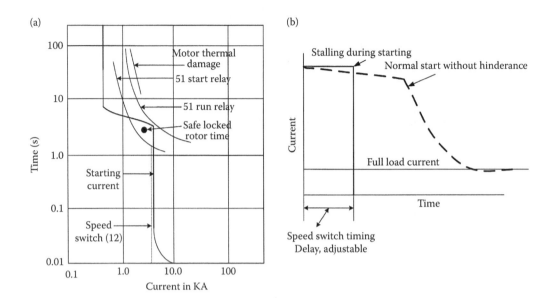

FIGURE 10.7
(a) Situation when the locked rotor time of motor is smaller than the motor starting time, speed switch characteristics shown. (b) Speed switch settings to sense rotation of the motor.

interrupting current of 8.5 kA for the 400 A vacuum contactor based on manufacturer data is shown in this figure. It is also pertinent to draw the contactor dropout line. It is recognized that the dropout time varies with the residual magnetism and is not a fixed number. A dropout time of 0.03 s is shown. In the area of lack of coordination between the fuse and the contactor interrupting rating, marked in this figure, the vacuum contactor will clear a fault beyond its interrupting rating.

The thermal damage curve of the motor is protected and the instantaneous setting with a time delay of 0.3 s protects the fuse and trips the contactor. This delay of 0.3 s can be further reduced.

The motor damage curve for the cold condition is plotted. The relay allows reducing the thermal capacity of the motor based upon the ratio of the locked rotor time in the hot and cold conditions. The manufacturer's recommendation for the particular relay type used should be followed.

The starting time is 15 s and with the coordination shown the motor is capable of two consecutive starts from cold, with motor coasting to rest between starts, per hour, or one start with the motor at the operating temperature, according to NEMA standards [6].

The transformer is protected with primary fuse of 200 E. It is assumed that there is no main secondary breaker for the 2500 kVA transformer.

We talked about coordinating the thermal model with overcurrent relay settings in the above discussions. *Modern MMPRs do not use conventional overcurrent relays at all.* Tracking the motor temperature with simple overcurrent relays is a problem when we consider cyclic loads like chipper motors and crushers. The cyclic overloads cause an overcurrent relay to false trip. The thermal model in the relay can accurately track the motor heating throughout the entire cyclic load cycle. The relay generates a thermal model based on the following:

- Locked rotor starts
- Running overload
- Unbalance current/negative-sequence current heating

Motor Protection

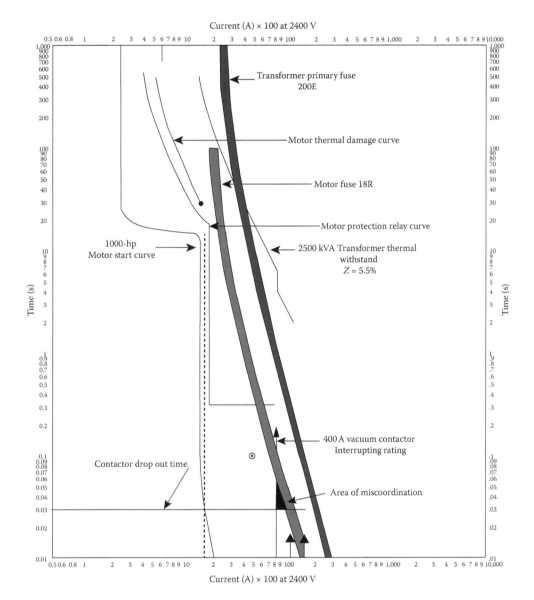

FIGURE 10.8
Coordination of motor protection, Example 10.1.

- Repeated or frequent starts

Run time constant (RTC) and cooling time constants can be automatically calculated by the relay for the thermal model. For example, for a rotor-limited motor, the relay calculates

$$\text{RTC} = \frac{(TD + 0.2)T_{\text{LR,hot}}}{60 \ln\left[\dfrac{I_{\text{LR}}^2 - (0.9 \times \text{SF})^2}{I_{\text{LR}}^2 - \text{SF}^2}\right]} \text{ min} \qquad (10.7)$$

and cooling time constant = 3RTC + 1 in minutes.
For a stator-limited motor,

$$\text{RTC} = \frac{T_{LR,hot}}{60\ln\left[\dfrac{I_{LR}^2 - 0.4^2}{I_{LR}^2 - SF^2}\right]}\,\text{min} \qquad (10.8)$$

where $T_{LR,hot}$ is the locked rotor time in hot condition, I_{LR} is locked rotor current, and SF is the service factor. *TD* is set = 1 when the driven load can accelerate within the locked rotor withstand time and a setting higher than 1.0, the speed switch is used, see Figure 10.7.

Note that during starting the motor uses a high thermal capacity unit (TCU) and after that the motor heat generated during starting falls and the TCU used during starting reduces.

The TCU when running can be approximated by

$$\text{TCU}_{run} = \left(\frac{I_{eq}}{OL \times I_f}\right)\left(1 - \frac{T_{stall,hot}}{T_{stall,cold}}\right) \times 100\% \qquad (10.9)$$

where $T_{stall,hot}$ and $T_{stall,cold}$ are the safe locked rotor hot and cold withstand times (generally of the order of 0.8–0.85), I_{eq} is the equivalent current that considers heating effect of negative-sequence current (approximately five to seven times that of positive-sequence current), and OL is the OL setting (1.10–1.15 for example). Figure 10.9 shows the TCU under running condition.

When the motor is tripped, it will cool down depending upon cooling time constant to zero TCU, see Figure 10.9. The motor will cool exponentially:

$$\text{TCU} = \left(\text{TCU}_{start} - \text{TCU}_{steadystate}\right)e^{-t/Tcrun}$$

where T_{crun} is the running cooling time constant.

Thus, a motor may fail to start if there is not enough TCU available, say due to more frequent starts, not enough time to cool down between starts, and overloads. The MMPR stores and remembers the TCU available at each instant of the motor operation.

With the thermal models and selected curves overcurrent elements are not used.

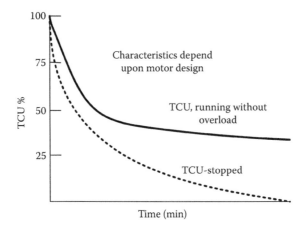

FIGURE 10.9
TCU % motor stopped and running.

10.5 RTD Biasing

RTD biasing is used to bias the motor thermal model based on the stator temperature as measured by embedded RTDs in the stator windings. It accounts for loss of motor cooling and high ambient temperatures, the model is in parallel with the current-based thermal model. It will only trip if the current is above the thermal overload pickup setting.

The minimum setting in Figure 10.10 is at the motor ambient temperature, the center setting is motor hot running temperature and the maximum setting is set slightly lower than the insulation temperature rating of the motor. If the cooling is lost, the RTD biasing will create a thermal capacity used model based on the motor actual temperature. This can influence motor thermal capacity for a stopped motor. The center bias point is given by

$$\left(1 - \frac{T_{stall,hot}}{T_{stall,cold}}\right) \times 100\% \tag{10.10}$$

10.5.1 Locked Rotor Protection Using Device 21

The locked rotor protection using 21 device is shown in Figure 10.11a. For the application of distance relays, see Chapter 2. The ratio of system voltage and starting current can be plotted in the R–X plane. The mho operating circle is chosen so that it encloses the impedance

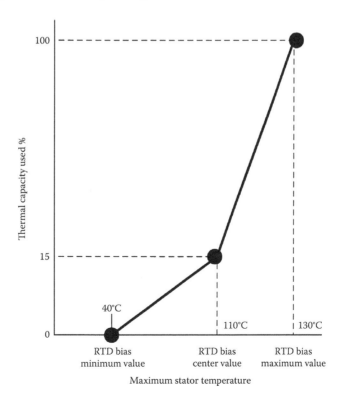

FIGURE 10.10
RTD biasing.

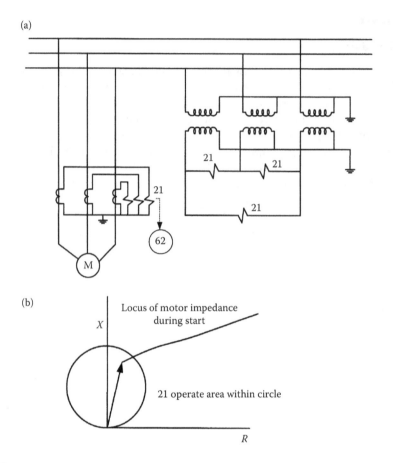

FIGURE 10.11
Application of a distance (mho) relay for motor starting, the starting time smaller than the locked rotor withstand time.

vector. On energization, distance relay will operate and energize a timer, see Figure 10.11b. For a successful start, the impedance phasor moves out of circle before timer contacts 62 close. For unsuccessful start, the impedance vector stays in the circle, the timer contact closes, and trip is initiated. The timer is set as determined by permissible locked rotor time curve from full voltage to 75% voltage. This protection does not cover failure to accelerate to full-load speed.

10.6 Medium-Voltage Motor Starters

We have discussed medium-voltage motor contactors briefly in Section 5.12. According to NEMA [6], the general purpose high-voltage contactors and class E for 50 and 60 Hz controllers are subdivided into two categories:

10.6.1 Class E1

These employ the contacts both for starting and stopping the motor and interrupting fault currents exceeding operating overloads.

10.6.2 Class E2

Class E2 contactors employ their contacts for starting and stopping the motor and employ fuses for interrupting short-circuit currents exceeding overloads.

The fuses used are generally R type, see Chapter 6.

Tables 10.1 and 10.2 from NEMA show the ratings of Class E controllers for nonplugging and nonjogging, reversing and nonreversing duty, when mounted in any enclosure with running overload protection or auxiliary devices. Class E controllers shall not be used for motors whose full-load currents or horsepower (HP) ratings exceed those given in these tables. Table 10.1 shows the maximum rms current rating that the controller can carry continuously without exceeding the temperature rise given in NEMA.

For the commercially available E1 and E2 contactor ratings, see Tables 6.4 and 6.5.

For motor ratings above those that can be started through E1 and E2 contactors, circuit breakers are applied. Note that contactors have higher capability of start/stops as compared to breakers.

Construction-wise three different forms of motor controllers are available.

TABLE 10.1

HP and Current Rating of Class E Controllers and Line Contactors

			HP Ratings					
			2200–2400 V, Three-Phase			4000–4800 V, Three-Phase		
Size of Controller and Contactor	Current Ratings Amperes		Induction Motors	Synchronous Motors		Induction Motors	Synchronous Motors	
	Continuous	Service Limit[a]		80% PF	100% PF		80% PF	100% PF
H2	180	207	700	700	900	1250	1250	1500
H3	360	414	1500	1500	1750	2500	2500	3600

Source: NEMA ICS 2-234. AC General Purpose HV Contactors and Class E Controllers, 50 Hz and 60 Hz.

[a] 1.15 times the continuous rating of the controller.

TABLE 10.2

Voltage and Interrupting Ratings of Class E Controllers

Size of Controller	Maximum Volts RMS	Range of Voltage at Which Interrupting Ratings Apply		Interrupting Rating, MVA	
		Maximum	Minimum	Class E1	Class E2
H2	2500	2500	2200	25&50	150	200[a]
H2	5000	5000	4000	25&50	250	350[a]
H3	2500	2500	2200	50	150	200[a]
H3	5000	5000	4000	50	250	350[a]

Source: NEMA ICS 2-234. AC General Purpose HV Contactors and Class E Controllers, 50 Hz and 60 Hz.

[a] For future design.

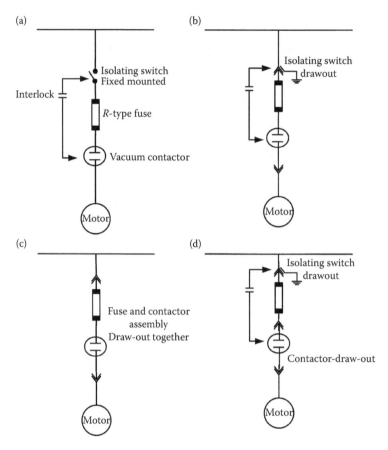

FIGURE 10.12
(a through d) Various commercial motor starter configurations.

- Fixed mounted: The contactor and its fuse are in non-draw-out construction, and an interlocked off-load switch is provided on the source side of the fuse, see Figure 10.12a.
- A manufacturer provides a draw-out, three-pole manually operated isolating switch mounted on the top of the assembly, rated up to 800 A continuous current and minimum 10,000 operations. Interlocks are provided from opening the switch when the contactor is closed, see Figure 10.12b.
- The fuse and contactor assembly is totally draw-out with safety shutters like metal-clad circuit breakers, see Figure 10.12c.
- Figure 10.12d is another variation.

10.6.3 Low-Voltage Magnetic Contactors up to 600 V

The controllers consist of magnetic contactors for starting and stopping of the motors. The controllers are equipped with thermal overloads (or more modern low-voltage protection relays) and loss of voltage is inherent with contactors when used with integral

Motor Protection

control supply and three-wire control circuits. However, this may not protect the motor from undervoltage operation. The contactors are maintained close by the potential taken from the motor line voltage or from a control power transformer. The line contactors must close at 85% of the line voltage according to ANSI/NEMA ICS2 [2], while the dropout point is not defined and may vary between 20% and 70% of the line voltage. The opening time is short, of the order of one cycle. This is an important consideration that during a starting or fault voltage dip, the low-voltage motor contactors may drop out fast and usher a process shutdown.

Figure 10.14 shows a three-wire control with time-delay loss of motor protection.

This is important, when it is desired that the motor should continue operating during voltage dips or voltage loss for a short duration. It consists of an undervoltage relay, which maintains a sealing contact for a definite time, depending on the time constant of the resistor and capacitor charging circuit.

The short-circuit protection is provided by the following:

- Fuses—but these can give rise to single phasing and consequent motor damage; these are not currently much in use
- By thermal magnetic circuit breakers (MCCBs) normally with fixed time delay characteristic and in large sizes with adjustable instantaneous magnetic characteristics, see Chapter 6

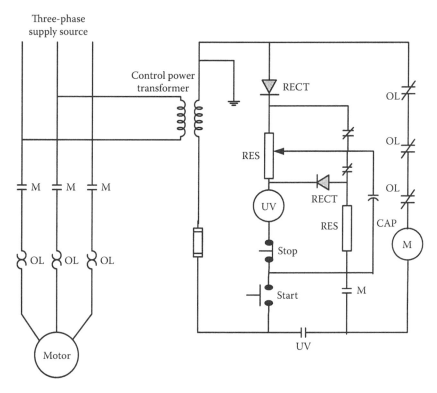

FIGURE 10.13
Stabilization of a motor contactor for voltage dip.

- With MCPs which have only instantaneous magnetic characteristics adjustable over a certain range; the MCPs do not have a short-circuit rating of their own and are tested with motor contactor starters

10.7 Two-Wire and Three-Wire Controls

In a three-wire control, the contactor is maintained through an auxiliary contact in parallel with the start push button. This is the most common control used in the industry. If the contactor opens due to a low voltage, the coil circuit is broken by the auxiliary contact of the contactor; and the motor cannot restart unless the start push button is again operated. This control is referred to as three-wire control with loss of voltage protection, see Figure 10.14a.

A synchronous motor on loss of voltage acts as a generator for a short period of time and tends to maintain the AC terminal voltage for a longer period than an induction motor. See undervoltage protection for further explanation.

Figure 10.14b shows a as two-wire control. Here, the auxiliary contact of the contactor is omitted. The start–stop push button is replaced with a toggle or latched switch, pressure switch, or other types of maintained contact sensing device. The contactor will drop out at a certain low voltage and the motor will automatically restart when the voltage is restored. This is referred to as two-wire control and loss of voltage release. This has the disadvantage that many motors provided on the same bus will restart simultaneously, which will depress the voltage to an extent that the motors will never be able to restart. When two-wire controls are used, overload relays with automatic reset should not be applied. This will make the motors restart when the relays are self-reset.

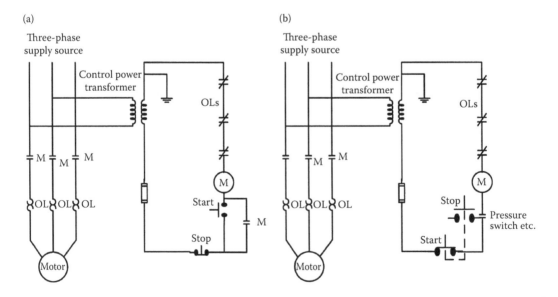

FIGURE 10.14
(a) Three-wire control circuit (b) Two-wire control circuit.

10.7.1 Schematic Control Circuit of Krondroffer Starter

We have discussed various starting methods for motors in Volume 2, not reproduced here; see Table 7.2 and Figure 7.6. The schematic circuit diagram of a reduced voltage starter is of interest, and Figure 10.15 shows the control circuit diagram of a Krondroffer starter. See description in Chapter 7 of Volume 2 for the sequence of operation of the contactors. The sequencing is controlled with auxiliary relays. The control power supply is derived from a CPT (Control Power Transformer) connected as shown. The motor protection relay contact MP should be of normally closed type, opening on a fault. A number of MMPR contacts (not shown in the figure) can be wired for remote indication, start/stop status, interlocks, ready to start, etc.

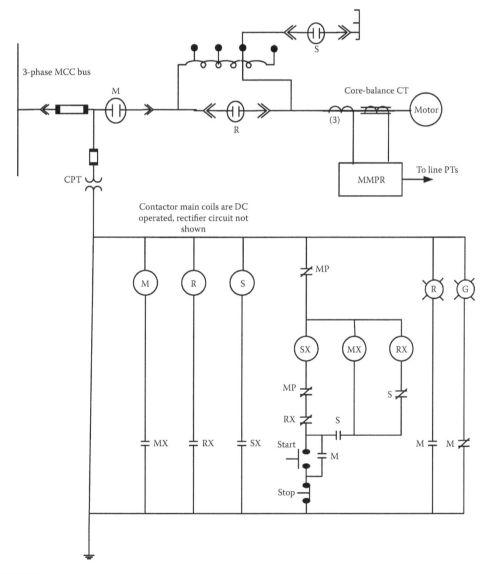

FIGURE 10.15
Schematic control circuit diagram of a Krondroffer reduced voltage starter.

10.8 Undervoltage Protection of Motors

Again in Chapter 7 of Volume 2, we have discussed the stability of induction and synchronous motors on voltage dips. The stability limits can be calculated based on the analytical analysis detailed in this volume. See also Example 7.3, which calculates the stability limits of 2000 hp and 1000 hp induction motors on voltage dips. The stability limit is a function of the following:

- Motor starting torque–speed characteristics
- Inertial constant H
- Load starting torque characteristics

The analytical calculation is complex and time-consuming. A transient stability study program with proper modeling of motor and load characteristics can be used for rigorous calculations.

In view of above, the settings on undervoltage relays cannot be recommended in general.

Generally, by proper system design, we limit the voltage dip on starting of the largest motor or impact load to 10% or less. Table 7.1 of Volume 2 shows the effects of voltage variations on the performance of induction motors. According to NEMA specifications, an induction motor should be capable of operation at:

- ±10% of rated name plate voltage at rated frequency
- A combined variation in frequency and voltage of 10% (sum of absolute values) provided the frequency variations do not exceed ±5% of the rated frequency
- ±5% of the rated frequency with rated voltage

ANSI C84.1 specifies the preferred nominal voltages and operating ranges A and B for utilization and distribution equipment from 120 to 34,500 V. For transmission voltages over 34,500 V, only nominal and maximum system voltage is specified. Range B allows limited excursions outside range A. For example, for 13.8 kV, nominal voltage range A is 14.49–12.46 kV and range B is 14.5–13.11 kV.

Overvoltages can occur in the motors connected in systems with cogeneration. On sudden load rejection, a generator voltage will rise.

Based on above, the general settings on the undervoltage and overvoltage protection can be:

- Undervoltage: 15%, time delay 5 s.
- Overvoltage: 15%, time delay 5 s.

It should be ensure under voltage settings override the starting voltage dip for the starting time.

10.9 NEC and OSHA Requirements

NEC articles are meant to safeguard persons, buildings, and their contents from hazards arising from the use of electricity, though they are not applicable to utility systems, and transmission and distribution systems.

Motor Protection

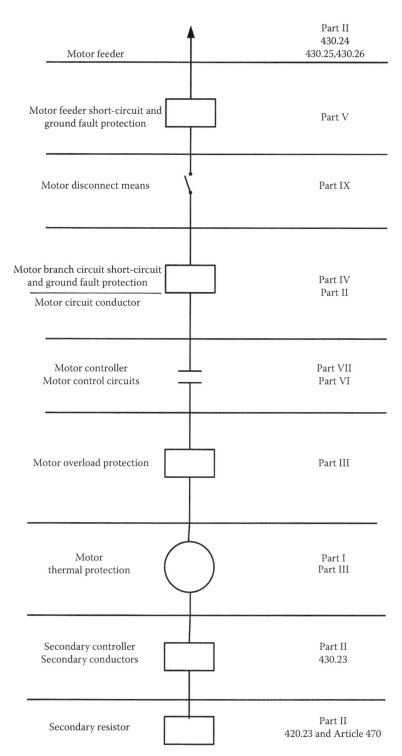

FIGURE 10.16
Exhibit from NEC for motor protection.

OSHA 29CFR, Chapter XVII, part 1910, is concerned with all establishments engaged in the manufacture of products for interstate commerce. It has adopted NEC and incorporated its requirements for electrical installations.

Figure 10.16 fundamentally shows the requirements of Article 430 for the motor circuits, branch circuit, controllers, and protection requirements. The relevant NEC Part numbers and articles are shown for each section of the motor circuit. Table 6.2 shows the maximum

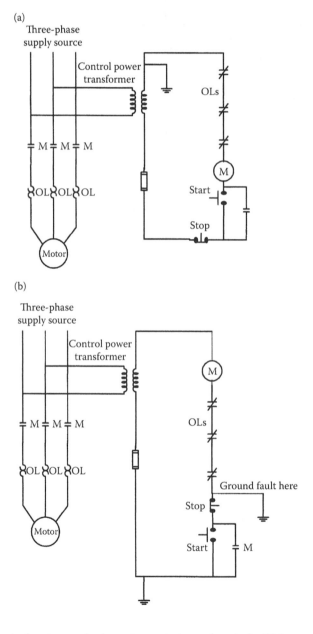

FIGURE 10.17
(a) Correct arrangement of motor overloads, motor contactor and controls. (b) Incorrect arrangement for a ground fault in the control circuit as shown the motor will start.

rating or setting or motor branch circuit short-circuit and ground fault protective devices and Table 6.3 shows the locked rotor indicating code letters. A motor short-circuit protector is permitted in lieu of devices in Table 6.2, if the short-circuit protector is part of listed combination motor controller having coordinated motor overload, and ground fault and short-circuit protection in each conductor, and it will open the circuit at currents exceeding 1300% of motor full-load current for other than design B high-efficiency motors and at 1700% of motor full-load current.

Figure 10.17 from NEC shows the control wiring of a 480/120 V CPT. Note that Figure 10.17b is incorrect. A ground fault in the control circuit shown at location can start the motor. Figure 10.17a is the correct order of connection of motor contactor, start–stop push buttons, and overloads.

10.10 Motor Insulation Classes and Temperature Limits

Deterioration of insulation systems of stator windings occurs due to a number of reasons:

- Overloads reduce the motor useful life.
- Impulse or switching surge overvoltages: It is usual to provide surge protection with surge capacitors and surge arrestors for large and critical motors, as close to the stator line terminals as possible. Generally, these devices are provided in the stator line side terminal box. The surge protection is not discussed in this handbook.
- Moisture penetration, mechanical stresses such as vibrations or distortion forces that can occur due to starting or misalignment cause deterioration.
- It can also occur due to restrikes in vacuum contactors, see Volume 1.

The physical and dielectric properties of the insulation deteriorates with age. And the process is elevated at higher temperatures. The life of insulation system is approximately reduced to 50% for 10°C incremental increase in temperature and approximately doubled for each 10°C incremental decrease.

10.10.1 NEMA Standards for Insulation Temperature Rise

Tables 10.3 through 10.6 show the temperature rise limits on induction and synchronous motors from NEMA [6, parts 20, 21] for various classes of insulations. The observable temperature rise under rated load conditions for each of the various parts of the induction machine, above the temperature of the cooling air, shall not exceed the values given in the tables. The temperature of the cooling air, the temperature of the external air as it enters the ventilating openings of the machine, and the temperature rise given in the tables are based on a maximum temperature of 40°C for the external air. For operation at higher ambient, the temperature rise of the machines shall be reduced by the number of degrees as the ambient temperature exceeds over 40°C.

Class A insulation systems are obsolete and class A insulated motors are no longer manufactured. Class B insulation systems are giving way to class F systems. The materials and construction of the insulation systems are not described.

TABLE 10.3

Temperature Rise Limits, Induction Motors with 1.0 Service Factor at Rated Load

Item	Machine Part	Method of Temperature Determination	Temperature Rise (°) Class of Insulation System			
			A	B	F	H
a	Insulated windings					
	All HP ratings	Resistance	60	80	105	125
	1500 hp and less	Embedded detector	70	90	115	140
	Over 1500 hp 7000 V or less	Embedded detector	65	85	110	135
	Over 7000 V		60	80	105	125
b	The temperatures attained by cores, squirrel cage windings, collector rings, and miscellaneous parts (such as brush holders and brushes etc.) shall not injure the insulation or machine in any respect.					

TABLE 10.4

Temperature Rise Limits, Induction Motors with 1.15 Service Factor at Service Power Factor Load

Item	Machine Part	Method of Temperature Determination	Temperature Rise (°) Class of Insulation System			
			A	B	F	H
a	Insulated windings					
	All HP ratings	Resistance	70	90	115	135
	1500 hp and less	Embedded detector	80	100	125	150
	Over 1500 hp 7000 V or less	Embedded detector	75	95	120	145
	Over 7000 V		70	90	115	135
b	The temperatures attained by cores, squirrel cage windings, collector rings, and miscellaneous parts (such as brush holders and brushes, etc.) shall not injure the insulation or machine in any respect.					

Note: For machines which operate under prevailing barometric pressure and which are designed not to exceed the specified temperature rise at altitudes from 3300 ft (1000 m) to 13,000 ft (4000 m), the temperature rise, as checked by tests at low altitudes, shall be less than those listed in the tables by 1% of the specified temperature rise for each 330 ft (100 m). Embedded detectors are located within the slot of the machine and can be either resistance elements or thermocouples.

TABLE 10.5

Temperature Rise Limits, Synchronous Motors with 1.0 Service Factor at Rated Load

Item	Machine Part	Method of Temperature Determination	Temperature Rise (°) Class of Insulation System			
			A	B	F	H
a	Insulated windings					
	All HP ratings	Resistance	60	80	105	125
	1500 hp and less	Embedded detector	70	90	115	140
	Over 1500 hp	Embedded detector	65	85	110	135
	7000 V or less Over 7000 V		60	80	105	125
b	Field windings Salient pole motors	Resistance	60	80	105	125
	Cylindrical rotor motors			85	105	125
c	The temperatures attained by cores, squirrel cage windings, collector rings, and miscellaneous parts (such as brush holders and brushes, etc.) shall not injure the insulation or machine in any respect.					

TABLE 10.6

Synchronous Motors with 1.15 Service Factor at Service Power Factor Load

Item	Machine Part	Method of Temperature Determination	Temperature Rise (°) Class of Insulation System			
			A	B	F	H
a	Insulated windings					
	All HP ratings	Resistance	70	90	115	135
	1500 hp and less	Embedded detector	80	100	125	150
	Over 1500 hp	Embedded	75	95	120	145
	7000 V or less	detector	70	90	115	135
	Over 7000 V					
b	Field windings	Resistance	70	90	115	135
	Salient pole motors					
	Cylindrical rotor motors			95	115	135
c	The temperatures attained by cores, squirrel cage windings, collector rings, and miscellaneous parts (such as brush holders and brushes, etc.) shall not injure the insulation or machine in any respect.					

Source: NEMA MG-1, Motors and Generators, Large Machines, (See in particular, Part 21).
Notes as for tables apply.

10.10.2 Embedded Temperature Detectors

The embedded temperature sensitive elements in the windings are as follows:

- RTDs are usually of platinum, having a resistance of 100 Ω at 0°C. RTDs of copper of 25 Ω have also been used. Platinum has a temperature coefficient of 0.385 Ω/°C in the range of interest. The resistance elements consist of thin bifilar wire wound around a carrier of insulating material and then further insulated and encased. The resistance elements shall have a minimum width of 0.25 in. and the detector length depends on the machine core length, see NEMA [MG-1, part 2]. The number of embedded detectors shall be at least equal to the number of phases for which the machine is wound. Usually two elements are provided per phase. The detector is located in the center of the slot (with respect to slot width) and in intimate contact with the insulation of both the upper and lower sides, where possible or otherwise it should be in contact with the insulation of the upper side, that is, the coils side nearest to the air gap.
- Thermocouples are usually of copper constantan type.
- Thermistors can have a negative or positive temperature coefficient (PTC) depending on the base material. PTC thermistors are used and the characteristics of abrupt change of resistance at Curie point are used. The discontinuous nature of thermocouples enables all embedded thermistors to be connected in series and, thus, the system is self-monitoring. Any fault in the circuit or open circuit can immediately be detected.
- The characteristics of thermistor sensors based on IEC 34-11-2 are shown in Figure 10.18.

The advantage of embedded temperature detectors is that a direct measurement of insulation temperature is possible and the motor can be tripped before potential damage to the insulation occurs. Generally, platinum RTDs are most common and provided for motors

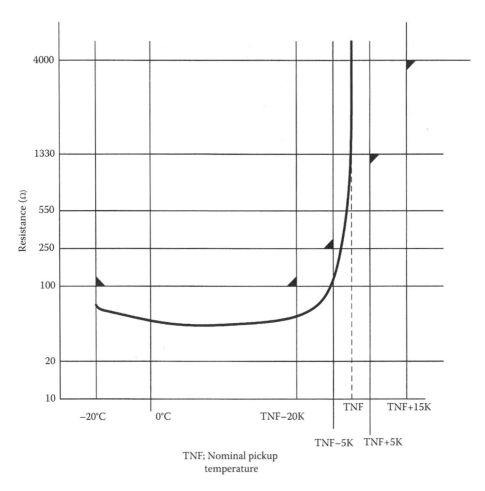

FIGURE 10.18
Temperature–resistance characteristics of a thermistor.

say above 500 hp rating. There are two set points available, alarm and trip, which can be set according to temperature rise limits shown in Tables 10.3 through 10.6.

It is estimated that the "hot spot" temperature of the insulation is 10°C higher than the temperature limits shown in these tables. Thus, for a class B insulated motor, permissible temperature 125°C, by embedded temperature detector for motors above 1500 hp and voltage rating less than 7000 V, hot spot temperature 135°C, it will be acceptable to set the RTDs for the following:

130°C trip
125°C alarm

10.10.2.1 Polling of RTDs

The modern MMPR relays provide for the polling of the RTDs, that is, out of the six RTDs for the stator windings, the temperature of each is continuously monitored and the RTD having the highest temperature rise will usher alarm and trip. Moreover, this temperature setting can be automatically adjusted by an ambient RTD, if provided.

10.11 Bearing Protection

10.11.1 Antifriction Ball or Roller Bearings

Ball and roller bearings transmit the rotor weight by direct contact with rolling action and have low starting friction. Failure of this type of bearing usually takes the form of fatigue cracks on the surface of races and rolling parts, leading to spalling and peeling, and destruction of the bearing follows rather quickly.

The bearings are grease lubricated designed for 130,000 h L-10 life for direct drive applications without axial thrust loading. For belt drives, a roller-type bearing with 40,000 h L-10 life is used. Motors with antifriction bearings have an external thrust capability of 500 pounds continuous and 1500 pounds momentary up through 1 hp per rpm and 1500 pounds continuous and 5000 pounds momentary for large motors.

10.11.2 Sleeve Bearings

Sleeve bearings and hybrid tilt pad bearings are split at the horizontal center line. Sleeve bearings are furnished with oil rings to provide coast down in the event of oil supply failure. Sleeve and HTP bearings are provided with fittings for flood or forced lubrication from external oil supply. Motors with sleeve bearings do not have external axial thrust capability. To protect the motor bearing, the usual procedure is to use a limited end float coupling so that with proper alignment the rotor is restrained axially within allowable limits. Belt-driven motors may be equipped with special roller bearings and a longer shaft for mounting the drive sheath as per recommendations of the manufacturer.

For motors that need to be lubricated from an external source of lubrication system, a system consisting of electric motor-driven pump, reservoir, and cooling arrangement is required.

In fluid film (sleeve) bearings, when the film of lubricant is destroyed, friction losses rise rapidly and metal-to-metal contact is likely to occur. Conditions leading to film failure are reduced lubricant viscosity (contaminated lubricant), increased loading or particles in the lubricant.

10.11.3 Bearing Failures

The bearing failures are attributed to the following:

10.11.3.1 Lubricant Problems

- Incorrect grade or viscosity of lubricant
- Inadequate cooling of bearing or lubricant
- Deterioration, saponification, or frothing oil
- Abrasive particles

10.11.3.2 Mechanical Problems

- Failure of oil supply due to stuck oil rings, lubricant pump failure, low lubricant reservoir level, fractured oil pipe

- Excessive radial loading due to misalignment of shaft and bearings and couplings
- Improper fit of the bearing
- Bent motor shaft, unbalanced rotor

10.11.3.3 Excessive Axial or Thrust Loading

- Improper leveling
- Improper axial alignment with respect to magnetic center
- Improper axial alignment of driven equipment reflected through double-helical gear drive

Other possible causes are as follows:

- Abrasive particles, shaft currents, fatigue cracks
- Phase current unbalance and harmonics

10.11.4 Bearing Protection Devices

- Bearing temperature RTDs, much alike stator RTDs
- Bearing temperature monitor can provide a continuous digital display of the bearing temperature and alarm and shutdown relays for each bearing
- Low oil level alarm in the reservoir, device 71
- Low oil pressure switch, device 63
- Reduced oil flow switch, device 80
- Rate of temperature rise
- Vibration detectors should be mounted with sensitive axis to coincide with the direction of displacement; such devices are usually deactivated during startup or shutdown
- Only the NDE bearing is insulated from shaft currents; however, with VFD drive motors and in presence of harmonics, both DE (Drive end) and NDE (Non-drive end) bearings should be insulated for shaft currents
- The derating of motors due to harmonics is discussed in Chapter 6, Volume 3
- The VFD motors should meet the requirements of insulation specified in NEMA MG-1, part 30, and 31. Also see Chapter 6, Volume 3
- Shaft slingers or bearing seals must be provided to prevent oil and grease leakage

10.11.5 End Play and Rotor Float for Coupled Sleeve Bearings Horizontal Motors

When induction machines are provided with sleeve bearings, the sleeve bearings and limited end float coupling should be applied as shown in Table 10.7.

To facilitate assembly of driven or driving equipment, the manufacturer should:

TABLE 10.7

End Float Couplings-Sleeve Bearings

Machine HP (kW)	Synchronous Speed	Minimum Rotor End Float, Inches	Maximum Coupling End Flat, Inches[a]
500 (400) and below	1800 rpm and below	0.25	0.09
300 (250) to 500 (400)	3600 and 3000	0.50	0.19
600 (500) and higher	All speeds	0.50	0.19

Source: NEMA MG-1, Motors and Generators, Large Machines (see in particular, Part 20).

[a] Couplings with elastic axial centering forces are usually satisfactory without these precautions.

- Indicate on the induction machine outline drawing the minimum machine rotor end play in inches
- Mark rotor end play limits on the machine shaft

10.12 Vibrations

NEMA Standard MG-1 may be referred for details. Curtailing the vibrations to acceptable limits is important; excessive vibrations can damage the bearings, bend the shaft, and result in misalignment and premature failure of the motor.

The measurement quantities are defined as bearing house vibrations—the greatest value measured at prescribed measuring points characterizes the vibrations of the machine.

10.12.1 Relative Shaft Vibrations

The vibrations are measured as peak to peak vibratory displacements.

Machine mounting is important. Smaller machines may be resilient mounted—the natural frequency of the suspension system shall be less than 25% of the system frequency corresponding to the lowest speed of the machine under test. The effective mass of the elastic support shall not be less than 10% of that of the machine, to reduce influence of mass and moments of inertia of these parts on the vibration levels. The rigidly mounted machines have a relatively massive foundation. A relatively massive foundation is the one that has a vibration in any direction or plane during testing to 0.02 in./s peak (0.5 mm/s peak). The horizontal and vertical natural frequencies of the complete test arrangement shall not coincide within ±10% of the rotational frequency of the machine, within ±5% of two times the rotational frequency, or within ±5% of the one-time and two-time line frequency.

10.12.1.1 Standard Machines

Unfiltered vibrations shall not exceed the velocity levels as shown in Figure 10.19 for standard (no special vibration requirements) machines resiliently mounted. The limits on rotational frequency are shown in Table 10.8.

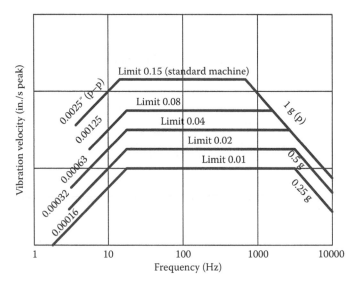

FIGURE 10.19
Unfiltered vibration levels, standard machines. (From NEMA MG-1, Motors and Generators, Large Machines, (See in particular, Parts: 1, 5, 6, 7, 20, and 21).)

TABLE 10.8

Unfiltered Vibration Limits

Speed rpm	Rotational Frequency Hz	Velocity in./s Peak (mm/s)
3500	60	0.15 (3.8)
1800	30	0.15 (3.8)
1200	20	0.15 (3.8)
900	15	0.12 (3.0)
720	12	0.09 (2.3)
600	10	0.08 (2.0)

Source: NEMA MG-1, Motors and Generators, Large Machines, (See in particular, Part 7).

10.12.1.2 Special Machines

For machines requiring vibration levels lower than that given in Figure 10.19 at limit 0.15, the vibration limits shall not exceed the limits shown by various curves. Machines to which these limits apply shall be by agreement between manufacturer and purchaser.

Banding is a method of dividing the frequency range into frequency bands and applying a vibration limit to each band. Banding recognizes that the vibrations at various frequencies are a function of the source of excitation. For further details, see Reference [6, Part 7].

10.12.2 Axial Vibrations

The evaluation of axial bearing housing or support vibrations depend upon bearing installation, bearing function, and bearing design, plus uniformity of stator and rotor cores.

TABLE 10.9
Limits of the Unfiltered Maximum Relative Shaft Displacement (Sp–p) for Standard Machines

Speed rpm	Maximum Relative Shaft Displacement (Peak-to-Peak)
3600	0.0028 in. (70 μm)
≤1800	0.0035 in. (90 μm)

Source: NEMA MG-1, Motors and Generators, Large Machines, (See in particular, Part 7).

TABLE 10.10
Limits of the Unfiltered Maximum Relative Shaft Displacement (Sp–p) for Special Machines

Speed rpm	Maximum Relative Shaft Displacement (Peak-to-Peak)
3600	0.0020 in. (50 μm)
1800	0.0028 in. (70 μm)
≤1200	0.0030 in. (76 μm)

Source: NEMA MG-1, Motors and Generators, Large Machines, (See in particular, Part 7).

In the case of thrust-bearing applications, axial vibrations correlate with thrust loading and axial stiffness. Axial vibrations of these configurations should be measured with axial thrust applied, when practical.

When bearings have no axial thrust capability, axial vibrations are judged as for standard or special motors as described earlier.

10.12.3 Limits of Relative Shaft Vibrations

Shaft vibrations are measured by noncontacting proximity probes. When specified, the limits of relative vibrations of rigidly mounted standard machines with sleeve bearings inclusive of electrical and mechanical runouts shall not exceed the limits specified in Table 10.9, for special machines limits as shown in Table 10.10 are applicable.

10.13 Motor Enclosure

Hitherto NEMA specified enclosure and cooling with descriptive words like open machine, drip proof, splash proof. IEC methods of designations have been adopted in NEMA. Details of protection are provided by two designations, letters "IP" for the enclosure and "IC" for the cooling. Table 10.11 is constructed based on these letters.

Weather-protected type I and II enclosures are popular choices for large machines. The type I machine is a guarded machine with its ventilating passages so constructed as to minimize entrance of rain, snow, and airborne particles into electrical parts. Type II machine is so constructed that the normal path of ventilating air, which enters the parts of the machine shall be so arranged by baffling or separate housing to provide at least three abrupt changes in direction; none of which shall be less than 90°. In addition, an area of low velocity not exceeding 600 ft/min shall be provided in the intake to minimize possibility of moisture or dirt being carried into the electrical parts of the machine.

TABLE 10.11

Enclosures and Cooling in Terms of IEC Designations

No	Description	IP	IC
1	Open machine	IP00	IC01
2	Drip-proof machine	IP12	IC01
3	Splash-proof machine	IP13	IC01
4	Semiguarded machine	IP10	IC01
5	Guarded machine	IP2	IC01
	Drip-proof-guarded Machine	IP22	IC01
	Open independently ventilated machine	As in 1–4	IC06
6	Open-pipe ventilated machine	As in 1–4	IC11, IC17
7	Weather-protected machine, type I		IC01
	Weather-protected machine, type II		
8	Totally enclosed nonventilated	IP54	IC410
	Totally enclosed fan-cooled machine		
	Totally enclosed fan-cooled guarded machine	IP54	IC411
	Totally enclosed pipe ventilated machine	IP44	
	Totally enclosed water-cooled machine	IP54	
9	Waterproof machine	IP55	
10	Totally enclosed air-to-water cooled machine	IP54	
11	Totally enclosed air-to-air cooled machine	IP54	
12	Totally enclosed air-over machine	IP54	IC417
13	Explosion-proof machine		
14	Dust ignition-proof machine		

Source: NEMA MG-1, Motors and Generators, Large Machines, (See in particular, Part 1).
Note: See NEC for various zone classifications and constructional features.

A totally enclosed water-cooled machine is cooled by circulating water, the water coming in direct contact with the machine parts. A totally enclosed air-to-water cooled machine is cooled by circulating air, which in turn is cooled by circulating water. It is provided with a water-cooled heat exchanger, integral (IC&-W) or machine mounted (IC8-W).

10.13.1 IP Designations

The designations used for the degree of protection consist of letters IP followed by two characteristic numericals, with conditions as indicated in Table 10.12. See Reference [6, Part 5] for the tests applicable to each characteristic numerical.

10.13.2 IC Designations

The IC designation method of cooling consists of numericals and letters representing the circuit arrangement, the coolant, and the method of movement of the coolant. See Figure 10.20 for complete and simplified designations. Table 10.13 shows the circuit arrangement and method of movement characteristic numerical and Table 10.14 shows the characteristic numerical for various coolant types.

A proper selection of the motor enclosure based on environmental conditions and rating of the motor go a long way in ultimate protection of the motor. For example, in the paper and pulp mill industry, the chipper motors are totally enclosed, and pipe ventilated because of wood chip dusty atmosphere.

Motor Protection

TABLE 10.12

Degrees of Protections Letters IP Followed by Two Numericals

First Characteristic Number	Description	Second Characteristic Number	Description
0	Nonprotected machine	0	Nonprotected machine
1	Machine protected against solid objects >50 mm	1	Machine protected against dripping water
2	Machine protected against solid objects >12 mm	2	Machine protected against dripping water, when tilted up to 15°
3	Machine protected against solid objects >2.5 mm	3	Machine protected against spraying water
4	Machine protected against solid objects >1 mm	4	Machine protected against splashing water
5	Dust-protected machine	5	Machine protected against water jets
6	Dust tight machine	6	Machine protected against heavy seas
		7	Machine protected against effects of immersion
		8	Machine protected against continuous submersion

Source: NEMA MG-1, Motors and Generators, Large Machines, (See in particular, Part 5).

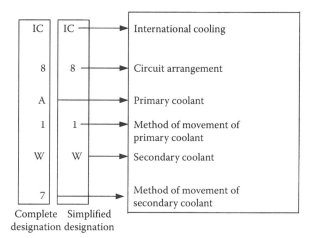

FIGURE 10.20
Cooling designations IC of motor enclosures.

10.14 Effect of Negative-Sequence Currents

The positive- and negative-sequence models of induction and synchronous machines are detailed in Volumes 1 and 2 and not repeated here.

A synchronous (as well as induction) motor has severe conditions imposed when there is unbalance in the supply system voltages. Under single phasing, a rotating flux is produced, which travels in a direction opposite to that of the rotor. This produces a 120 Hz

TABLE 10.13

IC Cooling Designations

Circuit Arrangement		Method of Movement	
0	Free circulation	0	Free Convection
1	Inlet pipe or inlet duct circulated	1	Self-circulation
2	Outlet pipe or outlet duct circulated	2	Reserved for future use
3	Inlet and outlet pipe or duct circulated	3	Reserved for future use
4	Frame surface cooled	4	Reserved for future use
5	Integral heat exchanger using surrounding medium	5	Integral independent component
6	Machine-mounted heat exchanger using surrounding medium	6	Machine-mounted independent component
7	Integral heat exchanger using remote medium	7	Separate and independent component or coolant system pressure
8	Machine-mounted heat exchanger using remote medium	8	Relative displacement
9	Separate heat exchanger using surrounding or remote medium	9	All other components

Source: NEMA MG-1, Motors and Generators, Large Machines, (See in particular, Part 6).

TABLE 10.14

Coolant for IC Designations

Characteristic Letter	Coolant
A	Air
F	Refrigerant
H	Hydrogen
N	Nitrogen
C	Carbon dioxide
W	Water
U	Oil
S	Any other coolant
Y	Coolant not yet selected

rotor current, which appears in the amortisseur bars and the rotor windings. A higher than the normal rotor resistance is experienced, giving rise to severe heating.

Starting of a motor with single phasing can give rise to negative-sequence current equal to approximately 50% of the normal three-phase starting current. Positive-sequence torque equals negative-sequence torque and negative-sequence impedance of the motor equals positive-sequence impedance. The starting current may fall to approximately 86% of the normal starting current in two phases.

The negative-sequence current is influenced by the combined series impedance of the negative-sequence and positive-sequence networks.

Figure 10.21 shows the ratio of the negative-sequence current/motor full-load current versus motor total load current.

Upon occurrence of a single phasing under running conditions, positive-sequence current will increase. The two phases will experience an increase of 73% or more.

Figure 10.22 shows the positive- and negative-sequence networks for a motor connected through a transformer.

Motor Protection

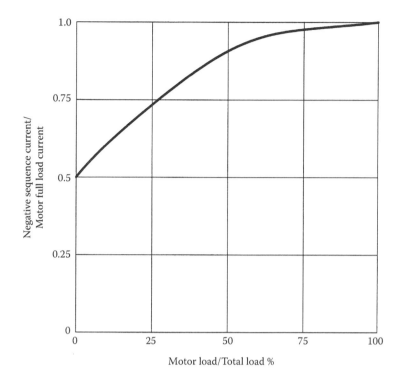

FIGURE 10.21
Negative-sequence current of motors with respect to load current on single phasing.

10.14.1 Protection for Negative-Sequence Currents

The negative-sequence protection should be provided for the normal running conditions as well as for the starting conditions. A 1% voltage unbalance causes approximately 6% current unbalance in the induction motors.

10.14.1.1 Phase Balance Protection

Almost all motor protection relays have protection function, which detects the phase unbalance conditions. The currents in the three-phase system are compared. The following settings are normally provided:

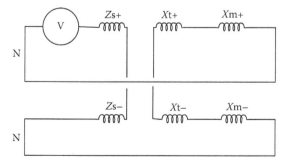

FIGURE 10.22
Equivalent negative-sequence impedance circuit of a motor connected to a transformer.

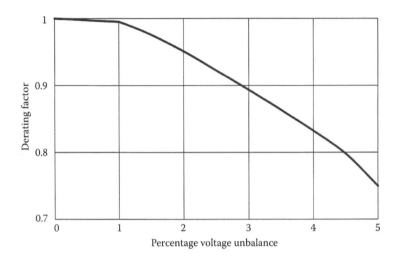

FIGURE 10.23
Derating of motor due to voltage unbalance in the supply source. (From NEMA MG-1, Motors and Generators, Large Machines, (See in particular, Parts: 1, 5, 6, 7, 20, and 21).)

- 15% unbalance for 15 s, trip—the 5% current unbalance level corresponds to 2.5% voltage unbalance. Sometimes the distribution systems may be inherently subjected to larger unbalances due to unbalances in the utility source voltages.
- 10% unbalance for 10 s, alarm.

A 5% voltage unbalance may give rise to 38% negative-sequence current with 50% increase in losses and 40°C higher temperature rise as compared to operation on a balanced voltage with a zero negative-sequence component. Also, the voltage unbalance is not equivalent to the negative-sequence component. The NEMA definition of percentage voltage unbalance is maximum voltage deviation from the average voltage divided by the average voltage as a percentage. Operation above 5% unbalance is not recommended [6]. Figure 10.23 shows the derating factor versus % voltage unbalance.

10.14.1.2 Time Delay and Instantaneous Negative-Sequence Protection

Figure 10.21 illustrates the negative-sequence current as a function of ratio of motor current to total current. This assumes that loads other than synchronous motor loads are connected to the same bus. The equivalent negative-sequence impedance circuit of a motor connected to a transformer is shown in Figure 10.22, and Figure 10.23 depicts derating of motor due to voltage unbalance in the supply source [6].

Unlike generators where I^2t capability is specified in appropriate standards, *no standard has been established for synchronous motors*. Gleason and Elmore [8] recommend that

$$I_2^2 t = 40 \tag{10.11}$$

TABLE 10.15

Parameters of 3250 hp, 0.8 PF Leading, 300 rpm, 4.0 kV, Three-Phase 60 Hz Chipper Motor

Parameter	Value
Full-load current	455 A
Full-load efficiency	96%
Starting torque	60%
Pull-in torque	60%
Pull-out torque	300%
Starting current	4.5 times If
X_s	1.0 pu
X'_d	0.284 pu
X''_d	0.193 pu
Full-load field current	122A
Maximum allowable time on cage winding, full voltage without injury to cage at zero speed	13 s
Enclosure	TEPV (totally enclosed, pipe ventilated)

Example 10.2

The motor negative-sequence withstand for 3250 hp chipper motor, data as shown in Table 10.15 are calculated based on Equation 10.11. This is shown in Figure 10.24. This negative-sequence withstand is coordinated with an inverse time characteristic of a protective relay. Trying various characteristics, the IEC extremely inverse characteristics, C curve type C3 provides the best coordination. The pickup setting is 250 A and time dial, 1.00 (maximum setting). Most MMPRs provide selection of ANSI or IEC characteristics.

Instantaneous negative-sequence protection should not be applied to motor protection schemes, because high-magnitude and low-duration negative transients exist due to circuit breaker or starter pole mis-coordination. Definite time negative-sequence protection can be enabled during starting phase, taking care that definite time delay is set long enough to ride through the CT saturation that may exist when the motor is energized. Nuisance trips can occur if the definite time is set too short [9]. A calculation of the negative-sequence withstand characteristics based on motor parameters and coordinating it with negative-sequence overcurrent element, as demonstrated in Figure 10.24, is recommended. This is a new concept, hitherto not discussed in any publication.

10.15 Differential Protection

The principles of differential protection have been amply discussed in previous chapters.

10.15.1 Flux Balancing Current Differential

In flux balancing current differential protection, both ends of each motor winding serve as primary winding of the current transformers. Each current transformer wraps around two winding conductors, one phase and another neutral; the neutral is formed above the CTs. The differential relays see the difference of internal fault currents. Extremely sensitive phase and ground fault protection can be obtained by using overcurrent relays. The relays can be set low, for example, 10 A primary current with six-cycle delay to override the starting transients. The cables from the starting device to the motor remain outside the zone of differential protection, see Figure 10.25.

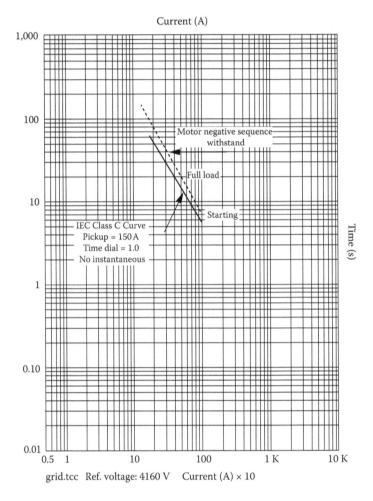

FIGURE 10.24
Calculation of a negative-sequence capability curve of a 3250 hp synchronous motor and its protection with an IEC class C curve.

10.15.2 Full Differential Protection with Ground Fault Differential

Figure 10.26 shows the full differential protection with added ground fault differential with a product type of relay. The current polarization can be done through the CT connected in the grounded system unit or voltage polarization can be done. See Chapter 9.

10.16 Ground Fault Protection

Apart from ground fault protection through differential relays, as illustrated earlier, the ground fault protection is provided with core balance CTs of 50/5 or some relays recommend special 2000:1 ratio core balance CTs. The motor ground fault protection is set at 5 A pickup and six-cycle time delay. Experience shows that this setting is appropriate even for large motors. See Chapter 6 also. It has been established that a ground fault current of even a few

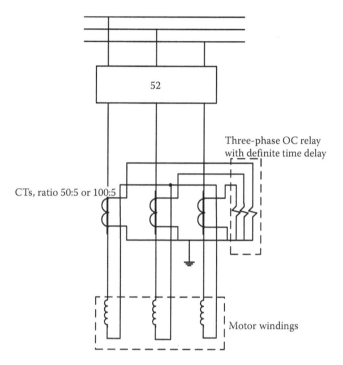

FIGURE 10.25
Flux balance differential protection scheme of motor.

amperes sustained for a long period of time can cause core damage. Rarely, the ground fault protection will be provided through residually connected ground fault relay in the phase CT circuit. The required sensitivity for the ground fault protection cannot be obtained.

10.17 Variable-Speed Motor Protection

The ASDs (Adjustable Speed Drives) and converter circuits are described in Volume 3.

Table 10.16 and accompanying Figure 10.27 provide basic variable-speed (ASDs) motor protection guidelines. Also see Reference [10].

10.18 Synchronous Motors Starting and Synchronization

The field circuit of the synchronous motor can be of brush type or brushless type. The starting characteristics of the synchronous motors and constructional features are described in Chapter 7, Volume 2.

The brush-type excitation system may consist either of the following:

- A shaft-driven DC exciter
- A separate motor generator set
- A separate static rectifier

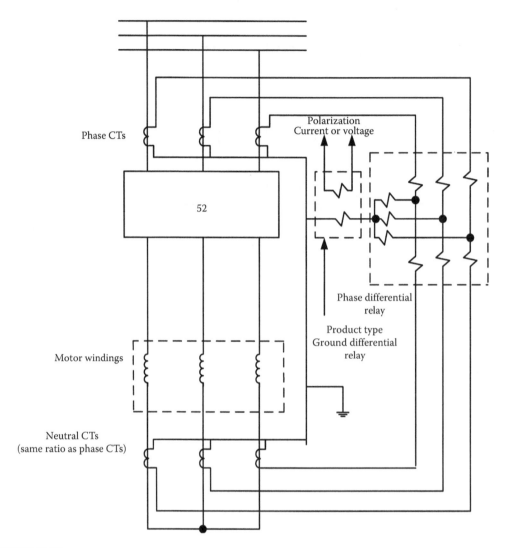

FIGURE 10.26
Full-phase fault differential and restricted ground fault differential protection scheme of a motor.

A brushless excitation system is generally of the permanent magnetic generator-type shaft mounted with rotating rectifiers and other control circuitry.

In either type:

- The field circuit must provide a discharge path for AC induced in the motor during starting. Without the discharge circuit, high voltages can be generated. The discharge circuit is opened when the field is applied.
- The field is applied at a suitable rate and at a favorable position of the rotor poles with respect to the rotating flux.
- The field circuit is automatically removed on an impeding PO (Pull-out).
- Automatic resynchronization is attempted on favorable conditions, or automatic synchronization is not adopted and shutdown is ushered.

TABLE 10.16
Variable Speed Motor Protection

Protection Zone 1	Protection Zone 2	Protection Zone 3
Apply typical transformer protection	Apply generic electronic drive protection	Apply typical motor protection
Differential, phase and time overcurrent relays, phase and ground instantaneous relays	Differential relays, phase and ground time and instantaneous relays	Differential relays
Transformer neutral relays	Transformer neutral relays, if employed in converter circuitry	Phase and ground time and instantaneous relays
Sudden pressure relays	DC overcurrent relay	V/Hz relaying
	Voltage controlled overcurrent relays	Thermal overload relays
	Motor speed protective circuitry	Negative-sequence relays
	Under/overvoltage relays	Overspeed protection relays
	Special semiconductor fuses	Speed switch or impedance relay supervision of time overcurrent relays
	Thyristor and thyristor gate protective circuitry	
	Microprocessor or control system protection circuitry	

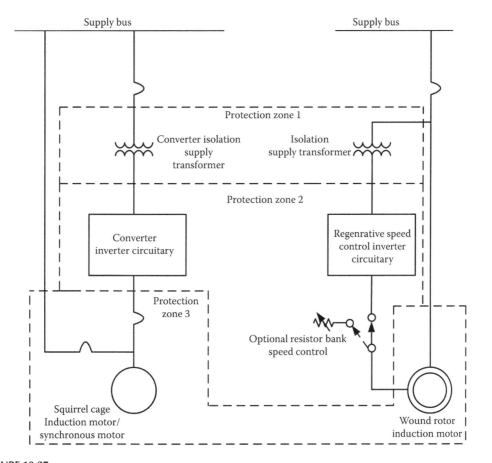

FIGURE 10.27
Protection of a VFD system, see Table 10.16.

10.18.1 Brush-Type Controllers

Figure 10.28 shows the circuit of a brush-type controller. The line contactor operated to connect the motor to the power supply for across the line starting. (Additional contactors are required for reduced voltage starting.) The field contactor is a DC-activated device, with one normally open and one normally closed power pole. A series dropping resistor is used to keep from exceeding rated coil voltage. The rated coil voltage is 50% of the supply to cover a wide range of field voltages.

The starting field discharge resistor is used to improve the motor starting torque and to limit induced filed voltage during starting or when the field excitation is removed.

The damper winding protection (SC) is provided in the event the motor fails to start and accelerate. It is thermally activated and operated to trip the main contactor. It must be hand reset. It is a transformer-operated relay with maximum sensitivity on initial starting when the induced field frequency is the maximum.

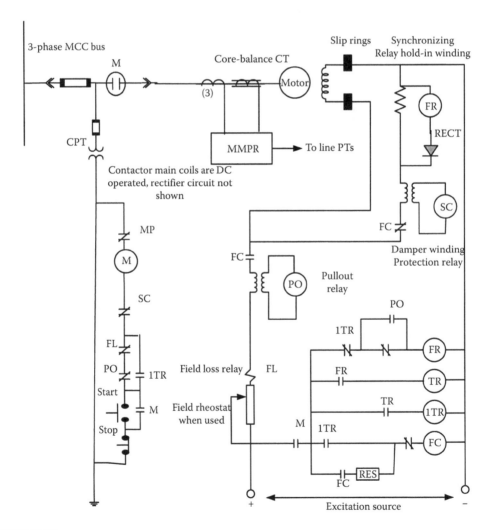

FIGURE 10.28
Control circuit diagram and protection of asynchronous motor brush-type excitation system.

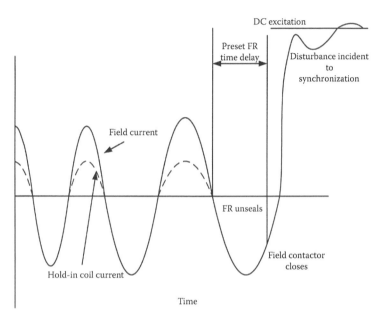

FIGURE 10.29
Correct instant for the application of field during synchronization.

The synchronizing relay FR controls the field contactor, so that it closes when the motor has reached sufficient speed and the poles are in favorable relation for synchronizing. The relay has two independent windings. The pickup winding energizes the relay, while the hold-in winding keeps it activated by the voltage drop across the starting and discharge resistor. The polarity of the two windings must be additive. The rectifier in the hold-in winding produces half wave direct current, and when the off time between pulses is longer than the inherent dropout time of the FR, relay drops out to energize the field contactor, see Figure 10.29. The rectifier rectifies the component of the field current thereby polarizing the hold-in circuit of the relay. See Equation 7.30 in Chapter 7 for the maximum slip at synchronization.

The PO relay operates on pullout of the synchronous motor to trip the line and field contactors, thus stopping the motor. Alternatively, it energizes the FR relay, thereby initiating a resynchronization sequence. The PO relay transformer is not impacted by DC but pulses of AC in the field are transferred to secondary to operate the relay.

The auxiliary sequence relays TR and ITR are off-delay timing relays. These control the sequence of the field application equipment and nullify the PO relay during synchronizing.

10.18.2 Brushless-Type Excitation System

Figure 10.30 shows a brushless excitation system. The AC output of the rotating armature exciter is converted into DC, but the control circuit blocks SCR1 from firing until the induced frequency in the rotor circuit is low. When the field terminal MF1 is positive, with respect to MF2, the diode D conducts and on the reverse cycle SCR2 conducts during starting and connects the discharge resistor FD across the motor field windings MF. This protects the motor windings and electronics from high induced voltages. As the induced frequency of the current in the field windings decreases, the frequency part of the control circuit blocks SCR2 and simultaneously SCR1 conducts to apply the field excitation. A zero slip circuit is included to apply excitation if the motor synchronizes on reluctance torque.

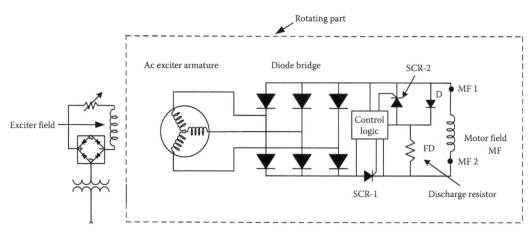

FIGURE 10.30
A brushless-type excitation system for a synchronous motor.

On first half-cycle after PO, the induced field voltage across SCR1 switches it off, removing the motor excitation and SCR2 conducts to reconnect the discharge resistor across the field windings. If the motor does not synchronize, it will slip a pole, the induced field voltage opposes the exciter voltage causing current to go to zero, turning off SCR1. The SCR2 is turned on only at a voltage higher than the exciter voltage and will not be conducting when SCR1 is conducting. There is positive interlocking between SCR1 and SCR2. Figure 10.31 shows the voltages during one cycle [11].

10.18.3 Types of Field Controllers

The simplified current and power diagram of a synchronous motor is depicted in Figure 10.32a. The derivation is arrived analytically in many texts [12–16]. The reactive power can be varied by control of excitation. The variation range of the excitation current I_e is limited by maximum operating temperature of the stator and rotor and by the stability limit of the motor. In this figure the stator temperature limit is shown by a dotted arc and the rotor

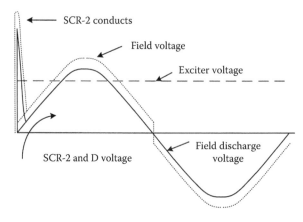

FIGURE 10.31
Operation of circuit in Figure 10.30.

temperature limit is shown by a solid arc. The underexcited stability limit is given by dotted line. The horizontal lines parallel to *x*-axis are constant power output lines. At point A, the motor operates at the maximum excitation current at its leading power factor φ. The stator current can be resolved into active and reactive components. The active power drawn from the system depends upon the load torque required and the reactive power can be varied by control of excitation.

The unregulated controllers are the simplest in terms of hardware and are brush type. A fixed rectifier transformer with off-load tap adjustments supplies excitation power through a field contactor. The reactive power output at no load will exceed that at the full load. The system voltage can fluctuate, depending upon the size of the motor and the Thévenin impedance as seen from the bus to which the motor is connected. For fluctuating loads, and large motors, this can give rise to instability on overloads, i.e., a crusher motor connected to a weak electrical system.

The four major types of regulated excitation systems are as follows:

1. *Constant field current*: This considers the variation of the field winding resistance as the motor is loaded; due to temperature rise, the resistance rises and the current falls. A constant field current is maintained by adjusting the conduction angle of SCRs, see Figure 10.32b.
2. *Constant power factor*: The kvar output rises as the motor is loaded, and this will tend to raise the motor bus voltage. An impact load will cause power factor to dip and the controller will respond with a boosted output. This will enhance the stability limit. In brushless excitation systems, the time constants associated with the pilot exciter and motor-field windings may not allow the excitation to respond quickly. Oscillations may occur on swinging loads and the stability may be jeopardized. The power factor controllers can be used to regulate the system power factor, rather than the individual motor, see Figure 10.32c.
3. *Constant reactive power output*: The motor bus voltage will be maintained constant from no load to full load. If a number of motors are connected to the same bus, the constant reactive power outputs of the running motor reduce the bus voltage dip when a motor is started, see Figure 10.32d.
4. *Constant voltage controller*: The characteristics are similar to that of var controllers, see Figure 10.32e.

10.19 Stability Concepts of Synchronous Motors

The output equation of a synchronous motor is as follows:

$$P = \left[\frac{EV}{X_d}\sin\delta + \frac{V^2}{2}\left(\frac{1}{X_q} - \frac{1}{X_d}\right)\sin 2\delta\right] \tag{10.12}$$

where E and V are the excitation and terminals voltages, respectively, X_d and X_q are the direct and quadrature-axis reactances, and δ is the torque angle. Mostly, the excitation is derived from the same source to which the motor is connected and a voltage dip in the

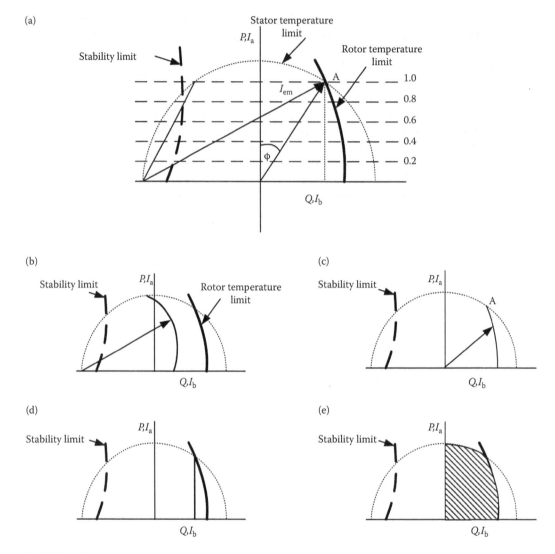

FIGURE 10.32
(a) Circle diagram of a synchronous motor, see text (b through e) constant field current, constant power factor, constant reactive power, and constant voltage excitation, respectively, control characteristics.

line voltage proportionally reduces the excitation voltage. If the excitation is derived from a separate stable source, the stability can be improved.

The second term of (Equation 10.12) shows that some torque (reluctance torque) is generated due to saliency. (Generally $X_q \approx 0.6\ X_d$.) If the excitation source is derived from the same power source to which the motor is connected, a voltage dip will reduce the excitation voltage also, proportionally. Therefore, it is desirable to provide excitation power from a separate stable source.

The transient stability of the synchronous motors has been widely discussed in the literature. An equal area a criterion of stability is briefly described which leads to calculations of PO power factor.

Equation 10.12 for the power output can be plotted as a power–angle relation, the second term in the power output relation has a sin 2δ term, which makes the power–angle curve peaky in the first half-cycle. The power–angle diagram of a generator is drawn in the first quadrant and that of a synchronous motor in the third quadrant. Referring to Figure 10.33, if the output shaft power demands an increase, from P_1 to P_2, the torque angle increases from δ_1 to δ_2. The limit of δ_2 is reached at the peak of the torque angle curve, δ_p, then at δ_p synchronizing power is zero. If the load is increased very gradually and there are no oscillations due to a step change, then the maximum load that can be carried is given by torque angle $\delta_p = 90°$ *for a sinusoid*. This maximum limit is much higher than the continuous rating and the operation at this point will be unstable—a small excursion of load will result in instability.

It is obvious that excitation plays an important role [1]. The higher the E, the greater the power output. On a simplistic basis, torque angle curves of a synchronous machine can be drawn with varying E. Neglecting the second frequency term, these will be half sinusoids with varying maximum peak.

Practically, the situation will not be as it is depicted in Figure 10.33. On a step variation of the shaft power (increase), the torque angle will overshoot to a point δ_3, which may pass the peak of the stability curve, δ_p. It will settle down to δ_2 only after a series of oscillations. If these oscillations damp out, we say that the stability will be achieved, if these oscillations diverge, then the stability will be lost, see Figure 10.34.

Note:

1. The inertia is fairly large.
2. The speed and power angle transients are much slower than electrical, current, and voltage transients.

In Figure 10.34, the area of triangle ABC translates into the retardation area due to variation of the kinetic energy of the rotating masses:

$$A_{\text{deaccelerating}} = \text{ABC} = \int_{\delta_1}^{\delta_2} (T_{\text{shaft}} - T_e)\,d\delta \tag{10.13}$$

The area CDE is the acceleration area:

$$B_{\text{accelerating}} = \text{CDE} = \int_{\delta_2}^{\delta_3} (T_{\text{shaft}} - T_e)\,d\delta \tag{10.14}$$

In accelerating and deaccelerating, the machine passes the equilibrium point given by δ_2 giving rise to oscillations, which will either increase or decrease. If the initial impact is large enough, the machine will be instable in the very first swing.

If the following condition is satisfied,

$$A_{\text{accelerating}} = B_{\text{deaccelerating}} \tag{10.15}$$

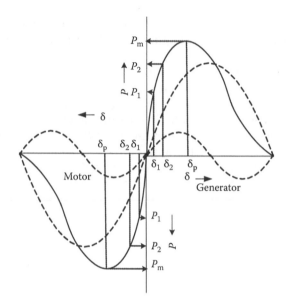

FIGURE 10.33
Torque–angle diagram of synchronous machines, generators, and motors.

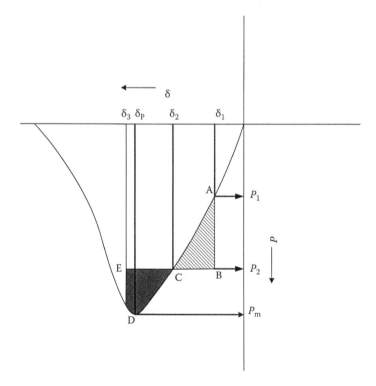

FIGURE 10.34
To illustrate equal area criteria of stability for a synchronous motor.

Motor Protection

Then, there are chances of the machine remaining in synchronism. The asynchronous torque produced by the dampers has been neglected in this analysis, also the synchronizing power is assumed to remain constant during the disturbance. This is the concept of "equal area criterion of stability."

10.19.1 Calculation of PO Power Factor

Figure 10.35a and b illustrates the synchronous motor phasor diagrams; Figure 10.35a is based on two-reaction theory of synchronous machines, which can be much simplified in Figure 10.35b. These diagrams are drawn with constant excitation and varying mechanical load, as we discussed in previous sections.

In these figures, φ is the power factor angle, δ is the torque angle, I_d and I_q are direct axis and quadrature-axis components of current I, I_a and I_r are active and reactive components of the current I, and X_{sd} and X_{sq} are the direct axis and quadrature-axis reactances.

A maximum torque angle setting of 90° is considered for setting the PO power factor. This is a conservative approach.

Example 10.3

First, we calculate the torque angle when the motor is operating at rated load and 0.8 leading power factor:

From the motor data in Table 10.15,

$X_s = 5.07\ \Omega$
Neglect motor resistance.
Motor full-load current is: $455(0.8 + j0.6)$ A.
Motor terminal voltage, phase to neutral = 2309 V.

Therefore, from Figure 10.35b,

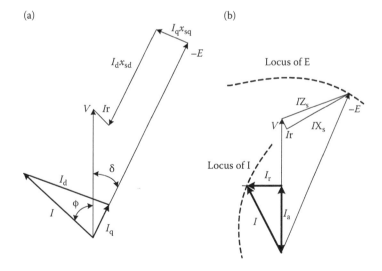

FIGURE 10.35
Phasor diagrams of a synchronous motor, (a) in direct and quadrature axis and (b) simplified; considering only direct axis.

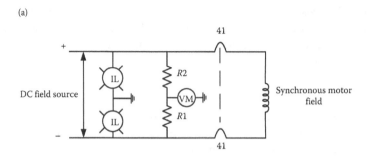

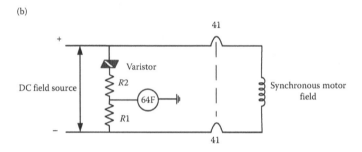

FIGURE 10.36
Detection of a ground fault in synchronous motor field windings, (a) a fault close to the center of the windings will not be detected and (b) preferred configuration with a varistor introduced.

$$E = 2309 - 455(0.8 + j0.6)(j5.07) = 4128 < -26.54°$$

Therefore, the torque angle of motor operating at full load, 0.8 power factor = $-26.54°$.
Assuming excitation constant, torque angle = 90° at PO, maximum current at PO is as follows:

$$I_m = \frac{E - V}{X_s} = \frac{4128 < -26.54° - 2309 < 0°}{j5.07}$$

$$= 454 < -36.7°$$

The angle changes from leading to lagging, and $\cos \varphi = 79°$.
Calculate the maximum PO torque:

$$T_m = \frac{3|V||E|}{5 \times 2\pi \times |X_s|} = 179.53 \times 10^3 \text{ NM}$$

This is 2.3 times the motor full-load torque. The motor data in Table 10.16 specify PO torque is equal to 3 times the full-load torque. This difference can be attributed to simplified calculations and neglecting saliency, and provides conservatism. The PO power factor setting provides redundant protection to PO protection on the field circuit.

10.20 Rotor Ground Fault Protection, Slip Ring-Type Synchronous Motors

Figure 10.36a shows a ground fault detection scheme using indicating lamps or voltmeter, connected as shown. A fault close to the midpoint of the windings will not be indicated.

Figure 10.36b shows a scheme which can detect ground faults even close to the middle point of the field windings. Without varistor, a ground fault at the center of the field windings will result in a zero voltage at device 64F. With varistor still there is a point in the field winding which results in zero voltage at 64F relay; however, any change in the field voltage moves the zero point because of varistor's nonlinear characteristic so the voltage at 64F is no longer zero and it could operate.

A second ground fault results in a short circuit in part of the rotor winding. This unbalances the air gap fluxes and magnetic field causing local heating and excessive vibration, and possibly disastrous rubbing between stator and rotor. Thus, the first ground fault should not be left in place, though it does not cause a disruption of operation. Studies [17–34] provide further reading.

References

1. EB Agamolh, S Peele, J Grappe. Response of motor thermal overload relays and phase monitors to power quality events. *Records on IEEE PPIC Conference*, Austin, TX, June 2016, pp. 14–21.
2. NEMA ICS 2-324. AC General Purpose HV Contactors and Class E Controllers, 50 Hz and 60 Hz.
3. NFPA70 National Electrical Code, 2017.
4. ANSI/IEEE C37.96. IEEE Guide for AC Motor Protection, 2012.
5. ANSI/IEEE 620. IEEE Guide for Construction and Interpretation of Thermal Curves for SCIM over 500-HP.
6. NEMA MG-1, Motors and Generators, Large Machines, (See in particular, Parts: 1, 5, 6, 7, 20, and 21).
7. ANSI/IEEE C37.112. IEEE Standard Inverse Time Characteristic Equations for Overcurrent Relays.
8. LL Gleason, WA Elmore. Protection of 3-phase motors against single phasing operation. *AIEEE Tran: Part III*, 77, 1112–1119, 1958.
9. SE Zocholl. *AC Motor Protection*. SEL, Inc., Pullman, WA, 2003.
10. CP LeMone. Large MV motor starting using ac inverters. *ENTELEC Conference Record*, Austin, TX, May 1984.
11. GE Publication GEH-5201. Synchronous motor control with CR192 microprocessor based starting and protection module, Mebane, NC, 1985.
12. AE Fitzgerald Jr., SD Umans, C Kingsley. *Electrical Machinery*. McGraw-Hill Higher Education, New York, 2002.
13. B Adkins. *The General Theory of Electrical Machines*. Chapman and Hall, London, 1957.
14. CV Jones. *The Unified Theory of Electrical Machines*. Pergamon Press, Oxford, 1964.
15. C Concordia. *Synchronous Machines, Theory and Performance*. John Wiley & Sons, New York, 1951.
16. AT Morgan. *General Theory of Electrical Machines*. Heyden & Sons Ltd., London, 1979.
17. OP Malik, BJ Croy. Automatic resynchronization of synchronous machines. *Proc IEE*, 113, 1973–1976, 1966.
18. JR Linders. Effect of power supply variations on AC motor characteristics. *IEEE Trans Ind Appl*, IA-8, 383–400, 1972.

19. JC Das. Effects of momentary voltage dips on operation of induction and synchronous motors. *IEEE Trans Ind Appl*, 36(4), 711–718, 1990.
20. JC Das, J Casey. Characteristics and analysis of starting of large synchronous motors. *Conference Record, IEEE I&CPS Technical Conference*, Sparks, Nevada, May, 1999.
21. JC Das, J Casey. Effects of excitation controls on operation of large synchronous motors. *Conference Record, IEEE I&CPS Technical Conference*, Sparks, Nevada, May, 1999.
22. JC Das. Application of synchronous motors in pulp and paper industry. *TAPPI Proceedings, Joint Conference Process Control, Electrical and Information*, Vancouver, BC, Canada, 1998.
23. GL Godwin. The nature of ac machine torques. *IEEE Trans Power Appar Syst*, PAS-95, 145–154, 1976.
24. M Canny. Methods of starting synchronous motors. *Brown Boveri Rev*, 54, 618–629, 1967.
25. J Bredthauer, H Tretzack. HV synchronous motors with cylindrical rotors. *Siemens Power Eng* (5), 241–245, 1983.
26. HE Albright. Applications of large high speed synchronous motors. *IEEE Trans Ind Appl*, IA-16(1), 134–143, 1980.
27. GS Sangha. Capacitor–reactor start of large synchronous motor on a limited capacity network. *IEEE Trans Ind Appl*, IA-20(5), 1984.
28. J Langer. Static frequency changer supply system for synchronous motors driving tube mills. *Brown Boveri Rev*, 57, 112–119, 1970.
29. K Albrccht, KP Wever. A synchronous motor with brushless excitation and reactive power control. *Siemens Rev*, (11), 577–580, 1970.
30. W Elmore. *Applied Protective Relaying*. Westinghouse (A Silent Sentinels Publication), Second Printing, 1979.
31. P Kundar. *Power System Stability and Control*, EPRI. McGraw-Hill, New York, 1994.
32. RH Park. Two reaction theory of synchronous machines, Part 1. *IEEE Trans*, 48, 716–730, 1929.
33. RH Park. Two reaction theory of synchronous machines, Part II. *AIEEE Trans*, 52, 352–355, 1933.
34. GL. Godwin. The nature of ac machine torques. *IEEE Trans*, PAS-95, 145–154, 1976.

11
Generator Protection

11.1 Ratings of Synchronous Generators

The world's largest single-shaft generator is rated at 1750 MW output. The microgrid and distributive power generators are small, which are of the order of 100 kW or even less. Also see Chapter 2 of Volume 1 for the largest thermal, hydro, wind, and solar plants. The generator construction, cooling arrangement, and capabilities vary according to their ratings and so also vary their protection.

The synchronous generator models, reactances, time constants, Park's transformation, capabilities, and reactive power capability curves are discussed in Chapter 9 of Volume 1 and also in Chapter 8 of Volume 2. These are not repeated here.

11.2 Protection of Industrial Generators

IEEE Standard 242 [1] recommends generator protection for bus-connected generators in industrial power systems.

- *Small generators*. Single isolated generators on LV systems are shown solidly grounded, which is not desirable due to third harmonic loading. The solidly grounded low-voltage power systems are no longer in use. The protective functions recommended are 51G and 51V only, which in the modern age of microprocessor-based relays will not protect the generator adequately. The generator is supposed to be operating isolated.
- *Multiple isolated generators on medium-voltage systems*. The generator is resistance grounded, with 51V, 32, 40, and 51G protective functions. The 87 function is shown in a flux balancing scheme. Again the generators operate in isolation, though in parallel with each other, on the same bus, as there is no synchronizing function shown. This is rather an unusual situation.
- *Medium-sized generator protection*. This is shown the same as that of multiple isolated generators as described above, except that the 87 function is now a full differential scheme and device 46 is optional. In terms of modern protective relaying, this protection is considered inadequate.
- *Large-sized generators*. The protection shown for a large-size generator adds 87G (ground differential), 49 (stator winding temperature detectors), 64 (field ground

fault protection), and device 60 (voltage selector relay between PT sources). Again in terms of modern protection, this is considered inadequate. Again no synchronization is shown.

Note that in industrial power systems, generators rated up to approximately 100 MVA, 13.8 kV are in service in cogeneration mode.

11.3 Functionality of a Modern MMPR for Generator Protection

Figure 11.1 shows the functionality (partial) of a modern MMPR for generator protection. Table 11.1 shows the description of the protective functions. The same relay type can be used for:

- Small to large generator protection
- Intertie or intertie/generator protection
- Multiple wind generator feeder protection

In addition, the following communication protocols are available:

- EIA-485/EIA-232/Ethernet Ports (single/dual, copper or fiber optic)
- Modbus TCP/IP Protocol

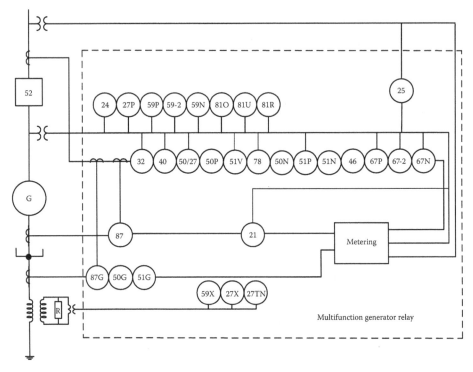

FIGURE 11.1
Functionality of a modern MMPR for generator protection.

Generator Protection

- Device Net
- IEC 61850 Communication
- DNP3 Serial LAN/WAN
- Simple Network Time Protocols

Also, the usual features of the MMPRs are data capture, storage, event recorder, fault reports and graphics, front panel displays, generator operating statistics monitoring, breaker wear monitoring, etc.

Using discrete electromagnetic relays to achieve even 50% of the functionality described above may require three panels, each of 3′ × 9′. Thus, electromagnetic relays are no longer in use.

It is a common industry practice to duplicate the MMPRs, say for generators of 20 MVA or above. Many a time, the redundancy is achieved by using relays of two different manufacturers with the same functionality.

We will describe the protective features and their settings as shown in Table 11.1. Also, the tripping logic is described.

TABLE 11.1

Functionality of an MMPR for Generator Protection

Device Number	Description
87	Phase differential, full, and also flux balance current differential
87G	Ground differential
REF	Restricted earth fault
64G	100% stator ground fault
40	Loss of field
49	Thermal overload
49-RTD	RTDs
46	Current unbalance
24	Volts/Hz
INAD	Inadvertent energization
21	Compensator distance
51V	Voltage-controlled or voltage-restrained TOC
51, 51Q, 51G	Phase time overcurrent, negative-sequence time overcurrent, ground time overcurrent
51N	Neutral time overcurrent
67, 67Q, 67G, 67N	Directional phase overcurrent, directional negative-sequence overcurrent, directional ground overcurrent, directional neutral overcurrent
50, 50Q, 50G	Phase inst. overcurrent, negative-sequence inst. overcurrent, ground inst. overcurrent
27	Undervoltage
59	Overvoltage (P, Q, G)
27S, 59S	Synchronizing under- and overvoltage
32	Directional power
81O, 81U	Over- and underfrequency
81R	Rate of change of frequency
BF	Breaker failure
60L	Loss of potential
25	Synchronization check
25A	Autosynchronizer

11.4 Voltage-Controlled and Voltage-Restraint Protection (51V)

Device 51V is for the generator backup protection to disconnect the generator from the system if a system fault has not been cleared by other protective devices after a sufficient time delay has elapsed. It protects the system components against excessive damage and prevents generator and its auxiliaries from exceeding their thermal limitations. Device 51V is used for bus-connected generators.

Where the output of the generator is stepped up to transmission voltage device 21 is used instead.

A simple overcurrent relay without voltage bias is unsuitable for this application. Most likely it will operate on generator load demand. If the settings are too low, it may not be able to protect the generator at all, and if the settings are high it may not be able to operate at all due to generator fault decrement characteristics.

There are two types of relays:

- In the voltage-restraint relay, the operating torque is proportional to voltage and effectively controls the pickup current of the relay over a ratio of 4:1.
- The voltage-controlled relay has its torque controlled by a high-speed voltage relay over a range of 65%–83% (adjustable). When the applied voltage is above the pickup setting, the relay contacts are held open. A clear distinction can, therefore, be made between normal no-fault conditions and abnormal fault conditions.

The settings on these two relay types are illustrated with an example.

Example 11.1

Figure 11.2 shows a distribution system with a 30/50 MVA, 138–13.8 kV transformer and a 50 MVA, 0.85 power factor generator operating in synchronism with the utility source at 13.8 kV. The load on the system is 40 MVA, and the largest feeder is shown for coordination and setting of 51V device.

Construction of the fault decrement curve of the generator is necessary before coordination is attempted. The reactances and time constants of the synchronous machines, and equivalent circuits during short circuit are covered in Chapter 9 of Volume 1 and are not repeated here, except the construction of fault decrement curve.

The expression for a decaying ac component of the short-circuit current of a generator can be written as

$$i_{ac} = \text{decaying subtransient component} + \text{decaying transient component}$$
$$+ \text{steady-state component} \tag{11.1}$$

$$= (i_d'' - i_d')e^{-t/T_d''} + (i_d' - i_d)e^{-t/T_d'} + i_d$$

The subtransient current is given by

$$i_d'' = \frac{E''}{X_d''} \tag{11.2}$$

where E'' is the generator internal voltage behind subtransient reactance:

$$E'' = V_a + X_d'' \sin \phi \tag{11.3}$$

where V_a is the generator terminal voltage and ϕ is the load power factor angle, prior to fault.

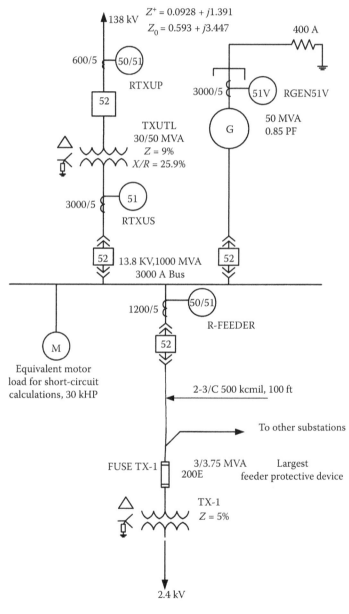

FIGURE 11.2
A system configuration for settings of generator 51V device.

Similarly, the transient component of the current is given by

$$i'_d = \frac{E'}{X'_d} \qquad (11.4)$$

where E' is the generator internal voltage behind transient reactance:

$$E' = V_a + X'_d \sin \phi \qquad (11.5)$$

The steady-state component is given by

$$i_d = \frac{V_a}{X_d}\left(\frac{i_F}{i_{Fg}}\right) \qquad (11.6)$$

where i_F is the field current at given load conditions (when regulator action is taken into account), and i_{Fg} is the field current at no-load-rated voltage.

The dc component is given by

$$i_{dc} = \sqrt{2}\, i''_d e^{-t/T_a} \qquad (11.7)$$

where T_a is the armature short-circuit time constant, given by

$$T_a = \frac{1}{\omega r}\left[\frac{2 X''_d X''_q}{X''_d + X''_q}\right] \qquad (11.8)$$

where r is the stator resistance.

The open-circuit time constant describes the decay of the field transient; the field circuit is closed and the armature circuit is open:

$$T''_{do} = \frac{1}{\omega r_D}\left[\frac{X_{ad} X_f}{X_{ad} + X_f} + X_{kD}\right] \qquad (11.9)$$

The quadrature-axis subtransient open-circuit time constant is

$$T''_{qo} = \frac{1}{\omega r_Q}\left(X_{aq} + X_{kQ}\right) \qquad (11.10)$$

The open-circuit direct-axis transient time constant is

$$T'_{do} = \frac{1}{\omega r_F}(X_{ad} + X_f) \qquad (11.11)$$

The short-circuit direct-axis transient time constant can be expressed as

$$T'_d = T'_{do}\left[\frac{X'_d}{(X_{ad} + X_l)}\right] = T'_{do} \frac{X'_d}{X_d} \qquad (11.12)$$

It may be observed that the resistances have been neglected in the above expressions. In fact, these can be included, i.e., the subtransient current is

Generator Protection

$$i_d'' = \frac{E''}{r_D + X_d''} \tag{11.13}$$

where r_D is defined as the resistance of the amortisseur windings on salient pole machines and analogous body of cylindrical rotor machines. Similarly, the transient current is

$$i_d'' = \frac{E'}{r_f + X_d'} \tag{11.14}$$

In Chapter 9 of Volume 1, a fault decrement curve of a generator of the following specifications is plotted.

Saturated subtransient reactance $X_{dv}'' = 0.15$ per unit

Saturated transient reactance $X_{dv}' = 0.2$ per unit

Synchronous reactance $X_d = 2.0$ per unit

Field current at rated load $i_f = 3$ per unit

Field current at no-load rated voltage $i_{fg} = 1$ per unit

Subtransient short-circuit time constant $T_d'' = 0.012$ s

Transient short-circuit time constant $T_d' = 0.35$ s

Armature short-circuit time constant $T_a = 0.15$ s

Effective resistance* = 0.0012 per unit

Quadrature-axis synchronous reactance* = 1.8 per unit

A three-phase short circuit occurs at the terminals of the generator, when it is operating at its rated load and power factor. It is required to construct a fault decrement curve of the generator for (1) the ac component, (2) the dc component, and (3) the total current. This is shown in Figure 9.6 of Volume 1. The dc component is given by Equation 11.7:

$$i_{dc} = \sqrt{2}\, i_d'' e^{-t/T_a} = 42.57 e^{-t/0.15}\, \text{kA}$$

This is also shown in Figure 9.6 of Volume 1, not reproduced here. At any instant, the total current is

$$\sqrt{i_{ac}^2 + i_{dc}^2}\, \text{kA rms}$$

Note that the short-circuit current with constant excitation is 50% of the generator full-load current. *This can occur for a stuck voltage regulator condition. Though this current is lower than the generator full-load current, it cannot be allowed to be sustained. Voltage-restraint or voltage-controlled overcurrent generator backup relays (ANSI/IEEE device number 51V) or distance relays (device 21) are set to pick up on this current.*

The computer programs for the relay coordination will plot the generator fault decrement curve based on the proper input data, and it is not necessary to construct the curve by hand calculations.

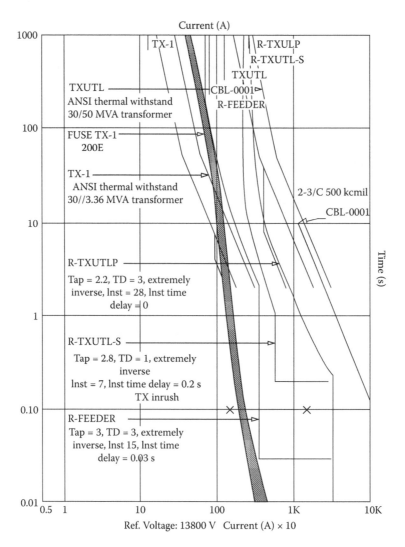

FIGURE 11.3
Coordination of protective devices in Figure 11.2 (The generator 51V device is not shown).

Continuing with the example, the 50 MVA, 13.8 kV, 0.85 power factor, and four-pole generator has the following particulars:

$$X''_{dv} = 10\%, X_{2v} = 9\%, X_0 = 7\%, X/R = 64, X'_d = 15\%, X_d = 160\%, T''_d = 15 \text{ ms}, T'_d = 400 \text{ ms}$$

The other major impedance data are shown in Figure 11.2.

In order to coordinate 51V device with the system relaying, first carry out the coordination in the entire power system to come up with the protective device with the maximum setting with which the coordination is required. Figure 11.3 shows the protective device coordination for the system.

The generator 51V protection should be set so that it coordinates with the system overcurrent device with the largest settings; in this case it will coordinate with the utility transformer primary overcurrent relay, with the following settings:

Generator Protection

Tap = 2.2, TD = 3, extremely inverse characteristics, Inst = 28 A, instantaneous time delay = 0 s.

51 V Settings on Voltage-Restraint Device

This device should be set with a pickup setting at full voltage equal to generator full-load current multiplied by a factor of 1.5 for possible overload operation. Thus, the pickup setting should be

$$\text{Generator full-load current} = 2092 \text{ A}$$

$$\text{Pickup} = \frac{2092 \times 1.5}{800} = 3.92$$

Note that 800 is the CT ratio. Thus, a tap setting of 4 can be selected at full rated voltage. Ensure that at zero voltage the minimum sustained generator current for the struck regulator condition, that is, at rated excitation, is protected. At zero voltage the pickup reduces to 800 A. This pickup value should be below the generator sustained short-circuit current of 1300 A. The coordination is shown in Figure 11.4. Complete coordination with the utility transformer primary *relay settings is not obtained*.

Settings on 51V, Voltage-Controlled Relay

For the voltage-controlled relay, first calculate the generator sustained short-circuit current. Here, it is 1300 A. Then the pickup setting is

$$\text{Pickup} = \frac{1300}{800} = 1.625$$

Provide a somewhat lower pickup setting to ensure that the relay pickup = 1.5 and TD = 15—very inverse characteristics, see Figure 11.5.

It is noted that the complete coordination with system relaying is not obtained. This is often the case. When the generator is relatively small compared to the system components, the lack of coordination will be more; the generator curves shown in Figures 11.4 and 11.5 will shift toward the left. *However, we do not sacrifice the generator protection and accept this lack of coordination.* The protection provided by 51V device should not normally operate, as the generator's other protections should take care of the generator faults.

Application of Device 21

In a unit configuration, the generator is directly connected to the high-voltage grid through a generator step-up transformer (GSU) and provided with phase-to-phase and three-phase fault protection. Zone 1 of device 21 is normally set to look through the generator impedance X_d plus 50% of the GSU winding impedance, with no appreciable time delay. Zone 2 can be set up to reach into the system and provide a longer time delay. Alternatively, zone 1 can be set to provide backup protection for a fault on the high-side bus with a coordinating time delay, and zone 2 is set up with a longer reach and longer time delay for breaker failure backup protection. The settings for zones 1 and 2 are as follows:

- Z1 reach in secondary ohms
- Z1 offset in secondary ohms

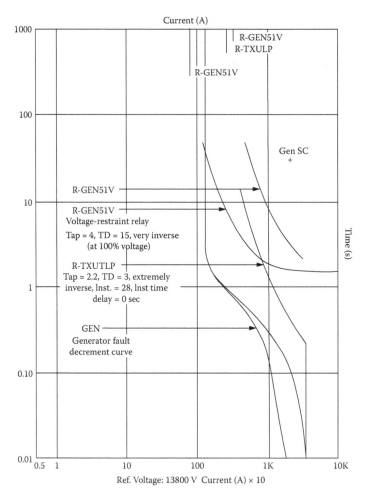

FIGURE 11.4
Settings of a 51V voltage-restrained device to coordinate with the system protection shown in Figure 11.3.

- Z1 time delay
- Z1 forward current in amperes
- Z1 positive-sequence angle in degrees
- Similar settings for zone 2

A fault study is required to determine the magnitude and angle of the apparent impedance seen by the relay during a system fault.

Example 11.2

A 60 MVA, 13.8 kV, 0.85 PF, 60 Hz generator with transient reactance X_d = 200% (on generator base) is connected to a 60/100 MVA step-up transformer with 13.8–230 kV, Z = 9%, and X/R = 33. The utility three-phase symmetrical short-circuit level is 20 kA at X/R = 13.28. Set zone 1 to look through 50% of the transformer windings and zone 2 to reach through a fault at a distance of 5 miles on the 230 kV line with the following parameters: R = 0.0982 Ω/mi, X = 0.778 Ω/mi, and XC = 0.1823 MΩ/mi. The CT ratio is 3000:5 and the PT ratio is 14400:120 V.

Generator Protection

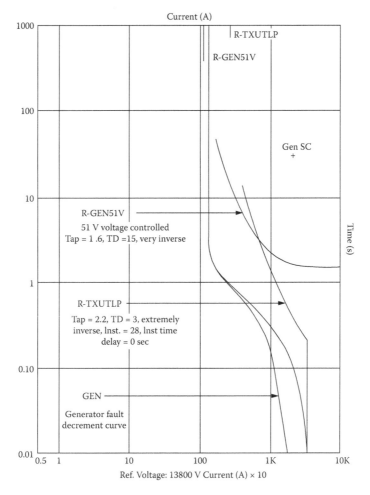

FIGURE 11.5
Settings of a 51V voltage-controlled device to coordinate with the system protection shown in Figure 11.2.

A fault study is required. It can be calculated that at a distance of 5 miles away, with the 230 kV line parameters, three-phase short-circuit level is 40.6 kA at $X/R = 49.58$.

Impedance to zone 1 fault in secondary ohms is equal to $0.071 + j355\ \Omega$
Impedance to zone 2 fault in secondary ohms is equal to $0.41 + j8.19\ \Omega$

A reader can verify these calculations, and zones 1 and 2 settings can be made.

11.5 Negative-Sequence Protection: Function 46

The negative-sequence capability of generator is discussed in Chapter 9 of Volume 1. The presence of power system harmonics can overload the generators with respect to negative-sequence currents (see Section 6.1.4 of Volume 3). These data are not repeated here, except

TABLE 11.2

Requirements of Unbalanced Faults on Synchronous Machines

Type of Synchronous Machine	Permissible $I_2^2 t$
Salient pole generator	40
Synchronous condenser	30
Cylindrical rotor generators	
Indirectly cooled	30
Directly cooled (0–800 MVA)	10
Directly cooled (801–1600 MVA)	10−(0.00625)(MVA-800)

TABLE 11.3

Continuous Unbalance Current Capability of Generators

Type of Generator and Rating	Permissible $I_{2\,(percent)}$
Salient pole, with connected amortisseur windings	10
Salient pole, with nonconnected amortisseur windings	5
Cylindrical rotor, indirectly cooled	10
Cylindrical rotor, directly cooled	
to 960 MVA	8
961–1200 MVA	6
1201–1500 MVA	5

that Tables 6.1 and 6.2 of Volume 3 are reproduced as Tables 11.2 and 11.3, respectively, for ready reference. These negative-sequence capabilities of the generators should never be exceeded. Also, continuous and short-time unbalanced capability of the generators is graphically shown in Figure 11.6 from ANSI standards [2,3].

Figure 11.7 shows the negative-sequence time overcurrent operating characteristics of a modern MMPR. The operating time is given by

$$t_{op} = \frac{46 Q_2 K}{\left(\dfrac{I_2}{I_{nom}}\right)^2} \tag{11.15}$$

Preferably the negative-sequence function is connected to trip the main generator breaker. This allows quick resynchronization after the unbalance conditions have been eliminated. If the machine conditions do not permit this operation, then the negative-sequence relay must also trip the prime mover and the field and transfer the auxiliaries.

Example 11.3

A 100 MVA has permissible $I_{2\,(percent)} = 10\%$ and $I_2^2 t = 30$. The appropriate settings for this generator are as follows:

Continuous unbalance current setting = 9% (1% is kept reserve)

Generator short-time negative-sequence current setting = 30. The curves in Figure 11.7 show that at this setting and 100% negative-sequence current, the operating time is 30 s.

Generator Protection

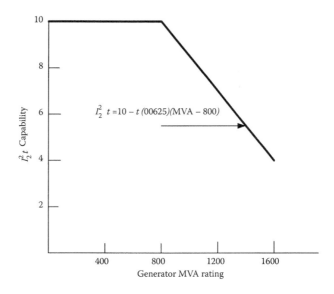

FIGURE 11.6
Negative-sequence capability of a generator based on its rating in MVA.

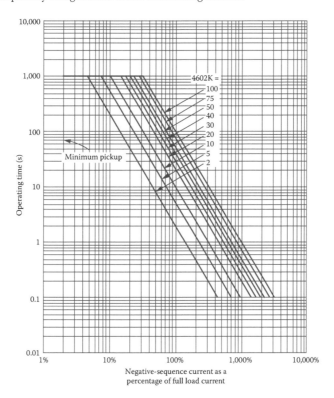

FIGURE 11.7
Negative-sequence operating time versus ratio of negative-sequence current to nominal current for various generator negative-sequence capabilities (capabilities in a modern MMPR).

11.6 Loss of Excitation. Protection: Function 40

The source of excitation to a generator can be completely or partially removed due to accidental tripping of the field breaker, field short circuit, or loss of supply to the excitation system. This presents a serious operating condition to the operation of the generator.

11.6.1 Steam Turbine Generators

On loss of excitation, the generator will over speed and operate as an induction generator. It will continue to supply some power into the system and receive its excitation from the system in the form of vars. The generator slip and the power output will be a function of the initial generator loading, generator, and system impedances and also govern characteristics. The severest condition for the generator and the system occurs when the generator is operating at full load. For this condition, the generator current can be 2.0 pu, its speed will be 2%–5% above normal, as the generator will lose synchronism. There will be a high level of currents induced in the rotor. These high currents can cause dangerous overheating of the stator windings and rotor within a short time. Also, as the loss of excitation means operating at a very low excitation, overheating of the end portions of the stator core may occur.

The var drain on the system can depress system voltage and may impact the operation of generators in the same substation or elsewhere in the system. The increased voltage reduction can cause tripping of transmission lines and impact stability of the system.

11.6.2 Hydro Generators

Due to saliency and reluctance torque, a hydro generator may be able to supply 20%–25% load on loss of excitation. Also with zero excitation operation, it is necessary to accept line charging current. If a loss of field occurs when the generator is supplying full load, the effect will be similar to that of a steam turbine generator. The high currents can damage the stator windings, rotor, and amortisseur windings, and the unit will impose a var drain on the system as in the case of steam turbine generators.

11.6.3 Protection

Distance relays are widely used for protection. These sense the variation of impedance as seen from the generator terminals. This variation of impedance will have the characteristics as shown in Figure 11.8. In this figure, curve (a) shows the variation of impedance with the machine operating initially at or near full load. The initial impedance point is C and the impedance locus follows the path C–D. The impedance locus will terminate at point D, to the right of the (–X) ordinate. It will approach impedance value somewhat higher than the average of direct- and quadrature-axis subtransient impedances of the generator. Curve (b) shows the profile when the machine is operating initially at 30% load and under excitation. The impedance locus follows the path EFG and will oscillate in a region between points F and G. For a loss of field at no load, the impedance as viewed from the machine terminals will vary between the direct- and quadrature-axis *synchronous* reactances (X_d, X_q). In general for any load the impedance will terminate on the dashed curve.

Generator Protection

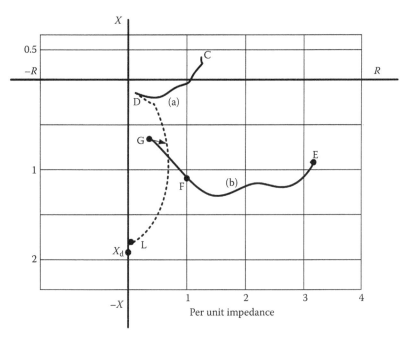

FIGURE 11.8
Impedance locus plotted on the R–X diagram for loss of excitation.

One type of protection is shown in Figure 11.9. The construction shows the required settings. A time delay of 0.5–0.6 s will be used with impedance diameter of X_d to prevent incorrect operation on stable swings. The 1.0 pu impedance diameter will detect a loss of field from full load down to about 30% load. In this impedance diameter no time delay is provided, thereby providing fast protection for more severe conditions in terms of possible machine damage.

The other type of protection is shown in Figure 11.10, which uses a combination of impedance unit, a directional unit, and an undervoltage unit. The impedance Z2 and directional units are set to coordinate with the generator maximum excitation limiter and its steady-state limit. During abnormally low excitation conditions, these units operate and sound an alarm, allowing an operator to correct the conditions. If a low-voltage condition exists indicating a loss of field condition, the undervoltage unit will trip with a time delay of 0.25–1.0 s. The Z1 setting trips without any time delay.

A hydro generator may be operated occasionally as a synchronous condenser, and the schemes described can operate unnecessarily when the generator is operated under excited. The undervoltage relay can be used to supervise distance relaying schemes—the dropout level of the undervoltage relay will be set at 90%–95% of the rated voltage.

In the past a number of electromagnetic relays were necessary to provide the protection shown in Figures 11.9 and 11.10 (not so in modern generator MMPRs).

The loss of field protection is normally connected to the trip main generator breaker and the field breaker and transfers the auxiliaries. The field breaker is tripped to minimize the damage to the rotor field in the case the loss of field is due to a rotor field short-circuit or flashover at the slip rings. If the loss of field could be easily corrected, a tandem-compounded generator could be resynchronized. This approach is not valid when sufficient auxiliary loads cannot be transferred to maintain boiler and fuel systems. In this case, turbine stop valves should also be tripped.

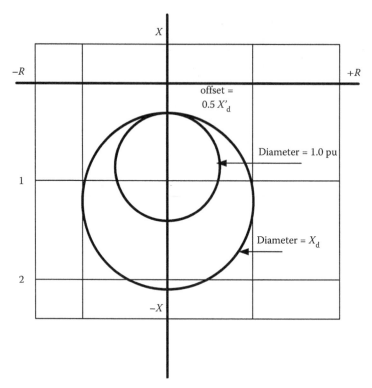

FIGURE 11.9
Settings for loss of excitation protection.

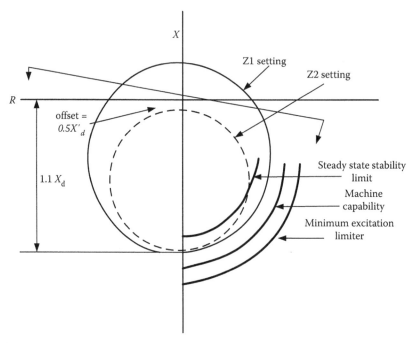

FIGURE 11.10
Settings for loss of excitation protection, alternative to Figure 11.9.

Example 11.4

A 13.8 kV 90 MVA, 0.85 steam turbine generator has $X_d = 175\%, X'_d = 20\%$. Provide suitable settings for the relay types shown in Figures 11.9 and 11.10. The generator is provided with CTs of 4000:5 ratio and open delta–connected PTs, 14400/120 V. The generator is directly connected to 138 kV system through a transformer of 60/100 MVA, with percentage impedance = 9% and $X/R = 28$: the three-phase short-circuit level at the 138 kV system is 30 kA at $X/R = 15$.

X_d in primary ohms = 3.797 Ω
X'_d in primary ohms = 0.434 Ω

In terms of secondary ohms:

$$X_d = \frac{3.797 \times 800}{120} = 25.31 \, \Omega$$

$$X'_d = 2.89 \, \Omega$$

$$X_{pu} = 14.46 \, \Omega$$

Settings for the protection as shown in Figure 11.9:

Zone 1 offset setting = –1.45 Ω
Zone 1 Mho diameter = 14.5 Ω
Zone 1 time delay = 0
Zone 2 Mho diameter = 25.3 Ω
Zone 2 offset setting = zone 1 offset setting = 1.45 Ω
Zone 2 time delay = 0.6 s

Settings for the protection as shown in Figure 11.10:

Zone 1 Mho diameter = $1.1 X_d + \frac{-X'_d}{2} = 1.1 \times 25.3 - 1.45 = 26.4 \, \Omega$
Zone 1 offset = –1.45 Ω
Zone 1 time delay = 0.25 s
Zone 2 diameter is set = $1.1 X_d + X_t + X_s \, \Omega$

where X_t and X_s are the transformer and system reactances.
  Calculated values of X_t and X_s in secondary ohms based on the data provided are as follows:

$$X_t = 1.89 \, \Omega$$

$$X_s = 0.0139 \, \Omega$$

Thus, zone 2 diameter is 29.734 Ω. Thus, set at 28.74 Ω.

Zone 2 offset is set = $X_t + X_s$ = 1.90 Ω
Zone 2 time delay = 60 s
Zone 2 directional angle: set at –20°

When undervoltage setting is used, the zone 2 time delay can be much reduced. Set undervoltage = 0.8% of the nominal voltage for single machine and 0.87% for multimachine buses.
  Then undervoltage setting AND gated with directional angle setting with a pickup time delay of 0.25 s and dropout time of zero seconds.

11.7 Generator Thermal Overload

The continuous output capability is given at the terminal of the generator at the specified frequency and power factor. For hydrogen-cooled generators, the capability is given at maximum and several lesser pressures. For combustion generators, this capability is given at an inlet air temperature of 15°C at sea level. Also, the synchronous generators can operate successfully at rated kVA, frequency and power factor for a voltage variation of 5% above and below rated voltage.

The short-time thermal capability of generators is given in Table 11.4 and also shown in Figure 11.11 (from ANSI C57.13).

Most generators are provided with embedded RTDs in the stator windings and provide thermal overload protection. The RTDs can be wired for alarms or trips or alarm and trip, depending on the settings. The settings on the RTDs are made according to the permissible temperature rise on generator insulation, as for motors (see Chapter 10). Currently, the class F insulation systems are most common.

Depending on ratings and design, stator core and windings may be cooled by air, water, or hydrogen. In direct-cooled (also called conductor-cooled) generators of large ratings, the coolant (usually water) is in direct contact with the heat-producing members such as generator stator windings. In indirectly cooled generators, the coolant cools the generators relying on heat transfer through the insulation. The generator manufacturer will provide

TABLE 11.4

Short-Time Thermal Capability of Generators

Time (s)	10	20	60	120
Armature current (percent)	226	154	130	116

Source: ANSI C50.13.

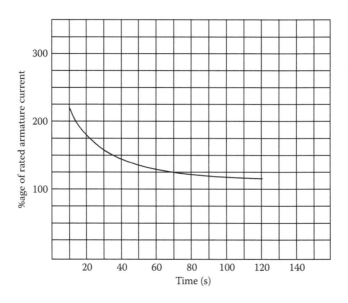

FIGURE 11.11
Overload capability of generators.

the protection for the loss of the coolant. This protection may consist of temperature monitors, pressure, and flow sensors.

The core hot spots can be produced by lamination insulation failure caused by foreign objects left in the machine, by damage to core during installation. The hot spots are the results of high eddy currents produced from the core flux that find conducting paths across the insulation between laminations.

For air-cooled machines, RTDs are the only way of protection against core hot spots. For hydrogen-cooled generators, the hot spots can be detected by a *core monitor*. It is an ion particle detector, which is connected to a generator in a manner that permits a constant flow of cooling gas to pass through the monitor. Under normal conditions, the coolant contains no particles. However on deterioration and decomposition of the organic material, core laminations and epoxy paint produce a large number of particles, which are of submicron size and are detected by the monitor. Such monitors are supplied only on large machines.

The thermal protection of the field includes the following:

- Protection of the main field circuit
- Protection of the main rotor body, wedges, retaining rings, and amortisseur windings.

The short-time limits of the field windings are shown in Figure 11.12. Under normal operating conditions, the field windings operate continuously at a current equal to or less than required to produce rated kVA at rated power factor. For power factors less than the rated power factor, the generator output kVA can be reduced. The generator reactive capability curve shown in Figure 8.6 of Volume 2 is derived on this basis.

It is not possible to embed RTDS in the field windings. The monitoring of the current and field voltage can give the measurement of the *average* temperature of the field windings, *but not of the hot spot, which is more important.* Further, this method cannot be applied to brushless exciters.

When the generator is equipped with core monitor, it will also detect overheating of the field windings and hot spots.

An overcurrent relay can be set to coordinate with the curve shown in Figure 11.12. On operation of this relay an alarm can be sounded, allowing time to take remedial action and adjust the field, and then after a time delay trip can be initiated. An inverse time relay to coordinate with the curve shown in Figure 11.12 is preferable than a fixed time relay.

The excitation systems have a voltage regulator, which controls the field in response to terminal changes in the generator voltage based on maximum and minimum excitation limits of the machine. See Figure 8.6 of Volume 2 for URAL and steady-state and transient stability limits superimposed on reactive capability curve.

Modern MMPRs can provide thermal protection of the generator: thermal model as described in IEC 60255-8 [4]. RTD biasing can be used, which uses the ambient temperature and the winding temperature measured by the RTD elements. The relay generates and continuously updates a thermal model, based on the actual operating conditions and the calculated generator thermal capacity used.

Figure 11.13 shows the generator overload curve. This is based on the pickup level of 1.05 pu, the generator cooling time constant of 10 min, and the generator preloaded to 95%. The time constant data must be obtained from the generator manufacturer. The curve does not consider negative-sequence currents.

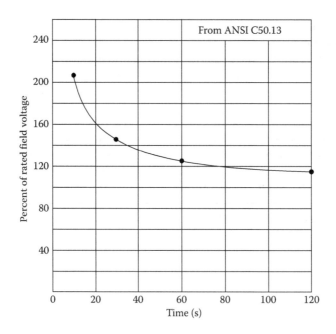

FIGURE 11.12
Percentage of rated field voltage versus time duration of withstand in seconds.

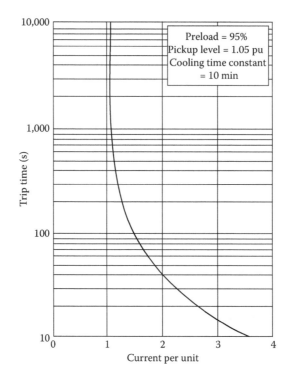

FIGURE 11.13
A current versus trip time characteristic generated for the generator preload of 95%, the pickup level 1.05 pu, and the cooling time constant of 10 min.

11.8 Differential Protection

11.8.1 Generator Winding Connections

The application of differential relaying depends on the generator winding connections.

Small generator windings may be connected in delta. The wye-connected generators are common (even up to 600 MVA ratings). Figure 11.14a shows a two-circuit, three-phase, six-bushing-type connection. When more than one circuit per phase is used, these circuits are connected in parallel inside the machine and two leads brought out for external connections. In some hydro generators, there may be a number of circuits per phase and each circuit may consist of a number of multiturn coils connected in series. In these machines, the parallel circuit may be formed in two groups that are paralleled and only two leads brought out for external connections. Figure 11.14b and c illustrates a wye-connected, double-winding construction sometimes used in large steam turbine generators. Each phase has two separate windings, which are connected externally to form two wye connections. The high-voltage terminals of each phase are connected in parallel to form a single three-phase output. Separate wye connections are formed at the neutral end of the windings, which may be physically at the opposite ends of the machine. This

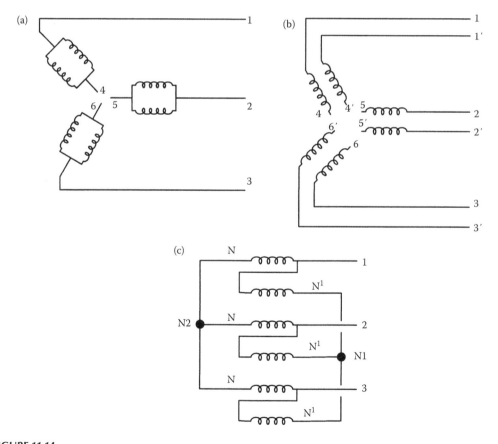

FIGURE 11.14
Generator winding connections: (a) two-circuit three-phase six bushings and (b and c) double-winding, one-circuit, three-phase 12 bushings.

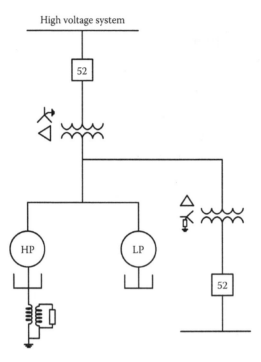

FIGURE 11.15
Cross-compounded generators.

arrangement is sometimes referred to as the double-ended 12 bushing machine. It is used when the total phase current exceeds the current carrying capability of single bushing.

The generator connections to the system vary. We have already discussed bus-connected generators and generators directly connected to high-voltage system through a GSU transformer. This configuration will be further discussed. Figure 11.15 shows connections of cross-compounded generators. The low-pressure and high-pressure units are bussed together at generator voltage and connected to the system through a grounded wye–delta unit transformer. Both the LP and HP units are usually wye connected, and it is the recommended practice to ground one of the neutrals.

11.8.2 Protection Provided by Differential Relays

Differential relaying will not detect turn-to-turn fault in the same phase since there is no difference in the current entering and leaving the winding. Where applicable, separate turn fault protection can be provided with split-phase relaying scheme. Differential protection will detect three-phase faults, phase-to-phase faults, double-phase-to-ground faults, and some single-phase-to-ground faults depending on how the generator is grounded. Figure 11.16 shows the percentage of stator windings that will be unprotected by differential relaying for phase-to-ground faults. In addition, the magnitude of the ground fault decreases linearly as the fault moves from the generator line terminals to the neutral end. When the ground fault is limited below the generator full-load current, a large percentage of the generator windings will remain unprotected. As generally the generator ground fault is limited to low values, a separate ground fault protection is required as discussed further.

Generator Protection

11.8.3 Self-Balancing Differential Scheme

Much like protection of motors, Chapter 10, Figure 10.25 self-balancing differential scheme can be used with generators also. This scheme is very sensitive for providing phase and ground protection (Figure 11.17).

11.8.4 Application Depending on Winding Connection

Most generators will have single stator windings, six bushing type. As noted previously, turn-to-turn protection cannot be provided in these machines. When the generator

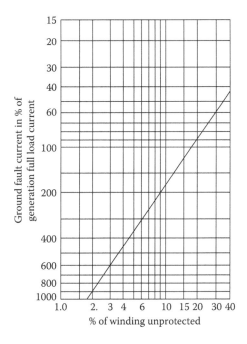

FIGURE 11.16
Percentage of stator windings protected for ground faults, based on generator ground fault current in percentage of generator full-load current.

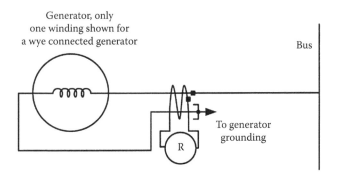

FIGURE 11.17
Self-balancing differential protection, one phase only shown.

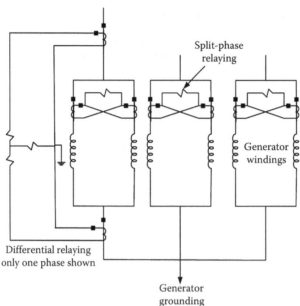

FIGURE 11.18
Application of split-phase and differential relaying.

windings have split phases, with two or more circuits per phase, the turn-to-turn fault detection is possible. Figure 11.18 shows the application of combination split-phase and differential relaying. The circuit of each phase is split into two equal groups and the currents in each group are compared. A difference in the currents indicates an unbalance caused by a turn fault. As there will be some unbalance in each group, the overcurrent relay is set so that it does not respond to this normal unbalance but will operate for unbalance on a single turn fault. Time delay is employed to ride through CT transients and normal unbalance. The problem of CT error can be eliminated by using a single-window or double-window CT. A single-window CT has a common core and the currents balance in the magnetic core, and only the difference current provides an output in the secondary circuit. In a double-window CT, the theoretical principle is the same but without physical restrictions of a single-window CT. See Figure 11.19a, which shows split-phase protection using a single-window current transformer and Figure 11.19b, which illustrates split-phase protection using double-primary, single-secondary current transformer.

11.8.5 Backup Protection Using Differential Relays

The backup protection is dependent on how the generator is connected to the system and also on generator rating. In a unit generator configuration for large machines, we have three zones of protection:

- GSU transformer differential protection
- Generator differential protection
- Overall protection covering the generator as well as GSU

This is shown in Figure 11.20. The differential relay for the transformer alone is of high impedance type, with harmonic restraint (see Chapter 12). The generator differential

Generator Protection

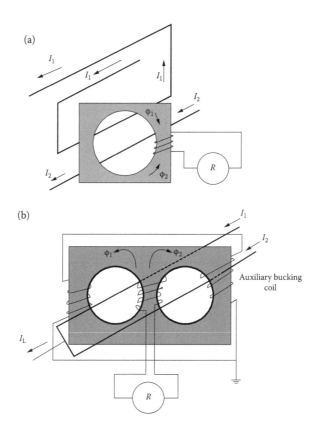

FIGURE 11.19
(a) Split-phase protection using a single-window CT and (b) split-phase protection using double-primary and single-secondary CT.

relay is the current differential variable slope percent differential relay. For this relay, harmonic restraint is not necessary. The overall differential relay for the transformer and the generator is the current differential variable slope relay but harmonic restraint is required.

Figure 11.21 shows overall differential protection for a cross-compounded generator.

11.8.6 Characteristics

The dual-slope percentage differential protection of an MMPR provides more sensitive and stable protection. The dual characteristics shown in Figure 11.22 compensate for steady-state, transient, and proportional errors, within the zone of protection. Steady-state errors are those that do not vary with loading. These include transformer magnetizing currents and unmonitored loads. Proportional errors are those that vary with loading. These errors include CT ratio errors, tap changing, and relay measuring errors. Transient errors are those that occur temporarily, say due to CT saturation. The four settings that define the characteristics are as follows:

O87P = minimum operate level required for operation. It is typically about 10 times the tap setting. The restraint elements determine whether the operating current quantity is greater than the restraint quantity by using the characteristics shown in Figure 11.22. For harmonic blocking the harmonic content of the differential current must exceed the set individual threshold levels. Magnetization currents of transformers contain large amount

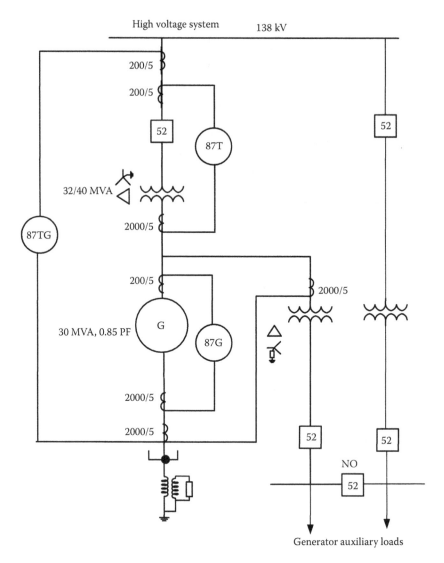

FIGURE 11.20
Three separate differential protections shown for a unit-connected generator.

of harmonics, see Volume 3. Uneven CT saturation during external faults can lead to nuisance operation. The high security mode can be programmed to obviate this problem.

SLP1 = initial slope beginning at the origin and intersecting O87P on IRT-restraint current IRT= O87P.100/SLP1

IRS1 limit of IRT for SLP1 operation, intersection with SLP2 begins

SLP2 = second slope, SLP1
U87P = unrestraint pickup

In unit connection schemes, the three-phase currents for phase angle and phase interaction effects are compensated by appropriate matrices, see Chapter 12.

Generator Protection

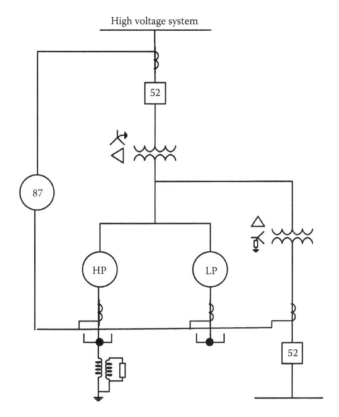

FIGURE 11.21
Differential protection of cross-connected generators.

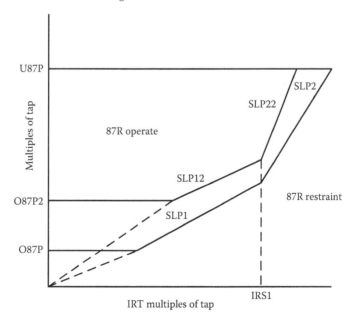

FIGURE 11.22
Two slope differential characteristics of a modern MMPR.

Example 11.5

This example does not include GSU transformer in the protection zone.

Consider a 13.8 kV generator of 30 MVA, and CT ratios on line and neutral side = 2000:5. CTs are connected in wye configuration. Then the tap is equal to

$$\frac{30 \times 1000}{\sqrt{3} \times 13.8 \times 400} = 3.1$$

In this case, dual-slope characteristics are used without harmonic restraint. The slope 1 equation is

$$\text{SLP1}_{\min}\% = \left(\frac{\text{Er}\%}{(200 - \text{Er}\%) \times k}\right) 100 \qquad (11.16)$$

where
$\text{SLP1}_{\min\%}$ = slope ratio that will accommodate error Er with no margin
Er = amount of error expected in normal operation
k = average restraint scaling factor = 1

If we consider the following:

Excitation current error = 4%
CT accuracy = 3%
No-load tap changer = 5%
Relay accuracy = ±5%

then the minimum slope from Equation 11.16 is 9.3%. Generally, a 25% slope setting is desirable to account for unforeseen errors. Thus, the settings are as follows:

O87P = 0.3
SLP1 = 25%
SLP2 = 70%
IRS1 = 6.0
U87P = 10
O87P2 = 1.25—high security mode operate current pickup in multiples of tap.
  Also, high security mode dropout time = 5.0 s.

Example 11.6

Consider that the generator of 30 MVA of Example 11.5 is connected in unit step-up configuration to 138 kV high-voltage system as shown in Figure 11.20. (Practically, this generator rating is too small for unit connection; however, this example is for the calculations of settings.) It is required to calculate the settings for the overall differential protection 87TG (shown in this figure). The CT ratios on the 138 kV side are 200:5. The 138 kV and 13.8 kV CTs are in wye configuration. The unit auxiliary transformer (UAT) is delta–wye connected and has a CT ratio of 2000:5. (This should be identical to the CT ratio on the generator neutral.) First use the matrix equations to compensate for the winding shift in 138 kV (wye-connected windings solidly grounded) and 13.8 kV windings delta connected. Referring to Chapter 12, on the 138 kV side the connection matrix is 0, while on the 13.8 kV side it is 11.

The taps on primary and secondary sides with the given CT ratios are identical, which are equal to 3.14.

Set:

O87P = 0.3
SLP1 = 30%
SLP2 = 70%
IRS1 = 6U87P = 10
Fourth harmonic threshold for blocking = 15%
Fifth harmonic threshold for blocking = 35%
Harmonic restraint and blocking enabled
O87P2 = 1.25
High security mode dropout time in seconds = 10

In addition, the relay has the restricted ground fault protection for 138 kV wye-connected, solidly grounded GSU transformer windings.

11.8.7 The Tripping Modes

It is common practice to have primary and backup protections to energize separate hand-reset multicontact relays:

- Trip main generator breakers
- Trip field and exciter breakers
- Trip prime mover
- Turn on CO_2 internal fire protection, if provided
- Operate alarm and enunciators
- Transfer station service to standby source

11.9 Generator Stator Ground Fault Protection

The stator ground fault protection and the %age of stator windings protected are a function of the generator grounding. Table 11.5 shows common generator grounding methods with notes on protection.

11.9.1 HR Grounded through Distribution Transformer Unit Generators

- The 100% stator winding ground fault protection using third harmonic voltage distributions across the generator stator windings from line-to-neutral terminals is discussed in Chapter 8 (see Figures 8.29 and 8.30).

11.9.1.1 Third Harmonic Differential Scheme

Figure 11.23 illustrates a third harmonic voltage differential scheme. This scheme compares the third harmonic voltage appearing at the generator neutral to that, which appears at the generator terminals. The ratio of these voltages is relatively constant for all loadings.

TABLE 11.5

Grounding Methods of Generator Neutral

Configuration	Grounding Arrangement	Application	Protection
Unit-connected generators through GSU transformer	High-resistance grounded through a distribution transformer	Practically all generators in the utility generation	100% stator winding protection can be provided. Ground fault differential relaying not required
Bus-connected generators	Hybrid grounding See chapter	Current trend in grounding industrial bus-connected generators	Ground fault differential is required to protect maximum stator windings.
	High-resistance ground through a distribution transformer	New grounding possibility.	Ground fault differential is not required. 100% stator winding can be protected
	Low-resistance grounded, grounding current limited to 100–400 A	Has been the most popular method of grounding of industrial generators	Ground fault differential is required. 100% stator windings cannot be protected. Ground overcurrent relays coordinated with system ground fault protection
	Reactance grounding	Not popular	
	Ungrounded generators	Current IEEE recommendations do not recommend that the generators be ungrounded. Possibility of subjecting the windings to overvoltage due to generator reactance capacitance couplings	Neutral displacement type of relays can be provided
	Solidly grounded	Small industrial generators—this practice is not recommended even for small machines, especially when two or more generators are connected to the same bus due to third harmonic loading	Neutral-connected ground overcurrent relays coordinated with system ground fault relays

A ground fault will disrupt this balance causing the differential scheme to operate. An advantage of the scheme is that it continuously monitors the grounding transformer primary and secondary connections and voltage transformer at the terminals of the machine. It operates for open circuits or short circuits that will prevent the overvoltage relay or other relays from operating. Thus, a problem could be detected before the stator ground fault occurs. This may favor an alarm rather than a trip.

11.9.1.2 Subharmonic Voltage Injection

Figure 11.24 illustrates a scheme with subharmonic voltage injection at the neutral terminals of the generator being protected. The injected signal returns to ground through

Generator Protection

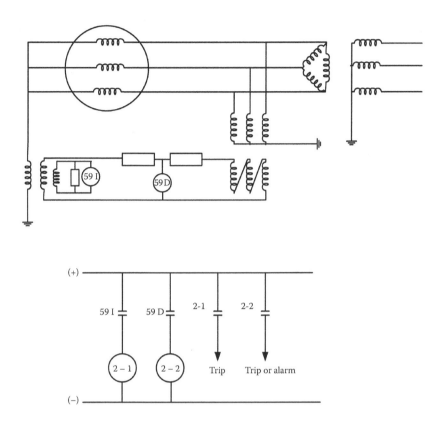

FIGURE 11.23
Hundred percent stator winding protection for ground faults using differential third harmonic sensing.

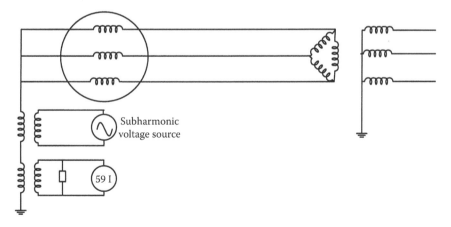

FIGURE 11.24
Hundred percent stator winding protection using subharmonic injection.

the stator winding shunt capacitances to ground. When a stator ground fault occurs, the shunt capacitances are short-circuited, and the magnitude of the injected signal increases. This change in the signal level is detected by the relay. This scheme provides protection whether the generator is operating or is at standstill.

When the generator is out of service, the grounding system associated with the generator also goes out of service. As the generator is connected through a GSU transformer with delta-connected windings on the generator side and UAT has delta primary windings, the system reverts to an ungrounded system. A neutral displacement relay is provided to take care of this condition, see Figure 8.11 in Chapter 8.

11.9.2 Generator Ground Fault Differential

The generator ground fault differential scheme with a current polarized directional relay is shown in Figure 11.25. The polarizing coil is energized through a CT in the generator neutral circuit, while the operating coil is in the residual circuit of the generator phase differential scheme. This provides sensitive ground fault protection. Also see Chapter 12 for transformer restricted fault protection. Modern MMPRs can accept a certain mismatch between CT ratios.

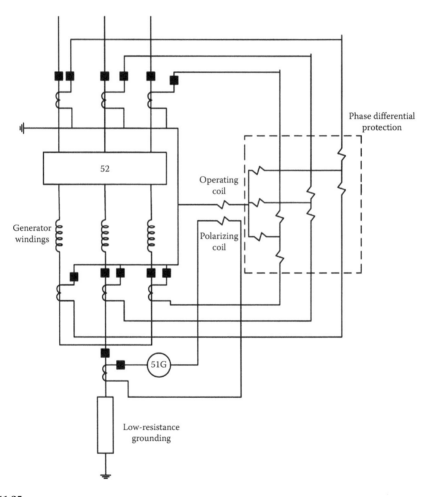

FIGURE 11.25
Combined phase differential and ground fault differential protection, with the ground fault relay polarized from a neutral-connected CT.

11.10 Rotor Ground Fault Protection

11.10.1 Slip Ring Machines

The field circuit of the generator is an ungrounded system. Thus, a single ground fault will not affect the operation of the generator. If a second ground fault occurs a portion of winding will be short-circuited, thereby producing unbalanced fluxes in the air gap of the machine. These may cause rotor vibrations which may quickly damage the machine. Also, unbalance temperatures caused by unbalance currents can give rise to similar damaging vibrations. The probability of a second ground fault increases, since the first ground fault establishes a ground reference for the voltages induced in the field by the stator transients.

Figure 11.26 shows a method similar to the one employed to detect a control battery ground using a voltage divider and a sensitive voltage relay connected to the midpoint. A maximum voltage will be impressed on the relay by a ground fault close to the positive or negative terminals. However, a fault at the midpoint of the windings will remain undetected. This problem is avoided by using a nonlinear resistor in one leg of the voltage divider. The resistance of the nonlinear resistor varies with the applied voltage. The divider is proportioned so that the field winding null point is at the field winding midpoint when the exciter voltage is at rated voltage. Changes in the exciter voltage will move the null point from the middle of the windings. This concept is similar as illustrated in Figure 10.36 for the ground fault protection of the field of synchronous motors, Chapter 10.

11.10.2 Brushless Type of Generators

This scheme is shown in Figure 11.27. There are three units:

- Transmitter mounted on the rotating part and connected to the rotating rectifier three-phase bridge as shown (between two phases).

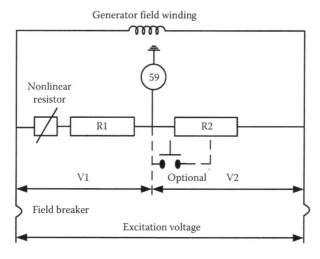

FIGURE 11.26
Generator field, brush-type generators, ground fault protection.

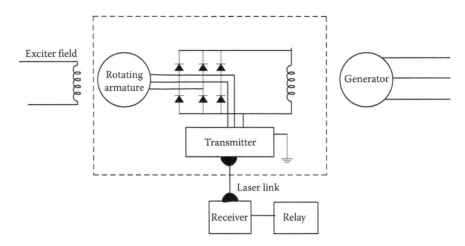

FIGURE 11.27
Generator field, brushless-type generators, ground fault protection.

- A receiving unit mounted outside the rotating part. The light information conveyed by the transmitter is interpreted by optical phototransistors in the receiver.
- A relay unit—the loss of LED light in the receiving unit will actuate this ground relay.

The circuit of the transmitter is shown in Figure 11.28. The transmitter has a transformer T1, which is powered by the connections to the three-phase rotating rectifier bridge as shown. This connection puts the rectifier in the transmitter, which includes the base circuit Q2 in series with the rotating rectifiers. Base current Q2 is determined by the field ground resistance and the location of the ground fault with respect to positive and negative bus.

With no ground fault, there is no Q2 base current and Q2 is off. Q1 is saturated and keeps LEDs on. If base current flows in Q2 due to low-resistance path external to the transmitter, transistor Q1 and LEDs are turned off, indicating the possibility of a ground fault.

The receiving unit, using infrared phototransistors, senses the passing pulse of infrared energy. The pulse period is of the order of 0.05 ms for each revolution for an 11 in. radius at 3600 rpm. The period is lengthened in the receiver by a monostable flip-flop to provide sufficient signal to ground relay, which has an adjustable time delay. The LED ground detection sensitivity curve is provided in Figure 11.29, which shows that fairly high-resistance ground faults can be detected over a wide range of exciter voltage. The time delay setting on the ground relay prevents operation during start up when the field voltage is building up.

As a backup to rotor ground fault protection, vibration monitoring is provided. This is organized to trip the main and field breaker if the vibration is above that associated with short-circuit transients for the faults external to the unit.

11.11 Volts per Hertz Protection

This section can be read along with volts per hertz protection for transformers in Chapter 12. Overexcitation occurs when volts/Hz applied to the terminals of the equipment exceeds 1.05 pu (generator base) for a generator and 1.05 pu (transformer base)

Generator Protection

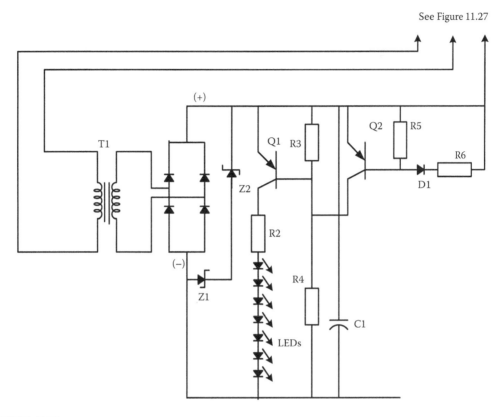

FIGURE 11.28
Schematic diagram of operation of the circuit shown in Figure 11.27.

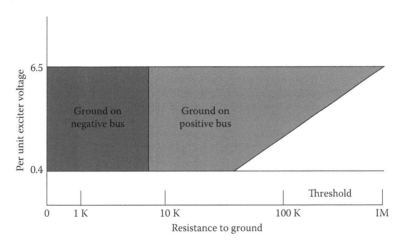

FIGURE 11.29
Field windings protected as a function of fault resistance to ground.

for a transformer at full load or 1.1 pu when the transformer is at no load. When these values are exceeded, saturation of the magnetic core of the generators and transformers can occur and stray flux can be induced in nonlaminated components which are not designed to carry flux. The field's current in the generator could also be excessive. This can cause severe overheating in the generator and transformer and breakdown of the insulation.

One of the major causes of overexcitation is operation at reduced voltage under regulator control at reduced frequencies during startup or shutdown. With the regulator maintaining rated voltage, while the unit is at 95% or lower speed will result in volts/Hz of 1.05 or higher resulting in possible damage.

Overexcitation can also occur during complete load rejection which leaves transmission lines connected to the generating unit. Under these conditions, volts/Hz may exceed 1.25 pu. With the overexcitation control, the excitation may be reduced in a couple of seconds and with excitation control out-of-service damage will occur. Failures in the excitation system can be another possible cause of overexcitation.

The excitation systems are provided with overcurrent, undercurrent, and volts/Hz limiter. Yet, it is a standard practice to provide additional dedicated protection for this condition.

Currently, there are no ANSI/IEEE standards describing safe limits of volts/Hz for short-term capabilities, but manufacturers supply these data on request.

The protection relays are connected to the voltage transformers. A PT fuse failure will give an incorrect indication. Redundancy can be achieved by connecting one set of relays to the PTs that supply voltage regulators and the other to PTs that are for metering and relaying functions. A voltage transfer device 60 is commonly used between two sets of PTs. Modern MMPRs have a logic for loss of potential.

There are two types of protection schemes illustrated in Figure 11.30. These are dual-level volts/Hz sensing (see Figure 11.30a) and inverse volts/Hz sensing (see Figure 11.30b).

It is possible that the differential protection may operate on volts/Hz. The delta winding of the GSU may produce a large 60 Hz component with little odd harmonics, and the harmonic restraint in differential relays may not be adequate. As discussed in differential protection, the modern MMPRs restrain on fifth harmonic and second harmonic.

Example 11.7

Referring to Figure 11.30a and b, the typical settings are as follows.

Level 1 setting 24D1P is set close to 1.05 or slightly greater with a time delay of 1 s to allow correction for overexcitation. This setting is for alarm.

Either a dual overexcitation protection or inverse definite time/inverse characteristics can be selected.

Dual-Level Volts per Hertz

Set 24D2P = 118% and time delay = 6 s, set 24D2P1 = 110% and time delay = 60 s. See Figure 11.30a.

Inverse Characteristics

A composite inverse time and definite time characteristics can be selected (see Figure 11.30b). Select 24IP = 108% and 24D2P2 = 118%, and time delay = 12 s. Select the settings to coordinate with the manufacturer supplied characteristics; the relay has various inverse time–current characteristic shapes. First plot the withstand characteristics supplied by the manufacturer and then try different curve shapes for coordination built into the MMPR.

The protection is generally connected to trip the main generator breaker and field breaker and transfer the auxiliaries.

Generator Protection

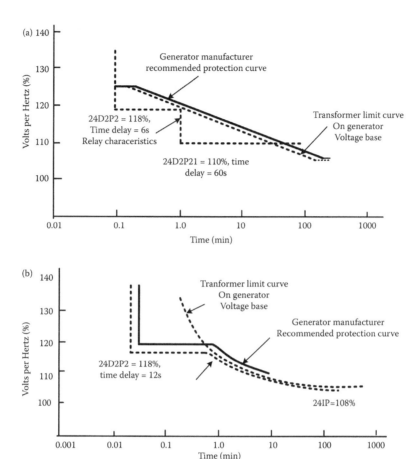

FIGURE 11.30

Volts per hertz protection: (a) two slope characteristics and (b) inverse time characteristics. Settings as shown for a specific case.

11.12 Over- and Underfrequency Protection

Generators and turbines are limited in the degree of abnormal frequencies that can be tolerated without damage.

At reduced frequency, there will be a reduction in the output capability of the generator. There are no ANSI/IEEE standards detailing this derating; however, manufacturers can supply useful data.

The steam turbines are more restricted with respect to abnormal frequency operation. This is because mechanical resonances can occur in many stages of the turbine blades. The departure from the normal frequency will bring stimulus frequencies closer to the natural frequencies of the various blades and there can be a resonant condition giving rise to excessive vibrations and ultimate damage. Manufacturers provide time limits for abnormal frequency operation. These data are supplied in the form of permissible duration of operation in a certain frequency band. Figure 11.31 from IEEE C37.106 [5] illustrates these limits supplied by five manufacturers. These capabilities apply to steam turbines. Combustion turbine generators (CTGs), in general, have greater capability than the steam

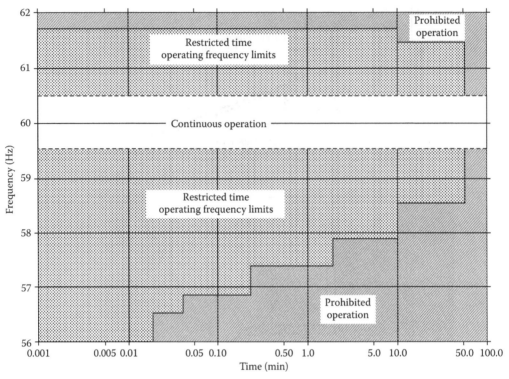

FIGURE 11.31

Under- and overfrequency operation of a steam turbine generator, restricted time and prohibited time, based on data of five manufacturers [5].

units. However, the CTGs are frequently limited by combustion instability and/or sharply reduced turbine output as the frequency drops.

11.12.1 Load Shedding

The primary protection is provided by automatic load shedding programs in the power system. These programs are designed so that sufficient load is shed to quickly restore the frequency to normal value. A frequency-dependent load shedding study is required. With under shedding of load, it may cause an extremely long time for frequency to return to normal or the frequency may bottom to a level lower than the system normal frequency. In either case, there is a possibility of damaging the steam or CTGs. In general, the underfrequency operation is more critical than the overfrequency operation.

11.12.2 Protection

The modern MMPRs record the total time of operation of the generator at off-nominal frequencies in as many as six bands. If the frequency is within a time accumulator band, the relay will assert an alarm bit and start a timer for that band. If the frequency remains in that band for the set timer delay, the relay will start adding time to that accumulator band timer. If the total time of operation exceeds the limit setting of that band, an alarm or trip is initiated.

Generator Protection

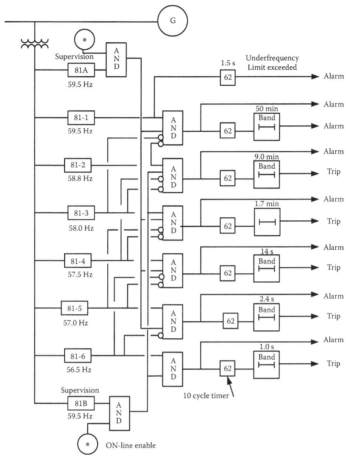

FIGURE 11.32

A multiple set-point scheme with frequency band logic and accumulating counters.

Example 11.8

This example is based on IEEE C37.106. A multiple set-point scheme with frequency band logic and accumulating counters is shown in Figure 11.32. The protection scheme is designed to closely follow the manufacturer's limit curves for underfrequency operation. In addition to the supervision steps, six underfrequency set points are rarely utilized. It takes into account the cumulative time spent in each underfrequency band. It accumulates the time spent in each band independently and stores it in nonvolatile memory. IEEE C37.106 recommends a time delay of 10 cycles to allow the turbine blade underfrequency resonance to establish to avoid unnecessary accumulation of time.

The dotted line in Figure 11.33 represents characteristics of the relay settings, and the solid line represents turbine abnormal frequency operating limit for time-restricted operation. Dotted lines represent the relay settings.

The sample settings shown provide some margin under the turbine damage limit for the trouble-free operation curve and maintain coordination with the system load shedding curve. Table 11.6 summarizes the frequency and timer setting used in this example.

Figure 11.34 shows the operation at over ranges of voltages and frequency from IEC 60034-3. The generators are required to deliver continuously rated output at rated power factor over the ranges ±5% in voltage and ±2% in frequency. The operation outside the

TABLE 11.6

Frequency and Time Settings (Example 11)

Frequency Band (Hz)	Time Delay	Comments
60.0–59.5	–	Continuous operation allowed
59.5 or below	1.5 s	Continuous underfrequency alarm. Time delay to avoid spurious alarms
59.5–58.8	50.0 min	Alarm "underfrequency limit exceeded". These bands may trip or alarm, depending upon individual practices. For alarms the operator has these respective times to shed loads or isolate the unit based on limits shown in Figure 11.31
58.8–58.0	9.0 min	
58.0–57.5	1.7 min	
57.5–57.0	14.0 s	
57.0–56.5	2.4 s	
56.5 or below	1.0 s	

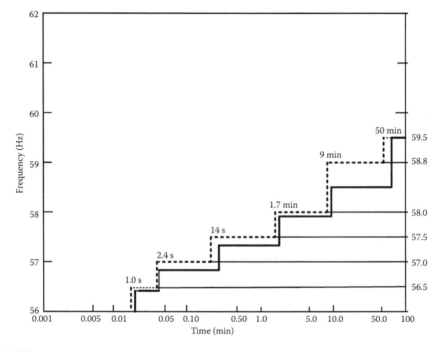

FIGURE 11.33
Characteristics of the relay settings. The solid line represents turbine abnormal frequency operating limit for time restricted operation. The dotted lines represent the relay settings.

area shown with bold lines should be restricted in extent, duration, and frequency of occurrence. A generator manufacturer will impose severe restrictions to avoid accelerated aging of generator and prime-mover mechanical components.

11.12.3 Rate of Change of Frequency

Frequency changes occur in the system due to load and generation unbalance. The generator regulator control action adjusts the active power and restores the frequency to normal value. Failure of this control action can lead to instability unless there is some remedial

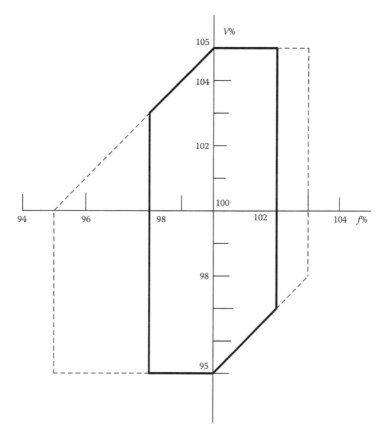

FIGURE 11.34
Operation at over ranges of voltages and frequency from IEC 60034-3.

action, such as load shedding. The rate of frequency change can detect and initiate this remedial measure. The modern MPPRs have a number of set points for the rate of change of frequency. A first frequency measurement is taken and then the second measurement after an adjustable time interval, time window dt. The element uses a hysteresis (dropout/pickup ratio of 95%) to prevent chattering. A minimum positive-sequence voltage is necessary for the operation of the element. Two levels of undervoltage and overvoltage elements are provided.

11.13 Out-of-Step Protection

The transient stability of the synchronous generator is discussed in Chapter 15 and also briefly it was touched upon in Chapter 10. The generator instability can be caused due to a variety of reasons, such as prolonged fault clearing times, low machine excitation, high impedance between the generator and the system, or some line switching operations. When a generator loses synchronism, the off-frequency operation caused winding stresses, pulsating torques, and mechanical resonances that are potentially damaging the generator. The generator should be tripped preferably during first half slip cycle on loss of synchronism.

The differential relays or system backup protection will not detect loss of synchronism. The loss of excitation relay may provide some protection but cannot be relied upon to provide protection for all system conditions.

The relaying approach is to analyze the variation of system impedance as viewed for the terminals of the system elements. It has been shown that on loss of synchronism between two system areas, the apparent impedance, as viewed from the generator or line terminals, varies as the impedance between the two systems, the system voltage, and angular displacement between the two systems.

The impedance loci are approximately circular characteristics that move in a counterclockwise direction.

There are two protection schemes in use for the detection of loss of synchronism:

- Single-blinder scheme
- Double-blinder scheme

We will discuss only single-blinder scheme as it is more popular. Referring to Figure 11.35, the mho element and the right and left blinders are shown. During short-circuit faults, the impedance moves from the load region to inside the mho element and between the two blinders instantaneously, preventing out-of-step function from pickup.

The relay picks up when the positive-sequence impedance trajectory moves from the load region to area A (also blinder pickups). When the impedance trajectory advances to area B, mho element and both blinders assert. If the impedance trajectory exits mho circle via area C, a timer and dropout relay start timing. The out-of-step function remains picked

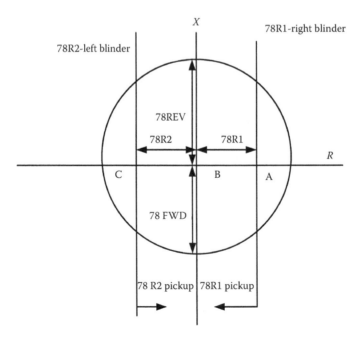

FIGURE 11.35
A single-blinder scheme for out-of-step protection.

Generator Protection

for the dropout delay after pickup delay time expires. For trajectories traveling from left to right the operation is symmetrical.

Example 11.9

The following data are required for the settings:

Generator transient reactance X'_d in secondary ohms
GSU impedance in secondary ohms
Impedance of the lines or lines beyond GSU

Convert all impedances on generator base kV.
Consider $X'_d = 24.5\%$, generator MVA = 125, generator rated voltage = 13.8 kV, three-phase = 60 Hz. CT ratio = 8000:5, and PT ratio = 14400:120 V.

$X_t = 10\%$ on generator base
$X_{sys} = 6.25\%$ on generator base

$$X'_d = 0.245 \text{ pu}$$

$$X'_{d,primary} = \frac{13.8^2 \times 0.245}{125} = 0.373 \, \Omega$$

$$X'_{d,sec} = \frac{0.373 \times 1600}{120} = 5.0 \, \Omega$$

Transformer:

$$X_{T,primary} = \frac{13.8^2 \times 0.1}{125} = 0.152 \, \Omega$$

$$X_{T,sec} = \frac{0.152 \times 1600}{120} = 2.03 \, \Omega$$

Similarly, $X_{sys,sec} = 1.27 \, \Omega$. Consider an $X/R = 15$ and $R = 0.847 \, \Omega$.

Normally set forward reach at two to three times the generator X'_d; here, we will set it at two times the generator $X'_d = 10 \, \Omega$. Set reverse reach 78_{REV} at 1.5 to 2.0 times the transformer reactance. To provide adequate coverage with a margin of error, add system Z_{sys} as shown in the geometric construction in Figure 11.36. Set blinders at 120° approximately, angles α and β. Separation angles of 120° between two sources result in loss of synchronism. Time delay = 8 cycles, pole slip counter = 1.0, pole slip reset = 120 cycles, and trip on mho exit = disable.

The scheme is connected to trip only the generator breaker and isolate the generator with its auxiliaries. In this way when the system conditions are favorable, the generator can be resynchronized.

11.14 Inadvertent Energization of Generator

When a generator is energized while on turning gear, it will behave and accelerate as an induction motor. The equivalent machine impedance can be expressed as a

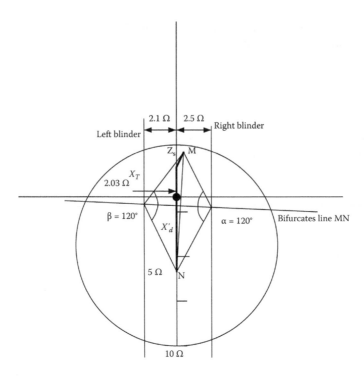

FIGURE 11.36
An example of out-of-step settings.

negative-sequence reactance X_2 in series with a negative-sequence resistance; the negative-sequence reactance is approximately equal to subtransient reactance.

The terminal voltage and current of the machine depend upon the impedance to the energization circuit.

If the generator and the transformer are connected to a source of infinite impedance, the machine current is high,—three to five times pu. If the generator and the transformer are connected to a weak system, the current can be 1–2 pu.

If the generator is accidentally energized through a station service transformer, the current will be much lower, of the order of 0.1–0.2 pu.

High currents will be induced in the rotor and the damage may occur in a few seconds. There are a number of protection schemes described in IEEE C37.102 [6], namely:

- Loss of excitation relays
- Reverse power relays
- System backup relays such as mho distance relays and 51V device
- Directional overcurrent relays
- Frequency-supervised overcurrent relays
- Distance relays

This reference may be seen for their limitations and applications. Modern MMPRs use the logic shown in Figure 11.37. The indicators that the generator has been removed from service include:

Generator Protection

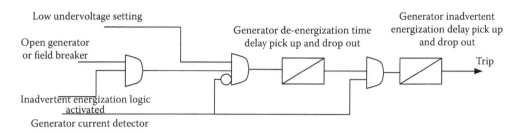

FIGURE 11.37
Logic circuit diagram of inadvertent energization of a generator.

- Low terminal voltage
- Field circuit breaker/generator breaker that are open
- No phase current

The undervoltage pickup value, as sensed by the PTs on the generator side, is set to a very low value to ensure that there is no voltage at the generator terminals. The open generator breaker and/or field breaker is an input to the AND gate. Settings generated de-energized pickup and dropout provide time delay in the logic. The dropout setting ensures that inadvertent protection remains armed for the setting duration after undervoltage element deasserts to allow for reapplication of the generator field. These settings add an arming delay to prevent the scheme from arming prematurely, when the generator is removed from service. The dropout time ensures that the scheme trips when the relay detects current during inadvertent energization. The generator low current is detected; the setting is 0.25 in terms of secondary CT current. The last AND gate is ready to activate inadvertent time as soon as it detects energization current. Its pickup must be set less than dropout.

The following settings are appropriate in most cases:

Generator de-energization pickup = 3 s
Generator de-energization dropout = 1 s
Undervoltage pickup = 0.1 pu
Generator current detector = 0.25 A in terms of CT secondary current (5 A secondary CT)
Inadvertent energization timer pickup = 0.3 s
Inadvertent timer dropout = 0.15 s

11.15 Generator Breaker Failure Protection

The scheme described is applicable to failure of any breaker—alternate upstream breaker/breakers must be tripped to clear the fault. Though the modern breakers are reliable and can be provided with duplicate trip coils, trip signal may be applied to the first trip coil and if the breaker is not tripped, it is applied to the second trip coil with some delay; yet the breaker may fail to trip (see Chapter 2).

A simple logic is shown in Figure 11.38. The protective relay operation will trip the breaker and simultaneously initiate breaker failure timer. If the breaker is not tripped, the trip output is utilized to trip other breakers and isolate the generator. The breaker auxiliary contact must be used, as there are protection conditions that will not result

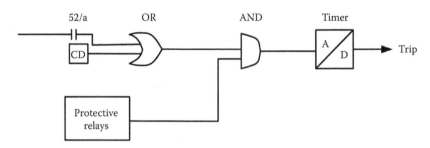

FIGURE 11.38
A simple generator breaker failure logic.

in sufficient current in the current detector CD—such as stator or bus ground faults, volts/Hz protection, and excessive negative sequence. When the generator is connected to a system through two circuit breakers, each must be equipped with breaker failure scheme.

Another breaker failure consideration occurs for an open breaker flashover to energize the generator. This can occur prior to synchronization or just after the breaker is removed from service, and the voltage across the breaker contacts can be twice the rated voltage. Though the breakers are rated for this condition and out-of-phase switching, the probability of a flash does exist.

The breaker failure scheme is modified as shown in Figure 11.39. An instantaneous relay (50N) is connected to the neutral of the step-up transformer. The relay output is supervised by generator 52/b contact. When the generator breaker is open and one or two poles flash over, the resulting transformer neutral current is detected by 50N. The current detectors must be set with sufficient sensitivity to detect the flash over condition.

Generator breaker flashover can also be detected by breaker pole disagreement relaying. This scheme monitors the three-phase currents and monitors whether any phase is below a certain low threshold, indicating an open pole, while the other phases have higher threshold level indicating a closed or flashed-over pole.

The trip is connected to a lockout relay, hand reset that will trip the necessary breakers to isolate the generator.

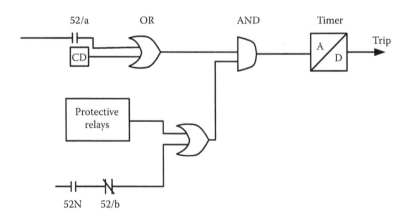

FIGURE 11.39
A modified breaker failure scheme, see text.

Generator Protection

11.16 Antimotoring Protection

If for any reason the energy supply to the prime mover is cut off while the generator is on line, the generator will act as a synchronous motor and drive the prime mover. The primary consideration is that the prime mover may be damaged during a motoring condition.

The rotation of the turbine rotor and blades in steam environment causes windage losses, which will be higher in the exhaust end of the turbine, where the diameter of the rotor disc and blade lengths is highest. Also, losses are a function of the steam density—any situation where the steam density is high will cause dangerous windage losses. For example, if the vacuum is lost on the unit, the density of steam increases and the windage losses will be many times greater than the normal.

Under normal conditions, the windage loss energy is carried away by the steam flow. With no steam flow, the turbine parts are heated as windage heat energy is not carried away and dissipated. Even when the unit is synchronized and operating practically at very little load, this phenomenon can occur. As temperature of the various parts of the turbine is controlled by steam flow, under motoring, the parts will heat or cool at uncontrolled rate. This can lead to unequal contraction and expansion of the turbine parts, and could even cause a rub between stationary and rotating parts—generating further heat.

There is a maximum safe operating time limit that can be supplied by the manufacturers. Windage probes depend upon the type of prime mover:

- Gas turbines may have gear problems when driven from the generator end.
- With hydro turbines, cavitations of the blades can be caused. If hydro turbines are being operated as synchronous condensers, motoring may be allowed.
- With diesel generator, there are chances of fire and explosion due to unburned fuel.

The reverse power protection is applied:

- With gas units, the large compressor load represents substantial power demand from the system, up to 50% of the rating of the unit.
- With hydro turbines with blades under tail-race water, the motoring power is high, and above tail-race water, it is low of the order of 2%–2% of the rated.
- Steam turbines under full vacuum and zero steam flow require 0.5%–3% of rating to motor.

Figure 11.40 shows the setting for a reverse power operation. The setting should be lower than the motoring power, minimum, for the specific prime mover as specified by the manufacturer.

Example 11.10

The minimum motoring power of a 100 MW steam turbine unit as specified by the manufacturer is 1.5%. Calculate the setting on the reverse power relay. CT ratio = 1600, and PT ratio = 120.

In terms of secondary watts, the power is 7.8 W.

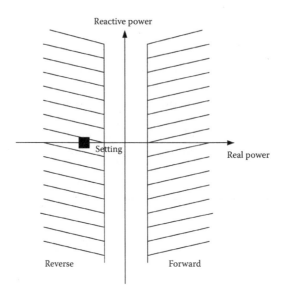

FIGURE 11.40
Reverse power relay settings for antimotoring protection.

        Set at 7.0 W
        Set fast time delay = 3 s
        Set slow time delay = 30 s

This requires some explanation. See Section 11.20 for the simultaneous and sequential trip logic. The fast time delay setting first closed the steam valve and then certain protective functions in the generator operate. This means that the generator breaker, field breaker, etc., are not tripped and *the turbine is allowed to purposely motor*. The longer time delay is used to trip the required breakers and auxiliaries as a standby.

11.17 Loss of Potential

Generally, two or more sets of PTs are used in the generator zone. PTs for voltage regulator and protective relays are dedicated PTs, normally connected grounded wye, grounded wye connection, with primary and secondary fuses. If one or more fuses operate, the secondary PT voltages applied to the relays and the regulator will be reduced in magnitude and shifted in phase angles. This can cause both the relays and the regulator to misoperate.

It is a common practice to apply a voltage balance relay, device 60, which compares the three-phase secondary voltages of the two sets of PTs as shown in Figure 11.41. If the fuses blow in one set of PTs, the resulting unbalance will cause the relay to operate. The relay is usually connected to remove the regulator from service and block possible incorrect operation of the relays, which have a potential input. Also see Chapter 2.

Ferroresonance can occur when grounded PTs are connected to an ungrounded system. Under these conditions, the voltage appearing in one or more PTs can be distorted 60 Hz or subharmonic voltages and the PTs will be operating in a saturated region. The PTs can fail thermally in a relatively short time. This is possible if the generator is disconnected and

Generator Protection

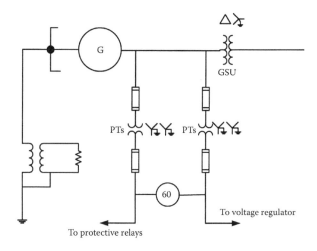

FIGURE 11.41
Connection of a voltage balance relay.

the PTs are left connected to a delta winding of the UAT. This possibility of ferroresonance can be minimized by using line-to-line rated PTs. To eliminate ferroresonance, it will be necessary to apply resistance loading. Figure 11.42 shows that a single nonlinear resistor is applied to two PTs. This is not recommended. When the single resistor fails, both PTs are left with zero or reduced voltage signals. A current-limiting resistor should be provided for each PT.

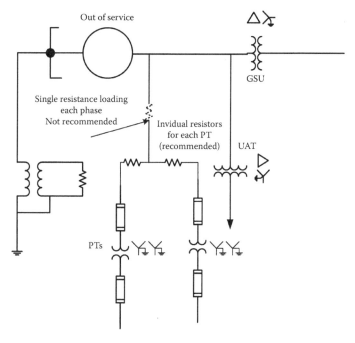

FIGURE 11.42
Resistance loading of PTs to avoid ferroresonance, see text.

11.18 Under- and Overvoltage Protection

Generators are usually designed to operate continuously at 95% of the rated voltage, while delivering rated power at rated frequency. Operation below 95% may result in undesirable effects such as reduction in stability limit and import of excessive reactive power from the grid. Undervoltage protection can be provided with a definite time or inverse time delay relay. A setting of 93% with a time delay of 2–3 s is appropriate.

On sudden load rejection, the hydro units will over speed which may be above 200%. In this mode, the volts/Hz may remain within the limits but the sustained overvoltages may be above permissible limits. This is generally not a problem with steam and gas units. The protection is provided with an instantaneous and time delay unit, frequency compensated or tuned to the fundamental frequency. The instantaneous setting is at 130%–150% of the pickup voltage while the inverse time delay unit is set to pickup at 110% of the normal voltage. The protection can also be applied to steam and gas units and connected alarm.

11.19 Synchronization

Unlike synchronization of synchronous motors, where the field is applied at a proper angle when the slip has reduced and the motor has reached a certain speed, the field voltage in the generator builds up as it accelerates and the field is continuously applied. When the generator has built up a terminal voltage and is operating almost at rated frequency, it is synchronized with the system. The criteria are as follows:

- Voltage mismatch should not be more than ±5%.
- The frequency mismatch should not be more than 0.067 Hz.
- The closing angle should be as close as possible within a maximum of ±10°.

There are two types of synchronization schemes:

11.19.1 Manual Synchronizing

The generator voltage and frequency are manually controlled and the breaker is manually closed when the voltage and frequency limits are within the range. The manual closing angle can be 10 electrical degrees or even more, depending upon the skills of the operator. The greater the closing angle, the greater the synchronization stresses both for the generator and the system. These stresses occur in the current surges, increased shaft vibrations, a change in bearing alignment, loosened stator windings and laminations, and the shaft and other mechanical parts. With automatic synchronization, much nearer angles of closing of the order of 5° electrical or even less can be achieved.

When the generator is connected to a step-up transformer with a generator breaker at generator voltage, there will be a phase shift between the secondary voltage obtained through generator transformer and the generator voltage. This phase shift can be corrected by using appropriate matrices, see Chapter 12.

Figure 11.43 shows that the voltages should be within a certain range. The voltage is matched to V_s. If the slip frequency is within acceptable bands, both voltages are within

Generator Protection

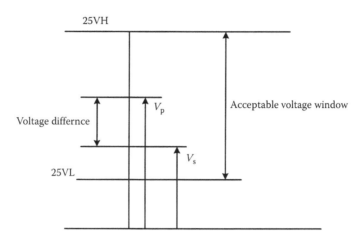

FIGURE 11.43
Acceptable voltage window for synchronization.

settable limits, the phase angle difference is within limits, and the relay issues a signal to close the breaker. The settings 25VH and 25VL define the acceptable system voltage Vs magnitude window prior to closing the generator breaker. If the system requires that the generator voltage should be higher than the system voltage, this can be set in the relay.

In Figure 11.44, the C< setting defines a target closing angle, generally set less than −5°. Positive angle indicates Vs lagging the synchronizing voltage. The relay will issue a close signal when the angle difference equals C<. The 25<1 and 25<2 define the acceptable generator breaker closing angles, generally set at 5° and 15°. The relay has the capability to account for generator breaker closing time and preset slip frequency. Also a closing angle, equal to 30°, can be set. Figure 11.45 shows the manual synchronizing scheme with synchroscope, voltage and frequency meters, and synchronizing selector switches.

11.19.2 Automatic Synchronizing

In automatic synchronization, the generator frequency is automatically matched with the system frequency. The synchronizer produces pulses which can be used to raise or lower the generator speed. In general, the generator frequency is adjusted higher than the system frequency for synchronization to prevent motoring or tripping on reverse power. The synchronizer also works on the generator voltage regulator for voltage matching. The automatic synchronizers provide much closer synchronizing angles. This reduces the stresses in the generator and the systems. Practically, all generators (except very small ones) are provided with autosynchronizer, digital display of the voltages, and angles and slip frequency.

Even when automatic synchronization is used, the synchronism check function is retained.

11.20 Tripping Schemes

On operation of various protective functions, the entire generating plant need not be tripped. The factors contributing to the selection of tripping logic are as follows:

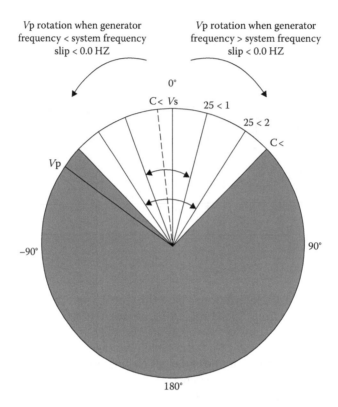

FIGURE 11.44
Acceptable closing angle for manual synchronization.

- Type of prime mover
- Impact of sudden loss of generation on an electrical system
- Management of unit auxiliary loads—the critical loads are often supplied from a diesel generator and UPS systems
- Extent of potential damage that can occur
- Need to restore the system operation as quickly as practical

Table 11.7 from [6] shows the tripping of a unit-connected generator. Practically, most of the trips are simultaneous trips.

11.20.1 Sequential Tripping

The simultaneous trip can expose the unit to potentially catastrophic overspeed conditions and stresses. The purpose of sequential trip is to avoid overspeed conditions. The method is also used for normal shutdown mode. Abnormal conditions which do not require immediate generator shutdown can be set for sequential trip. A sequential trip logic initiated by a reverse power relay is shown in Figure 11.46.

Reference works [7–20] provide further reading.

Generator Protection

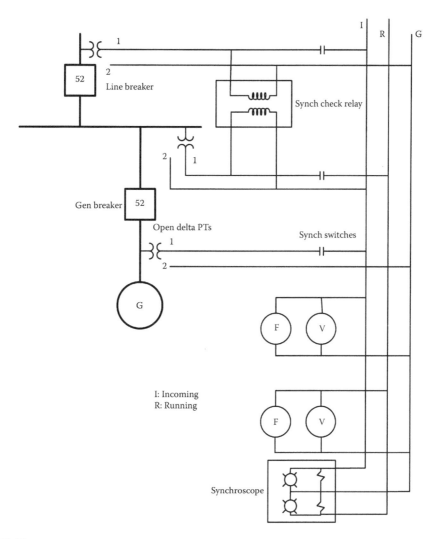

FIGURE 11.45
A manual synchronizing scheme with synchronizing switches, synchroscope, and voltage and frequency meters.

TABLE 11.7

Tripping Table of a Generator in Unit Connection, High-Resistance Grounded

Protective Function	Generator Breaker	Field Breaker	Transfer Auxiliaries	Prime Mover	Alarm
87G, 87GSU, 87UAT, 87 overall, 21 or 51V, 32, inadvertent energization, 100% stator winding fault, 40, 24, 46 (Note 1), 63, 63UAT, restricted ground fault GSU, 64F generator rotor fault	X	X	X	X	X
49, 60, 71, 71UAT, 59					X

Note: 46 trip only generator breaker if auxiliaries permit operation. 64F generator rotor ground fault—follow manufacturer's recommendations, 59 generally connected for alarm, follow manufacturer's recommendations.

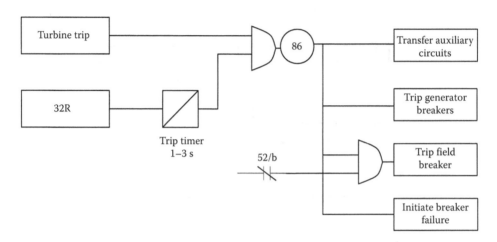

FIGURE 11.46
Sequential trip logic.

References

1. IEEE Std. 242. IEEE Recommended Practice for Protection and Coordination of Industrial and Commercial Power Systems, 1986.
2. ANSI C50.10. ANSI Requirements for Synchronous Machines, 1977.
3. ANSI C50.13. Approved Draft IEEE Standard for Requirements for Cylindrical Rotor 50 Hz and 60 Hz Synchronous Generators Rated 10 MVA and Above, 1989.
4. IEC 60255-8. Thermal Electrical Relays, 1990 (now withdrawn).
5. IEEE Std. C37.106. IEEE Guide for Abnormal Frequency Protection of Power Generating Plants, 2003.
6. IEEE Std. C37.102. IEEE Guide for AC Generator Protection, 1995.
7. MS Baldwin, WA Elmore, JJ Bonk. Improved turbine-generator protection for increased plant reliability. *IEEE Trans Ind Appl*, 99(3), 982–989, 1980.
8. J Berdy, ML Crenshaw, M Temoshoh. Protection of large steam turbine generator during abnormal conditions. Paper 11.05, CIGRE 1972.
9. SEL Inc. Instruction Manual for SEL 700G Generator Protection Relay.
10. Beckwith Electrical, Instruction Manual for M3425A Generator Protection Relay, Beckwith Electric.
11. CIGRE WG 34.05. Results of CIGRE survey on the electrical protection of synchronous generator. CIGRE SC34 Colloquium, Paper no. 108, Johannesburg, South Africa, October 1997.
12. IEEE Committee Report. Potential transformer application on unit-connected generators. *IEEE Trans Power Appar Syst*, PAS-94, 24–28, 1972.
13. IEEE/PSRC Working Group Report. Performance of generator protection during major system disturbances. *IEEE Trans Power Delivery*, 19(4), 1650–1662, 2004.
14. WA Elmore, Ed. *Protective Relaying, Theory and Applications*. Marcel and Dekker, New York, 2004.
15. IEEE Tutorial Course. The protection of synchronous generators. Publication 95 TP102, IEEE New York, 1995.
16. JW Pope. A comparison of 100% stator ground fault protection schemes for generator stator windings. *IEEE Trans Power Appar Syst*, 103, 832–840, 1981.

17. IEEE Committee Report. Out-of-step relaying for generators. *IEEE Trans Power Appar Syst*, 96, 1556–1564, 1977.
18. LJ Powell. The impact of generator grounding practices on generator fault damage. *IEEE Trans Ind Appl*, 34(5), 923–927, 1998.
19. Electrical Machinery-Dresser Rand Industries, Inc. Field Ground Detection System-Instruction Manual 324.
20. IEEE C37.101. IEEE Guide for Generator Ground Protection, 1993.

12

Transformer Reactor and Shunt Capacitor Bank Protection

12.1 Transformer Faults

Transformers in power systems represent large investments and are one of the important components of the power system which must be properly protected depending on their rating and importance. In a radial system, a transformer failure may result in a prolonged shutdown as the downtime for repair of a transformer can be large. Even if one similar transformer is available for immediate replacement, the removing of the faulty unit and inserting the new unit, and its testing and commissioning may optimistically take more than 48 h.

We may classify the transformer faults as internal or external to the transformer.

12.1.1 External Faults

12.1.1.1 Short-Circuit Faults

The most severe external faults are short circuits (including line-to-ground faults). ANSI/IEEE standards publish the transformer frequent and infrequent fault withstand curves, and the provisions of NEC [1] for transformer protection are followed in the industry. NEC does not apply to utility installations, generating stations and substations distribution, and metering designed, maintained, and operated by the utilities. The external faults can be cleared fast if unit protection systems like differential relaying is provided or these may be cleared by overcurrent and ground fault protective devices, coordinated with the transformer protection. Figure 12.1 shows that faults on the secondary side in the block area shown must be cleared by the primary protection. The ground faults in the secondary windings are of concern, and should not be cleared as phase faults. This could be an NEC contrivance, see Chapter 8.

12.1.1.2 Overloads

Overloads can occur if the downstream protection from the transformer is not properly coordinated. Within their thermal withstand capability, transformers can take short-time overloads. The derating in transformer life expectancy can be calculated using methodology detailed in IEEE standard, discussed in Section 12.4. Many utilities allow calculated overloads on transformers under certain contingency conditions. Prior to the overload, generally, the transformer may not be loaded to its continuous kVA rating.

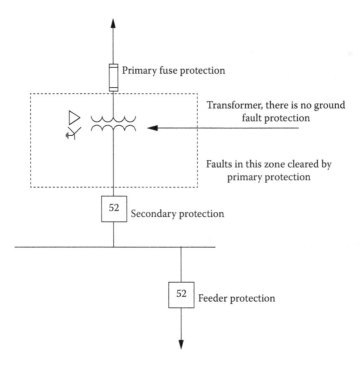

FIGURE 12.1
Faults in the secondary zone shown in dotted block cleared by primary protection device.

12.1.1.3 Overvoltage and Underfrequency

Overvoltages can be caused by sudden load shedding. Consider a transformer provided with underload tap changing (ULTC) and let it be operating at full load with taps providing 10% boost on the secondary side. On sudden loss of load, say due to collapse of the primary voltage, the taps will be stuck at 10% boost and the returning voltage will subject the transformer to overvoltage and overfluxing. Underfrequencies can be caused by a major mismatch between the generation and the load. The V/f protection, much alike generators, Chapter 11, is provided.

12.1.2 Internal Faults

12.1.2.1 Short Circuit in the Windings

The short circuit in secondary windings when reflected to primary windings can be different depending on winding connections. Phase-to-ground faults in wye-connected windings will depend on the grounding of the windings, solidly, resistance, or high-resistance grounded. For ground fault protection, see Chapter 8. The phase and ground fault differential relays provide the protection, backed up by overcurrent protection.

12.1.2.2 The Interturn Winding Faults

The interturn winding faults are difficult to detect; approximately 14% of windings must be short circuited for a current equal to full-load current to flow. Differential protections cannot detect low-level interturn faults. Fault-pressure relays and Bucholz protection relays are applied.

Transformer Reactor and Shunt Capacitor Bank Protection

For leakage in tanks and radiators and failure of cooling fans or circulating water pumps, oil level gauge and oil and winding temperature alarm relays are provided.

12.1.2.3 Faults in ULTC and Off–Load Tap Changing Equipment

Core-type faults are rare but can occur due to overfluxing and deterioration of insulation between laminations.

12.1.2.4 Bushing Flashovers

These can occur due to lightning or other switching overvoltages for high-voltage transformers. Surge arresters on primary and secondary windings can be provided. These help to mitigate the magnitude of the surge voltages and transients transferred through the transformer windings through inductive and capacitive couplings.

12.1.2.5 Part-Winding Resonance

Part-winding resonance can occur due to switching transients of vacuum breakers, see Volume 1.

The failure statistics of transformers for two periods is shown in Table 12.1.

12.2 NEC Requirements

NEC requirements for overcurrent protection of transformers and transformer vaults including secondary ties are contained in article 450.2 [1]. It is important to note that the transformer secondary overcurrent protection is not required, provided certain conditions of installations in NEC are met. The recommendations for protection in "supervised location" and "any location," part of NEC Table 450.3(A), are summarized in Tables 12.2 and 12.3.

Also for transformers of 600 V or less, secondary protection is not required provided the primary protection is set at no more than 125% for currents for 9 A or more and 167% for currents less than 9 A.

Secondary protection of transformers over 600 V can be omitted if there are no more than 6 secondary circuit breakers and the sum of the ratings of circuit breakers does not exceed 300% of the rated secondary current of the transformer. If there are secondary fuses and circuit breakers in combination, not exceeding 6, the sum of their ratings should not exceed 250%.

TABLE 12.1

Failure Statistics of Two Time Periods

Failures	Period 1955–1965		Period 1875–1982	
	Number	% of Total	Number	% of Total
Winding failures	134	51	615	51
Tap changer failure	49	19	231	19
Bushing failure	41	16	114	9
Terminal board failure	19	7	71	6
Core failures	7	3	24	2
Miscellaneous failures	12	6	72	13
Total	262	100	1217	100

TABLE 12.2

Maximum Rating or Setting of Overcurrent Protection for Transformers Rated over 600 V in Supervised Locations

Transformer-Rated Impedance	Primary Protection Over 600 V		Secondary Protection Over 600 V		Secondary Protection 600 V or Less
	Circuit Breaker (%)	Fuse (%)	Circuit Breaker (%)	Fuse (%)	Circuit Breaker of Fuse (%)
Any	300	250	Not required	Not required	Not required
Not >6%	600	300	300	250	250
>6% but not >10%	400	300	250	225	250

Source: NEC. National Electric Code, NFPA 70, 2017.

TABLE 12.3

Maximum Rating or Setting of Overcurrent Protection for Transformers Rated over 600 V in Any Location

Transformer-Rated Impedance	Primary Protection Over 600 V		Secondary Protection Over 600 V		Secondary Protection 600 V or Less
	Circuit Breaker (%)	Fuse (%)	Circuit Breaker (%)	Fuse (%)	Circuit Breaker of Fuse (%)
Not > 6%	600	300	300	250	125
>6% but not >10%	400	300	250	225	125

Source: NEC. National Electric Code, NFPA 70, 2017.

The overcurrent protection of primary and secondary conductors of the transformers is covered in NEC article 240. Again, it is permissible to omit the secondary protection of conductors provided certain specified conditions in NEC are met.

As a result of these provisions, many industrial systems do not have transformer secondary overcurrent protection. The secondary overcurrent devices and circuit breakers are eliminated as a cost saving measure.

12.3 System Configurations of Transformer Connections

Figure 12.2 shows some of the configurations for substation transformers in industrial environment.

12.3.1 Radial System of Distribution

A radial system of distribution is fairly common in the industry due to its cost saving advantages (Figure 12.2a). As many as 10 or more unit substation transformers are daisy chained on to a single 13.8 kV feeder circuit breaker. From protection and arc flash point of view this is not a desirable system. For any fault in the 13.8 kV primary feeder cables, for a ground fault on the secondary sensed through a ground fault relay, operation of a sudden

Transformer Reactor and Shunt Capacitor Bank Protection 447

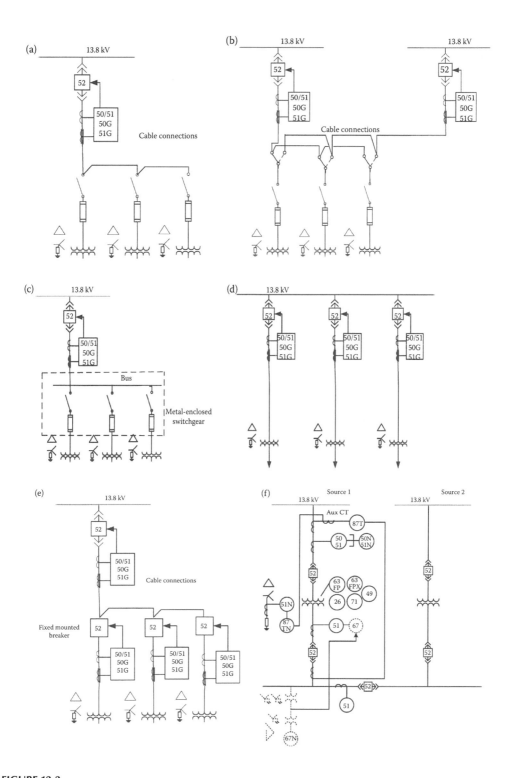

FIGURE 12.2
(a) A radial system of distribution, (b) primary selective system, (c) group feed system, (d) dedicated primary circuit breakers, (e) fixed mounted primary relayed breaker, and (f) secondary selective system.

fault-pressure relay in any of the transformers, and a fault on any primary load interrupter switch must *all* trip the single 13.8 kV feeder breaker; resulting in a complete shutdown of all the transformers fed from the single 13.8 kV breaker.

12.3.2 Primary Selective System

In a primary selective system with redundant sources of power, a substation transformer can be connected to any of the two sources through interlocked selector switches which are load break type (Figure 12.2b). Yet an entire shutdown of the load will occur, before a switchover can be made to the alternate source of power. The situation is identical to that of Figure 12.2a, except that the primary switchover of power for a cable fault allows bringing the system on line after a short interruption.

12.3.3 Group Feed System

In a group-fed system, the transformer primary protection is grouped in one location through fused load interrupter switchgear (Figure 12.2c). Again for all secondary fault types as discussed in Figure 12.2a, the primary circuit breaker has to be tripped, resulting in complete shutdown. The only difference is that for a number of transformers, primary fuse protection is grouped together.

12.3.4 Dedicated Circuit Breakers

Dedicated circuit breakers on the primary and secondary sides of each transformer are provided (Figure 12.2d). All the protective devices are shown in this figure, except that the differential protection is required to be provided to meet the requirements of Factory Mutual (FM) Global Property Loss Prevention Data Sheet [2] for transformers rated 1000–10,000 kVA located outdoors. Similar protection is required for transformers rated less than 1000 kVA when it creates a fire hazard. Though expensive in terms of providing dedicated circuit breakers, it limits the area of shutdown and transformer secondary and primary faults can trip out its respective circuit breaker. Addition of differential protection further enhances the protection and limits the damage.

12.3.5 Fixed Mounted Primary Circuit Breaker

The primary load break switch-fuse protection is replaced with fixed mounted (non-draw-out) or metal-clad draw-out circuit breaker and overcurrent relays (Figure 12.2e). At the additional cost of providing a circuit breaker instead of a fused switch, it has distinct advantages of fault reduction and minimizing the area of shutdown for a fault in a transformer. Only the faulty transformer will be isolated leaving rest of the daisy chained transformers in service.

12.3.6 Secondary Selective System

A further enhancement in protection and maintaining the continuity of power can be achieved through double-ended secondary selective systems (Figure 12.2f). The system is operated with bus section circuit breaker normally open, with each transformer supplying its bus load. All the protective devices shown in this figure are required to be provided for transformers rated above 10,000 kVA according to FM [2]. In case of failure of one of the transformers or primary source of power, the bus section switch can be closed, either manually, or an automatic bus transfer scheme can be arranged. In case of manual closing, the loads will be interrupted. With fast autotransfer of power, it is possible to maintain the motor running loads. It implies that each transformer must be rated to carry the entire system load when the bus section circuit breaker is closed.

A variation of this configuration is that the two transformers can be operated in parallel. This requires that both sources on the primary side of the transformers must have the same voltage and phase angle shift and must be in synchronism, and transformers must have similar winding connections, ratings, and percentage impedances. The limitation of this scheme is that the short-circuit levels on the low-voltage systems may increase beyond acceptable limits with the parallel running transformers. The protection system can be arranged so that only the faulty transformer is selectively isolated. The protective devices shown dotted in this figure will be required, when such parallel operation of the transformers is required.

12.4 Through-Fault Current Withstand Capability

Prior to 1978, the transformer through-fault protection involved protection of a single ANSI point. Fault currents that could be sustained at a lower level for a considerable period of time could result in transformer damage. The protection engineers were satisfied by protecting only one point as there was no published through-fault withstand curves.

The development of through-fault withstand characteristics of transformers has seen many changes and currently, for liquid immersed transformers, these are defined in IEEE standard [3] and are categorized as follows:

Category I: Single-phase 5–500 kVA, three-phase 15–500 kVA

Category II: Single-phase 501–1667 kVA, three-phase 501–5000 kVA

Category III: Single-phase 1668–10,000 kVA, three-phase 50,001–30,000 kVA

Category IV: Single-phase >10,000 kVA, three-phase >30,000 kVA

ANSI standard [4] for loading of oil-immersed distribution and power transformers provides the transformer short-time thermal overload capability as shown in Table 12.4.

TABLE 12.4

Transformer Short-Time Thermal Overload Capability

Time	Times-Rated Current
2 s	25.0
10 s	11.3
30 s	6.3
60 s	4.75
5 min	3.0
30 min	2.0

Source: ANSI/IEEE Standard. IEEE C57.109, IEEE Guide for Liquid Immersed Transformer Through-Fault Current Duration, 1993 (R2008).

Low values of currents of 3.5 or less times the normal base current of transformer may occur from overloads rather than faults, and for such cases, loading guides may indicate allowable time durations [4]. The per-unit short-circuit currents shown in through-fault curves are for balanced transformer winding currents. The line currents that relate to these winding currents depend upon transformer winding connections and the type of fault. The following explanations apply to each category.

12.4.1 Category I

The recommended duration limit is based on curve in Figure 12.3. This curve reflects both thermal and mechanical damage considerations. The dot-dash portions of the curve cover transformer varying short circuit withstand capabilities required by IEEE Std. [3], up to a maximum of 40 times the normal current.

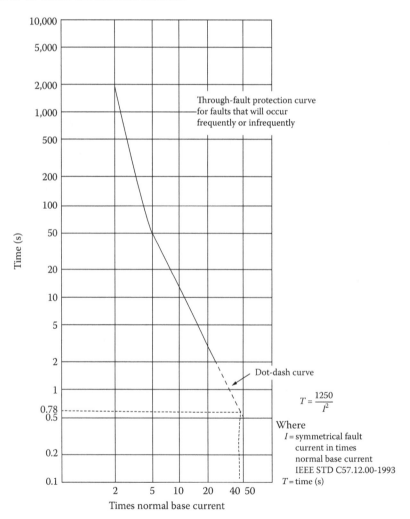

FIGURE 12.3
ANSI through-fault withstand curve, category I transformers, liquid immersed.

12.4.2 Category II

In Figure 12.4, the fault frequency refers to number of faults with magnitudes greater than 70% of the maximum possible. Figure 12.4a reflects both thermal and mechanical damage considerations. It is applied for faults that will occur frequently, more than 10 times in the life of the transformer. Part of the curve is dependent upon transformer short-circuit impedance for faults above 70% of the maximum possible and is keyed to I^2t of the worst case mechanical duty as shown by dashed curves for a few selected impedances. The remaining portion matches the thermal portion of curve for faults below the 70% level. Figure 12.4b, which is the solid portion of the curve in Figure 12.4a, reflects primarily thermal damage. This curve may be applied for less frequent faults typically no more than 10 in the life of the transformer.

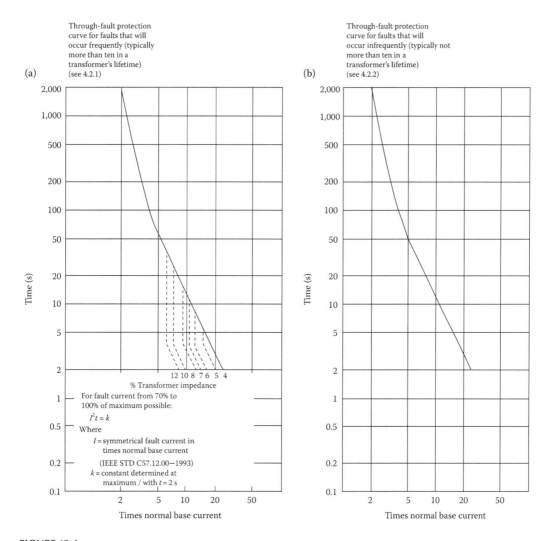

FIGURE 12.4
ANSI through-fault withstand curve, category II transformers: (a) frequent fault curve and (b) infrequent fault curve, liquid immersed.

12.4.3 Categories III and IV

For these categories, it is required that the system short-circuit reactance based upon the available short-circuit current is added to the transformer short-circuit impedance. Consider a three-phase transformer of 10 MVA, 138–13.8 kV, which falls in category III. If transformer impedance is 9% and the 138 kV source has a short-circuit level of 5000 MVA, then the combined impedance on transformer 10 MVA base is 9.2% for the construction of the curves. In the absence of available short-circuit data on the primary of the transformer, recommended values are provided in Reference [3].

12.4.3.1 Category III

In Figure 12.5a, which reflects both thermal and mechanical damage, part of the curve is dependent upon the transformer short-circuit impedance for faults above 50% of the

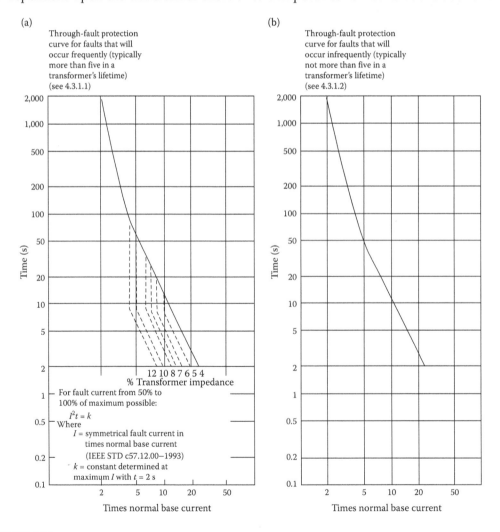

FIGURE 12.5
ANSI through-fault withstand curve, category III transformers: (a) frequent fault curve and (b) infrequent fault curve, liquid immersed.

maximum possible keyed to the I^2t of the worst case mechanical duty as shown by dashed curves for a few selected impedances. (For category II this transition point is 70%.) This curve is applied when typically more than 5 faults are expected in the life of a transformer. Again Figure 12.5b is applicable for faults <5 in the life of the transformer.

12.4.3.2 Category IV

For category IV transformers there is only one set of curves, as shown in Figure 12.6. This represents both thermal and mechanical damage and is applied for all transformers in this category.

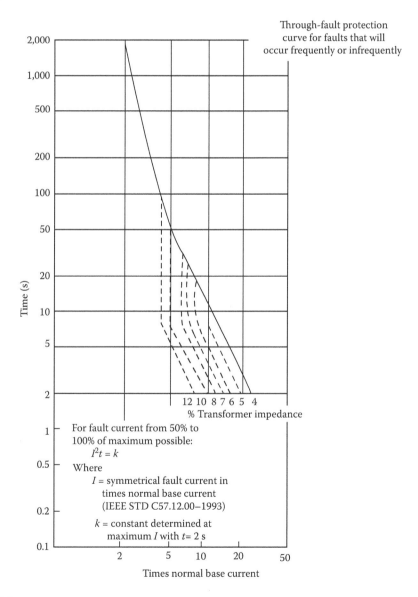

FIGURE 12.6
ANSI through-fault withstand curve, category IV transformers; liquid immersed.

12.4.4 Observation on Faults during Life Expectancy of a Transformer

It is rather difficult to project how many faults will occur in the life expectancy of a transformer, as it depends upon a number of factors; for example, how carefully the transformer was chosen for the required load duty, its protection and preventive maintenance. Generally, for transformers connected to overhead lines, frequent fault damage curve can be applied while for transformers connected through cables in an industrial environment less frequent fault damage curve can be applied. To be conservative, this book applies frequent fault damage curves of the transformers. It is also recognized that for some applications, the through-fault withstand of the transformer provided in ANSI/IEEE standards may be exceeded. An example will be unit auxiliary transformer (UAT) to serve auxiliary generation loads for generating systems, with generator and transformer connected in a unit configuration directly to the utility source. This is further discussed in Section 12.6.1.

12.4.5 Dry-Type Transformers

It is recognized that dry-type transformers differ considerably from liquid immersed transformers, with respect to their withstand characteristics.

Dry-type transformers can be designed for different temperature ratings, that is, 75°C, 90°C, 115°C, 130°C, and 150°C.

There is a significant variation in the construction of dry-type transformers. The transient heating of liquid immersed transformers is considerably buffered by the insulating medium, providing a relatively long thermal time constant as compared to dry-type transformers [5].

The following categories are applicable:

Category I: Single-phase 15–500 kVA, three-phase 15–500 kVA, Figure 12.7.

Category II: Single-phase 501–1667 kVA, three-phase 501–5000 kVA, Figure 12.8.

Dry-type transformers of higher rating of 10 MVA and sometimes larger by special designs are commercially available and are in operation. In the absence of any other data, category II curve can be applied for these transformers too.

Thermal short-time overload capability for both categories I and II is 2 s, at 25.0 times the rated current.

For category I transformers, the recommended duration limit is shown in Figure 12.7. This curve reflects both thermal and mechanical damage, and can be applied for faults that will occur frequently or infrequently.

For category II transformers, the fault frequency is 70%; same as for liquid immersed transformers, applicable for faults typically more than 10 in the life of the transformer. Figure 12.8 shows curves for category II dry-type transformers.

For dry-type transformers, the thermal curve with equation

$$I^2 t = 1250$$

where I is the symmetrical fault current in times the normal base current and t is the time in seconds and applies to $t = 102$ s in Figure 12.7. For dry-type transformers, the dash-dot portion shown in Figure 12.3 is absent.

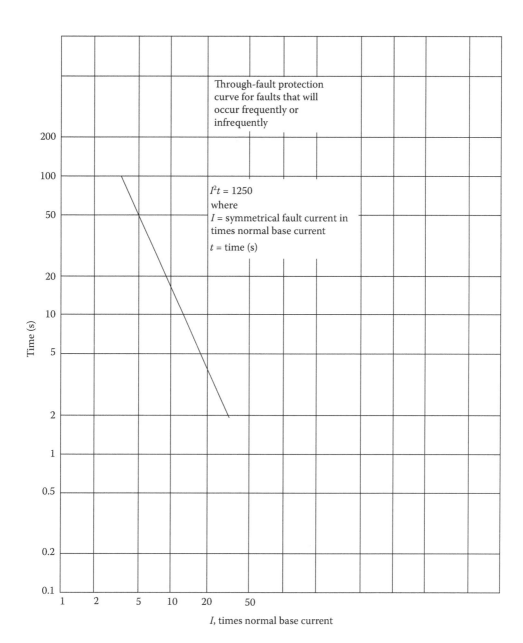

FIGURE 12.7
ANSI through-fault withstand curve, category I transformers, dry-type transformers.

12.5 Construction of Through-Fault Curve Analytically

Practically, all computer programs have inbuilt data to plot the transformer frequent or infrequent curves based upon the input data. The following example details the procedure for hand construction.

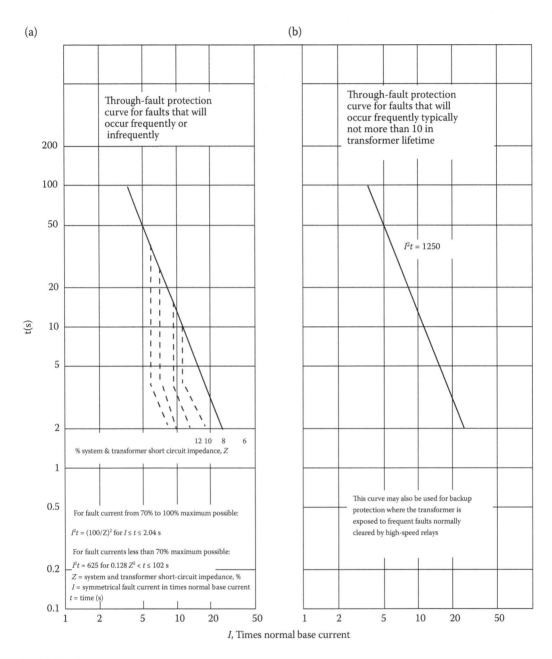

FIGURE 12.8
ANSI through-fault withstand curve, category II transformers, dry-type transformers.

Example 12.1

Construct a frequent fault withstand curve for a liquid immersed 2500 kVA, three-phase, 13.8–2.4 kV transformer of 5.5% impedance. The primary (13.8 kV) three-phase short-circuit current is 30 kA rms symmetrical.

This is done in the following steps:

Transformer three-phase through-fault current limited by transformer impedance only is 10.94 kA rms symmetrical at 2.4 kV. This is equal to 18.19 times the transformer-rated full-load current = 601.42 A.

If the source impedance is considered, based upon 30 kA rms at 13.8 kV, the through-fault current falls to 10.28 kA. However, the primary source short-circuit impedance is not considered for category II transformers and is therefore ignored.

Seventy percent of the calculated short-circuit current = 7.65 kA, and I^2t line is constant between 100% short-circuit current at 2 s and 70% short-circuit current:

$$(10.94)^2(2) = (7.65)^2(t_1)$$

This gives $t_1 = 4.09$ s. Thus, point C can be plotted in Figure 12.9. Also a current of 7658 A can be read on x-axis at this point.

Using $I^2 t = k$, where I is the current in terms of base current at 2 s, that is = 18.19 pu, gives $k = 661.7$. Therefore, t_1 can also be calculated as

$$(0.7 \times 18.19)^2(t_1) = 661.7$$

which gives the same value of $t_1 = 4.09$ s, as before.
According to Table 12.4, at 10 s, 11.3 times rated current = 6796 A.

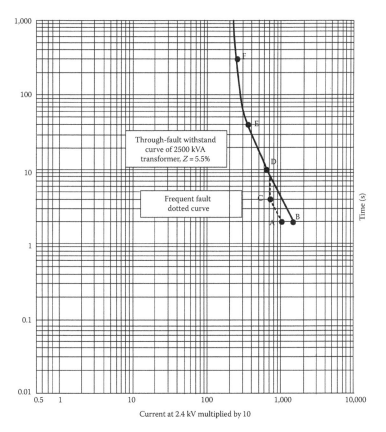

FIGURE 12.9
Analytical construction of a through-fault curve.

This gives point D in Figure 12.9. The infrequent fault curve is drawn parallel to line AB to 2 s.

Again from Table 12.4, at 6.3 times the base current (= 3789 A) $t = 30$ s (point E), at 4.75 times the base current (= 2856 A) it is 60 s, and at 3.0 times (= 1804 A) it is 5 min (point F).

Figure 12.9 shows the calculated curve.

12.6 Protection with Respect to Through-Fault Curves

Standards do not establish guidelines that to what extent the through-fault withstand curve of the transformers should be protected? The NEC primary protection settings as detailed above will not protect the through-fault withstand curve over its entire range. See Example 12.2 for further calculations. It is opined that fault currents in the range of 2–2.5 times are more of a temporary overload nature. A high-resistance fault will quickly break down, and the fault current will escalate. Consider, for example, NEC maximum permitted setting of six times the transformer full-load current for primary protection (Table 12.2). Practically, protection for much lower levels of fault currents is desirable, even when secondary protection is provided.

12.6.1 Withstand Ratings of UT (GSU) and UAT

The through-fault withstand ratings as described above are not adequate for the generator unit transformer (UT) also called, generator step-up transformer (GSU), UAT, and station service transformer (SST).

A typical generating station auxiliary power system one-line diagram is shown in Figure 12.10.

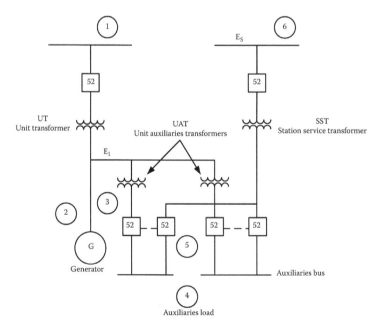

FIGURE 12.10
UT (GSU), UAT, and SST transformer connections in a unit-connected generator.

12.6.1.1 UAT

The standard thermal short-circuit requirements are generally not adequate for UAT; it may be subjected to higher through-fault duty than the network transformers.

- The short-circuit duration is not limited to 10–20 cycles, as in the case of network transformers, but may be as long as 10–20 s.
- Depending upon generator loading prior to fault, the UAT may have voltage up to 125% of its rated voltage imposed on it during short circuit.
  - The DC decrement is less than in network applications.

The total fault duration can be approximated by the equation:

$$T_F = t_T + 2T''_{do} \tag{12.1}$$

where t_T is the time delay between fault inception and full-load rejection that results due to tripping of system breaker and the generator exciter field. Typically, t_T is in the range 6–60 cycles. Six cycles represents high-speed breaker opening and 60 cycles represents backup clearing time for a failed station auxiliary in bus breaker during a three-phase fault on the station auxiliary's bus. This is shown in Figure 12.11a, while Figure 12.11b shows the asymmetrical fault current versus time of a typical transformer designed to meet ANSI/IEEE C57.12.00. T''_{do} is the generator open-circuit time constant.

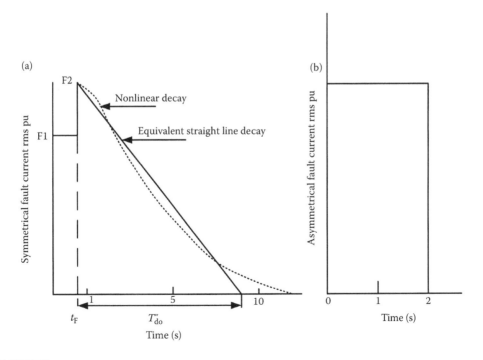

FIGURE 12.11
(a) Symmetrical fault current decay for a fault on station auxiliary bus and (b) the asymmetrical fault current versus time of a typical transformer designed to meet ANSI/IEEE C57.12.00.

The generator open-circuit time constant can exceed 2 s and users are advised to obtain the accurate value from the manufactures.

The specifications must include a plot similar to Figure 12.11a.

A fast transfer of load from the UAT to SST may result in circulating currents that could exceed the mechanical design capability of one of them:

$$|I_c| = \frac{E\sqrt{2 - 2\cos\phi}}{Z_u + Z_s} \qquad (12.2)$$

Consider, for example,

$Z_u = 0.04$ pu
$Z_s = 0.08$ pu

Then $Z = 0.12$ pu

$\phi = 180°$
$E = 1.0$ pu

Substituting in Equation 12.2 gives $I_c = 16.66$ pu.

The fault current capability of UAT = $1/0.4 = 25$ pu, while that of SST is 12.5 pu. Thus, the maximum allowable current for SST is exceeded by 0.33 pu. This requires a transformer of $(1.33)2 = 1.77$ times the mechanical strength of a standard transformer design. If the phase angle between the two voltages is zero degrees then the circulating current is zero.

12.7 Transformer Primary Fuse Protection

12.7.1 Variations in the Fuse Characteristics

We discussed power fuses in Chapter 5. Power fuses used for transformer primary protection, can be current-limiting or expulsion type. Class E power fuses rated 100E or less open in 300 s at a current level between 200% and 240% of their E rating. Fuses rated 100E or more open in 600 s at current level between 220% and 264% of their E rating [6]. The E rating also reflects 2:1 minimum melting current versus continuous current ratio, which is a design feature of the power fuse. The expulsion-type fuses are generally available in higher continuous ampere ratings, and current-limiting fuses are available in higher short-circuit ratings, Chapter 6. *We limit our discussions to current-limiting fuses.*

Only some typical operating parameters are fixed by the standards for the current-limiting class E fuses. Figure 12.12 shows the variations in the total clearing time characteristics of 100E fuses, based upon the data of four different manufacturers. A spread of 500–900 A is seen at 2 s. (For expulsion-type fuses this spread is even more.) These variations in the time–current characteristics of fuses of the same specifications and ratings do not easily permit a generalization of the through-fault protection of the transformers with fuses, though some guidelines are provided further.

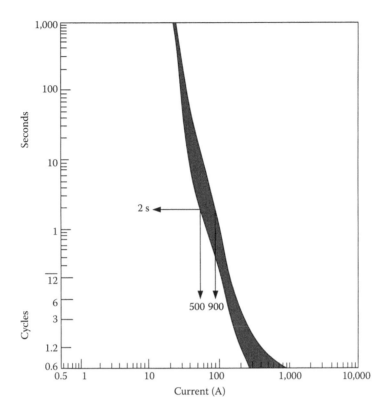

FIGURE 12.12
Variations in the total clearing time–current characteristics of 100E current-limiting fuses of four different manufacturers.

12.7.2 Single Phasing and Ferroresonance

Operation of a fuse in one of the phases can give rise to single phasing. For example, for a line-to-line secondary fault, the distribution of currents on the primary side is shown in Figure 12.13. The fuse in the phase carrying 1.0 per unit fault current will operate first, leaving two fuses in service. Thus, a phase of the transformer windings is energized through cable capacitance. This circuit can give rise to ferroresonance, as it involves excitation of one or more saturable reactors (transformer windings) through capacitance. When ferroresonance occurs at high peak voltages, irregular voltage and current waveforms and loud noise in the transformer due to magnetostriction may be produced [7,8]. Also see Appendix C of Volume 1.

12.7.3 More Considerations of Fuse Protection

- Fuses should not operate on transient inrush switching current of transformers. The integrated heating effect of the switching current is represented by 8–12 times the transformer full-load current for 0.1 s and 25 times the full-load current for 0.01 s. For transformers of rating 10 MVA and above, it is safer to consider inrush current equal to 15 times the full-load current to take account of residual trapped flux.

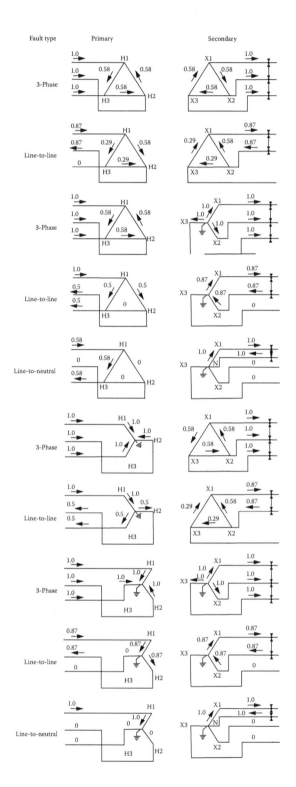

FIGURE 12.13
Line and winding currents in transformers for various secondary faults, different transformer winding connections.

TABLE 12.5

Maximum Ratings of Current-Limiting Fuses and Protected Transformers, Primary Protection

System Operating Voltage (kV)	Maximum Available Fuse Size	Maximum Rating of Protected Transformers, Z = 5.75% on Base kVA
2.4	750E, 5.5 kV	2228/2600
4.16		3862/4506
6.9	300E, 8.3 kV	2560/2987
7.2		2672/3117
12.47	300E, 15.5 kV	4450/5190
13.2		4898/5715
13.8		5127/5975

- The short-circuit rating of the transformer primary fused load break switch should be selected compatible with the calculated short-circuit results at the point of application. The current-limiting fuses generally have a short-circuit rating of 50 kA rms symmetrical and 80 kA rms asymmetrical. The switch short-circuit momentary (first cycle) and 10-cycle fault closing rating should be compatible.

- A current-limiting fuse generates an arc voltage much higher than the system voltage, see Volume 1. This is of concern for coordination with surge arresters, which are provided for protection of transformer windings.

- The fuses will operate at their rated ampacity for 40°C ambient temperature. In an enclosure and in an outdoor location exposed to direct sunlight, the temperature may exceed 40°C and derating will be required.

- Fuses tend to fatigue due to repeated inrush currents and transients below minimum melting operating limits. The time–current characteristics of a fatigued fuse may change, *and there is no way to ascertain these effects*. Proper margins are provided in the coordination.

- Table 12.5 shows the maximum available class E current-limiting fuse sizes and the largest transformer for which these can be generally used, considering transformer inrush currents.

12.8 Overcurrent Relays for Transformer Primary Protection

Some of the problems inherent in the application of fuses are absent when relayed circuit breakers are used. This is a desirable trend in the industry, with respect to arc flash protection and limiting the area of shutdown. Some obvious advantages are as follows:

- Single phasing is prevented.
- Possibility of ferroresonance is mitigated.
- Relay characteristics, properly calibrated and maintained, are not subject to change with time and, therefore, a consistent performance is inherent.
- Selections of a wide range of time–current characteristics, which can be closely coordinated with transformer through-fault, withstand curve, and downstream distribution.

A better through-fault protection can be provided with relays as compared to fuses. The time–current coordination in Figures 12.14 and 12.15 provides a comparison, fuses versus relays for transformer primary protection. These figures show three-step coordination, a molded case circuit breaker at the MCC, a LVPCB feeder circuit breaker provided with programmable trip device, a secondary main LVPCB and finally a delta–wye 2500 kVA, 13.8–0.48 kV transformer, which is solidly grounded on 480 V side. Its frequent fault and infrequent fault curves are shown in these figures.

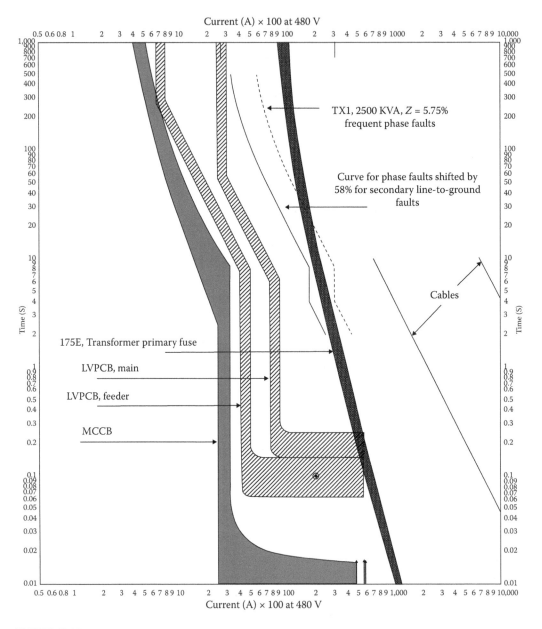

FIGURE 12.14
Through-fault withstand curve protection of a 2500 kVA, $Z = 5.75\%$ transformer with a primary fuse of 175E.

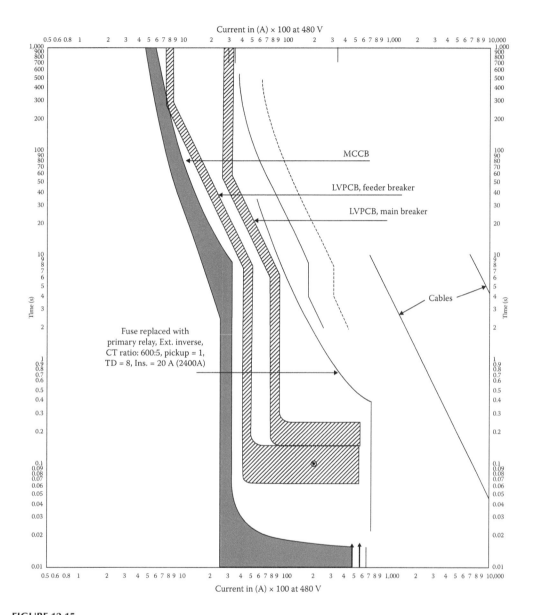

FIGURE 12.15
Through-fault withstand curve protection of a 2500 kVA, Z = 5.75% transformer with primary relayed circuit breaker.

Figure 12.14 shows a 175E class E fuse for the primary protection. There is a slight miscoordination with main LVPCB settings at the high level of short-circuit current; this is acceptable. The shifted through-fault curve for single line-to-ground faults on the 480 V side is not protected at all. Compare this with Figure 12.15, where all other settings and coordination remains the same as in Figure 12.14, except that the primary fuse protection is replaced with an overcurrent relay device 50/51. A better coordination is achieved, lack of coordination with the main LVPCB is entirely eliminated and the shifted 58% thermal damage curve of the transformer is fully protected. Also see Example 12.2.

12.9 Listing Requirements

Two listing agencies, FM Research Corporation and Underwriters Laboratories, Inc., list less-flammable liquids for transformers. These liquids have a fire point of not less than 300°C. All the listing requirements are not enunciated here. The FM Research listing is based on approved less-flammable fluid in the tank that meets certain criteria (the Envirotemp FR3 Fluid is approved by FM). Pressure-relief devices must be provided, enhanced electrical protection is required, spacing from adjacent combustibles must be provided, in event of a leak liquid confinement area and drainage is required [2]. The UL

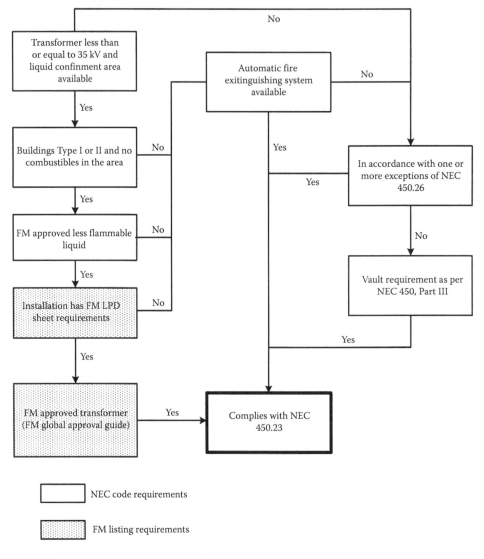

FIGURE 12.16
Less-flammable liquid immersed transformers compliance to NEC section 450.23 per FM listing, indoor installations.

Transformer Reactor and Shunt Capacitor Bank Protection

listing is based on the requirements that no tank rupture or noted fluid leakage occur during low and high current arcing faults test.

The transformers with less flammable liquids may be installed in type I and type II buildings [1] without a transformer vault, provided transformers are not rated above 35,000 V primary and other conditions specified in NEC 450.23 (A) are met. FM Global property loss prevention data sheets [2] may also be seen. These specify protections for indoor and outdoor transformers. As a specimen, a flow chart of compliance with NEC 450.23 and FM Listing is shown in Figure 12.16, *for indoor installations* using less flammable liquid-immersed transformers; Figure 12.17 shows the UL listing requirements.

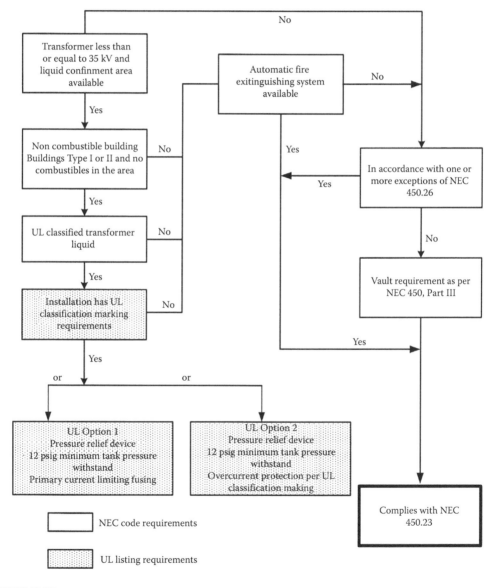

FIGURE 12.17
Less-flammable liquid immersed transformers compliance to NEC section 450.23 per UL listing, indoor installations.

TABLE 12.6

Requirement of I^2t Limitations and Minimum Pressure-Relief Capacity

Three-Phase Transformer kVA Rating	Required Current-Limiting Fusing Maximum I^2t (A²s)	Required Overcurrent Protection Maximum I^2t (A²s)	Minimum Required Pressure-Relief Capacity++ SCFM at 15 psig
45	500,000	700,000	35
75	500,000	800,000	35
112.5	550,000	900,000	35
150	600,000	1,000,000	50
225	650,000	1,200,000	100
300	750,000	1,400,000	100
500	900,000	1,900,000	350
750	1,100,000	2,200,000	350
1000	1,250,000	3,400,000	350
1500	1,500,000	4,500,000	700
2000	1,750,000	6,000,000	700
2500	2,000,000	7,500,000	5,000
3000	2,250,000	9,000,000	5,000
3750	2,500,000	11,000,000	5,000
5000	3,000,000	14,000,000	5,000
7500	3,000,000	14,000,000	5,000
10,000	3,000,000	14,000,000	5,000

The UL listing is based upon the limitation of the fault energy due to primary current-limiting fuses and transformer tanks capable of withstanding an internal pressure of 12 psig without rupture. Pressure-relief devices must be provided. Table 12.6 shows the limiting fusing maximum I^2t and minimum required pressure-relief capability in SCFM (Standard Cubic Feet per Minute) at 15 psig. The opening pressure is 10 psig maximum. With respect to protection without current-limiting fuses for the primary protection, it will be difficult to meet the requirements of maximum I^2t (A²s) in column 3 of Table 12.6. As an example consider a 2500 kVA transformer connected to a 13.8 kV primary system of short-circuit level 30 kA rms symmetrical on the transformer primary bushings. Considering an instantaneous relay with 0.5 cycle operating time and a 13.8 kV circuit breaker with interrupting time of three cycles, the I^2t released is 90×10^6 A²s. The permissible is 7.5×10^6 A²s. This means that unless the available three-phase fault level is 11.3 kA maximum, the labeling requirements are not met with relayed circuit breakers.

12.10 Effect of Transformer Winding Connections

A three-phase symmetrical fault on the secondary windings of a three-phase transformer gives rise to three-phase currents in the primary lines irrespective of the transformer connections. This primary fault current will be changed in the ratio of transformation. No shift in transformer through-fault withstand capability or in the primary or secondary protective device characteristics is required.

Figure 12.13 shows the currents in the primary lines for a fault on the secondary side of the transformer. The line currents and winding currents are dependent upon the type of unsymmetrical fault and the transformer winding connections. In this figure the line-to-line fault current is assumed equal to 0.87 times the three-phase fault currents, that is, the positive- and negative-sequence impedances are considered equal.

The general practice is to shift the transformer through-fault withstand curve, depending upon the transformer connections. For example, for a delta–wye transformer, wye windings solidly grounded, only two primary lines carry a fault current of 0.58 times the secondary line-to-neutral current, and therefore shift the transformer through-fault withstand curve by 0.58 toward left in the time–current coordination plot (that is, each current point on the curve is multiplied by a factor of 0.58).

Some authors are of the opinion that the transformer withstand curve is a fixed entity, and it should not be shifted. Rather than shifting the transformer withstand curve, shift the primary protective relay curve. As this protective device will operate slower, shift the curve by 1/058 (=1.73) times toward the right in the TCC plot.

Similar shifts for a line-to-line fault in a solidly grounded delta–wye transformer will be: (1) transformers withstand curve shift toward right by 16%, or the relay protective curve shift toward left, multiplied by a factor of 0.87 throughout.

Most computer programs for coordination shift the transformer withstand curves, rather than shifting the relay curves. Therefore, this method is adopted in this book.

12.11 Requirements of Ground Fault Protection

Many times the primary protection device is used to provide through-fault protection for secondary unsymmetrical faults also; as there may not be a secondary ground fault protection. This omission of secondary ground fault protection is in contrivance of NEC requirements for services, article 240.95, which states that ground fault protection will be provided for solidly grounded wye electrical services of more than 150 V to ground but not exceeding 600 V phase-to-phase for each service disconnect rated 1000 A or more; see also Chapter 8. This was first required in 1971, because of unusually high burnouts reported on these services. Further, it is specified that the ground fault protection will operate to open all ungrounded conductors (phase conductors) of the faulted circuit. The maximum setting is 1200 A, and the maximum time delay is 1 s for ground fault currents equal to or greater than 3000 A.

Listing Option B, FM Global *requires that all the following conditions must be complied with*:

- The transformer neutral ground fault relay as discussed above (device 51N in Figure 12.2d) should be provided.
- Indoor units >500 kVA and outdoor units >2500 kVA shall be equipped with alarm contacts on pressure-relief device and a rapid pressure rise relay.
- Three-phase pad mounted and substation transformers shall be equipped with an oil level gauge. All transformers >750 kVA shall be equipped with liquid temperature indicator and pressure–vacuum gauge.
- Transformer shall pass BIL (Basic Insulation Level) test at a minimum tilt of 1.5° from vertical.

For high-resistance grounded systems, no shifts in the thermal withstand curves for secondary ground faults are required.

12.12 Through-Fault Protection

12.12.1 Primary Fuse Protection

The extent of protection provided with respect to thermal damage curves of transformers using primary fuse protection is illustrated in Table 12.7. This table is constructed with the worst characteristics of the available class E fuses (Figure 12.12) and a number of coordination studies [9]. Delta primary and wye secondary transformers of various sizes connected to 13.8 kV primary voltages are considered. Transformers rated up to 2500 kVA have low-voltage secondary windings, solidly grounded and transformers of 3000 and 5000 kVA have RG (Resistance Grounded) medium-voltage secondary windings. The symbol "N" means no protection is provided by the selected fuse. The numbers "5–15" signify that transformer through-fault withstand is protected between 5 and 15 s only. The calculations are made with transformer impedances as shown.

This table shows that no protection is provided by the transformer primary fuses for secondary line-to-neutral faults. Therefore, a transformer neutral-connected ground fault relay is a must to protect the transformer. Also the effect of transformer impedance is clearly visible in this table with respect to the protection; a higher transformer impedance lowers the extent to which the through-fault curves can be protected.

TABLE 12.7

Protection Provided by Primary Fuses; Frequent Fault and Infrequent Fault Damage Curves

Transformer Three-Phase kVA	Maximum Current at Fan-Cooled Rating	%Z on Base KNAN	Fuse Rating Class E, Current limiting	Frequent Fault Withstand			Infrequent Fault Withstand		
				3-P	L-L	L-N	3-P	L-L	L-N
500/644	27	5.75	40	2-12	2-16	N	2-12	2-16	N
750/966	40	5.75	65	2	2-3 and 5-15	N	2-10	2-15	N
		8		N	N	N	2-10	2-15	N
1000/1288	54	5.75	80	2	2-3 and 5-15	N	2-10	2-15	N
		8		N	N	N	2-10	2-15	N
1500/1680	81	5.75	125	2	2-3 and 5-15	N	2-10	2-15	N
		8		N	N	N	2-10	2-15	N
2000/2576	108	5.75	150	2-20	2-28	N	2-15	2-25	N
		8		N	2-3 and 6-38	N	2-15	2-25	N
2500/3500	146	5.75	175	2-25	2-38	N	2-25	2-38	N
		8		2 and 9-25	2-3 and 6-38	N	2-25	2-38	N
3000/4200	176	5	200	2-50	2-60	RG	2-50	2-60	RG
		8		2-3 and 7-50	2-60	RG	2-50	2-60	RG
5000/7000	293	5	300	2-40	2-60	RG	2-50	2-50	RG
		8		2-40	2-60	RG	2-50	2-50	RG

See text for an explanation of symbols and abbreviations.

TABLE 12.8

Protection Provided by Fuses versus Relays for Transformer Through-Fault Withstand

Primary Protection	Frequent Fault			Infrequent Fault		
	3-P	L-L	L-N	3-P	L-L	L-N
Fuses	2	2-35	N	2-15	2-35	2-7
Relays	2-100	2-200	N	2-100	2-200	2-40

12.12.2 Primary Relay Protection

The through-fault protection calculated in Table 12.7 with fuses can be much improved with primary overcurrent relays. A comparison is shown in Table 12.8 [9]. Though the protection is improved, yet the need for a neutral-connected relay for secondary L–N faults is apparent. Also the listing requirements I^2t need to be simultaneously considered.

12.13 Overall Transformer Protection

The overcurrent protection discussed above does not provide thermal protection of the transformer. The conductor insulation in the transformers ages fast if the temperatures exceed the design temperatures. Winding temperature indicators with alarm, and sometimes for trip, are used, but these have limitations. With on-load tap changing (OLTC), the hot spot temperature may move from high- to low-voltage winding or vice versa. Winding temperature indicator consists of a temperature-sensing bulb immersed in a well in the top layer of the insulating oil, and it is heated by a replica of the load current of the transformer through a CT. It is calibrated for hot spot, but has limitations with respect to thermal protection of the transformer windings. Digital relays can be used for thermal winding protection, which can calculate the temperature on the primary, secondary, and tertiary windings. Transformers are provided with overexcitation, negative-sequence protection, voltage controlled overcurrent relays, mechanical detection of faults such as gas accumulator relay, gas detector relay, sudden gas/oil pressure relay, internal gas monitoring relay, oil temperature and oil level alarms, and the like. References [10–13] provide further reading on transformer protections.

12.13.1 System Configuration

A radial distribution system is shown in Figure 12.18. A single 1200 A 13.8 kV circuit breaker serves four substation transformers, three transformers of 2000/2240 kVA 480 V and one of 7500/9375 kVA, 2.4 kV secondary. All transformers have 13.8 kV windings in delta connection and the secondary windings in wye connection. The 480 V transformer wye windings are high-resistance grounded and 2.4 kV transformer wye windings are low-resistance grounded through a 200 A resistor. Downstream distribution for one 2000 kVA low-voltage transformer TX1 and 7500–2.4 kV transformer TX2 is shown. The 480 V transformer serves a lineup of low-voltage switchgear provided with LVPCBs having electronic trip programmers with LSI functions. At first consider that *there are no main secondary circuit breakers either on 480 V or 2.4 kV transformers*, that is, circuit breakers BK5 and BK6, shown in Figure 12.19, are not provided. The low-voltage switchgear serves a number of low-voltage motor control centers; one typical MCC is shown. For coordination purposes, we need to consider the largest motor/feeder on any MCC. In the industrial distribution system, generally, the 200 hp motor rating is the maximum that is applied, and higher rated motors are connected to medium-voltage distributions. The 2.4 kV transformer serves a lineup of 2.4 kV metal-clad switchgear, which in turn serves some medium-voltage MCCs. The major ratings of the equipment are shown in this figure. The interconnecting cable sizes are detailed in Table 12.9.

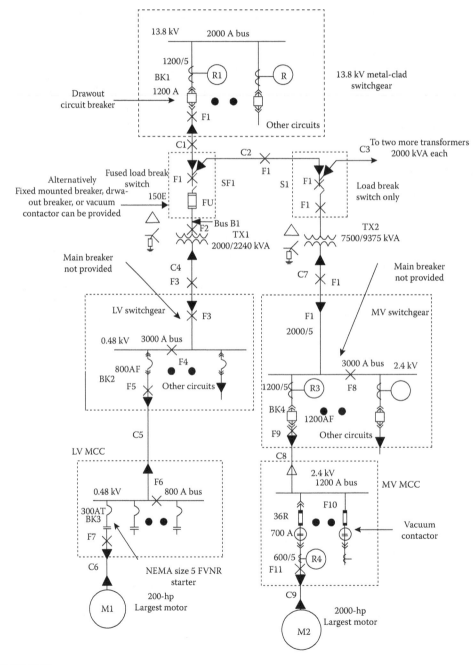

FIGURE 12.18
A low-voltage and medium-voltage distribution system for protective relay coordination.

Transformer Reactor and Shunt Capacitor Bank Protection

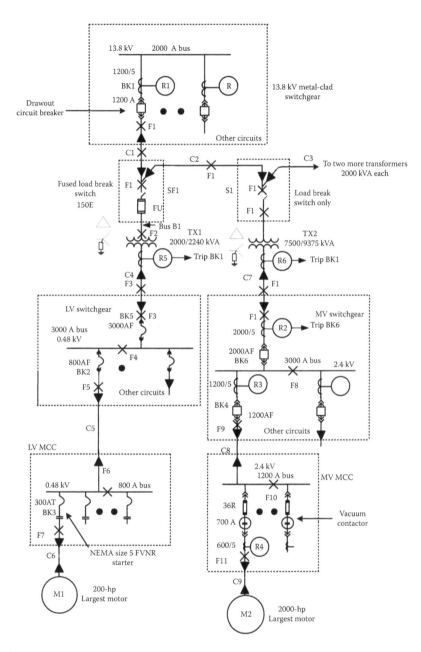

FIGURE 12.19
As in Figure 12.18, except that secondary main breakers BK5 and BK6 are added.

The 7500/9375 kVA transformer TX2 is protected by relay R1 on 13.8 kV circuit breaker BK1. The 2000/2240 kVA transformer TX1 is protected with 150E current-limiting fuse. The medium-voltage MCC is served from a relayed feeder circuit breaker, BK4. The 2000 hp motor has NEMA 2 fused starter, with 700 A vacuum contactor and 36R fuse. The system ground fault protection is not shown. All switching devices are applied well within their short-circuit ratings. The three-phase bolted short-circuit level at 13.8 kV switchgear is 30 kA rms symmetrical.

TABLE 12.9

Cable Sizes and Lengths, Figures 12.18 and 12.19

Cable Designation	Cable Description	Number in Parallel per Phase
C1, C2	15 kV grade, 3/C, 500 kcmil, 130% insulation level, 90°C temperature, XLPE	2
C3	15 kV grade, 3/C, 500 kcmil, 130% insulation level, 90°C temperature, XLPE	1
C7	5 kV grade, 1/C, 500 kcmil, 130% insulation level, 90°C temperature, XLPE	4
C8	5 kV grade, 1/C, 500 kcmil, 130% insulation level, 90°C temperature, XLPE	2
C9	5 kV grade, 1/C, 500 kcmil, 130% insulation level, 90°C temperature, XLPE	2
C4	600 V grade, 3/C, 500 kcmil, THHW, NEC 90°C temperature	7
C5	600 V grade, 3/C, 500 kcmil, THHW, NEC 90°C temperature	2
C6	600 V grade, 3/C, 500 kcmil, THHW, NEC 90°C temperature	1

This radial distribution system meets the requirements of NEC and is "well protected" with the protection devices and loads required to be served. Many industrial distribution systems implemented on this basis are in service.

12.13.2 Coordination Study and Observations

Let us consider fault locations shown from F1 through F11. Table 12.10 shows the fault locations and the protective devices which will clear the fault, when main circuit breakers BK5 and BK6 are not provided.

TABLE 12.10

Faults at Various Locations in Figure 12.18, and the Protective Devices Clearing These Faults, No Main Secondary Breakers

Fault at	Description	Fault Clearing Device
F1, F8	On the secondary side of the 13.8 kV feeder breaker in the 13.8 kV switchgear itself, in the cable circuits C1, C2, in the fused load break switch upstream of the 150E fuse, in the 7500/9375 kVA transformer load break switch, cable and to the 2.4 kV switchgear, in the primary cable terminations in the 2.4 kV switchgear and also in the 2.4 kV switchgear bus, as no main secondary beaker is present	Relay R1, Feeder Breaker BK1
F2	Load side of the fuse and up to the primary windings of 2000/2240 kVA transformer	Fuse FS1
F3, F4	On the secondary side of 2000/2240 kVA transformer, on the cable terminations in the low-voltage switchgear and also on the low-voltage switchgear bus, fault location F4; as there is no main secondary breaker	Fuse FS1
F5, F6	On the secondary side of the feeder beaker in low-voltage switchgear, in the incoming cable terminations at low-voltage MCC and on the MCC bus, fault location 5	Feeder breaker BK2
F7	On the secondary side of the motor feeder in the low-voltage MCC	Motor starter breaker BK3
F9, F11	On the load side of the feeder circuit breaker to medium-voltage MCC, on the incoming terminations of the medium-voltage MCC and also on the MCC bus, fault location F11	Relay R3, breaker BK4
F12	On the load side of the 2000 hp motor starter	Motor starter Fuse 36R

Figure 12.20 shows the TCC plot of the 2000/2240 kVA transformer protective devices, and Figure 12.21 shows the TCC plot for 7500/9500 kVA transformer. The following observations are of interest:

- The transformer ANSI frequent fault curves are protected.
- The system is well coordinated.
- The 150E fuse clears the transformer inrush point. There is a slight overlap with the feeder short-time delay setting, but this is acceptable. Transformer-rated current at

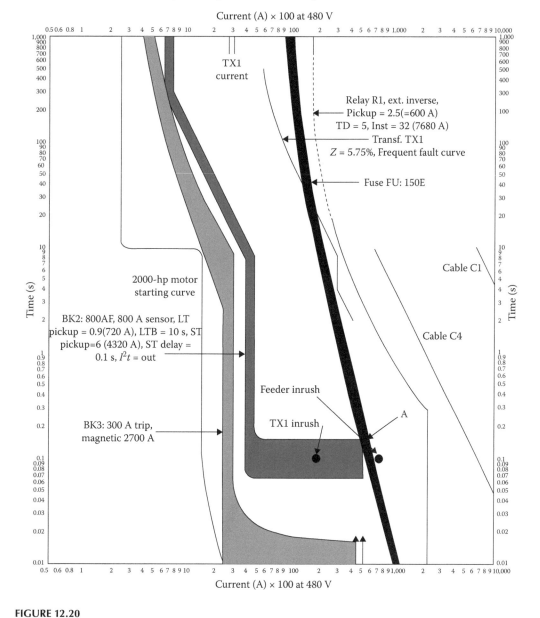

FIGURE 12.20
Time–current coordination plot for 480 V, 2000 kVA transformer protective devices in Figure 12.18, no main breaker.

13.8 kV is 93.7 A at its full fan-cooled rating of 2240 kVA. Thus, a lower rated class E fuse, for example 125E, could be used, but the overlap with feeder short-time characteristics will increase.

- To coordinate with 200 A MCCB, motor circuit breaker BK3, a setting higher than the minimum short-time delay setting available on the trip programmer of circuit breaker BK2 is not required.
- The motor starting curves are drawn for the actual motor starting time, driving its load inertia.

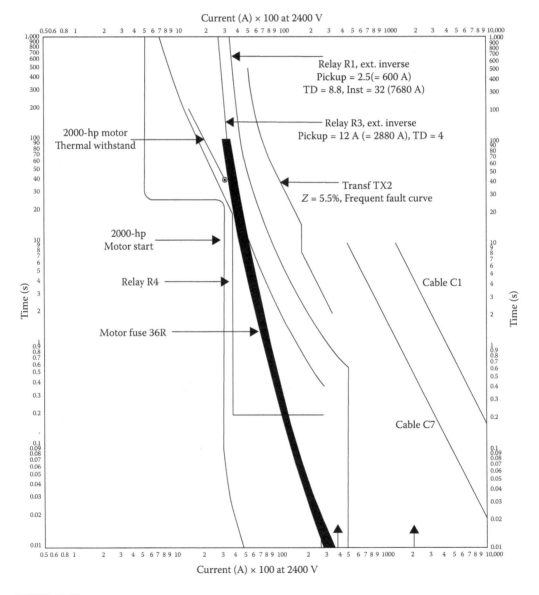

FIGURE 12.21
Time–current coordination plot for 2.4 kV, 7500 kVA transformer protective devices in Figure 12.18, no main breaker.

- The 2000 hp motor thermal curve is plotted based upon the manufacturer's data, and the multifunction relay R4 shown in Figure 12.18 admirably protects it.
- The settings on relay R1 protect the 7500/9375 kVA transformer as per NEC guidelines and allow the maximum system load equivalent to the total installed kVA rating of the transformer to be carried on continuous basis.
- The instantaneous settings on relay R1 take into account the total feeder inrush, that is, all the connected transformers will take their inrush currents on switching. This total inrush current is calculated as 6500 A at 13.8 kV.
- Curves for only a few of the cables are plotted as specimens.

12.13.3 Addition of Secondary Relay

The system design can be modified and the hazards on main circuit breakers are reduced with the modified configuration, as shown in Figure 12.19. This figure shows that additional overcurrent relays R5 and R6 are provided on the secondary of transformers.

The current transformers actuating relays R5 and R6 are located in the transformer tank or in the transformer secondary air terminal compartment. The relays R5 and R6 are located in their respective switchgears. By moving the sensing current transformers to the transformer tank or in the transformer secondary terminal compartment, these relays will respond to a fault in the secondary cables, cable connections and BK5 and BK6 circuit breaker primaries.

It is necessary that these relays trip an upstream circuit breaker, in this case circuit breaker BK1. This means that the entire loads of substations served from circuit breaker BK1 will be interrupted, which may not be acceptable.

Instead of fused switches, (1) fixed mounted circuit breakers with interlocked disconnect switches or (2) draw-out metal clad circuit breakers or (3) fused 13.8 kV vacuum contactors can be provided on the transformer primary in Figures 12.18 and 12.19. With this configuration, the entire loads need not be interrupted and the secondary relays in each substation will trip their respective primary circuit breakers.

The circuit breakers BK5 and BK6 can even be eliminated, retaining the relays R5 and R6. However, it is desirable to retain these. The switchgear bus faults will be cleared by these circuit breakers, and for such faults the tripping of primary circuit breaker is avoided. The relay R3 can be coordinated with circuit breaker BK6 settings, so that only a fault in the secondary cable connections from the transformer results in tripping of the upstream primary device. Furthermore, these cable or bus connections are of short length and mechanically protected—the likelihood of a fault occurring on these connections is small. Also provision of secondary main breakers is necessary to have acceptable arc flash levels on the LV or MV switchgear buses. See Chapter 17.

12.14 Differential Protection

The fundamentals of differential protection are discussed in Chapter 4. Generally, transformers rated above 5 MVA are provided with phase and ground fault directional protection.

(a)

Transformer type	Wdg #	Connection	Voltage phasors	Phase shift
D/d300°	1	Delta		300° lag
	2	Delta 300° lag	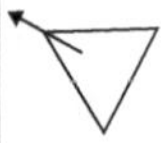	0°
D/y30°	1	Delta		0°
	2	Wye (gnd 1/2) 30° lag		330° lag
D/y150°	1	Delta		0°
	2	Wye (gnd 1/2) 150° lag		210° lag
D/y210°	1	Delta		0°
	2	Wye (gnd 1/2) 210° lag	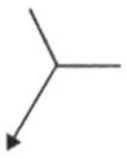	150° lag
D/y330°	1	Delta		0°
	2	Wye (gnd 1/2) 330° lag		30° lag
D/z30°	1	Wye (gnd 1/2)	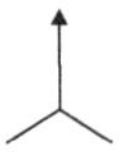	30° lag
	2	Zig-Zag (gnd 2/3) 30° lag		0°

Transformer type	Wdg #	Connection	Voltage phasors	Phase shift
Y/z150°	1	Wye (gnd 1/2)	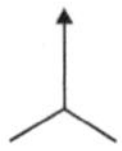	150° lag
	2	Zig-Zag (gnd 2/3) 150° lag	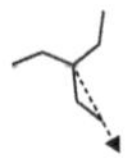	0°
Y/z210°	1	Wye (gnd 1/2)	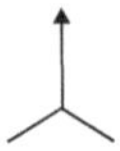	210° lag
	2	Zig-Zag (gnd 2/3) 210° lag	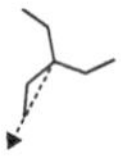	0°
Y/z330°	1	Wye (gnd 1/2)	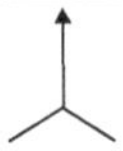	330° lag
	2	Zig-Zag (gnd 2/3) 330° lag	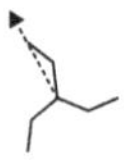	0°
D/z0°	1	Delta		0°
	2	Zig-Zag (gnd 1/2) 0° lag	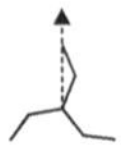	0°
D/z0°	1	Delta		60° lag
	2	Zig-Zag (gnd 1/2) 60° lag		0°
D/z120°	1	Delta		120° lag
	2	Zig-Zag (gnd 1/2) 120° lag	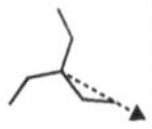	0°

FIGURE 12.22
(a and b) Transformer connections and phase shifts in winding connections. Some MMPRs will directly accept phase shift angles for differential relaying.

(*Continued*)

(b)

Transformer type	Wdg #	Connection	Voltage phasors	Phase shift
Y/y180°/d150°	1	Wye (gnd 1/2)	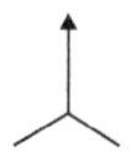	150° lag
	2	Wye (gnd 2/3) 180° lag		330° lag
	3	Delta 150° lag	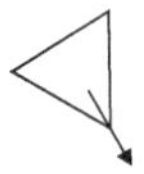	0°
Y/y180°/d210°	1	Wye (gnd 1/2)	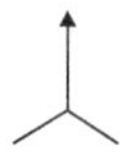	210° lag
	2	Wye (gnd 2/3) 180° lag	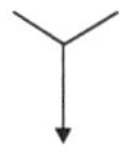	30° lag
	3	Delta 210° lag		0°
Y/y180°/d330°	1	Wye (gnd 1/2)	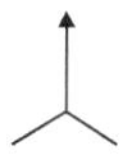	330° lag
	2	Wye (gnd 2/3) 180° lag	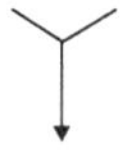	150° lag
	3	Delta 330° lag	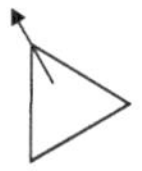	0°
Y/d30°/y0°	1	Wye (gnd 1/2)	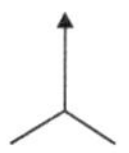	30° lag
	2	Delta 30° lag		0°
	3	Wye (gnd 2/3) 0°	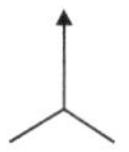	30° lag

Transformer type	Wdg #	Connection	Voltage phasors	Phase shift
Y/d30°/y180°	1	Wye (gnd 1/2)	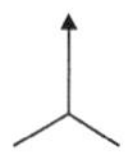	30° lag
	2	Delta 30° lag	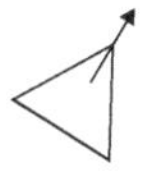	0°
	3	Wye (gnd 2/3) 180° lag		210° lag
Y/d30°/d30°	1	Wye (gnd 1/2)	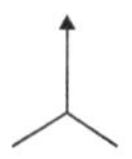	30° lag
	2	Delta 30° lag	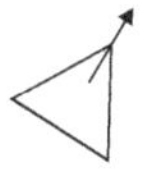	0°
	3	Delta 30° lag	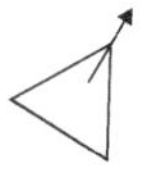	0°
Y/d30°/d150°	1	Wye (gnd 1/2)	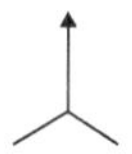	30° lag
	2	Delta 30° lag		0°
	3	Delta 150° lag		240° lag
Y/d30°/d210°	1	Wye (gnd 1/2)		30° lag
	2	Delta 30° lag	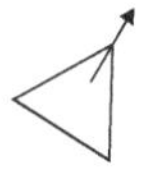	0°
	3	Delta 210° lag		180° lag

FIGURE 12.22 (CONTINUED)
(a and b) Transformer connections and phase shifts in winding connections. Some MMPRs will directly accept phase shift angles for differential relaying.

12.14.1 Electromechanical Transformer Differential Relays

When applying differential protection to transformers, additional considerations are as follows:

- There is a phase shift in three-phase transformer windings, see Volume 1 for full details. Again Figure 12.22a and b shows these phase shifts.
- The CT ratios on the primary and secondary sides of the transformers cannot be exactly matched with respect to primary and secondary currents.
- The transformer inrush currents contain harmonics, particularly second and fifth harmonics. The transformer inrush current profiles, harmonics, and the DC component produced are discussed in Volume 3 and not repeated here.

Figure 12.23 illustrates the circuit of a harmonic restraint electromechanical relay. Differential relays without harmonic restraint have also been used for protection of transformers, but are no longer in use.

Figure 12.23 shows the CT connections on the high- and low-voltage sides for a delta–wye-connected transformer. The CTs on the wye side are connected in delta and those at the delta side are connected in wye for proper phasing. In case the CTs on the wye-grounded side are connected in wye, then for an external ground fault, the current will flow in the relay operating coil and the relay will operate. Delta-connected CTs block this zero-sequence current. The phasing can be checked for current flows in the transformer and CT windings, as shown in this figure. A three-phase external fault on the high side (138 kV) is shown in this figure. The 13.8 kV side may be connected to a source like a generator. Next, a ratio check is made and the mismatch should be reduced to acceptable limits.

Example 12.2

A 40 MVA transformer, 138–13.8 kV, delta primary (138 kV), and wye-grounded secondary windings (13.8 kV), is assumed for this example.

Calculate the full-load currents on the primary and secondary sides. The 138 kV side full-load current is 167 A, and on 13.8 kV side it is 1673 A.

A CT ratio close to these currents is selected. Select a 200/5 CT on the 138 kV windings and 2000/5 CT on the 13.8 kV windings.

Calculate the currents in the CT leads. The CTs on the 138 kV side are wye connected and on the 13.8 kV side are delta connected. The currents in the secondary leads are 4.175 A for wye-connected CTs and $\sqrt{3} \times (1673/400) = 7.24$ A on the delta side CTs. The ratio of currents is 1.732.

The relay coils are provided with current taps. It is necessary to choose taps so that ideally the same ratio of 1.732 is obtained. The manufactures publish tables of taps provided on their relay coils.

If a tap of 8.7 is selected on the 13.8 kV side, then a tap of 5.02 is required on the high-voltage side to give exact 1.732 ratio. The nearest available tap is 5. This gives a ratio of 1.74, fairly close to the desired ratio. Mismatches can be calculated as

$$M = \left(\frac{\frac{7.24}{4.175} - \frac{8.7}{5}}{\frac{7.24}{4.175}} \right) \times 100 = -0.46 \quad (12.3)$$

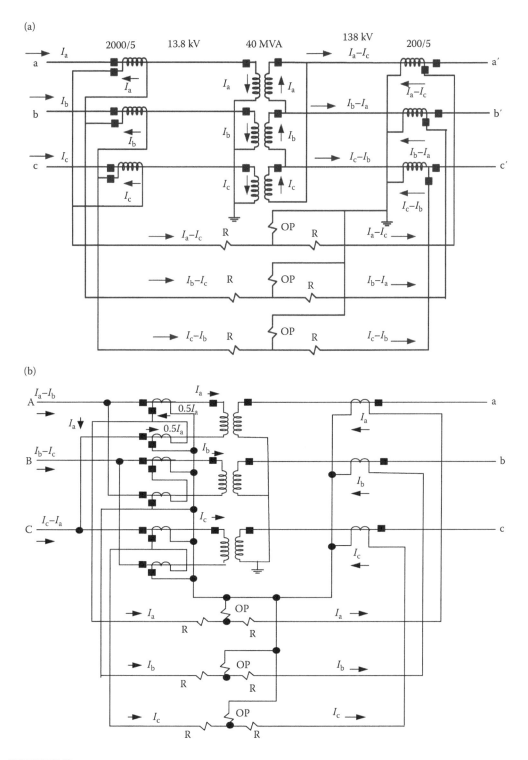

FIGURE 12.23
(a) Differential protection of a delta–wye transformer using an electromechanical relay. Note the CT connections on the delta and wye side and (b) single-phase transformers connected in three-phase configuration.

The manufacturers specify the acceptable limits of mismatch, which is 2.5%. This exercise indicates that a case-by-case analysis will be required and different CT ratios and taps can be tried to reduce the mismatch.

Considerations of CT accuracy are as follows:

With respect to CT accuracy, the following equation is provided by one manufacturer:

$$\left[\frac{N_p V_{CL} - (I_{ext} - 100)R_s}{I_{ext}}\right] > Z_T \qquad (12.4)$$

where
N_p = proportion of total CT turns in use for multiratio CTs, as the relaying accuracy is specified for the maximum turns ratio and should be proportionate if a tap on the CT winding is used
V_{CL} = CT C class accuracy
I_{ext} = maximum external fault current in CT secondary in rms, not < 100
R_s = CT secondary winding resistance
Z_t = total transformer secondary burden. This can be calculated for both primary and secondary side faults.

The expressions of Z_t, assuming that the CTs are dedicated for the differential protection, are as follows:

$Z_t = 1.13R_L + \text{relay burden} + Z_A$ Three-phase fault

$Z_t = 1.13(2R_L) + \text{relay burden} + Z_A$ Phase-to-ground fault   (12.5)

$Z_t = 2(1.13R_L + \text{relay burden} + Z_A)$ Phase-to-ground fault, delta-connected CTs

R_L is one-way lead resistance and Z_A is burden of any other device connected to the CT.

Example 12.3

In Example 12.2, the CTs have a C class accuracy of 100. Calculate if this is adequate for the application?

Continuing with Example 12.2 if the transformer has a percentage impedance of 9% on 40 MVA base, and the 138 kV source has a short-circuit current of 40 kA, then the through-fault current on 13.8 kV side is equal to 15 kA approximately. Reflected in the CT leads it is equal to 64.95 A (CT ratio is 400, delta connected).

As this is <100, we consider I_{ext} = 100. CT class C accuracy = 100, CT secondary resistance = 1 5 Ω. N_p = 1, as it is single ratio transformer of 2000:5. Substituting the values, Z_t should be <1 Ω.

When single-phase transformers are connected in three-phase configurations, the CTs provided on the transformer bushings are applied for differential protection. Two sets of CTs connected in parallel are required to provide protection for ground faults in these windings. Wye-connected CTs can be used on both sides. Figure 12.23b shows that there is a 2:1 ratio difference of the CTs on two sides, which can be adjusted by CT ratios or relay taps.

12.14.2 Harmonic Restraint

The harmonic restraint in the relay has a second harmonic blocking filter in the operating and restraint coils; the circuit is designed to hold the contacts open with second harmonic of 15% of the fundamental. The percentage characteristic varies from 20% on light loads to approximately 60% for heavy faults. The minimum pickup current is 30% of the tap value, for 30% sensitivity.

12.14.3 Protection with Grounding Transformer Inside Main Transformer Protection Zone

When a grounding transformer (zig-zag or wye–delta) is inside the main differential protection zone, a nuisance trip can occur due to zero-sequence currents. Figure 12.24a shows that there is a zig-zag grounding transformer to ground the secondary delta-connected winding. A zero-sequence filter is provided. The CT ratio can be 5:5. The flow of zero-sequence currents throughout the distribution system is illustrated in this figure. An alternate zero-sequence trap is also shown in Figure 12.24b. This requires a 1:3 CT ratio. These connections present relatively high magnetizing impedance to all but zero-sequence currents.

12.14.4 Protection of an Autotransformer

The protection of wye-connected autotransformer with unloaded tertiary is illustrated in Figure 12.25. Note that the CTs on both sides are connected in delta. The matching of the CT ratios is required as illustrated before, so that the currents in the relay coils are almost zero.

12.14.5 Protection of Three-Winding Transformers

When protection of three-winding transformers is involved, each secondary winding CT is connected to a restraint coil, as illustrated in Figure 12.26. Paralleling of CTs for connection to single restraint coil should be avoided for the most effective restraint action.

12.14.6 Microprocessor-Based Transformer Differential Relays

The microprocessor-based relays have simplified installations, CT connections, and added additional protection features, such as overcurrent instantaneous and time delay for phase and ground faults, functions such as 24, 80, 27, 59, 64, etc. These relays provide internal mathematical calculations for phase angle compensation and zero-sequence filtering. Built-in metering displays such as secondary current magnitudes and phase angles, and restraint and operate quantities are very useful in commissioning and fault diagnosis.

The dual-slope characteristic is shown in Figure 12.27. The lowest portion of the trip threshold represents the minimum sensitivity of the relay. Some relays provide options to set as low as 5%; however, this may give rise to nuisance trips. Settings of 10% have held without nuisance trips with matched CTs. Some manufacturers recommend a minimum setting of no less than 30%. Also the minimum settings are sometimes specified in terms of the calculated primary and secondary tap currents. The CT errors due to saturation are likely to be a problem at higher currents and the steeper slope 2 characteristics adds security for heavy external faults.

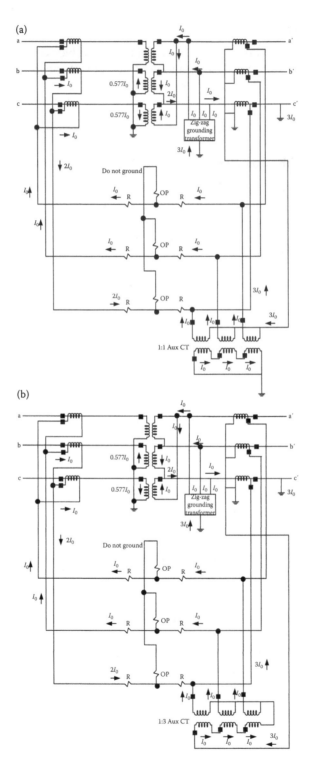

FIGURE 12.24
(a and b) Differential protection, when a grounding transformer is included in the differential zone of protection, external fault conditions are shown.

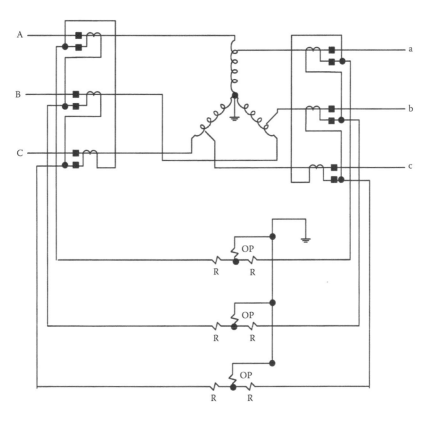

FIGURE 12.25
Differential protection of a three-phase autotransformer.

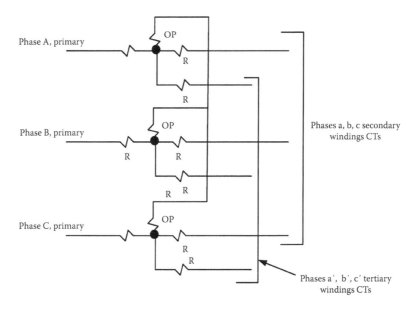

FIGURE 12.26
Differential protection of a three-winding transformer.

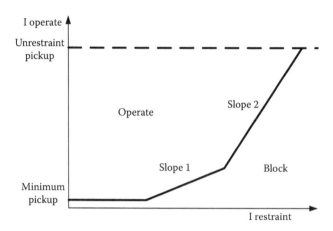

FIGURE 12.27
Two-slope differential characteristics in a modern MMPR.

12.14.6.1 CT Connections and Phase Angle Compensation

Considering a delta–wye transformer, the CTs on the delta and wye can be both wye connected. This requires that the phase shift in the transformer windings should be corrected and also the zero-sequence current that can flow for an external ground fault should be blocked.

Some advantages of connecting CTs in wye connection are that these present a lower burden than the delta-connected CTs, as much as three times less depending on fault type. With wye CTs it is possible to use zero-sequence elements, like ground fault protection.

However, for the differential protection external ground fault results flow through CTs without corresponding current flows through the transformer delta winding side CTs. To prevent relay operation, this zero-sequence current should be removed. This can be done by selecting correct phase angle compensation, which simultaneously corrects phase angle as well as zero-sequence source errors.

With the modern MMPRs for the transformer protection, this can be done by two methods:

1. A differential relay will accept the actual winding connection phasors directly. Figure 12.22a and b illustrates some phasor diagrams and vector groups for two-winding and three-winding transformers. Depending upon the transformer vector group it is adequate to input that group.

    In a three-phase two-winding transformer manufactured according to ANSI/IEEE standards the low-voltage side, whether in wye or delta connection, has a phase shift of 30° lagging with respect to phase-to-neutral voltage vectors on the high side or the high-side phasors lead the low-side phasors by 30° These transformers are called Dy1 or Yd1. The letter 1 indicates that low-side lags the high side by 1h on the clock. See Volume 1. A user can directly program the right connection into the relay.

2. For some other differential relays, a connection matrix is used, which corrects the phase shift and compensates for the zero sequence. Some connection matrices are shown in Figure 12.28a and b. Phase rotation must be considered while selecting the appropriate matrix for the compensation.

Transformer Reactor and Shunt Capacitor Bank Protection

(a)

$$\overline{M}_1 = \frac{1}{\sqrt{3}} \begin{vmatrix} 1 & -1 & 0 \\ 0 & 1 & -1 \\ -1 & 0 & 1 \end{vmatrix}$$

$$\overline{M}_2 = \frac{1}{3} \begin{vmatrix} 1 & -2 & 1 \\ 1 & 1 & -2 \\ -2 & 1 & 1 \end{vmatrix}$$

$$\overline{M}_3 = \frac{1}{\sqrt{3}} \begin{vmatrix} 0 & -1 & 1 \\ 1 & 0 & -1 \\ -1 & 1 & 0 \end{vmatrix}$$

$$\overline{M}_4 = \frac{1}{3} \begin{vmatrix} -1 & -1 & 2 \\ 2 & -1 & -1 \\ -1 & 2 & -1 \end{vmatrix}$$

$$\overline{M}_5 = \frac{1}{\sqrt{3}} \begin{vmatrix} -1 & 0 & 1 \\ 1 & -1 & 0 \\ 0 & 1 & -1 \end{vmatrix}$$

$$\overline{M}_6 = \frac{1}{3} \begin{vmatrix} -2 & 1 & 1 \\ 1 & -2 & 1 \\ 1 & 1 & -2 \end{vmatrix}$$

$$\overline{M}_7 = \frac{1}{\sqrt{3}} \begin{vmatrix} -1 & 1 & 0 \\ 0 & -1 & 1 \\ 1 & 0 & -1 \end{vmatrix}$$

$$\overline{M}_8 = \frac{1}{3} \begin{vmatrix} -1 & 2 & -1 \\ -1 & -1 & 2 \\ 2 & -1 & -1 \end{vmatrix}$$

(b)

$$\overline{M}_9 = \frac{1}{\sqrt{3}} \begin{vmatrix} 0 & 1 & -1 \\ -1 & 0 & 1 \\ 1 & -1 & 0 \end{vmatrix}$$

$$\overline{M}_{10} = \frac{1}{3} \begin{vmatrix} 1 & 1 & -2 \\ -2 & 1 & 1 \\ 1 & -2 & 1 \end{vmatrix}$$

$$\overline{M}_{11} = \frac{1}{\sqrt{3}} \begin{vmatrix} 1 & 0 & -1 \\ -1 & 1 & 0 \\ 0 & -1 & 1 \end{vmatrix}$$

$$\overline{M}_{12} = \frac{1}{3} \begin{vmatrix} 2 & -1 & -1 \\ -1 & 2 & -1 \\ -1 & -1 & 2 \end{vmatrix}$$

FIGURE 12.28
(a and b) Application of matrices for correction of phase shifts in transformer winding connections for differential protection in a modern MMPR.

Consider that we have a Dy1 transformer, phase rotation ABC. First, the phasors with proper phase rotation on the high side of the transformer to represent high-side currents are drawn. Also consider phase-to-bushing connections. When we choose appropriate matrix from Figure 12.28, the compensated currents produce zero operate current for external faults. If the primary winding phase current is taken as a reference, the compensation matrix is a unity matrix, M_0, because the transformer primary winding is in delta and there is no zero-sequence source current between the CT and transformer windings. Next, move the low-side vector in counterclockwise direction till it coincides with the vector on the primary side. A Dy1 or Yd1 connection requires a 30° movement in the counterclockwise direction. Thus matrix M_1 is selected.

$$\bar{I}_{\text{abc-comp}} = \frac{1}{\sqrt{3}} \begin{vmatrix} 1 & -1 & 0 \\ 0 & 1 & -1 \\ -1 & 0 & 1 \end{vmatrix} \begin{vmatrix} I_a \\ I_b \\ I_c \end{vmatrix}$$

or

$$I_{\text{a-comp}} = \frac{(I_a - I_b)}{\sqrt{3}}$$

$$I_{\text{b-comp}} = \frac{(I_b - I_c)}{\sqrt{3}} \quad (12.6)$$

$$I_{\text{c-comp}} = \frac{(I_c - I_a)}{\sqrt{3}}$$

This produces compensated secondary currents which are 180° out of phase with the primary currents. Operate or difference current is therefore zero. Matrix M_{12} filters the zero-sequence currents, without any phase corrections. Also see Reference [14].

While in MMPRs the zero-sequence components are trapped by appropriate choice of a connection matrix, in electromagnetic relays, no longer in use, the method of zero-sequence trap was auxiliary CTs connected, as shown in Figure 12.24a and b.

12.14.6.2 Dynamic CT Ratio Corrections

We noted that for electromechanical relays that the CTs on the primary and secondary sides must be carefully selected and also proper taps on the relay must be selected to minimize mismatch. *With the MMPR this is not required.* A differential relay will accommodate automatically a certain mismatch in the ratio, and the only input required is the CT ratios on the primary and secondary sides. A particular relay can accommodate a mismatch in the primary and secondary CT ratios, *up to a mismatch factor of 16*. The CT currents are scaled with respect to the rating of the transformer to reflect the load currents in the windings.

12.14.6.3 Security under Transformer Magnetizing Currents

The MMPR takes advantage of harmonics and DC component of the current waveform during inrush. The inrush current of a transformer is rich in second harmonic which is of the order of 60% of the fundamental. The overexcitation of the transformer produces third, fifth, and seventh harmonics, of the order of 49%, 23%, and 8% of the fundamental, respectively. The third harmonic is eliminated by the delta connection of the transformer windings. Thus, overexcitation can be detected with fifth harmonic.

Figure 12.29 depicts that the input currents from the CT secondary are converted into voltage signals. Low-pass filters remove high-frequency components from the voltage signals. Digital filters extract the fundamental, second, and fifth harmonics from the digital signals. These are then scaled and the zero-sequence components are removed. The currents are also compensated for transformer phase shift. The winding 2 compensated currents are similarly obtained. These are inputted to differential elements and the second and fifth harmonic blocking. The magnitude of the compensated fundamental frequency currents determines the operating quantity of the restrained differential element.

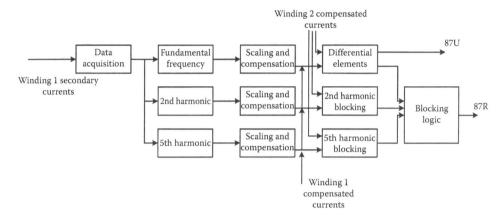

FIGURE 12.29
Blocking logic in an MMPR for second and fifth harmonics on transformer inrush currents.

The relay compares the operating quantity to the restraint quantity. An 87R trip is actuated if the operating quantity is less than the restraint quantity. The relay also calculates the second and fifth harmonic content of the differential current. It compares the user settable percentage with the operating quantity. If the harmonic percentage is greater than the operating quantity, the harmonic blocking element asserts to block the percentage restraint differential element. The unrestraint element compares the operating quantity against a settable threshold; and 87U element is asserted if operating quantity is bigger than the unrestrained element threshold. Some typical settings are as follows:

87P restraint element pickup = 0.50 of per unit tap setting
Slope 1 = 25%
Slope 2 = 50%
Second harmonic blocking = 15%
Fifth harmonic blocking = 8%

12.15 Sensitive Ground Fault Differential Protection

Figure 12.30 shows a sensitive ground fault differential protection for transformer/generator, device 87GN. [This is also called restricted earth fault (REF) protection.] The lowest setting on electromagnetic-type relays is 16 A. Consider that the transformer is grounded through a 100 A resistor. With a sensitivity of 16 A, approximately 84% of the winding from the line end is protected, assuming a linear distribution of voltage from the line-to-neutral terminals. For the ground fault toward the neutral end, neither the 51G nor 87GN will operate for 16 A sensitivity. If the transformer is low-resistance grounded through 400 A resistor, approximately 96% of the winding from the line-to-neutral end will be protected, which is a common practice. Thus, reducing the ground fault current compromises some protection. If the fault does occur on the unprotected windings, it may persist for a long time and may cause core damage unless cleared by other devices like fault-pressure relays. However, this winding section can be protected by special means as described in Reference [11].

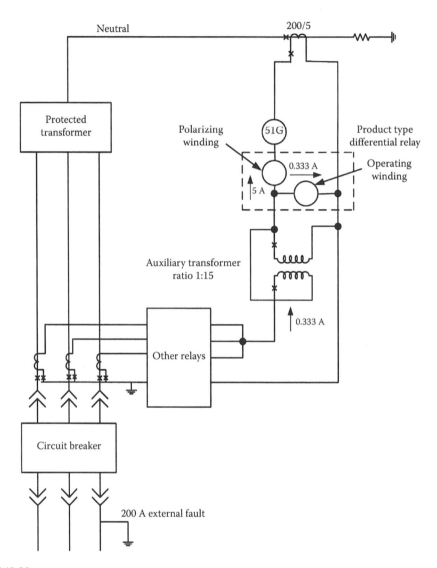

FIGURE 12.30
Sensitive restricted ground fault differential protection using a product type of electromechanical relay. In an MMPR ratio matching auxiliary CTs are not required and sensitivity of pickup is much lower.

Figure 12.30 shows the stability on external ground fault of 200 A, the wye-connected transformer neutral is grounded through 200 A. The neutral CT ratio 200:5 is matched with 3000:5 phase CTs through an auxiliary CT of 1:15. Note the distribution of the ground currents. This circuit of a differential relay called a "product type" of relay. The 0.33 A current is in the reverse direction of the operating current in the operating winding and gives some amount of stability for external faults. For an external fault the two currents in the differential relay coils are 180° out of phase and the relay has no operating torque. Note that in modern MMPRs, the ratio matching CT can be eliminated which is a source of errors and nuisance trips have occurred due to inappropriate accuracy selection of this CT, see Chapter 3. The line-end CTs must be connected in wye configuration to allow comparison of zero-sequence currents.

In an MMPR, the principle of operation is much similar. A directional element is used that compares the direction of the operating current derived from line-end CTs, and polarizing current obtained from neutral CT. A zero-sequence current threshold and positive-sequence restraint supervise tripping. The directional element asserts if the magnitude of neutral CT secondary current is greater than a pickup setting. A blocking function asserts if any of the winding residual currents used in REF function are less than a positive-sequence restraint factor. This safeguards against CT saturation during heavy three-phase faults.

The operating time is small of the order of 1.5 cycles or so, and the sensitivity of pickup is much better compared to electromagnetic type of relay described above. The residual current sensitivity threshold can be set at 0.05 times the CT secondary current = 0.25 A. Further security constraints are applicable, One, the minimum pickup should be set no less than:

$$0.05 \times 5 \times \frac{\text{CTR}_{max}}{\text{CTR}_{neutral}} \tag{12.7}$$

Also load unbalance should be considered. For example, with 10% load unbalance, the setting $>0.1 \times 5 = 0.5$ A.

12.15.1 Protection of Zigzag Grounding Transformer

The grounding transformers are either wye–delta or zigzag transformers. The theories of zigzag transformer and flow of ground fault currents for an external fault are shown in Volume 1, and not repeated here.

For overcurrent protection, the CTs must be connected in delta to block the zero-sequence currents. The turn-to-turn fault protection is provided by sudden-pressure or fault-pressure relay though the protection provided may be marginal, as the fault may be limited by the magnetizing impedance of the unfaulted phase. A grounding transformer may be selected for short-time ground fault withstand like 10 s or withstand continuously its ground fault current rating. In any case, coordination with the system ground fault relays is required, and if the fault is not cleared within a short time, the grounding transformer can be protected with a 51N device as shown in Figure 12.31a. The phase overcurrent relay will pick up for internal overcurrent faults. A better protection for internal faults is ground differential relay as illustrated in Figure 12.31b. This scheme can also be applied to wye–delta grounding transformers. The positive- and negative-sequence currents cannot flow, as these are 120° out of phase. See Volume 1 for further discussions.

12.16 Protection of Parallel Running Transformers

Figure 12.32 shows a scheme of parallel running delta–wye transformers. The fundamental aspects of parallel running transformers and load sharing are discussed in Appendix C of Volume 1 and are not repeated here. The parallel running transformers raise the short-circuit level on the secondary buses interconnected through a normally closed bus section breaker. Protection should ensure that only the fault bus or transformer is taken out of service keeping the loads partially powered. Figure 12.32 shows transformer primary phase

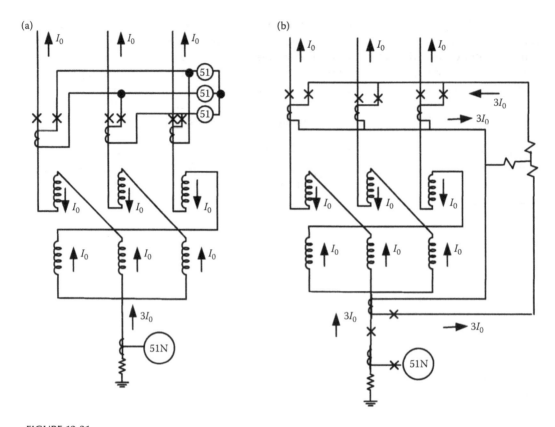

FIGURE 12.31
(a) Protection of a zigzag grounding transformer, the phase CTs must be connected in delta. (b) A more sensitive ground fault protection using differential principle.

overcurrent and ground fault protection, phase differential and ground fault differential protections. The connection of 51 device connected to circuit breakers 51-2 and 52T circuits forms a partial differential scheme, see Chapter 4. On operation of transformer phase and ground fault differentials (in one of the sources), the bus section remains closed and there is no load interruption. A bus fault is cleared by 51 device (a full bus differential scheme is normally adopted), which trips bus section breaker 51T and also 51-2, resulting in 50% interruption of the load. The loads on the healthy bus continue to be fed. 87G must trip bus section breaker 52T and also secondary breaker 52-2. Sometimes another 51G device is added to first trip bus section breaker 52T.

Now consider a system fault on the bus marked source 1 in this figure. An upstream device will operate, but the fault continues to be fed from source two through the parallel running transformers. Phase directional overcurrent protection will operate and trip breaker 52-2. The ground directional function 67N is not so much required and acts as a backup to 87TG.

If the transformers are protected with primary fuses and not the circuit breakers the protection shown can be modified. It is possible to remove phase differential and ground differential protection and relay on 67 and 67N directional phase and overcurrent relays for selective tripping. Alternatively, the differential protections can be retained and intertripping of an upstream beaker will be required.

Note that all the protective functions shown in this drawing are available in a single MMPR transformer protection relay.

Transformer Reactor and Shunt Capacitor Bank Protection

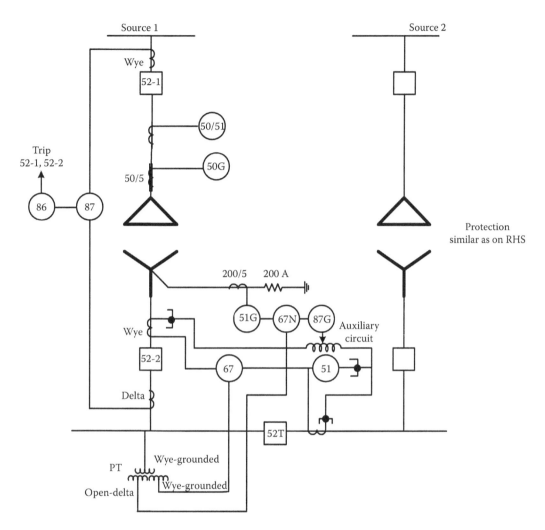

FIGURE 12.32
Protective relaying for parallel running transformer with bus section breaker closed.

12.17 Volts per Hertz Protection

Overexcitation is of concern for direct-connected generator transformers. See Chapter 11 for protection of unit-connected transformers and UAT. ANSI/IEEE standard C57.12.00 requires that the transformers should be able to operate continuously at 10% above rated secondary voltage at no load without exceeding the limiting temperature rise. The requirement applies at any tap at rated frequency.

Though some guidelines are provided in the literature, the volts per hertz withstand curve should be obtained from the manufacturer. With generator UT, the generator and transformer curves are plotted together for proper protection.

Figure 12.33 shows a two-step protection curve superimposed on the transformer overexcitation curve. Time delay inverse characteristics are also available for coordination, see Chapter 11.

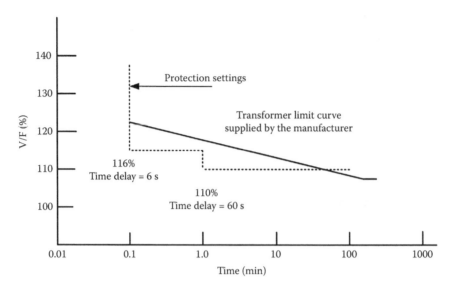

FIGURE 12.33
Volts per hertz protection of transformers.

12.18 Shunt Reactor Protection

The shunt reactors may be dry type or oil immersed. Dry-type reactors are generally limited up to 34.5 kV and are applied to the tertiary of a transformer which is connected to the transmission line being compensated. See Volume 2 for series and shunt compensation of transmission lines, and Volume 1 for the switching transients of reactors.

In the air core reactors, as there is no housing or shielding, a high intensity magnetic field is produced, and appropriate magnetic clearances are required. Close magnetic loops are to be avoided and shielding may be required in some cases.

The iron-core oil-immersed reactors may be coreless type or gapped-core type. On de-energizing the low-frequency currents persist for a long time depending upon time constants, determined by reactance and line capacitance. The gapped iron-core reactor is subject to more severe energizing inrush currents than the coreless type. Most coreless reactors have a magnetic shield to confine the magnetic field within the reactor tank. In the gapped iron-core reactor, the construction is similar to the power transformer except that gaps are introduced in the core to improve the linearity of inductance and reduce residual flux.

Dry-type reactors are generally connected to the tertiary delta of a transformer bank. See Figure 12.34.

A grounding transformer is used to provide a limited amount of grounding current. The grounding transformer and resistor are usually sized for a continuous zero-sequence current flowing through tertiary circuit capacitance to ground underground fault conditions. The types of faults are as follows:

- Phase-to-phase faults on tertiary bus
- Phase-to-ground faults on tertiary bus
- Turn-to-turn faults

Protection for phase-to-phase faults are overcurrent relays, negative-sequence current relays, and differential relays. Figure 12.34 shows the phase and negative-sequence current protection. See Figure 12.35 for differential protection of an autotransformer. In Figure 12.35 the reactor is a part of autotransformer tertiary.

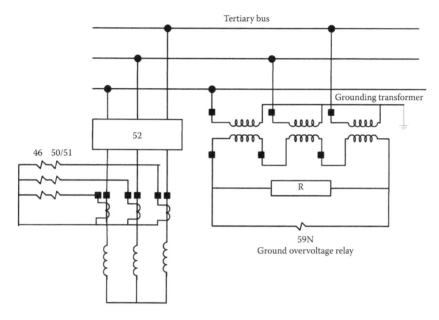

FIGURE 12.34
Typical dry-type shunt reactor connections with grounding transformer, overvoltage ground relay, and phase relays.

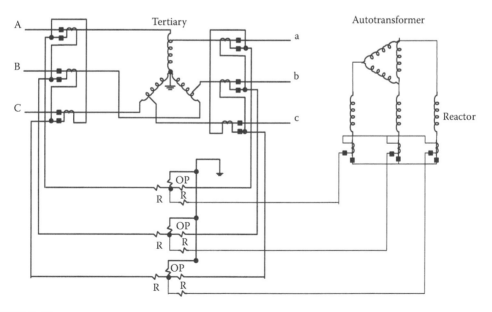

FIGURE 12.35
Differential relay protection for dry-type reactors. Reactor is a part of zone of autotransformer tertiary.

The overvoltage ground fault protection is also shown in Figure 12.34. Note that the relay has a third harmonic filter. Generally, the situation is alarmed.

Interturn faults are difficult to protect. The current and voltage changes for an interturn fault are the same as expected in normal service, and the voltage unbalance scheme illustrated in Figure 12.37 has come into use. The voltage between the reactor bank and ground can be due to the following:

Reactor bank unbalance due to a faulted reactor

Unbalance due to manufacturing tolerances

Unbalance of tertiary voltage with respect to ground

The scheme shown in Figure 12.36, with summing amplifier output, is representative of the fault in the reactor, and can distinguish the turn-to-turn fault and other sources of unbalance. The 59N relay shown in Figure 12.36 can provide only an alarm and the reactors can be kept in service. It should be remembered that during a ground fault the neutral voltage signal and the grounding transformer broken delta signals will have high levels. These signals must cancel in the summing amplifier. The voltage used to supply the phase-shifting network can be affected by tertiary bus ground fault. Then the compensation for unbalance may be changed in magnitude and phase angle leading to a false trip. This can be avoided using phase-to-phase rather than a phase-to-ground voltage as the phase-shifting network.

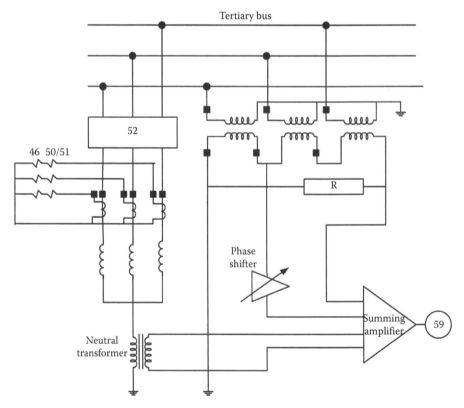

FIGURE 12.36
Voltage balance relay protection of dry-type reactors.

12.18.1 Oil-Immersed Reactors

Oil-immersed reactors are connected to one end or both ends of a transmission line and are wye connected with a solidly grounded neutral. These may be switched or remain permanently connected, see Volume 1. Oil-immersed reactors can also be connected to a substation bus. The typical tripping scheme applied is to trip local line breaker and transfer trip the remote line breaker. A dual channel is generally used for security, and both breakers are locked out for a reactor fault.

The distributed capacitance of the line can form a resonant circuit with the reactor, the resonance frequency being close to 60 Hz. Moreover, two transmission lines, one de-energized and the other energized with reactors can have mutual couplings due to proximity, and it is possible to impress higher voltage across de-energized reactor. This problem can be obviated by means of a dedicated switching device at the same time when de-energizing the line.

One major consideration for relay protection is that false operation of the relays can occur, due to DC offset and low-frequency components of the reactor energizing and de-energizing currents. High-impedance differential relays are recommended compared to low-impedance differential relays [15]. Figure 12.37 shows biased differential protection for oil-immersed reactors.

Figure 12.38 shows distance relay protection. Distance relays have been applied to detect shorted turns in iron-core reactors. This is so because under turn-to-turn fault there is a significant reduction in the 60 Hz impedance of the reactor. The sensitivity is limited by the apparent impedance seen by the relay during inrush period when the reactor is energized. The distance relay reach must be set below the reduced impedance on inrush currents and also that the relay will not operate incorrectly on natural frequency oscillations which occur when a compensated line is de-energized. Note that differential relay schemes cannot detect such faults. Sudden pressure relay can provide sensitive means to detect these faults.

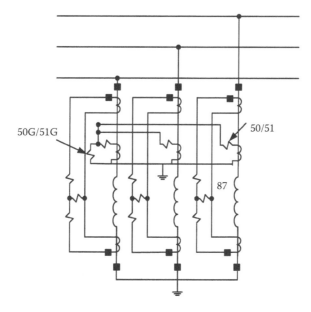

FIGURE 12.37
Differential protection with restraint and also overcurrent and ground fault protection through separate CTs.

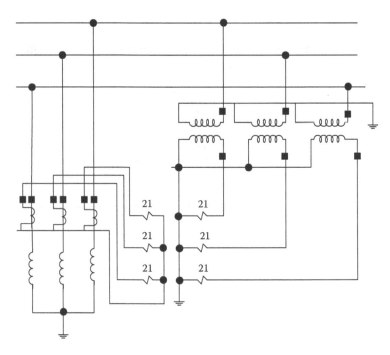

FIGURE 12.38
Distance relay protection, dry-type reactors.

In the applications for EHV lines and buses, the switching equipment usually consists of single-pole devices which are not mechanically linked and each pole has an independent operator. With such an arrangement, a pole disagreement can occur and all the poles may not close coincidentally. This may create imbalance in voltages or a faulty reactor may not be removed from service. A pole disagreement relay is shown in Figure 12.39, designed to compare the currents in each reactor connected to the transmission system. A spare reactor is provided which can be switched to replace any of the normal phase reactors. Two trip outputs with separately adjustable time delays are provided. The shorter delay trips the reactor switches in the event of current disagreement, and the longer delay trips local and remote line breakers if the first trip fails to clear the pole disagreement condition. Also see References [16,17].

12.19 Transformer Enclosures

While relay protection discussed above is important, equally important are the accessories and external devices that are provided on the transformers depending upon their ratings, and constructional features. Transformer enclosures depend on their ratings and also industrial practices in a country. The water ingress is very detrimental to the life of the transformer. Figure 12.40 shows breakdown voltage kV rms with respect to the water content in ppm. The limits of acceptable water content and other dissolved gasses are provided in References [18].

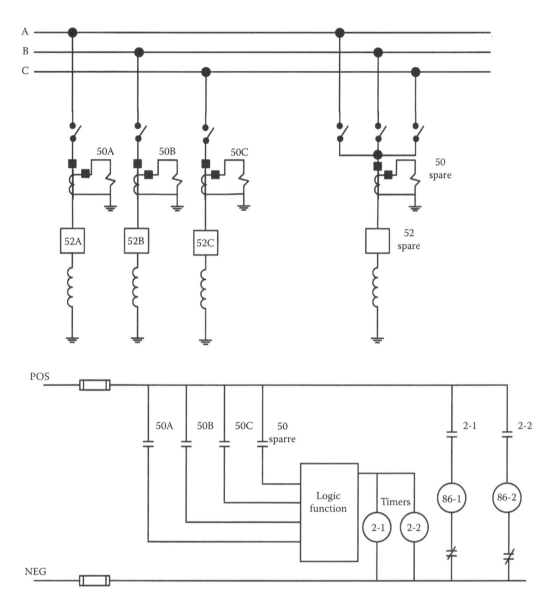

FIGURE 12.39
Pole disagreement protection for three-phase reactor installation with switchable spare reactor.

12.19.1 Dry-Type and Cast Coil Transformers

The dry-type transformers may be natural cooled, located inside or outside in totally enclosed nonventilated enclosures. Alternatively, forced ventilation may be provided and intake filters may be installed depending upon the environment. A temperature monitoring system is a must, with probes installed in the transformer winding ducts, which can get clogged due to contamination. The two or more stages of contacts can be set for ventilation fans, temperature alarm and trip. Generally the transformers in dry-type construction are available up to 34.5 kV and a rating of 10 MVA.

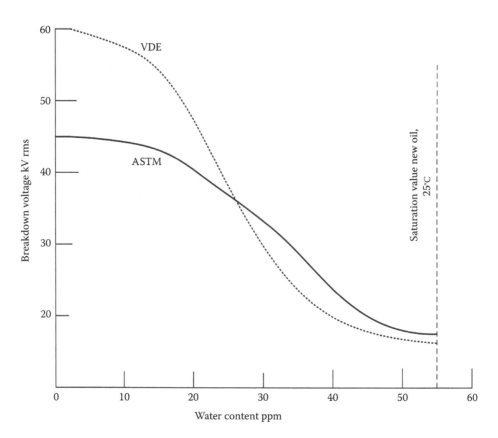

FIGURE 12.40
Breakdown voltage of transformer oil with respect to water content in ppm.

12.19.2 Liquid-Filled Transformers

12.19.2.1 Sealed Tank Construction

This design is commonly used for transformer up to 230 kV and ratings extending to 100 MVA. The transformer tank is sealed isolating it from the outside atmosphere. A gas space equal to about one-tenth of the liquid volume is maintained to allow for thermal expansion. This space is generally nitrogen filled.

A pressure–vacuum gauge and bleeder device are furnished to monitor the internal pressure or vacuum and to relieve any static pressure build up that could damage the enclosure and operate the pressure-relief device.

The tank is designed to withstand full vacuum at 15 psi.

12.19.2.2 Positive Pressure Inert Gas

A variation to the sealed tank design is the addition of a nitrogen gas pressurizing assembly. It provides a slight positive pressure in the supply line to prevent air from entering the transformer during operating mode and temperature changes. Transformers above 7.5 MVA can be equipped with this system.

12.19.2.3 Conservator Tank Design

This is diagrammatically shown in Figure 12.41. It has a conservator, a second oil tank above the main oil tank. The interconnecting pipe has a relay called a Buchholz relay. It was first developed by Max Buchholz in 1921 and has been in use since the 1940s. On slow accumulation of gas, due to say overloads, the gas produced by decomposition of insulating oil accumulates in the top of the relay and forces the oil level down. A float switch in the relay is used to initiate an alarm signal.

On an internal arcing fault or turn-to-turn faults which are difficult to detect by protective relaying, the gas accumulation is rapid and the oil flows rapidly into the conservator. The flow of oil operates a switch attached to a vane located in the path of the moving oil. This is wired to trip an upstream circuit breaker to isolate the transformer. The relay has a test port, and the accumulated gas can be withdrawn for testing. Flammable gas indicates internal faults such as arcing and overheating, while the air found may indicate a low-level leak.

A transformer is subject to temperature variations, not only because of varying loads on the transformer but also the variations in the ambient temperature. As the volume in the conservator tank changes, the air on the top of the oil will be breathed in and out of the transformer. This can cause ingress of moisture and dust particles, damaging the transformer operation. Desiccators are provided. The earlier desiccator had silica gel which will change color from blue to pinkish blue when it has absorbed moisture. It requires reheating to remove moisture and refilling. Modern desiccators may use long-life electronic devices to trap moisture and containments.

As a further development, the breathing can be into a plastic inflatable balloon floating in the conservator tank.

European practice is more in favor of conservator type of oil preservation system. Larger transformers say above 100 MVA or so are conservator type.

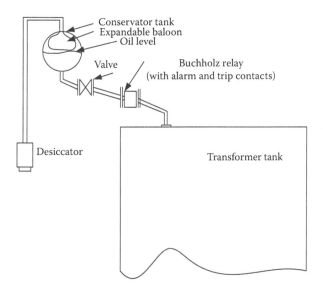

FIGURE 12.41
A transformer construction with conservator tank and Buchholz relay.

12.20 Transformer Accessories

12.20.1 Pressure–Vacuum Gauge and Bleeder Valve

It is specific for the sealed tank designs and indicated the difference between the transformer internal gas pressure and atmospheric pressure. The pressure in the gas space depends on thermal expansion and contraction. Large positive or negative pressure indicates an abnormal condition, and limit switches may be provided to detect excessive vacuum or positive pressure, which could cause tank rupture or deformation. These can be wired for alarm. Usually the pressure–vacuum gauge and a pressure–vacuum bleeder valve are mounted together. It automatically adjusts to control the internal vacuum or positive pressure in excess of 5 psig. It also prevents operation of pressure-relief device in response to slowly increasing pressure due to overloads or ambient conditions. Through a hose connection to bleeder valve, the transformer can be connected to an external source of gas pressure, and leaks can be checked.

12.20.2 Liquid Level Gauge

Liquid level gauge is a standard accessory even for a small liquid-filled transformer. It measures the level of the insulating liquid in the transformer tank with respect to the predetermined level usually at 25°C. A low level indicates the loss of insulting liquid, and alarm contacts are provided for remote indication and sometimes trip.

12.20.3 Pressure-Relief Device

This is a standard device on all liquid immersed transformers, except transformers of only some kVA ratings. It can relieve minor and serious internal pressures, though the main purpose is to prevent tank rupture due to excessive pressure. When the pressure exceeds the tripping pressure, the device snaps open, allowing excess gas or fluid to escape; the device is normally automatic-resetting and self-sealing and requires parasitically no adjustments or maintenance. Alarm contacts for remote alarm can be provided.

12.20.4 Rapid Pressure Rise Relay

The pressure rise relay is an important device to isolate the sealed type of transformers from the electrical system when there is an internal fault such as winding-to-winding fault, shorted turns, and ground faults. The bubbles of gas formed in the insulating fluid create abrupt pressure waves that activate the relay. After the transformer is taken out of service, it must be thoroughly checked for extent of damage due to faults.

The relay is mounted below the transformer oil level. The pressure built-up is transmitted to the relay bellows, which transmits it to a piston that actuates the relay contacts. Nuisance trips have been observed, sometimes. It seems that the relay can operate due to vibrations. The actuation time of the contacts is small of the order of four cycles. Figure 12.42 shows an auxiliary FPX switchboard type of relay with hand reset lockout relay to obviate this problem. Its coil is shorted by a normally closed contact of the transformer pressure relay, FP, to prevent the relay operation for a transient surge that could flash over the normally closed contacts of the pressure relay.

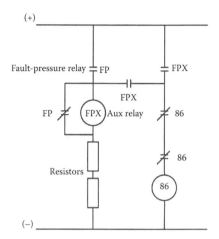

FIGURE 12.42
Schematic of a switchboard-type fault-pressure relay.

12.20.5 Liquid Temperature Indicator

Liquid temperature indicator measures the temperature of the insulating liquid at the top of the transformer. When winding temperature indicator is not provided, the contacts on the liquid temperature indicator also control the cooling fans. As the hottest liquid is less dense, the indicator can be supposed to measure the temperature of the hottest liquid. However, there can be a considerable difference between the winding hot spot temperature and the liquid temperature due to considerably large time constant. Thus, the temperature warning may be too optimistic or too pessimistic depending upon the loading cycle of the transformer.

12.20.6 Winding Temperature Indicator (Thermal Relays)

To replicate the winding temperature, a current transformer mounted on one of the three phases of the transformer bushings supplies current to the thermometer bulb heater coil (Figure 12.43). This simulated the transformer hot spot temperature.

The temperature indicator is a bourdon gauge connected through a capillary tube to the thermometer bulb. Coupled to the shaft of the gauge are cams that operate individual switches. Four (4) contacts are provided, two for the cooling fan control, one for alarm and the other for trip.

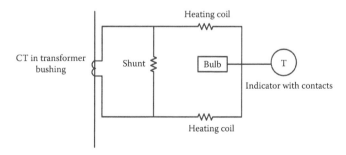

FIGURE 12.43
Schematic of a winding temperature indicator.

12.20.7 Combustible Gas Relay

A combustible gas relay will continuously monitor the combustible gas coming from the transformer. Combustible gases can be formed inside the transformer due to low-level faults or corona. These faults are not detected till these escalate into more damaging ones. The relay can be used on transformers with positive pressure inert gas preservation system. The sample of the gas is passed over a heated wire, and if the comestible gases are present these will add more heat, and change the resistance, which can be detected by a bridge network. It is rather an expensive device provided on larger substation transformers. Note that portable gas detection equipment can be used. Figure 12.44 shows a schematic diagram. Also see Reference [19].

12.20.8 Underload Tap Changing Equipment

ULTC equipment has a variety of constructional features not discussed. A reactor-type ULTC with vacuum switching is shown in Figure 12.45. The tap changer employs the following mechanisms: (1) reversing switches, which selectively raise or lower connection of the tapped winding connection, (2) selector switches, which connect to the proper tap from the winding, (3) vacuum interrupters, which perform the arcing duty of the switching operations, (4) by-pass switches, which operate in conjunction with vacuum interrupters to shunt current around the vacuum interrupter during a tap change when they select which leg of the four-preventive autotransformer will have its current interrupted. Figure 12.45a shows the schematic diagram; the tapped section of the winding is shown between terminals 3 and 2 with taps 4 through 11 connected to stationary contacts of the selector switches. Taps 3 and 2 are connected to reversing switch stationary contacts. Figure 12.45b depicts a sequence of tap change in steps.

Due to contacts which can degrade the main transformer oil, the ULTC equipment is mounted in a separate tank. It generally provides ±16 steps, with a voltage adjustment of ±10%. The tap changer is motor operated and has a number of mechanical parts such as springs, rollers, and levers. The tank is provided with the following:

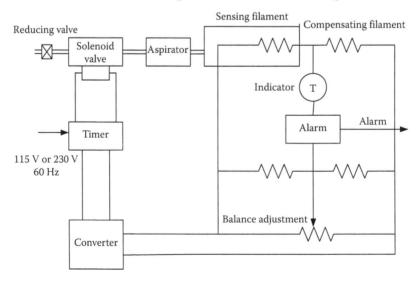

FIGURE 12.44
Schematic of a gas detector relay.

Transformer Reactor and Shunt Capacitor Bank Protection

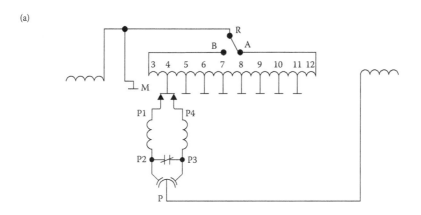

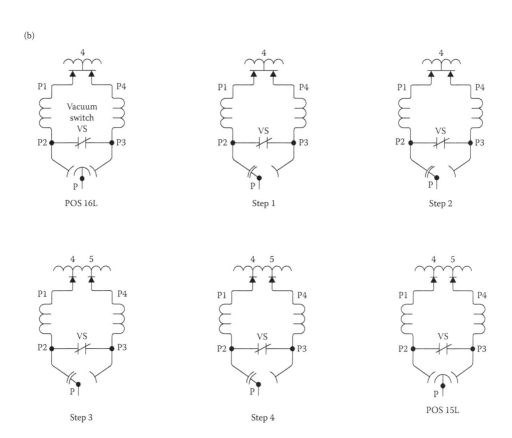

FIGURE 12.45
(a) Schematic diagram of an underload tap changing equipment and (b) sequence operation of tap changing.

- Local tap position indicator
- Liquid level gauge
- Oil compartment vent plug
- Automatic resetting pressure-relief device with alarm contacts
- A separate dehydrating breather

12.20.9 Surge Protection

Surge protection is not discussed in this book. Primary surge protection is provided even for small transformers of 1 MVA, the surge-rated voltage is selected according to system grounding. The surge arresters are located as close as possible to the transformer primary windings. For high-voltage transformers, it is usual to select transformer winding BIL one level below that of substation insulators and buses and protect the transformers with primary surge arresters. Secondary surge protection is needed for large transformation ratios as the surges can be transferred through transformer windings through capacitive and inductive couplings, and these can easily exceed the secondary (low-voltage) windings BIL, see Reference [20].

Figure 12.46 shows overall alarm and protection scheme for a transformer.

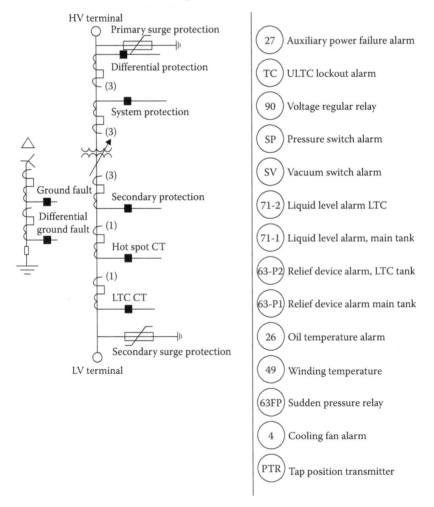

FIGURE 12.46
Overall schematic of transformer protection with underload tap changing and surge protection.

12.21 Shunt Capacitor Bank Protection

We have discussed formation of shunt capacitor banks, their fuse protection, and neutral unbalance protection in Volume 3, and not repeated here.

References

1. NEC. National Electric Code, NFPA 70, 2017.
2. FM Global, Property Loss Prevention Data Sheets, 5–4, January 1997.
3. IEEE C57.109, IEEE Guide for Liquid Immersed Transformer Through-Fault Current Duration, 1993 (R2008).
4. IEEE Standard. C.57.91 (Also Cor.1-2002). IEEE Guide for Loading Mineral-Oil Immersed Transformers, 1995.
5. IEEE. C57.12.59. Guide for Dry-Type Transformer Through-fault Current Duration, 2015.
6. ANSI Standard C37.46, American National Standard for High-Voltage Expulsion and Current Limiting Type Power Class Fuses and Fuse Disconnect Switches, 2010.
7. RH Hopkinson. Ferroresonance during single phase switching of three-phase distribution transformer banks. *IEEE Trans*, PAS-84, 289–293, 1965.
8. DR Smith, SR Swanson, JD Borst. Overvoltages with remotely switched cable fed grounded wye-wye transformers. *IEEE Trans PAS*, PAS-94, 1843–1853, 1975.
9. JC Das. Overcurrent relays versus current limiting power fuses for transformer primary protection, *Conference Record, 48th Annual Georgia Tech Protective Relaying Conference*, Atlanta, 1994.
10. IEEE. C57.12. General Requirements for Liquid Immersed Distribution, Power and Regulating Transformers, 2000.
11. IEEE C37.91. IEEE Guide for Protective Relay Applications to Power Transformers.
12. ANSI/IEEE Std. 242. IEEE Recommended Practice for Protection and Coordination of Industrial and Commercial Power Systems, 1986.
13. IEEE Committee Report, Protection of power transformers. *IEEE Trans PAS*, PAS80, 1040–1044, 1963.
14. SEL Inc. Instruction manual for 387E.
15. ANSI/IEEE C37.109. IEEE Guide for Protection of Shunt Reactors, 2006.
16. JW Cooper, LW Eiltes. Relay for ungrounded shunt reactors. *IEEE Trans PAS*, PAS-92, 116–121, 1973.
17. JJ LaForesst et al. Resonant voltages on reactor compensated extra-high-voltage lines. *IEEE Trans PAS*, PAS-91, 2528–2536, 1972.
18. IEEE Standard C57.104. IEEE Guide for Interpretation of Gases Generated in Oil-immersed Transformers, 2008.
19. PS Pugh, HH Wagner. The detection of incipient faults on transformers by gas analysis. *AIEE Trans PAS*, 80(pt. III), 189–195, 1961.
20. JC Das. Surge transference through transformers. *IEEE Ind Mag*, 9(5), 24–32, 2003.

13
Protection of Lines

Chapter 2, Section 2.8 for transmission systems and Section 2.10 for distribution systems, in Volume 1 may be seen. These discuss radial, loop, parallel, and grid systems, and subtransmission and transmission system voltages. Maintaining acceptable voltage in the power systems according to ANSI C84.1, with respect to their nominal voltages, is discussed in Table 8.1 (Section 8.1, Chapter 2, Volume 2). The calculations of transmission line parameters are discussed in Appendix A, Volume 2. The load flow over transmission lines and their models are discussed in Chapter 3, Volume 2. These concepts are important to proceed with line protection of transmission and distribution lines and are not repeated here.

13.1 Distribution Lines

As discussed in Chapter 2, Volume 2, the distribution systems vary, the lines serving the loads may operate at 4–34 kV, 13.2 kV being more popular, and up to 34.5 kV has been used for areas of high load density. In a radial line feeder, the short-circuit current is supplied from only one end, the effect of some induction motors loads at the consumer's premises can be ignored, while in loop or network configuration the short-circuit currents can be supplied from both ends. Generally, the lines are connected to substations, which are solidly grounded in multigrounded system configurations. See Chapter 7 for multiple grounded systems. Thus, the terminals must be tripped for both ground and phase faults. The protection of distribution lines should consider the following:

- Radial lines have the limitation that a fault at the source will result in a wide area of interruption and the time to repair and bring the service back on line is not acceptable.
- The area shutdown for a fault condition must be minimized by proper coordination of protective devices. Nuisance trips are not acceptable.
- Automatic restoration of power should occur. As approximately 80% of faults on overhead lines are of temporary nature, autoreclosers and sectionalizers are used.

A description of fuses, automatic circuit reclosers, and sectionalizers is provided in Chapter 5. The spot networks are discussed in Volume 1.

13.2 Transmission and Subtransmission Lines

The transmission lines are at voltages of 69–800 kV, and the subtransmission lines are at voltages from 12.47 to 138 kV; see Chapter 2 of Volume 1.

Transmission lines do not directly serve loads to the consumers, but there are exceptions. The largest utility substation in the state of Georgia, USA, serves a newsprint paper mill from 230 kV transmission line. Transmission lines have relayed circuit breaker, and so also subtransmission lines. It is desirable that all transmission line faults are cleared instantaneously. The subtransmission systems and transmission systems are solidly grounded and the line-to-ground fault current is high, sometimes approaching close to the three-phase fault currents. Again the criteria for protection are as follows:

- Transient stability is important, see Chapter 15. Even a reduction of fault-clearing time by a couple of cycles can make a profound difference in the stability or instability.
- The basic principles of protective relaying are applicable without exceptions; for example, the selectivity—the faulty section only should be removed. The power to the area that was shutdown should be quickly restored (see Chapter 2, Volume 2).
- Autoclosing is used for transmission lines as well, much like distribution lines, as the fault may be of a temporary nature.
- The subtransmission lines serve consumers at various voltage levels, depending on the load demand. The concepts of protective relaying are no different.

13.3 Protective Relays

The protection devices consist of the following:

- Nondirectional instantaneous phase and ground fault relays
- Directional instantaneous and phase fault and ground relays
- Nondirectional time delay overcurrent phase fault and ground relays
- Directional time delay overcurrent phase and ground fault relays
- Current balance relays
- Directional distance—instantaneous, or inverse time relays
- Pilot relaying systems, see Chapter 14

Figure 13.1a shows the protection zones and time coordination for typical loop lines. For radial lines fed from one end, the current coordination is required for only one direction. Figure 13.1b shows the coordination with directional overcurrent relays, line fed from either ends or Figure 13.1c with directional distance relays. CTI is defined in Chapter 6; here it is defined as the time interval between the protection devices at near station and the protection device at the remote station or the remote faults that the near-station device overreaches. Therefore, for the remote fault, the near-station device operating times must not be less than the sum of the remote operating time and CTI. Faults at the remote line should be cleared by the remote protection device and backed up by the near-station devices. The considerations for CTI are discussed in Chapter 6, the breaker interrupting or fault-clearing time, relay over travel (for electromechanical relays), safety margin, etc.

Protection of Lines

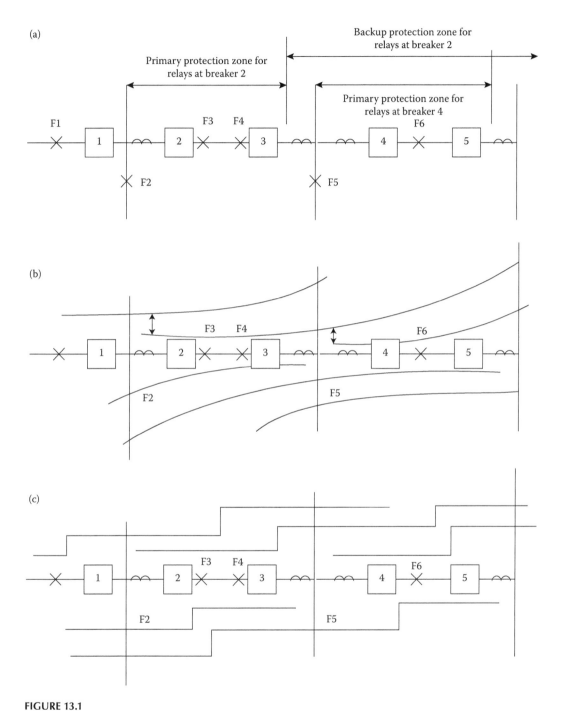

FIGURE 13.1
(a) Loop-type line, (b) coordination with inverse directional overcurrent relays, (c) coordination with distance relays.

Example 13.1

Coordination in a radial distribution feeder is investigated. Consider a system configuration as shown in Figure 13.2. This shows three line sections at 12.47 kV, each of 3 miles long, and each section serves a load of 3 MVA at a power factor of 0.85 lagging. The total feeder length is 9 miles (14.4 km). The radial system is served from a 10 MVA, 34.5–12.47 kV transformer. The short-circuit levels at the 34.5 kV source are shown in this figure. The impedance data of the system configuration are shown in Table 13.1.

As the first step, a load flow study is carried out (see Volume 2 for details). As the lines have considerable impedance, shunt capacitor banks of 2.3 Mvar (two) and 2.875 Mvar (one) are required at locations shown in Figure 13.3. With the application of shunt capacitor banks, the load flow picture is depicted in Figure 13.3. The voltages are at acceptable levels. In addition to the capacitor banks, the taps of the 10 MVA transformer, off-load type, are set to provide 2.5% secondary (12.47 kV) voltage boost at no load.

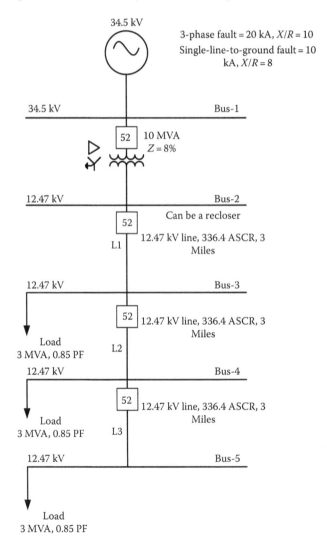

FIGURE 13.2
A radial distribution system for coordination study with three load buses.

TABLE 13.1

Impedance Data for Example 13.1

Equipment	Impedance Data
Utility Source	
Transformer, 34.5–12.47 kV, 10 MVA	$Z+ = Z- = 0.4783 + j7.986$ on 10 MVA base $Z_0 = 0.4783 + j7.986$ on 10 MVA base
12.47 kV line 1, 336.4 kcmil, ACSR, flat conductor formation, 3 miles	$Z+ = Z- = 0.278 + j0.6761$ Ω/mile $Xc = 0.1554$ MΩ/mile $Z_0 = 0.5642 + j2.887$ Ω/mile, $Xc_0 = 0.39294$ MΩ/mile
12.47 kV line 2, 336.4 kcmil, flat conductor formation, ACSR, 3 miles	$Z+ = Z- = 0.278 + j0.6761$ Ω/mile $Xc = 0.1554$ MΩ/mile $Z_0 = 0.5642 + j2.887$ Ω/mile, $Xc_0 = 0.39294$ MΩ/mile
12.47 kV line 3, 336.4 kcmil, flat conductor formation, ACSR, 3 miles	$Z+ = Z- = 0.278 + j0.6761$ Ω/mile $Xc = 0.1554$ MΩ/mile $Z_0 = 0.5642 + j2.887$ Ω/mile, $Xc_0 = 0.39294$ MΩ/mile

The 10 MVA wye-connected 12.47 windings are solidly grounded, which is the normal system grounding practice for distribution, subtransmission, and transmission systems.

Next, conduct short-circuits calculations for three-phase and line-to-ground faults. The results are shown in Table 13.2. The short-circuit levels are low, which is normally the case for distribution circuits.

The location of the circuit breakers and protective relays R-1 through R-4 is shown in Figure 13.3.

The coordination using inverse time characteristic overcurrent relays is shown in Figure 13.4. The pickup setting is made to consider the short-time or transient overloads. It is common practice to select pickup at 150% of the load current. The coordination in Figure 13.4 is achieved by using extremely inverse characteristics. Starting from downstream, all the relays are properly coordinated on time–current basis, and each setting provides a backup to the upstream setting. Proper CTI is applied at the maximum available short-circuit current. If the short-circuit level is reduced, say due to changes in the system short-circuit level at 34.5 kV, the coordination throughput will be retained. The transformer is adequately protected with respect to its through-fault ANSI withstand characteristics.

As an option, rather than using inverse time characteristics, definite time characteristics with appropriate time delay can be used. This is shown in Figure 13.5. Note that there is adequate CTI between the settings. The pickup current values can be much closer as compared to coordination with extremely inverse time–current characteristics.

Finally, the ground fault coordination is shown in Figure 13.6. Here, while the settings on relays R-2, R-3, and R-4 should coordinate with each other, this coordination is not required for relay at R-1. This relay does not see the secondary ground fault current, due to primary windings of 10 MVA transformer connected in delta; which are a sink to the zero-sequence secondary currents. Generally, a neutral-connected CT and overcurrent ground fault relay will be provided for winding ground faults in the secondary windings of 10 MVA transformer, not shown. See Chapter 11 on transformer protection. The pickup setting for the ground fault current may not be more than approximately 15% of the load current—the sensitivity and the minimum pickup setting available on the relay is also a factor.

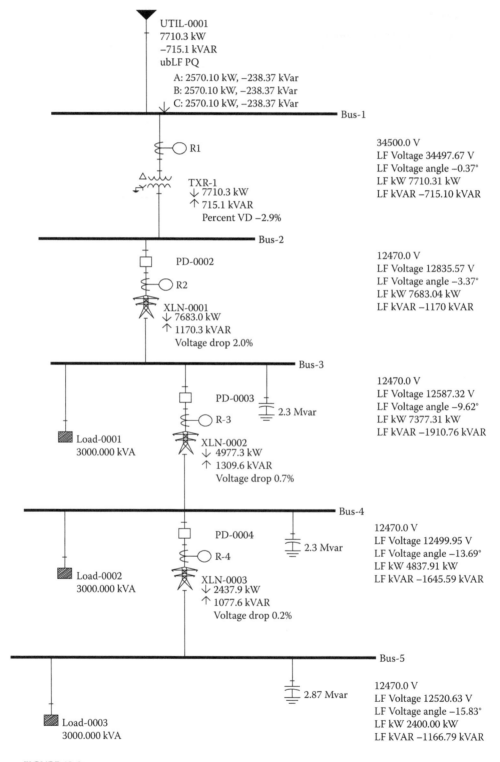

FIGURE 13.3
Computer calculations of the load flow in Figure 13.2.

TABLE 13.2

Short-Circuit Data, Example 13.1

Bus ID	Three-Phase Fault		Single-Line-to-Ground Fault	
	kA rms sym	X/R	kA rms sym	X/R
Utility Source, 34.5 kV	20 kA	10	10	8
Bus 2, 12.47 kV	5.240	16	5.410	16
Bus 3, 12.47 kV	2.044	4	1.264	5
Bus 4, 12.47 kV	1.262	3	0.714	4
Bus 5, 12.47 kV	0.912	3	0.498	4

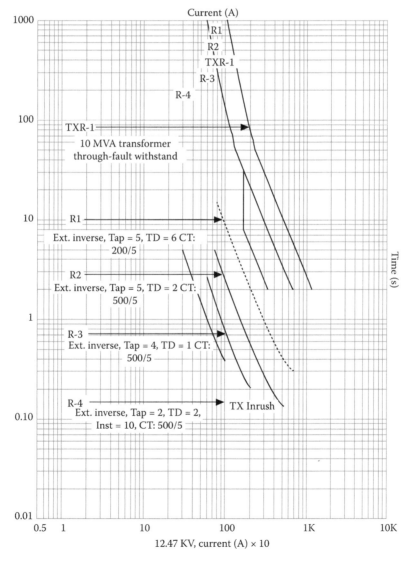

FIGURE 13.4
Coordination for phase faults in Figure 13.2, using inverse characteristic relays.

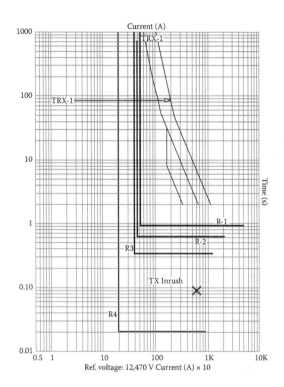

FIGURE 13.5
Coordination for phase faults in Figure 13.2, using definite time overcurrent relays.

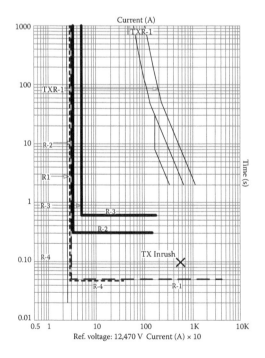

FIGURE 13.6
Coordination for ground faults in Figure 13.2.

13.3.1 Application of Directional Overcurrent Relays

Figure 13.7a shows parallel transmission lines interconnecting to systems with generation. Here for a fault in one of the lines, only the faulty line must be tripped from both ends, retaining the unfaulted line in service.

Figure 13.7b illustrates that simple overcurrent relays are not suitable for selective tripping. Consider a fault at F2 in line 1 (Figure 13.7a). This fault has three fault contributions: first from bus A, second from bus B, and the third from bus A through unfaulted line. Though

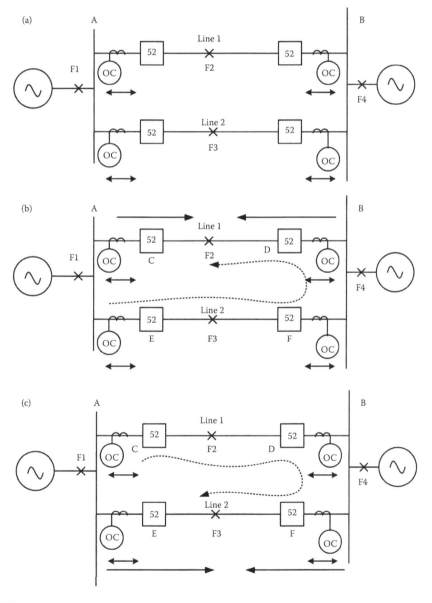

FIGURE 13.7
(a) Parallel transmission lines interconnecting two systems with generation, (b) and (c) simple overcurrent relays are inappropriate for line protection.

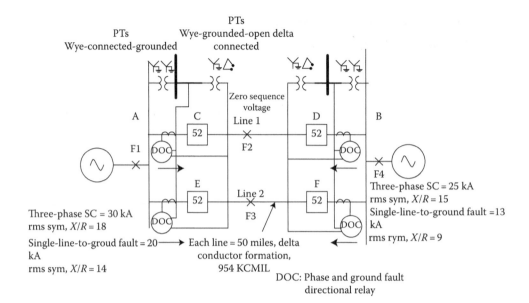

FIGURE 13.8
Application of directional overcurrent relays for protection of parallel lines.

the magnitudes of the fault current will vary depending upon the location of fault F2 in line 1, for the selective tripping, breakers C and D must trip faster than the breakers E and F.

Now consider fault at F3 on line 2 (Figure 13.7c). Then, the breakers E and F must trip faster than breakers C and D. This contradicts with the selective fault clearance requirement of line 1.

Furthermore, the relays at B, C, D, and F will be operative for faults at buses A and B; faults F1 and F4, jeopardizing the selectivity.

Directional relays at C, D, E, and F will obviate this problem, as shown in Figure 13.8. The direction of operation of the directional overcurrent relays is as shown. Also we need not consider the system faults at locations F1 or F2, because the directional relays will behave like unit protection to the protected lines. Note that wye–wye-grounded PTs are required with auxiliary wye-grounded and open-delta PT for polarizing the ground directional element of the relays.

Example 13.2

Consider the data in Figure 13.8. Each 138 kV line is 50 miles (80 km) long, and the source three-phase and single-line-to-ground fault currents are as shown in this figure. Normally, each line carries 300 A current, and in case one line is isolated due to a fault or any other reason, the line is loaded to 600 A.

The coordination with directional overcurrent relays for phase faults is illustrated in Figure 13.9. Consider a fault at 25 miles (middle of the line 1). First carry out the short-circuit calculations. Depending upon the location of fault in the line, the short-circuit currents supplied from the two ends will vary. See Table 13.3 for the relays which will be in operation and maximum and minimum fault currents supplied from the two ends of the line.

Note that at source 2 end, the settings on relays D and F are same, and at source 1 end, also the settings are same.

The same strategy is applicable to ground faults; the settings will be much lower than phase overcurrent protection minimum setting approximately 20% of the load current and the maximum to coordinate.

Protection of Lines

We should also consider faults at source buses A and B. These faults will not be cleared by the directional overcurrent phase and ground fault relays at C, D, E, and F. A fault, say on bus A, will remove the source 1, and if there are any taped loads on the lines, these will be supplied by the source at the other end. The fault on buses A and B should be cleared preferably by differential protection wrapped around each bus. Alternatively nondirectional overcurrent relays can be used. These must coordinate with the maximum settings of the directional overcurrent relays at either end of the transmission lines.

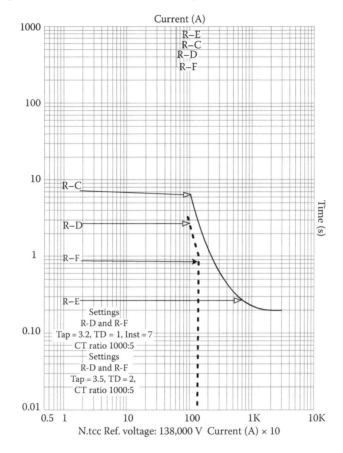

FIGURE 13.9
Coordination for phase faults in Figure 13.8, using directional overcurrent relays.

TABLE 13.3

Example 13.2 Short-Circuit Currents and the Operation of the Direction Relays

Short-Circuit Location	Relay C	Relay D	Relay E	Relay F
Line 1 is faulty	O	O	O	X
Short-circuit currents, max/min	30/1.9 kA	25/1.8 kA	1.9 kA	—
Line 2 is faulty	O	X	O	O
Short-circuit currents, max/min	1.9 kA	—	30/1.9 kA	25/1.8 kA

O, operate; X, does not operate due to direction of current flow.

13.3.2 Loop System with One Source of Fault Current

Figure 13.10 shows a loop system with one source of fault current. Any fault on bus A will interrupt the entire load; this is not desirable. Single-source loop systems are therefore not in much use. Note that at locations 1 and 10, the relays need not be directional overcurrent type, as there is only one source of power. All other relays shown at locations 2, 3, 4, 5, 6, 7, 8, and 9 are directional type. It is desired that fault in any section A–B, B–C, C–D, D–E, or E–A is selectively cleared, while the other sections remain in service. Loads are shown at buses B, C, D, and E. A tripping matrix is shown in Table 13.4; this describes which relays are nonoperative for fault in each section and which relays will be operative. Thus, coordination is required between all those relays which operate for fault in each section. Once the fault in a section is cleared, the loop is broken and the system returns to a radial system. With the overcurrent phase fault and ground fault directional relays, the time to clear a fault in a zone can be minimized.

Example 13.3

As an example of calculation, consider the single-source-fed loop system as shown in Figure 13.11. Each line section is of same configuration and length, which may not be practical situation. Similarly, the four loads shown are each 10 MVA at a power factor of 0.85. A load flow study is required to establish the voltage profile, and also the tap settings on the protective devices must ensure that the maximum loads are supported. Note that the load flow shows no load in the section C–D. This is so because each side is

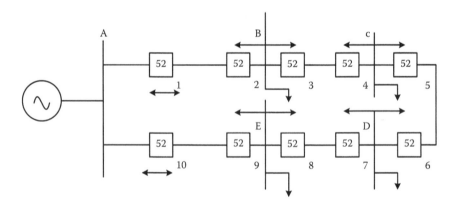

FIGURE 13.10
Loop system with one source of fault current.

TABLE 13.4

Operation of Relays for Selective Tripping

Fault in Section	Relay 1	Relay 2	Relay 3	Relay 4	Relay 5	Relay 6	Relay 7	Relay 8	Relay 9	Relay 10
A–B	O	O	X	O	X	O	X	O	X	O
B–C	O	X	O	O	X	O	X	O	X	O
C–D	O	X	O	O	X	O	X	O	X	O
D–E	O	X	O	X	O	X	O	O	X	O
E–A	O	X	O	X	O	X	O	X	O	O

O, operate; X, does not operate due to direction of current flow.

Protection of Lines

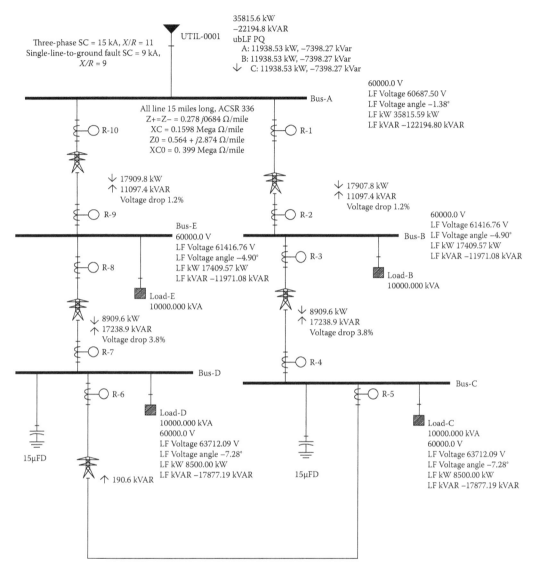

FIGURE 13.11
Load flow calculations in the loop system of Figure 13.10.

symmetrical and has the same impedance to the load flow. To maintain acceptable voltage profile on the buses, shunt capacitor banks at buses C and D are shown.

Next a computer-based short-circuit calculation is carried out to establish the maximum/minimum current flows for faults in each section under consideration. This calculation is shown in Table 13.5. For example, consider a fault in section A–B. If the fault is close to the circuit breaker A, the short-circuit current contributed by bus A is high, a maximum of 15 kA. As the fault location moves toward bus A, the short-circuit current supplied from bus A decreases due to line impedance.

A combination of instantaneous, time delay overcurrent only and time delay plus instantaneous settings provides the required coordination. The completed coordination is shown in Figure 13.12.

TABLE 13.5

Example 13.3 Maximum and Minimum Three-Phase Short-Circuit Currents, kA rms sym

Fault in Section →	Bus B	Bus C	Bus D	Bus E	Bus A
Fault kA rms sym From breaker 1 side	2.51	1.35	1.34	0.63	15
Fault kA rms sym From breaker 1 side	0.63	0.9	0.9	2.51	

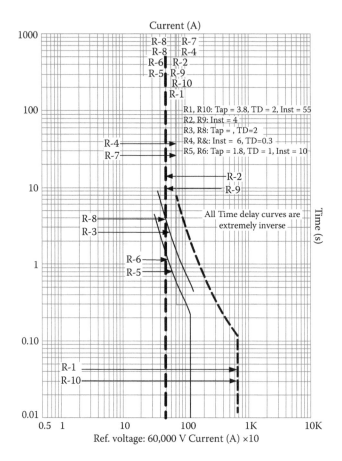

FIGURE 13.12
Coordination in the loop system of Figure 13.10.

13.4 Distance Protection

We discussed the fundamental of distance protection in Chapter 2. Also distance protection was applied as system backup protection, loss of excitation protection, and out-of-step protection for generators, in Chapter 11.

The distance protection has a major advantage that the relay zone of operation is a function of impedance of protected line though fault impedance and line length play a distinct role. The line impedance is relatively independent of current and voltage magnitudes.

Protection of Lines

Thus, distance relay has a fixed reach, as opposed to overcurrent relays, for which reach varies with the system changes.

13.4.1 Zoned Distance Relays

Consider Figure 13.13a showing a line section with two sources. Z_L is the impedance of the line to be protected from bus A to B, Z_s is the source impedance up to bus A, and Z_u is the source impedance at bus B, Z_e could be the impedance of a parallel line equal to Z_{IL}. Consider one line for which distance protection is to be applied (Figure 13.13b). The

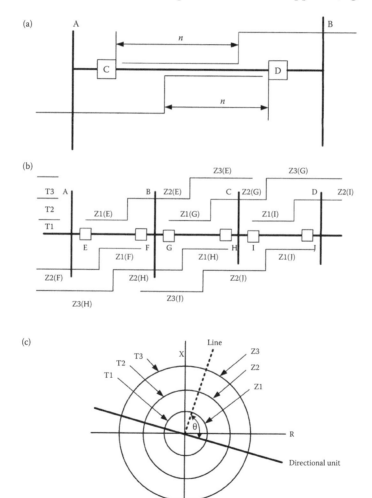

FIGURE 13.13
(a) A radial line interconnecting buses A and B, distance relay settings, (b) step times and impedance zones for distance relays, (c) operating characteristics of zoned distance relays.

(Continued)

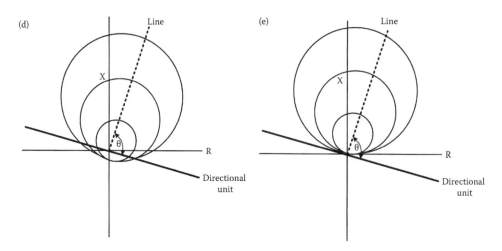

FIGURE 13.13 (CONTINUED)
(d) modified operating characteristics of zoned distance relays, (e) operating characteristics of zoned mho relays.

system can be plotted on R–X diagram as shown in Figure 13.13c. The impedance angle θ of the protected line is shown. The directional operating characteristics perpendicular to θ indicate the trip direction. The relay has contacts for each zone and in addition a set of time delay contacts. The time T_1 for zone 1 is set to zero.

The modified impedance zones may be arranged as shown in Figure 13.13d. This is same as discussed in Figure 13.13c except the location of the circles in the Z plane. Figure 13.13e shows the mho-type three-zone relay. The relay is inherently directional and therefore no directional element is required.

The impedances are reflected to the secondary side considering appropriate CT and PT ratios. Z_E is assumed to be very large.

Based on the above description, the settings for distance relays at either end are shown in Figure 13.13b. More than one distance element can be provided in the same relay. Each element is time delayed based on CTI. Here three time delays and three zones are shown, Z1, Z2, and Z3. Zone 2 can be set at 100% of the first line plus 50% of the adjacent line. Zone 3 is set at 100% of the impedance of the first two lines plus 25% of the impedance of the third line. Note that in Figure 13.13d, the characteristics are tilted toward the right, which may be the region of load encroachment, discussed further.

13.4.2 Distance Relay Characteristics

We discussed some distance relay characteristics in Chapter 2. Some of these are again shown in Figure 13.14

(a) Impedance
(b) Mho
(c) Off-set mho
(d) Reactance
(e) Blinder
(f) Quadrilateral for a radial system

Protection of Lines

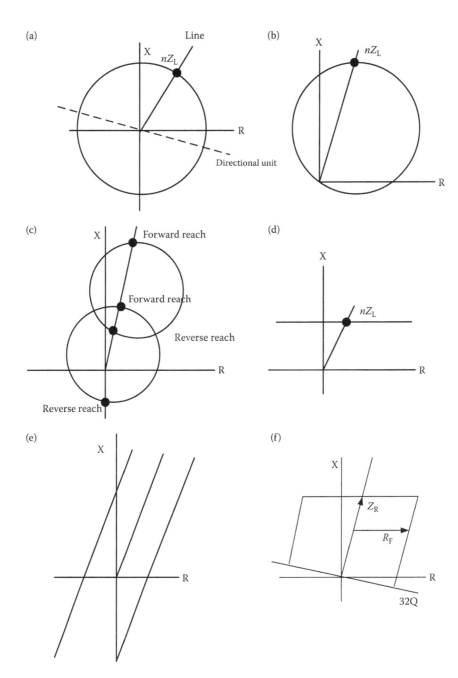

FIGURE 13.14
Distance relay characteristics, (a) impedance, (b) mho, (c) offset mho, (d) reactance, (e) blinder, (f) quadrilateral.

The operating zone is within the circles in (a), (b), (c) or within the quadrilateral in (f). The load can be represented as impedance phasors. The phasor is toward the right when flowing into the protected line to the bus and to the left when flowing out. Avoidance of the nuisance trips to heavy load encroachment is further discussed in a section to follow.

As faults are at much higher currents and much lower voltages, the load phasor usually falls outside the operating circles. This is not true for reactance characteristics, which are nondirectional. The reactance element without supervision will operate for faults behind the relay. The blinder characteristics are essentially two reactance-type units shifted to the line angle. The right unit operates for a wide area to the left, and the left unit operates for a wide area to the right of the X-axis. Together these provide operation in the band shown. Three-phase relays are used, which operate for all faults, regardless of the specific phases involved.

In Figure 13.15a, consider the distance protection of the lower line. We want the relay at C to trip for faults within a certain distance between C and D (Figure 13.15b). This distance

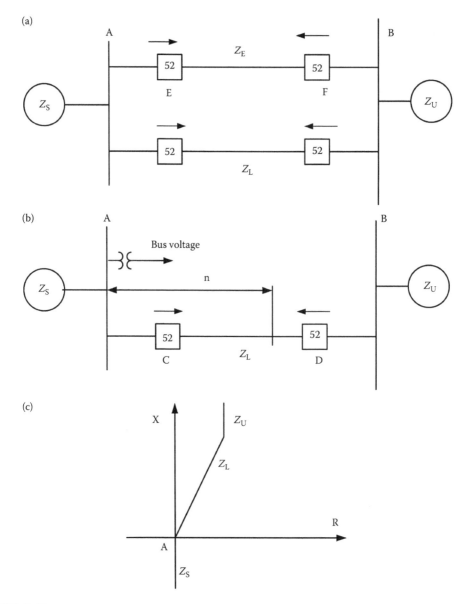

FIGURE 13.15
(a) Parallel transmission lines, (b) to illustrate reach of the relay, (c) equivalent circuit diagram for fault at n.

is called the "reach" of the distance relay. A little reflection will show that this cannot be 100% of the line length as an overreach beyond bus B is not acceptable. Generally, the reach is set at 80%–90% of the line length. In this zone, the distance protection will operate without any time delay, practically instantaneously. Consider a solid three-phase fault, then the voltage at relay C is

$$V_C = nZ_L I_C \tag{13.1}$$

where V_C is the voltage at the relay, n is the reach of the distance relay, Z_L is the line impedance, and I_C is the current. Then the impedance seen at the relay is

$$Z_C = \frac{V_C}{I_C} = nZ_L \tag{13.2}$$

The relay can be set to operate at nZ_L. As the fault location moves toward C, the current increases and the voltage decreases; that is, the impedance seen by the relay changes and is $<nZ_L$.

This method of fault detection is inherently selective with mho characteristics. If the fault is beyond n, the relay will not operate. The tripping criteria are independent of the system conditions and are the same for variations in generation and different system conditions. The impedance relay and the modified impedance relay are inherently nondirectional (Figure 13.14a,c) and directional element must be added to make these directional.

Figure 13.15c shows the equivalent circuit diagram for fault at n. The mho relay characteristic passing through the origin is set to give a reach of 90% of the total line length. The mho circle of magnitude 0.9 Z_L is plotted and a line can be drawn at the appropriate impedance angle.

It should be noted from the equivalent circuit diagram in Figure 13.16 that the reach can somewhat vary depending upon the fault impedance Z_F and also with Z_S and Z_U. The fault resistance tends to reduce n.

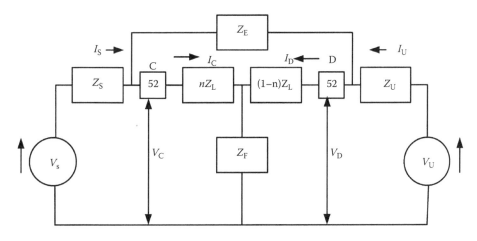

FIGURE 13.16
Equivalent circuit diagram to show variations in reach.

13.4.3 Operating Time in the First Zone

The modern MMPR have an operating time in the subcycle range in the first zone without intention time delay. The operating time for the ground faults and phase-to-phase faults is slightly higher than the three-phase faults in the subcycle zone.

13.4.4 Effect of Arc Fault Resistance

The impedance of the transmission lines has high X/R ratio. The impedance of the fault has been ignored in the previous figures. This impedance is given by the expression [1]

$$R_{arc} = \frac{8750(s + uT)}{I^{1.14}} \Omega \tag{13.3}$$

where
 s is conductor spacing (ft)
 u = wind velocity (miles/h)
 T = time (s)
 I is rms current (A)

The effect of the arc resistance is to move the end-of-reach impedance to the right in the Z plane. From Equation 13.3 the arc resistance is more predominant at the end of the reach as the current reduces and the current is in the denominator of Equation 13.3. This effect is shown in Figure 13.17. This means that the reach of the distance relay is reduced because of arc resistance. This may impact zone 1 but it is not good to have zone 2 fail to reach the end of the line and clear the fault with a greater time delay in zone 3. This should be checked for any relay application. The reach of zone 2 can be written as

$$R_{Zone\ 2} = Z_L + Z_K = d \tag{13.4}$$

This corresponds to the diameter of the circle = d. Referring to Figure 13.18a,

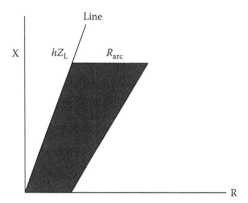

FIGURE 13.17
The effect of arc resistance on reach of the distance relay.

Protection of Lines

$$d^2 = a^2 + b^2 \tag{13.5}$$

We could write a and b as

$$a = \left[R_{arc}^2 + Z_K^2 - 2R_{arc}Z_K \cos\phi \right]^{1/2}$$
$$b = \left[(Z_L \cos\phi + R_{arc})^2 + (Z_L \sin\phi)^2 \right]^{1/2} \tag{13.6}$$

From these two equations, we can write

$$R_{arc} = \frac{Z_L}{2}\left[\sqrt{\left(1 - \frac{Z_K}{Z_L}\right)^2 \cos^2\phi + \frac{4Z_K^2}{Z_L^2}} \mp \left(1 - \frac{Z_K}{Z_L}\right)\cos\phi \right] \tag{13.7}$$

This value of the arc resistance is the maximum value that will ensure fault-clearing time T_2 or less for all line faults on the protected line.

One solution will be to rotate the mho circle to the right as shown in Figure 13.18b

The arc resistance for a line supplied from both ends is not as simple as illustrated for the radial line above. Let Z_1 and Z_2 be the impedances to the left of the fault and right of the fault, respectively, and let I_1 and I_2 be the fault current contributions from the left and right sources, respectively. Then the total impedance seen by the relay on the left is

$$Z_R = \frac{V_R}{I_1} = Z_L + \left(\frac{Z_1}{Z_2} + 1\right)R_{arc} \tag{13.8}$$

where Z_L is the actual impedance from the relay to the fault:

$$Z_1 = Z_L + Z_S \tag{13.9}$$

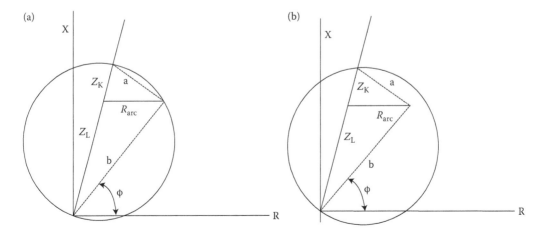

FIGURE 13.18
Loss of reach due to arc resistance with mho characteristics, (a) normal orientation, (b) circle rotated.

The apparent impedance is increased from its actual value to that as shown in Equation 13.9. Since source impedance is a function of generation level, the ratio of impedances in Equation 13.9 will depend on the system conditions, and apparent arc resistance will be most close to the relay. Note that phase angles of impedances apply.

13.5 Load Encroachment Logic

Load encroachment logic (Figure 13.19) prevents operation of the phase distance elements under high load conditions. The load region encroaches on the mho characteristics. This feature of modern MMPRs permits load to enter a predefined area of the phase distance characteristic without causing a trip. Calculations of the load encroachment are required, as will be illustrated further in an example.

13.5.1 Communication-Assisted Tripping

The modern MMPRs provide communication-assisted tripping schemes, with fiber-optic interfaces and communication protocols, sometimes specific to a relay type with transreceivers for relay to relay transmission times:

- POTT: permissive overreaching transfer tripping for two or three terminal lines
- DCBU: directional comparison unblocking for two or three terminal lines
- DCB: directional comparison blocking

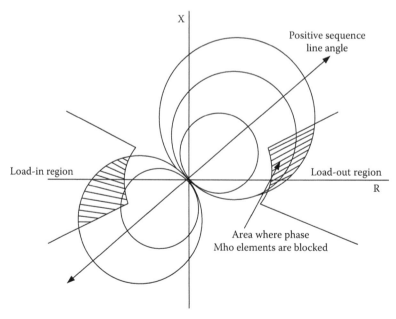

FIGURE 13.19
Load encroachment logic.

The logic can accommodate

- Current reversals
- Breaker open at one terminal
- Weak-in-feed conditions at one of the terminals
- Switch-on-to-fault conditions
- Step distance and time overcurrent protection provides backup protection should the communication channel be lost

This requires the presence of a fiber-optic channel between the terminals. With the multimode fiber, the limit is between 5 and 10 km, and with single-mode fiber, say 9 μm, the distance limit is approximately 160 km.

13.6 Ground Fault Protection

A reader is advised to familiarize with the calculation of sequence impedance of transmission lines as illustrated in Appendix A, Volume 2, also with sequence components and unsymmetrical fault current calculations as illustrated in Volume 1. Also see calculations for open conductor faults in Chapter 5, Volume 1. These aspects are not repeated.

The ground fault detection and relaying schemes are as follows:

- Directional zero-sequence overcurrent
- Directional negative-sequence overcurrent
- Quadrilateral ground distance
- Mho ground distance

Transmission lines are generally looped systems. Directional relays play an important role in transmission line protection. A ground relay must detect all phase-to-ground faults within the protected zone, and also under the fault conditions which produce the minimum ground fault current. The ground relay zone of protection can be defined as the current threshold or the measured impedance. The classical method has been to use directional overcurrent relays that measure zero-sequence current [2,3]. Many MMPRs now offer negative-sequence current polarized elements. Selecting the correct directional polarization technique will improve relay performance and sensitivity [4]. The developments in polarization techniques have improved the directional security and performance [5–9]. The distance elements measure the apparent impedance to the fault, and the most common used characteristic is the quadrilateral of mho elements. As discussed with directional overcurrent elements, the polarization is required [5–8]. Also see Chapter 8 for directional polarization of overcurrent relays.

13.6.1 Zero-Sequence Overcurrent

Sensitive ground fault protection is obtained using a relay that responds only to the zero-sequence current of the system. See Chapters 4 and 5, Volume 1, for zero-sequence currents. It should, however, be remembered that there can be some zero-sequence currents in the unbalanced three-phase systems, say, caused by nontransposed transmission lines or unbalanced loading. These can severely impact the sensitivity of the relay. A zero-sequence relay measures the three-phase currents, then

$$I_r = 3I_0 = I_A + I_B + I_C \tag{13.10}$$

The zero-sequence fault quantities are readily obtained from a short-circuit study. The pickup thresholds are simply to determine from the results of the study.

The majority of system studies available today model intercircuit zero-sequence couplings. The zero-sequence overcurrent elements can provide very effective ground fault coverage. The elements can be used independently or with time delays or in pilot tripping schemes. However, these can be impacted by changes in power system source, zero-sequence mutual couplings, and normal load unbalance.

Figure 13.20a shows phase and zero-sequence vectors for a single-phase-to-ground fault, and Figure 13.20b illustrates phase and negative-sequence vectors for a single-phase-to-ground fault. The torque equation for a zero-sequence voltage polarized element is

$$T_{V0} = |V_0||I_r|\left[\cos((<-V_0)-(<I_r+<\tau_{max}))\right] \tag{13.11}$$

Similarly for negative-sequence polarized element

$$T_{VQ} = |V_2||I_2|\left[\cos((<-V_2)-(<I_2+<\tau_{max}))\right] \tag{13.12}$$

where τ_{max} is the maximum torque angle.

A newly developed negative-sequence directional element [4] measures the negative-sequence impedance at the relay; see Figure 13.21 showing that negative-sequence

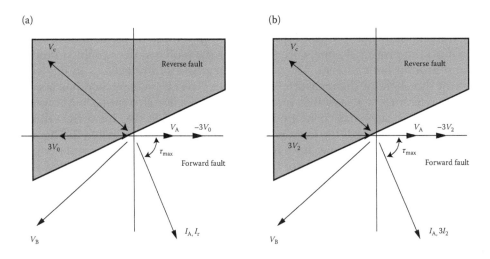

FIGURE 13.20
(a) Phase and zero-sequence phasors for a single-line-to-ground fault, (b) phase and negative-sequence phasors for a single-line-to-ground fault.

Protection of Lines

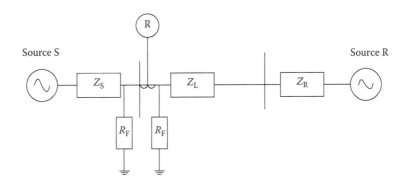

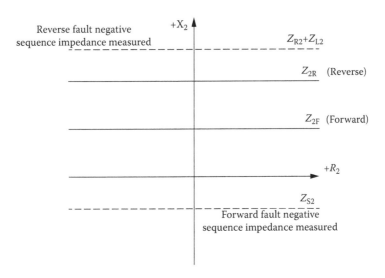

FIGURE 13.21
Measured negative-sequence impedance yields direction.

impedance yields fault direction. The measurement is then compared to forward and reverse impedance thresholds, which give the settings. The fault direction is determined as follows:

- Forward: If the negative-sequence impedance is less than the forward threshold setting.
- Reverse: If the measured negative-sequence impedance is greater than the reverse threshold setting.

The impedance-based directional element is more secure and reliable when compared to a conventional negative-sequence directional element [4].

13.6.2 Quadrilateral Ground Distance and Mho Ground Distance Characteristics

In Figure 13.14f, the top line is the reactance element, and right- and left-side lines are the positive and negative resistance boundaries. The bottom line is the directional element.

Quadrilateral characteristics operate if the measured impedance is within the box defined by the four elements mentioned above. The resistive element can be set independent of the load flow on the line. It is a good choice for protection of underground cables; these have low zero-sequence impedance angles, i.e., the zero-sequence impedance is mainly resistive, and low positive to zero-sequence ratios. Mho characteristics have already been discussed for phase faults.

13.6.3 Effect of Nonhomogeneous System on Reactance Elements

The discussion in this section is restricted to reactance elements that use zero-sequence polarization. A system is nonhomogeneous when the source and line impedance angles are not the same, this will be applicable to most systems. In such systems the angle of the total current in the fault is different from the angle of current measured in the relay. For a bolted fault, that is fault resistance is zero, it is not a problem. With fault resistance, the relay can under or overreach.

In a two source line, shown in Figure 13.22a, the phase voltage at bus S is

$$V_p = V' + V''$$
$$V' = nZ_{1L}(I_p + k_0 I_r) \quad (13.13)$$
$$V'' = R_F I_F$$

where
- I_r = the ground current measured by the relay = $3I0$
- I_p = phase current associated with the faulted phase voltage
- Z_{1L} = positive-sequence impedance of the line
- k_0 = the ground distance relay compensation factor:

$$k_0 = \frac{Z_{0L} - Z_{1L}}{3Z_{1L}} \quad (13.14)$$

The voltage V'' gives rise to the errors. When the system is homogeneous or radial, the voltage drop across the fault resistance is purely resistive and in phase with the polarizing current I_r. Thus, the term I_F is effectively removed from reactance element measurements.

When the system is nonhomogeneous, the voltage drop across the fault is no longer in phase with the polarizing quantity and includes both real and imaginary terms. The reactance error is defined as $A < T°$, and the reactance element underreaches or overreaches (Figure 13.22b,c).

This error can be simply defined as follows:

$$A < T° = \frac{I_{0F}}{I_{0S}} \quad (13.15)$$

That is the total zero-sequence fault current is divided by the zero-sequence current seen by the relay.

Protection of Lines

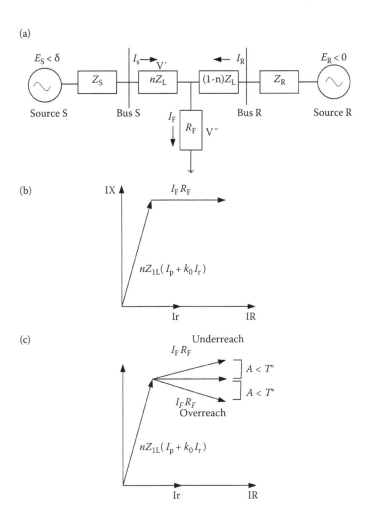

FIGURE 13.22
(a) Two-source system single-line diagram, (b) voltage vector diagram for a resistive fault in homogeneous radial system, (c) voltage vector diagram for a resistive fault in a nonhomogeneous system.

The fault induced error can be defined as

$$\Delta x = R_F \left[\frac{I_r A \sin <T}{I_p (\sin <Z_{1L} + <I_p - <I_r) + k_0 I_r \sin(<Z_{1L} + <k_0)} \right] \quad (13.16)$$

This error can be calculated in a fault study and properly set the quadrilateral reactance element to prevent over- and underreach. Adjusting the polarizing element is equivalent to setting the variable $T = 0$, thus removing the error introduced by the fault resistance.

13.6.4 Zero-Sequence Mutual Coupling

With transmission lines in close proximity, a zero-sequence mutual coupling exists. The ground distance to the fault can be falsified when there are zero-sequence couplings. It causes an increase or decrease of the current measured at the relay and the relay can overreach or underreach. Overreaching means that the measured distance is less than the distance to fault impedance. Consider parallel lines. A zone 1 distance element can overreach if the parallel line, grounded at both ends is removed from service. A ground distance relay can also overreach in cases where the zero-sequence current in the unfaulted line is a large percentage of the zero-sequence current in the faulted line [5,7,8]. An equation developed in Reference [7] can be used for setting overreaching zone 2 of distance elements:

$$\frac{Z_{app}}{Z_{iL}} = 1 + \frac{\frac{Z_{0M}}{Z_{1L}}}{\left[\frac{2K_1}{K_0}\right] + p} \qquad (13.17)$$

where
 Z_{app} = apparent loop impedance
 Z_{1L} = positive-sequence line impedance
 Z_{0M} = Zero-sequence mutual coupling impedance

$$K_1 = \frac{I_{1,\text{relay}}}{I_{1,\text{fault}}}$$

$$K_0 = \frac{I_{0,\text{relay}}}{I_{0,\text{fault}}}$$

$$p = \frac{Z_{0L}}{Z_{1L}}$$

This equation works for lines which are terminated at common buses and coupled with only one another. When the lines are not terminated at common buses or the line is coupled with more than one circuit, an apparent impedance calculation is recommended.

$$Z_{app} = \frac{V_p}{I_p + k_0 I_r} \qquad (13.18)$$

where V_p is the phase voltage measured by the relay, I_p is the phase current measured by the relay, I_r is the ground current measured by the relay, and k_0 is as defined before.

13.6.5 High-Resistance Fault Coverage and Remote Infeed

The amount of resistive fault coverage is influenced by the following:

- Distance element reach
- Directional sensitivity
- Remote infeed and, thus, the impedance behind relay location
- Line length
- System unbalance and load flow

Protection of Lines

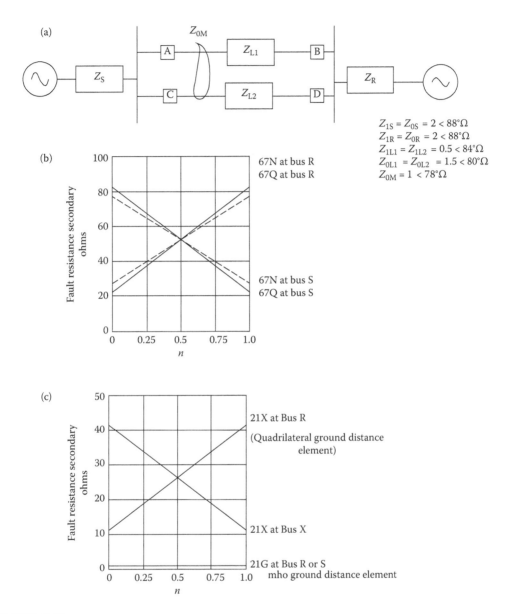

FIGURE 13.23
Example of a long line, showing mutual coupling between the lines, (b) long line high fault resistance detection using 67N and 67Q functions, (c) long line high fault resistance detection using 21G and 21X functions.

Roberts et al. [9] discuss the directional element sensitivity and its ability to detect high-resistance faults and how to determine the resistive fault coverage. Figure 13.23a shows two parallel lines with mutual couplings. Figure 13.23b shows resistive fault coverage using ground overcurrent (67N) and negative-sequence (67Q) elements on a short line. As the fault location moves away from the bus, the amount of fault resistance detected by zero-sequence element 67N decreases, while the negative-sequence element (67Q) increases. This is due to mutual couplings and differences in the line zero- and negative-sequence impedances.

TABLE 13.6
Application of Ground Fault Detecting Elements

Description of the Application	67N	67Q	21X	21G
Short-line applications	Best selection	Best selection	Satisfactory	Satisfactory
Long-line applications	Best selection	Best selection		Satisfactory
Parallel-line application	Satisfactory	Best selection		
Channel-independent direct tripping—instantaneous	Satisfactory	Satisfactory	Best selection	Best selection
Channel-independent direct tripping—time delay	Satisfactory	Satisfactory	Best selection	Best selection
Pilot scheme	Best selection	Best selection	Satisfactory	Satisfactory

TABLE 13.7
Specific System Conditions; Application of Ground Fault Detecting Elements

Description of the application	67N	67Q	21X	21G
System nonhomogeneity	Best selection	Best selection		Satisfactory
High fault resistance	Best selection	Best selection	Satisfactory	
Strong source	Best selection	Best selection	Satisfactory	Satisfactory
Weak source	Satisfactory	Satisfactory	Satisfactory	Satisfactory
Zero-sequence mutual coupling	Satisfactory	Best selection		
Load flow	Best selection	Best selection		Best selection
In-line load switching			Best selection	Best selection
Nontransposed transmission lines			Best selection	Best selection
Unbalanced loading			Best selection	Best selection

Note: Improvements in the MMPRs impact this table.

Figure 13.23c is the comparison of the resistive fault coverage using ground distance (21G and 21X) functions on a short line. In this figure, these elements are applied in a pilot scheme.

Table 13.6 summarizes the application performance of ground fault detecting elements. Distance elements provide fair fault resistance coverage and are more tolerant to system unbalance and in-line load switching. The mho element provides little or no fault resistance coverage and can be adversely impacted by zero-sequence mutual coupling. The quadrilateral element, while providing better fault resistance coverage than mho elements, can be impacted by zero-sequence couplings.

Quadrilateral elements are sensitive to system nonhomogeneity, if corrective measures are not taken. See also Table 13.7 for the applications in the presence of specific system conditions.

13.7 Protection of Tapped 345 kV Transmission Line

This will be illustrated with an example based on Reference [10]. The example uses communication-assisted tripping to provide high-speed protection. The relay uses distance elements and residual ground directional overcurrent elements in this scheme. A schematic of the tapped line is shown in Figure 13.24.

Protection of Lines 539

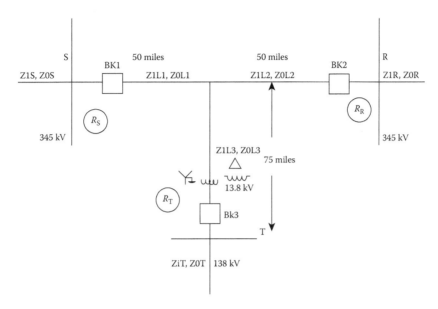

FIGURE 13.24
The 345 kV tapped transmission line.

System data and the data converted to secondary ohms are shown in Table 13.8. DCB is used. Three zones of mho phase and ground distance protection, two levels of zero-sequence directional overcurrent protection, inverse time directional zero-sequence backup protection, load encroachment logic, and finally, SOFT (fast tripping when the circuit breaker closes) are used.

Source T is a weak source; set positive-sequence impedance magnitude to 4.0 Ω and its angle to 84.7° ($Z_{1L1} + Z_{1L2}$) so that the fault locator corrects results for the faults not located on the tap. Also Z0MAG is set at 12.88 Ω and angle at 73°.

The relay has transient detection logic. It prevents incorrect operation of direct tripping (zone 1) distance element. It determines the source impedance ratio (SIR) and acts to inhibit zone 1 for only those conditions where a transient condition exists. This protection is disabled if SIR is >5. Calculate the ratio based on zone 1 reach, as we do not want distance protection to overreach during an external fault.

TABLE 13.8

System Data 345 kV Tapped Transmission Line

Parameter	Value, Primary	Value, Secondary
$Z_{1L1} = Z_{1L2}$	29.67 Ω < 84.7°	2 Ω < 84.7°
$Z_{0L1} = Z_{0L2}$	96.65 Ω < 73°	6.44 Ω < 73°
Z_{1L3}	44.5 Ω < 84.7°	3 Ω < 84.7°
Z_{0L3}	144.98 Ω < 73°	9.65 <73°
Transformer impedances: XHM	1.6% on 100 MVA	1.6% on 100 MVA
XML	40% on 100 MVA	40% on 100 MVA
XHl	60% on 100 MVA	60% on 100 MVA
Source impedance Z1S = Z0S	10 Ω < 87°	0.67 Ω < 87°
Source R impedance Z1R = Z0R	35 Ω < 87°	2.33 Ω < 87°
Source T impedance Z1T = Z0T	0.656 Ω < 87° pu	0.656 Ω < 87° pu

Here,

$$\text{SIR} = \frac{|Z_{1S}|}{0.8|Z_{1L1} + Z_{1L2}|} = \frac{0.67}{0.8(2+2)} = 0.209 \quad (13.19)$$

Mho phase distance element reach

Set zone 1 phase distance protection equal to 80% of the positive-sequence impedance from station S to station R:

$$Z1MP = 0.8(2+2) = 3.2 \ \Omega$$

Set zone 2 to include tapped autotransformer. A fault analysis is required for fault impedance distance element measures for fault at 13.8 kV terminals of the autotransformer. These measurements are used to set distance reach settings. A line-to-ground and three-phase fault is placed at the autotransformer, and the secondary voltage and current at relay at substation S is calculated. Consider these measurements to be as follows:

- Phase A to ground fault = 7.77 Ω = Z_{AG}
- Three-phase fault = 8.8 Ω.

$$Z_{AG} = \frac{V_A}{I_A + k_0 \times 3I_0} \quad (13.20)$$

where
  V_A = phase to neutral voltage
  I_A = A phase current
  k_0 = zero-sequence compensation factor
  $3I_0$ = zero-sequence current

As before

$$k_0 = \frac{Z_{0L1} - Z_{1L1}}{3.Z_{1L1}}$$

$$3I_0 = I_A + I_B + I_C$$

Thus, set zone 2 reach as

$$Z2MP = 1.25 \times 8.8 = 11.0 \ \Omega$$

where 1.25 is a safety factor.

Zone 3 phase element reach

The zone 3 phase distance protection is reverse looking. Zone 3 at substation S must have sufficient reach to prevent unwanted tripping by relays at R and T for external faults behind local terminal. The zone 3 reach at substation S must cover overreach from furthest reaching remote zone 2 for reverse faults when there is no infeed from other remote

TABLE 13.9

Apparent Impedance Measurements for Remote Faults, Tapped 345 kV Transmission Line

Station	ZAG	ZABC
Relay at R, fault at T	152.7 Ω (0.128 pu)	196.65 Ω (0.165 pu)
Relay at T, fault at S	79.605 Ω (0.418 pu)	76.845 Ω (0.404 pu)
Relay at T, fault at R	103.86 Ω (0.545 pu)	115.86 Ω (0.608 pu)

terminal. Faults are successively placed AG and ABC and the apparent impedances measured. This is shown in Table 13.9.

Figure 13.25a shows the coordination issue. The zone 2 reach at T is set to account for infeed during faults beyond the tap on a 345 kV system. However, when one 345 kV station is out of service, the zone 2 at substation T overreaches for faults on the other side of the tap on the 345 kV system. Figure 13.25b is an impedance diagram of the system.

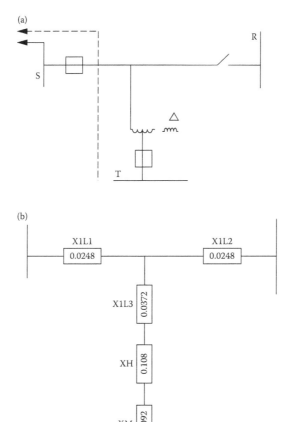

FIGURE 13.25
(a) Reverse zone 3 coordination, (b) impedance diagram.

Relay at T measures the largest apparent fault impedance for faults at substation R, because the source at S is stronger than the source at R. Therefore zone 2 at station T must be set to 115.86 Ω plus safety margin; so that relay can detect faults at R.

The overreach from a remote terminal during reverse faults with respect to substation S is calculated as follows, based on Table 13.9

Overreach at R:

$$|Z_{app}| - X_{1L1} - X_{IL2}$$
$$= 0.165 - 0.0248 - 0.0248$$
$$= 0.115 \text{ pu}$$

Overreach at substation T:

$$|Z_{app}| - X_M - X_H - X_{1L3} - X_{1L1}$$
$$= 0.608 - (-0.092) - 0.108 - 0.372 - 0.0248$$
$$= 0.53 \text{ pu}$$

Note that these calculations and Table 13.9 is in terms of primary quantities. Set zone 3 phase distance element reach

$$Z3MP = \frac{CTR}{PTR} \times 0.53 \times Z_{base} \times 120\%$$

where

$$Z_{base} = \frac{(345 \text{ kV})^2}{100 \text{ (MVA)}} = 1190.25$$

and 120% is the safety factor. This gives Z3MP = 50.50 Ω.

Mho ground distance element reach

Zone 1 has instantaneous underreaching tripping and zones 2 and 3 have a DCB tripping scheme.

Set zone 1 reach at 3.2 Ω, zone 2 reach at 11.0 Ω, and zone 3 reach at 50.5 Ω. Ground elements should measure fault impedance in terms of positive-sequence only. The zero-sequence current compensation factor is calculated as follows:

For zone 1,

$$k_{01} = \frac{Z_0 < Z_0 - Z_1 < Z1}{3Z_1 < Z_1} = 0.75$$

Zones 2 and 3 use the same compensation.

The common time delays for the distance elements are as follows:

Zone 1: instantaneous

Zone 2: 20 cycles

Zone 3: instantaneous as the distance protection is reverse looking

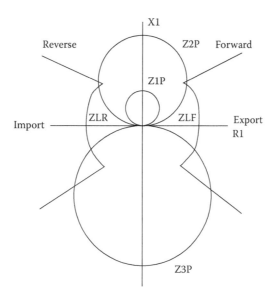

FIGURE 13.26
Load encroachment function.

Load encroachment permits load to enter predefined areas of phase distance characteristics without causing unwanted tripping, as explained before. The load impedance and its angle need to be calculated. Define the load encroachment characteristics with load impedance settings in the forward (ZLF) and reverse directions (ZLR). Define two load sectors export and import with angle settings (Figure 13.26). Considering that substation S can supply the entire load (4.2 A), and that the line-to-neutral voltage of the PT is 65.7 V, the load impedance is 15.6 Ω. This is inside the zone 3 setting of 50.5 Ω. As a safety factor, the load impedance can be considered 12.5 Ω. Then the forward and revise load encroachment can be set at 12.5 Ω, and forward load positive angle conservatively at 45.0°, forward load negative angle at −45°, reverse load positive angle at 135°, and reverse load negative angle at 225°. Local zone 2 measurements show the following:

Fault: A phase to ground, ZAG = 7.77 Ω
Fault ABC, ZBC = 8.8 Ω
Set zone 2 phase distance element reach Z2MP = 1.25 × 8.8 = 1.0 Ω.

Residual ground instantaneous and time delay settings; level 1 residual ground instantaneous element is disabled. Level 2 element for DCB tripping is set at a pickup of 20% of the nominal current of % A = 1.0. The ground level 3 residual ground overcurrent element is enabled to send blocking signals for out-of-section faults, at station S to 50% of the forward looking ground overcurrent element = 0.05 A. The level 2 element is forward looking and level 3 element is reverse looking, without intentional time delays.

An overcurrent element is selected to provide backup protection for high impedance ground faults; a setting of 0.96 pickup will sense 1037 Ω primary fault resistance. A very inverse characteristic and a time dial of 2 are selected.

Directional elements are selected to supervise the ground distance elements during ground fault conditions. Negative-sequence voltage polarized and zero-sequence voltage polarized functions can be simultaneously selected.

Unbalanced faults: The forward and reverse directional decisions during unbalanced faults are made based on the following:

- If negative-sequence impedance measured by the relay is less than Z2F, forward directional threshold, the unbalanced fault is declared forward and if Z2 > Z2R, reverse directional threshold, it is declared reverse.
- Similarly, if the zero-sequence impedance measured by the relay is Z_0, which is less than Z_{0F}, the forward directional Z_0 threshold unbalance fault is forward, and if $Z_0 > Z_{0R}$, reverse directional Z_0 threshold, it is declared reverse.

$$Z2F = 0.5Z1\text{MAG} = 2.0\,\Omega$$

$$Z2R = Z2F = 05/I_{\text{nom}} = 2.10\,\Omega$$

$$Z0F = 0.5.Z0\text{MAG} = 0.5 \times 12.88 = 6.44$$

$$Z0R = 6.44 + 0.5/I_{\text{nom}} = 6.54$$

32QG reverse directional check

Figure 13.27a shows a negative-sequence network for 345 kV tapped line. Assume that the negative-sequence impedances are equal to the positive-sequence impedances.

Also,

$$z2 = Z_{1L1} + Z_{2P} \tag{13.21}$$

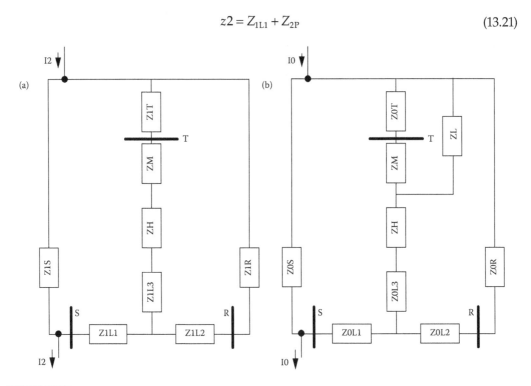

FIGURE 13.27
(a) 345 kV tapped transmission line equivalent negative-sequence impedance, (b) 345 kV tapped transmission line equivalent zero sequence impedance.

Protection of Lines

$$Z_{2P} = j\left[\frac{(X_{1L3} + X_H + X_M)X_{1L2}}{X_{1L3} + X_H + X_M + X_{1L2}}\right] = j0.017 \text{ pu} = 1.35\ \Omega, \text{ secondary} \qquad (13.22)$$

For correct operation

$$Z2F < |Z_{1L1}| + |Z_{2P}|$$
$$2\ \Omega < 2\ \Omega + 1.35\ \Omega \qquad (13.23)$$

Reverse directional check
Similarly, the zero-sequence impedance parameters can be calculated. Figure 13.27b shows the zero-sequence network.

$$z_0 = Z_{0L1} + Z_{0P} \qquad (13.24)$$

where Z_{0P} is the parallel combination of line 3 impedances.

$$Z_{0P} = (Z_{0L3} + jX_H + Z_{0PP}) \text{ in parallel } (Z_{0L2} + Z_{0R}) = 3.8\ \Omega \qquad (13.25)$$

where Z_{0P} is parallel combination of the transformer low side and bus T impedance in parallel with the transformer tertiary impedance = $-j0.113$ pu.
For the correct operation of the voltage polarized element,

$$Z_0 < |Z_{0L1}| + |Z_{0P}|$$
$$6.44 < 6.44 + 3.8\ \Omega \qquad (13.26)$$

Thus, this condition is satisfied.

DCB trip scheme:
It consists of three sections:

1. Starting elements: Nondirectional or directional elements or both can be selected to detect out-of-section faults behind the local terminal. These elements will send a blocking signal to station R, to prevent unwanted tripping during out-of-section faults. The directional elements are as follows:
   - Zone 3 phase distance element
   - Zone 3 ground distance element
   - Level 3 residual ground directional element

   An output contact IN102 is assigned if the nondirectional or the directional element picks up. When energized, it sends a blocking signal, say if nondirectional element 50G3 picks up. However, the STOP has priority, if Z2P, Z2G, or 67G2 assert. See Figure 13.28.

2. *Coordination timers*: The forward-looking elements that provide high-speed tripping at station S must be delayed for a short-duration, so that local circuit breaker does not trip for faults beyond R. This time delay provides time for nondirectional and reverse looking blocking elements at station R to send a signal to station S

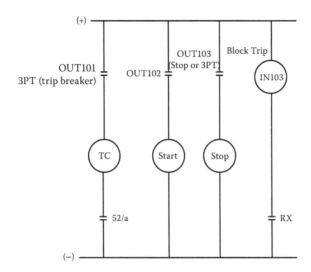

FIGURE 13.28
DC schematic for the DCB trio scheme.

during out-of-section faults. This is the coordination time for DCB scheme and depends upon control input recognition time, remote zone 3 distance protection maximum operating time and maximum channel communication time. A time delay of 1.64 cycles is adequate for zone 2 distance and also for level 2 overcurrent short delay.

3. *Blocking signal extension*: An output, IN103, is programmed to recognize when the local terminal receives a blocking signal from the remote terminal during external faults. The DCB trip scheme uses an on/off carrier signal to block high-speed tripping at stations S and R for out-of-section faults.

The communication tripping is enabled for delayed zone 2 mho phase and ground distance protection and delayed level 2 negative-sequence residual ground directional overcurrent element.

13.8 Series Compensated Lines

The distance protection cannot be applied as the directional distance mho element will operate undesirably due to the impedance of the series capacitor. These will not see the fault in the capacitor and part of the line. The relay protecting a series compensated line is presented with changing information following a fault depending upon the location of the series capacitors. These may be at the ends of the lines or in the middle of the line.

As discussed in Chapter 8, Volume 2, that on a fault the capacitors are bypassed with the capacitor protective gaps (Figure 8.16). These gaps are fast acting. So, if we assume that under a fault condition, with some time delay, the capacitors will be all bypassed, the line will revert to a conventional noncompensated line and the mho elements with proper time delay will operate. However, after the fault is cleared the capacitors are reinserted within a few cycles to enhance stability. Thus, the distance protection is not applied to the series

Protection of Lines

compensated lines and the preferred method of protection is phase comparison pilot wire systems discussed next in Chapter 14.

13.8.1 Subsynchronous Resonance

We also eluded to subsynchronous resonance in Chapter 8, Volume 2, that can occur because of series compensation of transmission lines. Subsynchronous resonance can occur due to the interactions caused by the following:

1. *Induction generator effect*: The resistance of rotor to subsynchronous currents is negative and network presents a resistance which is positive. If the negative resistance of the generator is greater than the positive system resistance, there will be sustained subsynchronous currents.
2. *Torsional interaction*: This occurs when induced subsynchronous torque in the generator is close to one of the natural modes of turbine generator shaft. When this occurs, generator oscillations will build up and this motion induces armature voltage components at subsynchronous and supersynchronous frequencies. The mechanical damping is always positive increasing from no load to full load, but it is small.
3. *Transient torques*: These result from a system disturbance, which causes changes in the network, resulting in sudden changes in the current that will oscillate at the natural frequency of the network.

13.8.2 Steady-State Excitation

Consider that a steady-state excitation of frequency f_0 is applied to a system shown in Figure 13.29a. A torsional vibration will be excited (Figure 13.29b), and it may continue to grow in magnitude, till the energy loss per cycle is equal to the energy that the small disturbance adds to the system during a cycle. The amplification curves of the resonant system are in Figure 13.30. If the excitation frequency varies at a certain rate the torsional vibrations will be amplified as the system passes through the resonant point.

13.8.3 Models

IEEE first benchmark model (FBM) was created by IEEE working group on subsynchronous resonance in 1977 for use in computer program comparison and development. It consists of a single-series capacitor compensated line connecting a synchronous generator to

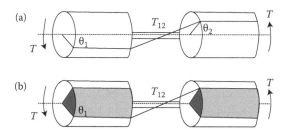

FIGURE 13.29
(a) Torsional model with two shaft coupled rotating masses under steady state, (b) free oscillations with the torque removed.

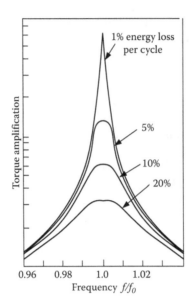

FIGURE 13.30
Amplification curves of the resonant system.

a large system. It describes a synchronous generator model, with defined parameters and shaft model with inertial constant H and spring constant K in pu T/rad [11].

IEEE second benchmark model (SBM) is similar to FBM; the systems are small and easy to implement in a computer simulation. It deals with parallel resonance and interaction between turbine generators that have a common mode of oscillation. In system 1, capacitance reactance is not specified explicitly, but as a variable—values between 10% and 90% of series inductance of line. System 2 has two generators with common torsional mode of oscillation. The SBM generator, circuit, and shaft data are provided [12].

The Corpals benchmark model is based on a system of more practical nature with 29 buses, 39 branches, and 5 generators.

13.8.4 Analysis

An n-spring-connected rotating mass can be described by the following equation:

$$J\frac{d\bar{\omega}_m}{dt} + \bar{D}\bar{\omega}_m + \bar{H}\bar{\theta}_m = \bar{T}_{turbine} - \bar{T}_{generator} \qquad (13.27)$$

The moment of inertia and damping coefficients are available from design data. The spring action creates a torque proportional to the angle twist.

Figure 13.31a shows a torsional system model for the steam turbine generator. The masses will rotate at one or more of the turbine mechanical natural frequencies called torsional mode frequencies. When the mechanical system oscillates under such steady state at one or more natural frequencies, the relative amplitude and phase of individual turbine-rotor elements are fixed and are called *mode shapes* of torsional motion (Figure 13.31b).

Torsional mode damping quantifies the decay of torsional oscillations. The ratio of natural log of the successive peaks of oscillation is called logarithmic decrement. The decrement factor is defined as the time in seconds to decay from the original point to $1/e$ of its value.

Protection of Lines

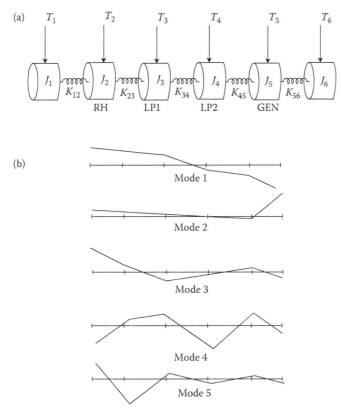

FIGURE 13.31
(a) Rotating mass model of steam turbine generator, (b) oscillation modes.

13.8.5 An Example with EMTP Simulations

Consider a 500 MVA, 22 kV generator, connected to a step-up transformer of 600 MVA, delta–wye (star)-connected, 22–500 kV, wye windings solidly grounded, which feeds into a 640 km long 500 kV line. A CP model of the transmission line is modeled in EMTP. EMTP permits a number of transmission line models for transient studies. The constant parameter (CP) model is for wave equations of the transmission line with distributed parameter, not discussed here. A series capacitor compensation of 50% at the terminal point of the transmission line is provided. For subsynchronous oscillations, the shaft mass system of steam turbine generators is modeled with four masses of certain inertia constants connected together through spring constants (HP and LP sections of turbine, rotor, and exciter). Again, synchronous generators in EMTP are modeled with Park's transformations in $dq0$ axis. The line serves receiving end loads. External torques can be applied to each of the masses, for example turbine, generator, and exciter masses.

An EMTP simulation of the frequency scan at the 500 kV side of the step-up transformer is shown in Figure 13.32. This shows one resonance at 19 Hz and the other close to the fundamental frequency. A three-phase fault occurs at the secondary of the transformer at 1 s, and cleared at 1.1 s, fault duration = 6 cycles. The resulting torque transients in the 500 MVA synchronous machine mass 1 is shown in Figure 13.33, total simulation time of 5 s. It is seen that these transients do not decay even after 5 s and diverge, imposing stresses on the generator shaft and mechanical systems. The angular frequency of mass 1 (zero external torque which will give maximum swings) is plotted in Figure 13.34. This shows violent speed variations. The frequency relays or vibration probes may isolate the generator from the system. The generator parameters for the EMTP model are as follows:

Field current at rated voltage = 1200 A,

$$R_a = 0.0045, X_0 = 0.12, X_d = 1.65, X'_d = 0.25,$$

$$X''_d = 0.20, X'_q = 0.46, X''_q = 0.20$$

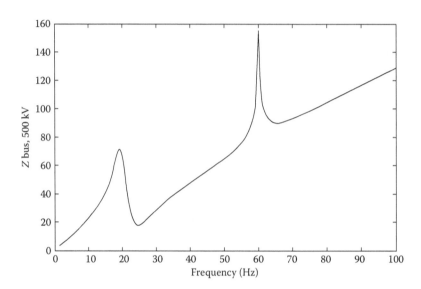

FIGURE 13.32
Frequency scan of transmission line showing resonant frequencies.

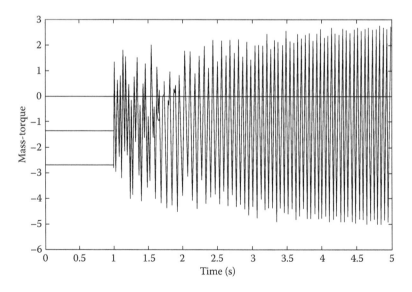

FIGURE 13.33
Diverging torque oscillations, EMTP simulations.

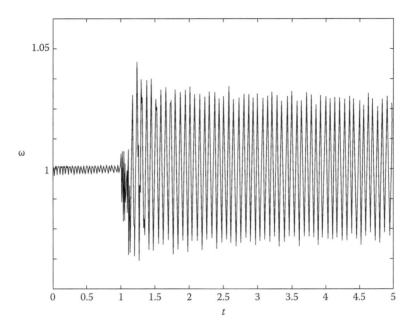

FIGURE 13.34
Angular speed transients, mass 1, EMTP simulations.

all in pu (per unit on generator MVA base).

$$T'_{qo} = 0.55, T''_{qo}T''_{qo} = 0.09, T'_{do} = 4.5, T''_{do} = 0.04 \quad \text{all in seconds}$$

The generator is modeled with automatic voltage regulator (AVR), and power system stabilizer (PSS). The transformer is rated 22–500 kV, 600 MVA, % Z = 10%.

13.9 Mitigation of Subsynchronous Resonance in HV Transmission Lines

13.9.1 NGH-SSR

NGH-SSR counteracts subsynchronous resonance. The principle of operation can be explained with reference to Figure 13.35a. The subsynchronous voltage V_C, when combined with 60 Hz voltage wave V_f, produces a voltage $V\alpha$ shown in dotted lines in this figure. Note that some half-cycles of $V\alpha$ become longer than the normal half-cycle period of 8.333 ms. These distorted wave shapes represent the voltage across the capacitor. This unbalance charge on the capacitor interacts with system inductance to produce oscillations. If this unbalance charge is eliminated, the system will be detuned to any other frequency other than the fundamental.

Figure 13.35b shows impedance in series with thyristor switch, connected across the capacitor. The impedance is a small resistance determined from the peak current capability of the thyristor switch in series with an inductor, the value of which is dictated by di/dt limit of the thyristors. The thyristor switch is controlled so that when a current

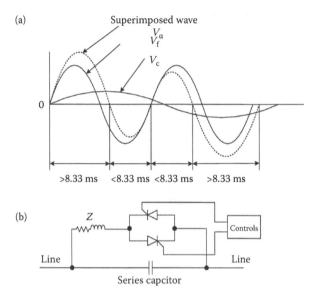

FIGURE 13.35
(a) Superimposition of 60 Hz wave on subsynchronous oscillation, (b) basic circuit of Narain Hingorani subsynchronous resonance suppressor.

zero is detected, the following half-cycle period is timed. If the half-cycle exceeds the set time of 8.33 ms, the corresponding thyristor is turned on to discharge the capacitor and bring about a current zero sooner. The switch stops conducting when the current zero is obtained. At each capacitor voltage zero, a new count starts.

13.9.2 Thyristor-Controlled Series Capacitor

A series capacitor can be controlled by parallel thyristors/GTOs (gate controlled series capacitor) in series with a reactor, a scheme proposed in 1986 by Vithayathil, which is shown in Figure 13.36. The steady-state impedance can be written as

$$X_{TCSC} = \frac{X_c X_L(\alpha)}{X_L(\alpha) - X_c} \tag{13.28}$$

$X_L(\alpha)$ varies with the firing angle:

$$X_L(\alpha) = X_L \frac{\pi}{\pi - 2\alpha - \sin\alpha}, \quad X_L \leq X_L(\alpha) \leq \infty \tag{13.29}$$

As the impedance is varied, the degree of series compensation is changed, and it can become inductive. A parallel resonance can occur, giving very high impedance, dividing the operating zone into inductive and capacitive regions depending on α. If the TCR is to be turned on at angle α, measured from the negative peak of the capacitor voltage; at this instant the capacitor voltage is negative and the line current is positive. This charges the capacitor in the positive direction. The thyristor switch can be viewed as an ideal switch, closing at α, and stopping condition as the current crosses zero. The charge of the capacitor is reversed during resonant half cycle of the LC circuit by closing the thyristor switch. This

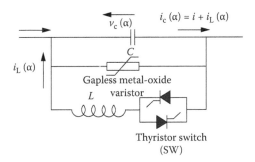

FIGURE 13.36
Thyristor-controlled series capacitor scheme (TCSC).

produces a DC offset for the next positive half cycle of capacitor voltage as illustrated in Figure 13.37. In subsequent negative half cycle, the DC offset can be reversed by maintaining same angle α.

The reversal of capacitor voltage by parallel TCR is shown in Figure 13.37. This reversal of capacitor voltage in the region near to end of each half cycle corresponding to

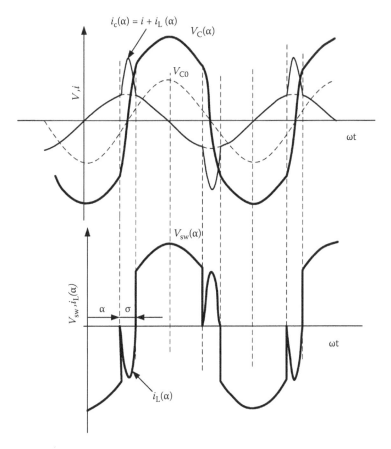

FIGURE 13.37
Capacitor voltage and current waveforms; also TCR voltage and current waveforms, TCSC in capacitive region under steady state.

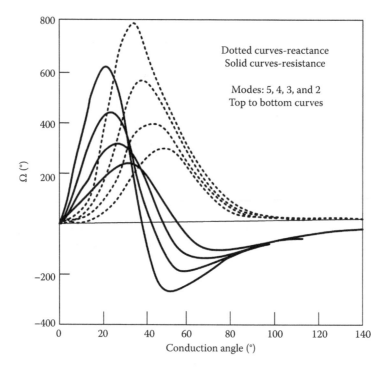

FIGURE 13.38
Subsynchronous frequency impedance, IEEE benchmark model FBM.

power frequency is the key to suppression of SSR. It is analogous to synchronous discharge of capacitor in NGH-SSR scheme with the additional benefit of no dissipation of energy, unlike NGH scheme. Similar curves for inductive mode can be drawn; however, when the subsynchronous frequency reactance is inductive, the system is free from SSR.

When the subsynchronous frequency reactance is capacitive, thyristor-controlled series capacitor (TCSC) should meet SSR requirements under each torsional mode. The conduction angle plays a major role; the selection must consider the resistive and reactive modes. Figure 13.38 shows derivation curves of TCSC equivalent impedance at five different torsional frequencies, IEEE FBM model. The control and setting of TCSC is complex.

In a practical installation, a number of cells are used in series with bypass disconnect and breaker. A TCSC installed at Bonneville Power Authority (BPA) at 500 kV has six modules in series, a rated current of 2900 A, and a three-phase nominal compensation of 202 Mvar with 10s compensation of 404 Mvar, designed for 60 kA peak fault current in the thyristor valve. References [13–21] provide further reading.

13.9.3 Supplemental Excitation Damping Control

The current literature describes many new technologies for SSR suppression. Supplemental excitation damping controls (SEDCs) consist of modulating the field voltage of the generator at torsional frequency. The speed of each torsional mode being controlled is inputted to the SDEC. A phase shift and gain is applied to each torsional speed and the summed signal is used to modulate excitation voltage.

TABLE 13.10

SSR Mitigation and Protection Examples

Plant	Units, MVA	KV	% Comp	SSR Mitigation
Navajo	3 × 892	500	70	SSR blocking filters SEDC, T-relays
Mohave	2 × 909	500	26	T-relay
Colstrip	2 × 377, 2 × 819	500	35	T-relays
Boardman	1 × 590	500	29	TCSC, T-relays
Jim Bridger	4 × 590	345	45	SEDC, T-relays
San Juan	2 × 410, 2 × 617	345	30–34	T-relays

13.9.4 Torsional Relay

A torsional relay can be designed to continuously monitor the turbine generator shaft for torsional oscillations; and trip the machine when shaft fatigue reaches predetermined levels. SSR mitigation and protection examples are shown in Table 13.10. This Table shows that torsional relay (abbreviated at T-relay in the Table) finds wide applications.

The measurements on the spinning shaft can be carried out in a number of ways. For example, toothed wheels with magnetic pickups provide speed and acceleration measurements, but these must be an integral part of the generator shaft. Strain measurements along the shaft can measure torque and indicate stress to shaft. However, shaft-mounted strain gauges require slip rings for wireless telemetry to transmit the measurements.

A synchronized rotor angle measurement concept is depicted in Figure 13.39. A laser is mounted on the generator and pointed at the shaft. Reflective material is applied to a part of the circumference of the shaft. The reflected light beam is directed to an optical pickup. These data are then input to a relay where the rising and falling edges are detected and then time-stamped with microsecond accuracy. In addition, the relay can acquire synchronized measurements of the field current and voltage and output of the AVR and PSS.

Synchronized measurements of phasors can give indications of SSR. The filter response of PMU units has to be properly designed; the narrow band response of IEEE C37.118.1 [21] is not adequate. The frequency response of filter must have minimum attenuation below 60 Hz, and even wide band filter may have some attenuation.

Combining electrical and mechanical measurements in a synchronized manner will improve the accuracy of SSR detection.

References [22–27] are added for further reading on this chapter.

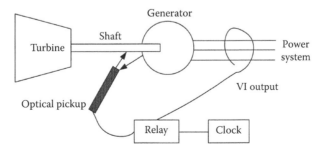

FIGURE 13.39
Rotor angle measurement system.

References

1. AR Warrington. *Protective Relays Their Theory and Practice*, Vol. 1. John Wiley & Sons, New York, 1962.
2. CF Henville. Combined use of definite and inverse time overcurrent elements assists in transmission line ground relay coordination, *20th Annual Western Protective Relay Conference*, Spokane, Washington, October 19–21, 1993.
3. WZ Tyska. Polarization of ground relays, *Western Protective Relay Conference*, Spokane, Washington, October 21–23, 1986.
4. EO Schweitzer III, J Roberts. Distance relay element design, *19th Annual Western Protective Relay Conference*, Spokane, Washington, October 19–22, 1992.
5. AT Giuliante, JE McConnell, SP Turner. Coordination for design and application of ground distance relays, *22nd Annual Western Protective Relay Conference*, Spokane, Washington, October 24–26, 1995.
6. A Guzman, J Roberts, D Hou. New ground directional elements operate reliably for changing system conditions, *23rd Annual Western Protective Relay Conference*, Spokane, Washington, October 15–17, 1996.
7. WA Elmore. Zero-sequence mutual effects on round distance relays and fault locators, *46th Annual Protective Relaying Conference at Georgia Institute of Technology*, Atlanta, Georgia, April 29–May 1, 1992.
8. GE Alexander, JG Andrichak. Ground fault relaying: Problems and principles, *18th Annual Western Protective Relay Conference*, Spokane, Washington, October 20, 1991.
9. J Roberts, EO Schweitzer III, R Aroroa, E Poggi. Limits to the sensitivity of ground directional and distance protection, *22nd Annual Western Protective Relay Conference*, Spokane, Washington, October 24–26, 1995.
10. SEL 421 Instruction Manual, SEL Inc.
11. IEEE Committee Report. First benchmark model for computer simulation of subsynchronous resonance. *IEEE Trans*, PAS-96, 1565–1570, 1977.
12. IEEE Committee Report. Second benchmark model for computer simulation of subsynchronous resonance. *IEEE Trans*, PAS-104, 1057–1066, 1985.
13. NG Hingorani. A new scheme for subsynchronous resonance damping of torsional oscillations and transient torques—Part I. *IEEE Power Eng Rev*, per-1, 56–57, 1981.
14. RA Hedin, KB Stump, NG Hingorani. A new scheme for subsynchronous resonance damping of torsional oscillations and transient torques—Part II. *IEEE Trans*, PAS-100, 1856–1863, 1981.
15. MS El-Moursi, V Khaddikar. Novel control strategies for SSR mitigation and damping power system oscillations in a series compensated wind park, *38th Annual Conference on IEEE Industrial Electronics Society*, pp. 5335–5342, 2012.
16. R Hooshmand, M Azimi. Investigations of dynamic instability of torsional modes in power systems compensated by SSSC and fixed capacitor. *IREE*, 4(1), 129–138, 2009.
17. Y Lu, A Abur. Improved system static security via optimal placement of thyristor controlled series capacitors (TCSC), *IEEE Power Engineering Society Winter Meeting*, Columbus, OH, pp. 516–521, 2001.
18. L Gyugyi. A solid state approach to the series compensation of transmission lines. *IEEE Trans Power Delivery*, 12(1), 406–417, 1997.
19. H Khalilinia, J Ghaisari. Sub-synchronous resonance damping in total variation ranges of operating conditions using a STATCOM. *IREE*, 4(1), 94–101, 2009.
20. KK Sen. SSSC-static condenser series compensator: Theory modeling and applications. *IEEE Trans Power Delivery*, 13(1), 241–246, 1998.
21. JC Das. Subsynchronous resonance-series compensated HV lines and converter cascades. *Int J Eng Appl*, 2(1), 1–9, 2014.

22. ANSI/IEEE Std. C37.95. Guide for Protective Relaying for Consumer-utility Interconnections, 1973.
23. Power System Relaying Committee. Distribution line protection practices, industry survey analysis. *IEEE Trans,* PAS 102, 3279–3287, 1983.
24. Power System Relaying Committee. EHV protection problems. *IEEE Trans,* PAS 100, 2399–2406, 1981.
25. WA Lewis, LS Trippett. Fundamental basis for distance relaying on three-phase systems. *AIEEE Trans,* 66, 694–708, 1947.
26. VF Cook. *Analysis of Distance Protection.* John Wiley & Sons, New York, 1985.
27. EB Davison, A Wright. Some factors affecting accuracy of distance type protective equipment under earth fault conditions, *Proc IEE,* (110), pp. 1678–1688, 1963.

14
Pilot Protection

In pilot relaying, a communication circuit is used to compare the system conditions at the terminals of the line. This provides selective high-speed fault clearance. High-speed instantaneous clearance of all types of faults has many advantages. It prevents thermal burnouts and improves transient stability. Pilot wire relaying requires metallic or fiber circuit over which the information is transmitted. The system involves transmission of a wave replica and phase angle; we can say it is much analogous to the differential protection of buses, transformers, and generators. The system is applicable to a large range of voltage levels.

With hard-wired differential schemes, a limitation of the distance to which the protection zone can be covered is soon reached depending upon the fault currents, the CT burdens, and the long CT leads. Also intertripping is required. Generally, the schemes can be used for distances no more than approximately 200–350 ft.

14.1 Pilot Systems

A classification of pilot wire systems is shown in Figure 14.1. This may seem complex, but the fundamental of each system and its application will be discussed further. Wave deflection from a fault is relatively a new technology, for high-speed fault clearance on extra-high voltage (EHV) lines.

1. Fault Detection Method

   Directional comparison based on power flow can be further subdivided as follows:
   - Pilot wire
   - Single-phase comparison blocking
   - Dual-phase comparison unblocking
   - Dual-phase comparison transfer trip
   - Segregated phase comparison

2. Phase comparison based on relative phase angles of the currents

   Wave deflections based on wave deflection at the fault point is relatively a new technique.

3. Current-based Systems
   - Current differential
   - AC pilot wire
   - Digital current differential
   - Charge comparison

4. Directional Comparison Systems
    - Directional comparison blocking
    - Directional comparison unblocking
    - Overreaching transfer trip
    - Underreaching transfer trip—can be permissive or nonpermissive

A much simplified classification based on CIGRE working group (consortium for large power system) [1] is shown in Figure 14.2. Many pilot schemes are referred to as permissive schemes. This is a general term which signifies that functional cooperation of two or more relays is required before control action is taken [2–4].

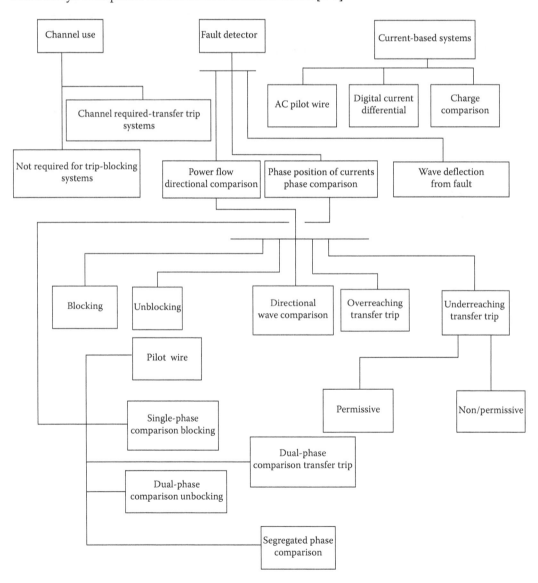

FIGURE 14.1
An overall functional diagram of pilot relaying.

Pilot Protection

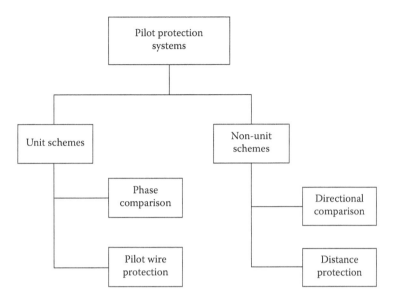

FIGURE 14.2
CIGRE—common forms of pilot wire schemes.

Transfer trip or blocking schemes

In a transfer trip scheme, the relay at each end of the line, recognizing a fault within a designated protection zone, will send a trip signal to the relay at the remote end of the line [5]. The blocking scheme just does the opposite. It sends a blocking signal continuously that prevents remote relay from tripping. The remote relay can operate only when the blocking signal is removed. Within transfer trip and blocking schemes, there are also underreaching and overreaching schemes, both of which imply the integration of distance measuring equipment within the logic. There are also direct tripping and permissive tripping schemes, which will be discussed further.

14.2 Signal Frequencies

The following signal frequencies have been used:

- DC—not in much use.
- Power frequency, sometimes called AC pilot relaying.
- Audio frequencies: These range between 30 and 20,000 Hz (human ear response). Much of telephone industry equipment operates in this frequency range and can be used for control signaling in power systems.
- Power line carrier frequencies: The communication system that couples the high-frequency carrier signal onto power conductors, frequency range 30–600 kHz or lower.
- Radiofrequencies: These lie between 10 kHz and 100,000 MHz. This medium is seldom selected because of the possibility of interference and requirements for licensing.

TABLE 14.1
Selection of Communication Methods

Method	Characteristics
Pilot wire	Generally the distance is limited to 10–25 km. The pilot wire consists of metallic wires. The line should remain a two-terminal line. If it is a three-terminal line, each leg should be less than about 4 km.
Power line carrier	Long transmission lines, also see Table 14.5
Microwave	Protection of long transmission lines, where power line carrier does not have enough channel capacity or a backup to power line carrier
Fiber optic	Protection of short transmission lines, depending on fiber type, multimode or single mode, and also data transmission protocols

- Microwave frequencies: The term is loosely applied to frequencies from 1000 MHz upward. This is characterized by line of sight between microwave antennas.
- Fiber optics: The visible spectrum of light is between a range of 0.4 and 0.7 µm. In practice, a much broader spectrum of 0.3–30 µm is applicable.

Though there are no rigid rules, some characteristics for general guidelines are shown in Table 14.1.

14.3 Metallic Pilot Wire Protection Using Electromechanical Relays

The earlier pilot wire scheme using electromechanical relays and metallic pilot wires for short lines up to approximately 20–30 miles is shown in Figure 14.3. It is a current-only system requiring no PT connections. These provide fast fault clearing for the 100% length of the protected line.

The three-phase currents at each terminal are converted into single-phase voltage, and this single-phase voltage is compared to the remote terminal via a pilot wire, whether the fault is within the protected zone or it is outside the protected zone. For the through-fault

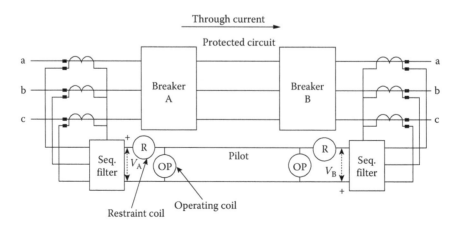

FIGURE 14.3
A pilot wire scheme using metallic pilots and electromechanical relays.

conditions, the filter output voltages V_A and V_B have the polarity as shown. This allows them to support a circulating current flow through the restraint coils and the pilot wires and little current flows through the operating coils. When an internal fault occurs, one of the sequence filter output voltages reverses. For the connections shown, V_B will reverse its polarity, and most of the current flows through the operating coils and the relays at both the terminals will trip. The pickup sensitivity for phase faults is of the order of 4 A for phase-to-ground faults, 0.5 A for phase-to-ground faults, and 2.5–7 A for phase-to-phase faults.

A shielded #19 or larger twisted pair pilot wire is required. There is a limitation of the series resistance and capacitance of the pilot wires. Extraneous voltages are of concern. These voltages arise from a rise in station ground potential by induction from power circuits, which can be reduced by properly shielding the pilot wires. Extraneous voltages will be maximum during a ground fault, as the zero-sequence currents return to the earth; mutual impedance Z_M exists between the power line and the pilot wire. For protection of the pilot wire gaps/arresters, carbon blocks and protector tubes are employed. Also drainage reactors, neutralizing reactors, and insulating transformers are required. To detect pilot wire faults, continuous DC supervision current is applied to the pilot wire. The drainage reactor and the gas tube control the induced voltages on the pilots from neighboring power stations. The neutralizing reactor is connected in series with the pilot wires. The two small capacitors permit magnetizing currents to flow from station ground to remote ground. Most of the station rise voltage drop occurs across the neutralizing reactor, and the pilot wires remain at remote ground potential. A circuit diagram of a practical pilot wire relaying scheme is shown in Figure 14.4. For one terminal end, the other terminal end is symmetrical.

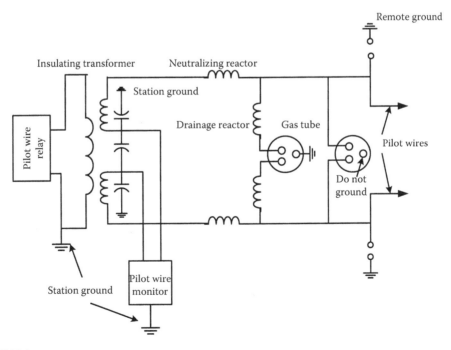

FIGURE 14.4
Details of equipment required at the line one end using pilot wire protection with electromechanical relays.

This pilot wire scheme can be used with point-to-point fiber-optic interface at each end and fiber-optic pilots, which obviates the necessity for drainage reactors and surge arresters at either end.

14.4 Modern Line Current Differential Protection

Unit protection systems for line differential protection are phase comparison, charge comparison, and current differential—the most widely used system is directional comparison. All these determine whether the fault is in the protected zone. The pilot wire scheme discussed above is a form of phase comparison system. Modern digital communication channels provide implementation of segregated phase comparison systems that enhance response to complex faults. Charge comparison systems are an alternate form of current differential protection intended to reduce communication channel bandwidth. This performs a numeric integration of samples of phase and residual currents over half a cycle. The sample integration process takes place between current zeros. Current differential protection combines phase and magnitude of current information in a single comparison. Microprocessor-based line current differential schemes utilize digital communication channels. These can be as follows:

- Digital microwave
- Direct fiber connections
- Synchronous optical network (SONET)
- Synchronous digital hierarchy (SDH)

The areas that impact the security of a line current differential system are as follows:

- The robustness of relay hardware and firmware
- The robustness of applied algorithms and logic
- The ability to deal with channel impairment
- Self-monitoring

14.4.1 Differential Protection of Two-Terminal and Three-Terminal Lines

An MMPR has five current differential elements as shown in Figure 14.5: one for each phase and one for negative sequence and one for ground fault current.

The time sampled and synchronized phase currents are exchanged between two-terminal and three-terminal lines through communication protocols which may be vendor specific over fiber-optic links. Each relay circulates $3I_2$ and $3I_0$ for the line terminals. The differential element phase currents, negative-sequence currents, and zero-sequence currents are compared from each line terminal. The relay performs differential calculations in a peer-to-peer architecture to avoid transfer trip delays. The operating time is of the order of less than a cycle. The phase elements detect three-phase faults. The negative-sequence element detects internal unbalance faults and is restrained from operation when all three of the phase currents exceed three times the nominal current. See an example to follow.

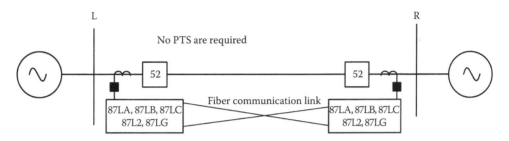

FIGURE 14.5
Line current differential elements in a modern relay.

14.4.2 The Alpha Plane

The alpha plane demonstrates the complex ratio I_R / I_L. It compares individual magnitudes and phase angles of the currents. For a two-terminal line and underbalanced conditions, I_L (local end) and I_R (remote end) are equal in magnitude and opposite in phase. This gives an operating point of $k = 1 < 180° = -1$. This neglects the line charging current. Practically, all the following parameters must be considered:

- Line charging current
- CT saturation
- Channel time delay compensation errors
- Tapped loads

Suppose the current entering the line is $5 < 0°$; then the current leaving the line is $5 < 180°$. Then the ratio of local to remote currents is

$$\frac{I_{AR}}{I_{AL}} = \frac{5 < 180°}{5 < 0°} = 1 < 180° \tag{14.1}$$

$$\frac{I_{BR}}{I_{BL}} = \frac{5 < 60°}{5 < -120°} = 1 < 180° \tag{14.2}$$

$$\frac{I_{CR}}{I_{CL}} = \frac{5 < -60°}{5 < 120°} = 1 < 180° \tag{14.3}$$

On the alpha plane, this can be shown as in Figure 14.6. All through-fault currents regardless of angle with respect to system voltages can be represented as shown in the figure.

If we define the operating current and restraint current as

$$I_{op} = |I_L + I_R|$$

$$I_{RT} = k|I_L - I_R| \tag{14.4}$$

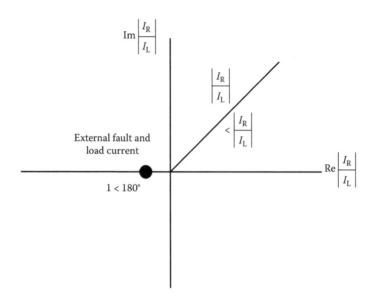

FIGURE 14.6
Alpha plane representation; complex ratio of remote to local currents.

and consider

$$I_{op} > kI_{RT} \tag{14.5}$$

with the relation

$$\left|\frac{I_R}{I_L}\right| = a + jb \tag{14.6}$$

Equation 8.9 can be written as

$$|1 + a + jb| \geq |1 - a - jb| \tag{14.7}$$

Expanding,

$$a^2 + b^2 + 2\frac{1+k^2}{1-k^2}a + 1 \geq 0 \tag{14.8}$$

This is the equation of a circle in the alpha plane, the operating characteristics of the differential relay (Figure 14.7). The relay trips when the alpha plane ratio travels outside the restraint region and the difference in current is above the settable threshold. The relay restrains when the alpha plane ratios remain inside the restraint region or when there is insufficient current.

This shape is determined by two settings. The angle setting determines the angular extent of the restraint region. The radius determines the outer region of the restraint region. The reciprocal of R gives the inner radius. All three types of elements qualify for trips with appropriate differential settings.

Pilot Protection

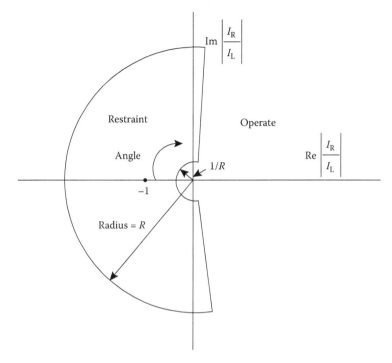

FIGURE 14.7
Characteristics of a new differential scheme in alpha plane.

14.4.3 CT Saturation

According to one manufacturer, the maximum CT burden that avoids saturation is given by

$$Z_B < \frac{V_S}{I_f \left(\dfrac{X}{R}+1\right)} \tag{14.9}$$

where Z_B is the permissible CT burden, I_f is the fault current, and V_S is the C voltage class of the CT. Also see Chapter 3.

The practical settings are as follows:

Set 87 angle at 195°.

The 87L pickup is set at 6 A conservatively, for a 5 A CT.

The outer radius is set at 6 and the inner radius, therefore, is 1/6.

Note that 87 angle and radius are common to all differential elements like negative sequence and ground fault.

The negative-sequence line pickup is set at 0.5 A, that is, 10% of system nominal current. The setting should be above maximum line charging current unbalance.

Set a ground fault differential setting also equal to 0.5 A. For load currents <1/3 of nominal current, the ground fault sensitivity is determined by minimum ground fault and

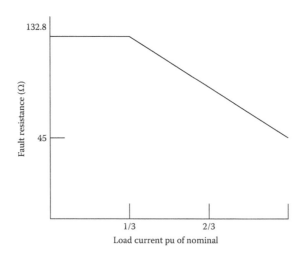

FIGURE 14.8
Ground fault sensitivity of differential elements in terms of secondary ohms, see text.

negative-sequence differential settings. Above 1/3 of nominal current, the ground fault sensitivity is determined by ratio of magnitude of negative-sequence current to positive-sequence current for negative-sequence setting and by ratio of zero-sequence to positive-sequence current for 87G setting. A negative-sequence element enables when the ratio is >0.05 from at least one terminal. Figure 14.8 shows the sensitivity in terms of secondary ohms. To prevent misoperation of negative-sequence element for an external fault due to CT saturation at one line terminal, it is blocked if all three-phase currents from any terminal exceed 300% of nominal current. The operating time for each type of fault is in subcycle region.

14.4.4 Three-Terminal Protection

The currents are vectorially compounded from two of the terminals to produce the remote current. This means that effectively the line is converted into a two-terminal line. For internal faults with no out feed and with no CT saturation, any possible combination of combining local and remote currents is acceptable, see Figure 14.9 and Table 14.2.

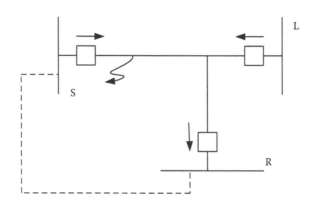

FIGURE 14.9
Internal fault on a three-terminal line may produce out feed on one terminal.

Pilot Protection

TABLE 14.2

Three-Terminal Line, See Figure 14.9

Options (All are Okay)	1	2	3
Remote current	Current from L+S	Current from R +S	Current from L and R
Local current	Current from R	Current from L	Current from S

Figure 14.9 shows that for an internal fault, the relay at R can experience an out feed. The direction of the fault current flow is as shown. Consider a ground fault. All three relays process the ground fault differential elements using the combinations shown in Table 14.2. Consider the relay at R. It has three trip/restraint decisions to make based on this table:

- Trip/restraint decision based on current from R as local current and the current from terminals L and S as the remote current.
- Trip/restraint decision based on current from terminal L as the local current and the currents from terminals R and S as the remote current.
- Trip/restraint decision based on current from S as the local current and the current from terminals L and R as the remote current.

The relay then selects the decision produced by the processing method that used the largest ground fault current as the local current.

Similar description applies to phase differential and negative-sequence differential elements.

14.4.5 Enhanced Current Differential Characteristics

The key factors in defining the required shape of the differential relay in alpha plane are as follows:

- Power system impedance nonhomogeneity
- CT saturation
- Low-frequency oscillations in compensated lines

The effect of charging currents and system power angle can be eliminated by using negative- or zero-sequence currents. Figure 14.10 shows the interconnection of relays at two terminals through direct fiber and also direct transfer trip (DTT). Each relay samples its analog input current (A/D converter); the sampling rate varies and in this respect various line current differential relays differ considerably. The channels traditionally used for line current differential protection are limited in bandwidth; 64 kbps is the typical value [6,7].

When working with phasors, twice the bandwidth is needed to send real and imaginary parts of the current. The phasor exchange rate is not high enough to facilitate phasor interpolation and the relay sampling clocks must be synchronized (through GPS signals). The implementations may use communication channels to force synchronization of the sampling clocks. This is one of the key elements of line current differential system.

The local current data are communicated to remote terminals i_{TX}–i_{RX} in Figure 14.10. Each relay then aligns the data and runs its differential trip equations. If a relay does not receive all the data from its remote terminal—say because of lack of communication—it is said to operate in a slave mode and DTT from the master terminal allows slave relay to

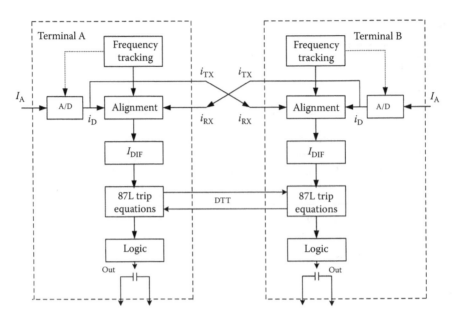

FIGURE 14.10
Simplified architecture of a modern line differential system.

issue trip commands to their circuit breakers. Dedicated point-to-point channels remove any third-party devices and the failure modes associated with these.

Blocking and operating functions are shaped, as in Figure 14.11 (the new differential element characteristics for protection of transmission lines). The relay restraining region is the area between two circular arcs and two straight lines and includes $a = -1$ point. Amplitude comparison provides the circular part of the characteristics with independent

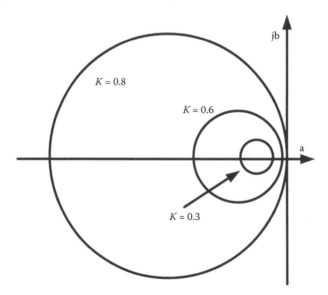

FIGURE 14.11
Alpha plane characteristics of differential relay described by Equation 14.8.

settings R and $1/R$ circle radii. Phase comparison provides the linear part of the characteristics and defines the angular setting α. The characteristics are symmetrical with reference to a-axis.

The traditional circular characteristics of the differential relay have low tolerance to the channel asymmetry. The new characteristics provide better stability, reliability, and sensitivity. The relays have the capability of single-phase reclosing and tripping, self-monitoring, communication channel loss, and a host of ancillary protective functions such as backup distance, circuit breaker failure synchronization, and directional overcurrent. The robustness of relay hardware and firmware is of importance.

An operating time in the subcycle range is obtained, and low settings of 0.5 A pickup for differential current, two-phase, and line-to-ground fault are possible.

14.5 Direct Underreaching Transfer Trip

Direct underreaching transfer trip is also called direct or nonpermissive underreaching transfer trip scheme. Transfer trip pilot wire protection provides high-speed tripping of line faults throughout the length of the line and also backup protection of adjacent sections.

Referring to Figure 14.12a, note that zone 1 reach is set at 80% of the total line length. Consider a zone 1 fault detected by relay R in Figure 14.12a. Then both zone 1 and zone 3 relays pick up instantaneously at R. Zone 3 pickup starts timers for zones 2 and 3 (Figure 14.12b). The zone 1 relay pickup sends an immediate signal to a local breaker trip and simultaneously sends a trip signal to relay Q at Bus K. At breaker Q, a tripping signal is received that actuates tripping relay UT. This completes the line trip before T2 times out. A drawback is that possibly a noisy communication channel will cause a transfer trip receiving circuit to detect a signal without its real transmission.

The security can be improved as shown in Figure 14.12c. The only difference between Figure 14.12b and c is that a normally closed contact of a guard relay, designated GD, is introduced in the circuit. The transmitter sends a tone to the remote terminal that picks up the guard receiver and contact GD is held open. When a fault is detected, the transmitter shifts the transmission frequency from the guard frequency to trip frequency, thus closing the GD contacts.

14.6 Permissive Underreaching Transfer Trip

Here, the configuration is the same as in direct underreaching transfer trip except that a second set of instantaneous elements, set to overreach the far end of the line, are employed. This is shown in Figure 14.13a.

Zone 1 is underreaching transfer trip as before. The instantaneous overcurrent relays are overreaching and fundamentally fault detectors. These are the permissive devices and must operate to permit a received trip signal to trip the circuit. The control logic is shown in Figure 14.13b.

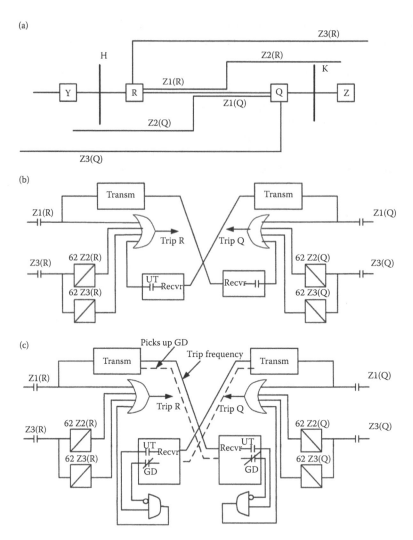

FIGURE 14.12
(a) Distance relay reach and zoned time delay for transfer trip, (b) logic for direct underreaching transfer trip, and (c) logic modified with guard frequency security.

14.7 Direct Overreaching Transfer Trip

DTT overreaching scheme is shown in Figure 14.14a, and the trip logic is shown in Figure 14.14b. Note that zone 1 is still underreaching as before but zones 2 and 3 are overreaching. The directional relay of zone 2 supervises the transfer tip relay UT. Consider a close in fault at R. The zone 2 relay picks up and sends a signal to Q and the transfer trip relay UT actuates clearing the fault instantaneously from both ends. Now consider a fault to the left of R. The zone 2 relays at Q pick up, but the transfer trip relay is not energized since zone 2 at R refuses to recognize this fault and send a trip signal to Q.

If a transfer trip scheme is used with PLC, the transfer trip signal may fail over a faulty transmission line.

Pilot Protection

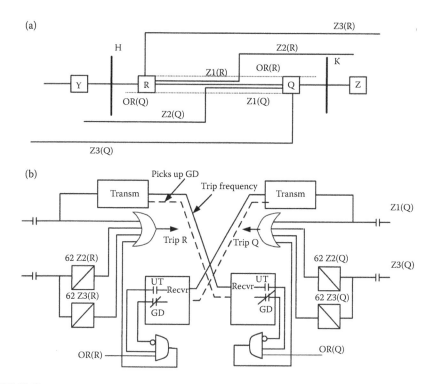

FIGURE 14.13
(a) Permissive underreaching transfer trip and (b) logic representation of a permissive underreaching transfer trip.

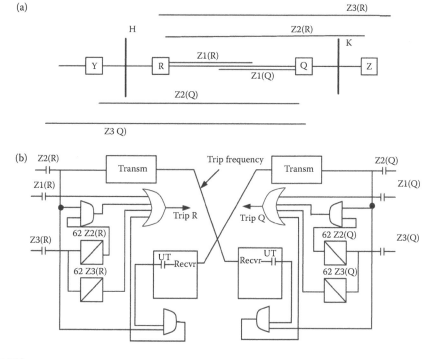

FIGURE 14.14
(a) DTT overreaching scheme and (b) logic diagram.

14.8 Blocking and Unblocking Pilot Protection

As described before, a blocking scheme is reverse of transfer trip schemes. In a blocking scheme, the sending relay monitors the protected line, and it also monitors the region behind the sending relay. Any fault behind the protected line triggers a signal to be transmitted to the remote need to prevent tripping, by opening the circuit of the remote relay.

14.8.1 Direct Blocking Scheme

A direct blocking scheme is shown in Figure 14.15a and its logic of tripping in Figure 14.15b. The relays normally applied are distance elements in the forward and reverse directions.

Consider a fault to the left of R, not in the protection zone of the line. This fault is not seen at R, but it is seen at Q and the elements Z2(Q) and Z3(Q) will pick up. But these are prevented from tripping Q and the relay at R sends a blocking signal to relay at Q. It makes it attractive for a PLC protection, as the signal is transmitted over an unfaulted line. ZR(R)

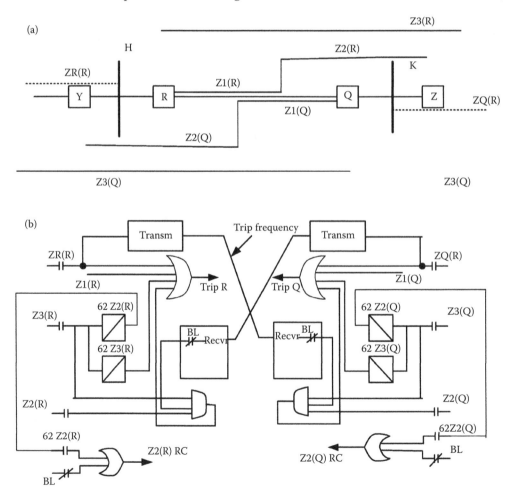

FIGURE 14.15
(a) Distance time blocking scheme and (b) logic diagram.

Pilot Protection

is the reverse blocking relay that sends a signal to Q. The reverse looking ZR is faster than the zone 2 relays.

It is clear that a fault in zone 1 is cleared instantaneously at both ends R and Q. Now consider a fault that is close to R. Such a fault will be cleared by zone 1 relay at R. The fault is also picked up by zone 2 and zone 3 relays at Q, which will operate after a time delay. Methods have been developed to accelerate the fault time in zone 2. For no blocking signal, the BL, blocking contacts, remain closed. This energizes the reach change coil Z2(R) RC, which changes the reach of zone 2.

14.8.2 Directional Comparison Blocking Scheme

The scheme is shown in Figure 14.16 and the logic of operation for external and internal faults is described in Table 14.3. Here, the protective relays for the phase and ground protection are designated as P(R) and P(Q). These are set to overreach so that these will pick up on internal faults. Usually, the overreach can be set 125%–150% of the line length. S(R) and S(Q) are the restraint units. These are set with different reach as shown. Both the protective units and the starting units must be of the same type.

The pilot signaling used for this type is a simple on–off PLC carrier. No signal is transmitted under normal conditions since the starting units operate only under fault conditions.

The directional comparison scheme is widely used because of flexibility and reliability of communications. Note that a communication channel is not required for tripping, so the problems with incorrect signaling are eliminated. Overtripping can occur if the signal channel fails or the communication fails to operate within the starting unit reach settings.

14.8.3 Directional Comparison Unblocking Scheme

This scheme is shown in Figure 14.17a. A continuous blocking signal is transmitted except for internal faults. The communication usually uses frequency shift PLC signal. The blocking or guard frequency is continuously transmitted under no fault conditions. On an internal fault, the frequency is shifted to a trip or unblocking frequency. The transmitted power may be increased during unblocking frequency transmission.

The logic for internal and external faults is shown in Table 14.4. The phase and ground detectors must overreach both ends. The reliability and security makes their application to PLC. The scheme is also applicable to multiterminal lines. The logic is shown in Figure 14.17b.

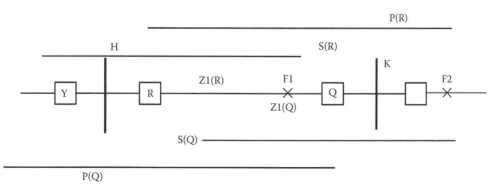

FIGURE 14.16
Directional comparison blocking scheme.

TABLE 14.3

Operation of Directional Comparison Blocking Scheme for External and Internal Faults

Fault	Operations at H	Operations at K	Remarks

Internal F1	P(R) operates
	S(R) may or may not operate
	P(R)operation prevents
	   sending blocking signal to Q
	Breaker R tripped
		P(Q) operates
		S(Q) may or may not operate
		P(Q) operation prevents
		   sending a blocking signal to R
		Breaker Q tripped
			Blocking signal not required for tripping breakers R and Q, which trip without time delay

External fault F2	P(R) operates
	S(R) does not see a fault
	Blocking signal received at R
	   from Q
	No trip
		S(Q) operates
		Transmitter keyed in
		Blocking signal sent to R P(Q)
		   does not see a fault
		No trip	

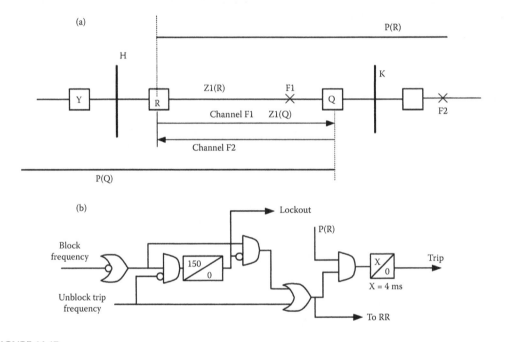

FIGURE 14.17
(a) Directional comparison unblocking scheme and (b) logic diagram.

TABLE 14.4

Trip Logic for Directional Comparison Unblocking

Type of Fault	At H	At K

Internal	P(R) operates
	F1 channel shifts to unblock
	Loss of block and/or receipt of
	   unblock (F2) trips R
		P(Q) operates
		F2 channel shifts to unblock
		Loss of block and/or receipt of unblock (F1)
		   trips Q

External	P(R) operates F1 channel shifts to unblock F2 channel continues to block No trip	P(Q) does not see a fault Loss of block and/or receipt of unblock F1 results in no trip

14.9 Phase Comparison Schemes

This is the most common type of unit protection for the transmission lines. The five schemes are as follows:

- Single-phase comparison blocking
- Dual-phase comparison blocking
- Dual-phase comparison transfer trip
- Segregated phase comparison
- Power line carrier protection

The fundamental concept can be gathered from Figure 14.18a. Square waves are generated based on current crossing through zero, or through a positive and negative threshold value. First consider that there are no filters. The square waves thus generated are shown in Figure 14.18a. These current crossing patterns are transmitted to the other end of the

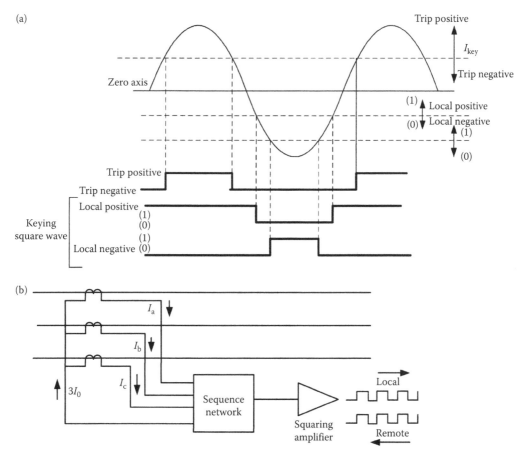

FIGURE 14.18
(a) Square wave processing from a phase current waveform and (b) square wave generator with sequence network combining three-phase currents.

line and compared. If it is determined that the currents at both ends are entering the line, an internal fault will be indicated.

In some phase comparison systems, the currents are processed through a filter, as shown in Figure 14.18b, which generated square waves at either end. The transmission delays are compensated. If the two currents are out of phase, this means that the currents from both ends are flowing into the lines, indicative of an internal fault.

14.9.1 Single-Phase Comparison Blocking

This is a current-only comparison system and requires an on–off PLC channel (Figure 14.19a). Two fault detectors, one at each end, are used. FD1 is called a carrier

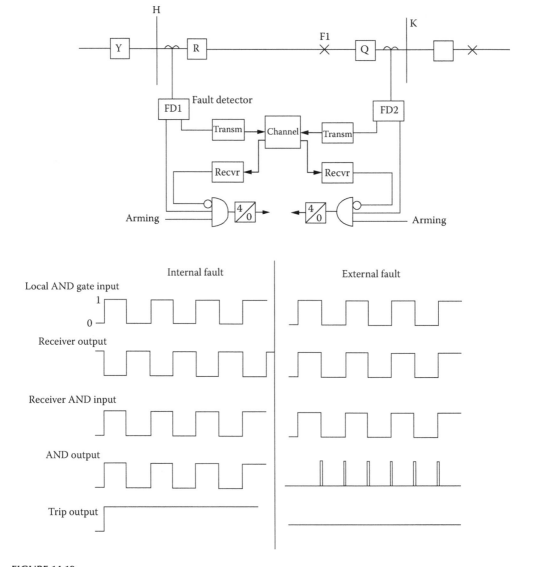

FIGURE 14.19
A single-phase comparison blocking scheme.

start and is more sensitive than FD2. It keys the on–off transmitter. FD2 has a higher pickup and armed to trip. Once FD2 operates, phase comparison can begin.

For an internal fault, the fault detectors at both ends of the line operate. A flip-flop is energized if the input to AND gate continues for more than 4 ms. This provides a continuous trip output which is supervised by FD2 operation. For this condition, the square wave inputs from the local and receiver outputs are in phase, and tripping occurs. A trip will occur if the currents at two ends are up to 90° out of phase.

For an external fault, blocking is continuous due to the phase of the receiver output being out of phase with the local square wave. Generally, the through-fault currents are approximately in phase.

14.9.2 Dual-Phase Comparison Blocking

The dual-phase comparison makes comparison on both cycles of the current wave, positive and negative. It requires a dual communication channel with one frequency for each line terminal.

This is shown in Figure 14.20. One frequency is called the mark and the other, space. As mark or space is transmitted continuously, it provides a continuous channel monitoring. The transmitter is keyed to its mark frequency when the square wave goes positive and is keyed to space frequency when the square wave goes negative.

For internal faults:

- The single-phase outputs of the sequence current networks are in phase.
- A delay circuit is tuned to delay the local square wave by a time equal to channel delay (not shown).
- The network develops two waves: local positive and the local negative.
- The positive wave is compared with receiver mark output in AND1.
- The local negative wave is positive during negative half cycle of sequence current and is compared with receiver's space output in AND2.
- For internal fault, the local positive and receiver mark waves are in phase, resulting in an output from AND1.
- Similarly the negative and receiver(s) space waveforms are in phase, giving an output at AND2.
- If an arming signal is received from the fault detectors or protective relay, AND3 has an output.
- If this signal persists for 4 ms, the output initiates breaker trip.

For external faults, there is 180° reversal of one of the currents shifts in the square waves and the other AND2 gate does not have an output.

14.9.3 Segregated Phase Comparison

We discussed that in single-phase comparison we generate only one current replica and convert it to square waves (Figure 14.18b). In dual-phase comparison, a comparison is made both on positive and negative cycles of the current. In a phase segregated system (Figure 14.21), each current is measured individually. Thus, these can be of a frequency other than the fundamental frequency; sub- and supersynchronous harmonics can be measured. *The*

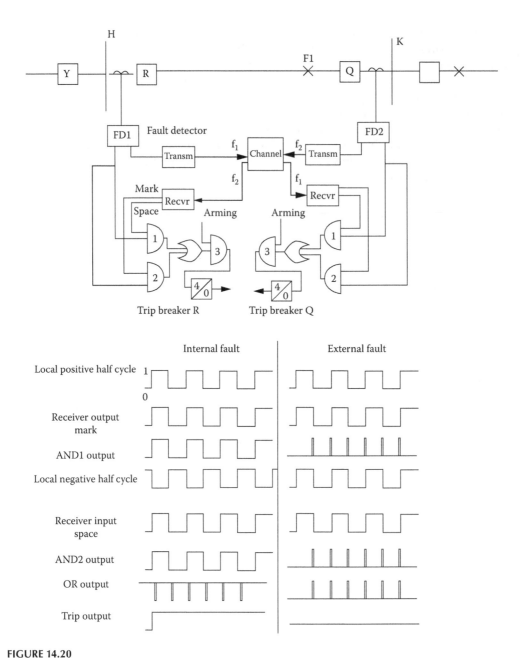

FIGURE 14.20
A dual-phase comparison blocking scheme.

system was primarily developed for series compensated transmission lines. The direct comparison of currents of each phase provides a reliable method of fault detection and selectivity. On series compensated lines, the problems are as follows:

- Voltage reversals caused by negative reactance to the series compensation
- Phase unbalance
- Abnormal frequencies of 200–400 Hz during the fault and postfault period

Figure 14.22 shows phase and ground currents (indicative) that can occur at the two ends of a series compensated line.

The system can operate successfully irrespective of the location of series capacitors and degree of compensation. A disadvantage is that it requires four pilot signals per terminal. One rectangular wave signal required 1 kHz bandwidth, so the pilot signals can be transmitted using 4 kHz voice frequency channel. When microwave links are available, several high bandwidth signal channels are also available, and phase segregated phase comparison is favored. It can also be used for protection of paralleled transmission lines.

The system can be divided into two types: a two-subsystem scheme and a three-subsystem scheme. The three subsystems have a subsystem for each phase current (Figure 14.23). The voltage output to the squaring amplifier is exactly proportional to the primary currents. The outputs of these amplifiers are used to key individual channels and through local delay timers to provide local square waves for comparison. The square wave comparison is made independently for each current in separate subsystems.

A two-subsystem scheme operates from delta currents (I_a-I_b) for phase faults and $3I_0$ for ground faults. Both systems incorporate off-set keying. Three subsystems are illustrated in Figure 14.24. See References [8–12] for further reading of the various schemes described above.

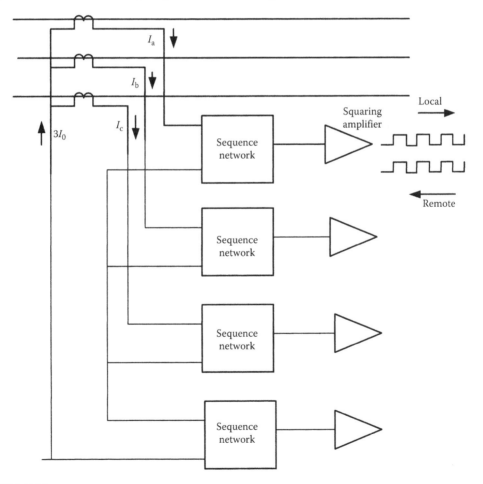

FIGURE 14.21
The segregated phase comparison waveform generation.

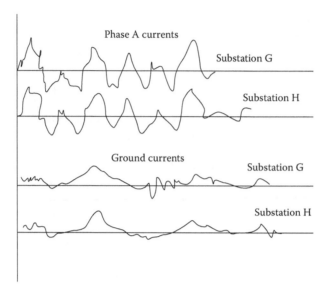

FIGURE 14.22
Phase *a* currents and ground currents at the terminals of a series compensated line or an external fault.

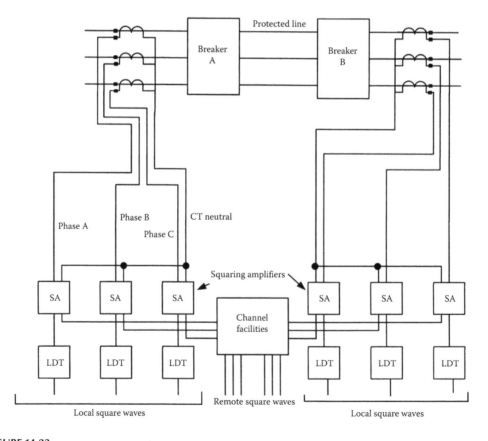

FIGURE 14.23
Three-substation segregated phase comparison system.

Pilot Protection

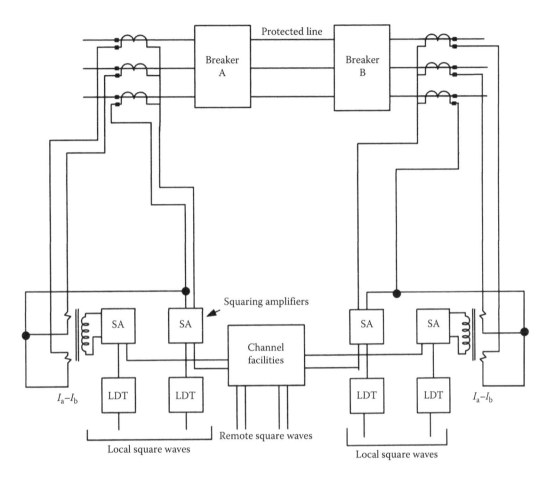

FIGURE 14.24
Two-substation segregated phase comparison system.

14.10 Power Line Carrier

Power line carrier utilizes the power conductors themselves at frequencies between 30 and 300 kHz, coupled to the transmission line with relatively small losses. Below 30 kHz, coupling is impractical and above 300 kHz, the line losses are higher and the radiations into space with unwarranted interferences are of consideration. Table 14.5 summarizes the advantages and disadvantages of this type of communication channel.

This application is fairly old, first used in the 1930s. It is still used and versatile for multiterminal lines. Simultaneous high-speed tripping is permitted at the line terminals. For an external fault, the information that the current flows out at one of the terminals is used to block tripping. It can be termed as directional comparison blocking system.

The power line carrier may use an on–off power line carrier (most common) frequency shift, or single-sideband (SSb) operation. The on–off signal is used to block tripping in a section not in the protection zone, and frequency shift is used for either blocking or transfer trip. When functions are combined into SSb channel, audio tones are used to modulate the carrier as explained further.

TABLE 14.5

Power Line Carrier—Advantages and Disadvantages

Advantages	Disadvantages
Reliable, as the power conductors themselves are a carrier	Limited frequency spectrum in some areas
Economical for long lines, where wire-line channels can be expensive	Susceptible to power line noise
Easy maintenance of terminal equipment	Line traps and coupling equipment required
Channel can extend over several line sections	
Relaying channel may include voice maintenance channel	

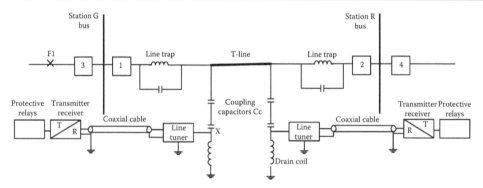

FIGURE 14.25
Main carrier components for protective relaying channel.

The major carrier components are illustrated in Figure 14.25. The description of each item is as follows:

The line traps: These consist of a parallel tuned resonant circuit that presents high impedance to the carrier frequency, but negligible impedance at power frequency. Thus, the carrier frequency is confined to the terminals and power frequency currents can have unabated flow. The line traps are designed to continuously carry the power frequency current and also fault currents without overheating or mechanical damage. These are often suspended from the line conductors themselves. We have already discussed these in Section 3.15.

14.10.1 Coupling Capacitor and Drain Coils

The coupling capacitor C_c is a series of stacked capacitors mounted inside an insulator and has a capacitance value of the order of 0.002–0.2°F, depending on the system voltage. For a system voltage of 115 kV, an acceptance of 0.025 µF, and for a system voltage of 765 kV, the value reduces to 0.0036 µF.

The drain coil mounted on the base of the coupling capacitor provides a low impedance path for flow of power frequency current through the capacitor and drain oil to ground. This minimizes the ground potential developed from point X to ground. At carrier frequencies, the drain coil has a high impedance, minimizing r–f losses to ground.

14.10.2 Line Tuner

The line tuner is a low-loss coupling between the coaxial cable and transmission line. Both resonant and broad-band line tuners are used. Single-frequency and double-frequency tuners are available. In a single-frequency tuner, the inductive reactance of coil cancels the capacitive reactance of coupling capacitor Cc and isolating capacitor Cs at one frequency. This provides a low-loss coupling circuit for carrier frequency energy.

The two-frequency resonant tuner provides low-loss coupling at two frequencies from a single coaxial cable to the power line. For two frequency tuners, the minimum separation is 25% of the lower frequency.

Figure 14.26a,b shows single-frequency and double-frequency tuner configurations. And Figure 14.27a through c shows typical response curves of typical single-frequency, double-frequency, and resonant link tuners.

14.10.3 Coaxial Cable

Low-loss concentric cable, type RG8/U, is used to connect the cable line tuner with transmitter receiver assemblies. It has a characteristic impedance of 52 Ω and an attenuation of about 0.4–0.9 dB/1000 ft, which increase gradually over the 30–300 kHz frequency range.

14.10.4 Transmitter/Receiver

Carrier transmitters and receivers are used for on–off operation or for the frequency shift operation (FSK). The transmitter wattage is 1–10 W, and the receiver bandwidth is

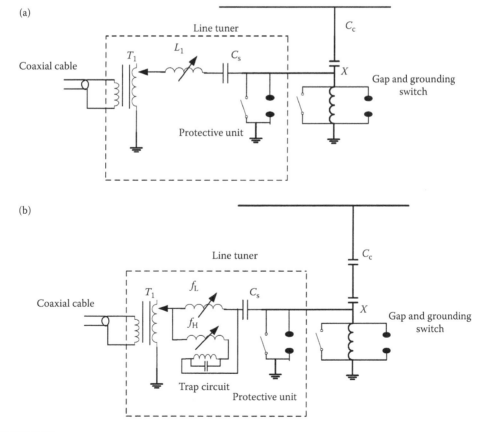

FIGURE 14.26
(a) Single-frequency tuner and (b) double-frequency tuner for a single coaxial cable.

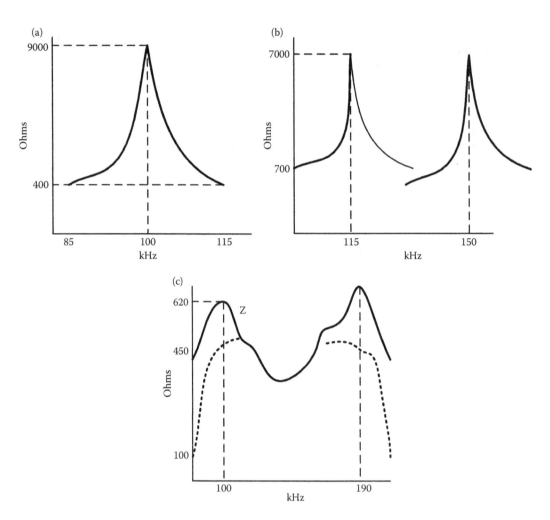

FIGURE 14.27
(a through c) Typical response characteristics of single-frequency, double-frequency, and resonant line traps.

500 Hz or 1500 Hz. The main applications of on–off transmitters are directional comparison blocking and phase comparison blocking. Transmitters of 10 W have been used for directional comparison with electromagnetic relays on high loss or high noise lines. The on–off carrier is usually operated at a single frequency, as only one station is transmitting at a time.

Since carrier signal is transmitted only under fault, the carrier channel is available for voice communication for maintenance. Table 14.6 shows carrier terminal transmitters.

14.10.4.1 On–Off Carrier

On–off operation is used for relay blocking systems, and prevents tripping of breakers in unfaulted section during an external fault. Power levels of 1–10 W are used. The transmitter consists of a crystal oscillator, amplifier stages, a power amplifier, and an output filter to keep harmonics at low levels. Voltage regulators keep the transmitter output and receiver sensitivity constant over expected range of battery voltage variations. Superheterodyne

TABLE 14.6

Carrier Transmitters

Operation	Transmitter (W)	Receiver Bandwidth	Main Applications
On–Off	1	500 Hz	EM (electromechanical-type relays) directional comparison blocking
On–Off	10	1500 Hz	EM and SS (solid-state type of relays) directional and phase comparison blocking
On–Off	10	500 Hz	EM directional comparison relays on high loss or high-noise lines
FSK	1/10	Narrow band (220 Hz)	All DTTs and EM transfer trip line relaying
FSK	1/10	Wide band (500 Hz)	SS directional comparison blocking
FSK	10/10	Wide band (500 Hz)	Dual-phase comparison line relaying

receivers have fixed filters that provide a constant bandwidth over the carrier band. Since a carrier is transmitted only during fault, the relaying carrier channel is suitable for other functions; a modulator and an amplitude modulated receiver can be added.

14.10.4.2 Frequency Shift Carrier: TCF and FSK

Power line carrier can be used to protect transformers with no primary breaker. The tripping signal for transformer faults is transmitted over the line to remote breaker. The receiver output relay directly trips the breaker; thus, no spurious output from the receiver can be tolerated. A frequency shift carrier is used where such security against incorrect operation is required. In this mode, a carrier signal is continuously transmitted at a guard frequency of 100 Hz above the nominal (30–300 kHz range) frequency. The receiver at the remoter end has a discriminator and limiter stages. Reception of the guard frequency saturated the receiver, making it insensitive to most noise voltages over the line. To transmit a trip signal, the transmitter frequency is shifted down 200 Hz producing a discriminator output of opposite polarity. This trips a breaker at the receiving end of the line.

When the protection requires duplex frequency shift, a considerable saving in a channel spectrum can be made by using a double-shift transmitter and two separate receivers. For example, a directional comparison unblocking (wide-band FSK receiver) can be used for line protection and a transfer trip channel (narrow-band FSK receiver) can be used for breaker failure or transformer band protection.

With a blocking on–off carrier, there is one transmitter and one receiver. With frequency shift carrier, there are two or more transmitters connected through a single tuner and coupling capacitor. All transmitters and receivers cannot be connected to the same coaxial cable as the transmitter outputs could overload the receiver input circuits and impair their ability to receive remote signals, and intermodulation distortion can occur. To overcome this problem, hybrid circuits are used. Figure 14.28a shows a reactance type hybrid circuit. The output load on a hybrid is a combination of coaxial cable, line tuner, coupling capacitor, and the surge impedance of the line. To obtain a satisfactory balance over a frequency range, the hybrid circuit is used. A tunable series resonant L–C circuit is added to balance resistor R. This scheme produces an acceptable balance over 6% bandwidth through most of the carrier spectrum. This circuit should be used when a single hybrid unit is required to separate two transmitters or a transmitter and a receiver. If more than one hybrid is needed, then the hybrid connected to a coaxial

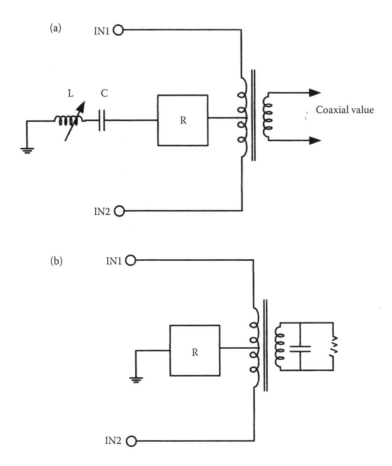

FIGURE 14.28
(a) Reactance type hybrid and (b) resistance type hybrid.

cable must be reactance type (Figure 14.28a), while the others can be of resistance type (Figure 14.28b).

14.10.5 Audio Tone Channels

Audio tones transmitted over wire-line or microwave communication circuits provide a second type of channel for protective relaying systems. These are used to transmit relay intelligence from one terminal to another:

- On relatively short lines, due to economical advantage
- In metropolitan areas with UG cables, that preset high carrier frequency attenuation
- Where pilot wire resistance and capacitance exceed limits for AC pilot wire relaying
- Where metallic pairs cannot be leased
- In areas with crowded power line carrier spectra

Table 14.7 shows types of tones and their applications. Also see Reference [13].

Pilot Protection

TABLE 14.7

Audio Tone Channels

Type	Frequency Range	Application
TA3	1200–3000 Hz, bandwidth 170–340 Hz	Transfer trip application with EM or SS
DIT-1	Two closely spaced, narrow-band frequency shift, voice band tone subchannels for transmission over telephone lines, microwave or single-sideband power line carrier circuits	Transfer trip applications. Dependable and good security
TA2.2	1500–2800 Hz, with available bandwidth of 340 and 600 Hz. Bandwidth of 600 Hz is used in all cases, except where a three-terminal line is protected with a two-wire audio channel, whereas 340 Hz bandwidth is required for additional frequencies	Phase comparison relaying
DIT-4	It is a multichannel frequency shift tone system. Each system is composed of four subchannels, each of which is a basic frequency shift tone channel	Segregated phase comparison relaying

14.10.6 Microwave Channels

A microwave communication channel for protective relaying offers many advantages:

- It provides a channel independent of power line. Thus, disturbances on the power line such as arcing faults, noise, and switching surges do no impact the integrity.
- Once the RF path is established, the incremental cost of adding channels is low.
- Due to large modulation bandwidth at microwave frequencies, many wide-band channels are available for high-speed solid-state relaying.

Microwave technology can be applied to any of the pilot wire relaying schemes using power line carrier or audio tone channels.

14.11 Modal Analysis

The performance of the carrier frequencies must be carefully analyzed, even before the line is built. Modal analysis provides such an approach. We discussed modal analysis of transmission lines in Section 3.10, Volume 2.

- Any set of phase currents or voltages at a point can be resolved into three sets of natural mode components.
- At any point mode components add up to the phase quantities.
- The mode characteristic impedance, which is the ratio of the mode voltage to the mode current, is constant.
- Each mode propagates with specific attenuation per unit length.
- A set of phase components corresponding to one mode only cannot be resolved into other modes.

Mode 3 is a high attenuation mode, Mode 2 is a medium attenuation mode, and Mode 1 is the least attenuated of the three modes. It makes carrier channels possible on long EHV

TABLE 14.8

Mode Attenuation and Phase Velocities

Mode	Attenuation (dB/mile)	Phase Velocity Relative to Mode 1
	30–300 kHz	
1	0.01–0.03 to 0.07–0.09	1.0
2	0.09–0.1 to 0.4–0.5	0.98–0.995
3	1.5-3.0 at 100 kHz	0.9

lines. For a horizontal line configuration with two ground conductors, the mode attenuation and phase velocities are shown in Table 14.8.

- Mode 1 has equal currents flowing via phases *a* and *c*; two units of currents returning through phase *b*.
- Mode 2 has current flowing out via phase *a* and returning via phase *c*.
- Mode 3 has equal currents in all three phases and ground return.

Considering mode 1 only, the efficiencies vary with respect to couplings.

- Center-to-outer phase has the maximum efficiency.
- Center to ground, outer to outer, and outer to ground for short lines only are the couplings with progressively lower efficiencies.

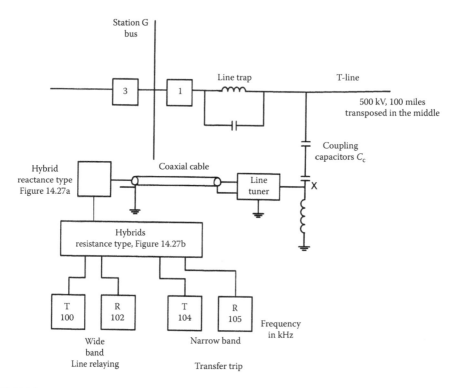

FIGURE 14.29
Terminal equipment at one end of a 500 kV line.

Other Losses

The other losses which must be considered are as follows:

RF hybrid loss

The conventional RF hybrid dissipates one-half or 3 dB of power at any carrier transmitter connected to it.

Coupling loss

Coupling loss is caused by resistive component of the line tuner and capacitor impedance. Depending on the single-frequency, double-frequency, or wide-band tuner, the losses are of the order of 0.5–1 (single-frequency), 1.0–2 (double-frequency), and 1.0–3 dB (hi-coupler, wide band).

Shunt loss

Shunt loss is the carrier loss through the line trap impedance and any shunt path to ground. Loss depends on the surge impedance of the line, the trap impedance, and the impedance of the local station bus. Again typical losses are of 1–3 dB, depending on single-frequency, double-frequency, or wide-band tuner.

The terminal equipment (at one terminal) for a 100 miles 500 kV line is shown in Figure 14.29. The losses in this configuration are calculated in Reference [10]. The line is transposed in the middle, which also causes some losses.

References

1. G Zigler, ed. Application guide on protection of complex transmission network configurations, CIGRE SC34-Working Group, 1991.
2. IEEE Power Line Carrier Working Group Report. Power line carrier practices and experience, IEEE Paper 94 SM 428-3, PWRD, 1994.
3. DWP Thomas, C Christopoulous. Ultra high speed protection of series compensate lines. *IEEE Trans Power Delivery*, PWRD-71, 139–145, 1992.
4. VM. Ermolenko, VF Lachugin et al. High speed wave directional relay protection of EHV lines, CIGRE Paper 34-11, *Presented at 1988 CIGRE Conference, Paris*, August 28–September 3, 1988.
5. M Chamia, S Liberman. Ultra high speed relay for EHV/UHV transmission lines-development design and applications. *IEEE Trans*, PAS-97, 2104–2016, 1978.
6. G Benmouyal. Trajectories of line current differential faults in the Alpha plane, *Proceedings of the 32nd Annual Western Protective Relay Conference*, Spokane, WA, October, 2005.
7. SEL 311L-0,6 Relay, Instruction Manual 20040412, SEL Inc.
8. GEC. The Art of Protective Relaying, Transmission and Subtransmission Lines. Publication GET 7206A, Switchgear Division, Philadelphia, 1971.
9. L Lohage et al. Protective systems using telecommunications, CIGRE WG 34/35-05, 1987.
10. Westinghouse Electric Corporation. *Applied Protective Relaying*. Westinghouse Electric Corporation, Coral Springs, FL, 1979.
11. IEEE Committee Report. Pilot relaying performance analysis. *IEEE Trans*, PWRD-5, 85–102, 1990.
12. IEEE Std. 487. Guide for the Protection of Wire Line Communication Facilities Serving Electric Power Stations.
13. ANSI/IEEE Std. C37.93. Guide for Protective Relay Applications of Audio Tones over Telephone Channels.

15
Power System Stability

Power system stability is a vast subject amply covered in the current literature. This chapter provided an introduction of and concentrates on the effect of the following:

- Fast protective relaying
- Load shedding
- Fast response excitation systems

on enhancement of power system stability limits. It is important to have knowledge of synchronous machine models, the induction, and synchronous machine models, which are discussed in Volume 1 of this series and are not repeated here. Also, the computer simulation of a power for the stability analysis starts with a converged load flow, see Volume 2.

A power system is highly nonlinear and continuously experiences disturbances. From the stability point of view, these can be classified into two categories:

1. Contingency disturbances due to lightning, short circuits, insulation breakdowns, and incorrect relay operations. These can be called *large perturbations or event disturbances.*
2. Load disturbances because of random variations in the load demand.

There are many definitions of the power system stability in the literature; however, with respect to fault disturbances and an initial (prefault) steady-state equilibrium point, it explores whether the postfault trajectory will settle down to a new equilibrium point in an acceptable steady state. The definition of the final steady state is important. For example, if the postfault state has periodic oscillations, it will not be acceptable. Even small fluctuations in the voltage and frequency will be undesirable. Nor subsynchronous resonance, which may occur due to conversion of mechanical energy into electrical energy associated with subharmonic mode and electromagnetic oscillations (due to interactions between magnetic fields) can be tolerated.

15.1 Classification of Power System Stability

Classification of the stability is necessary in view of the following:

1. The size of disturbance
2. Correct modeling and analysis of the specific disturbance
3. The time span for which the disturbance lasts
4. The system parameter which is most affected

The power system stability is classified into the following categories.

15.1.1 Rotor Angle Stability

This is large disturbance stability and concerned with the ability of interconnected synchronous machines to remain in synchronism after being subjected to a perturbation. This can be further subdivided into two categories:

1. Small-signal stability (sometimes termed dynamic stability)
2. Large disturbance rotor angle stability (termed transient stability)

As the nomenclature suggests, the small angle stability considers that the disturbances are small and the system equations can be linearized. Sometimes, the small rotor angle stability is termed dynamic stability. In modern systems, it is largely a problem of insufficient damping of the systems. The study is conducted for 10–20 s and sometimes even for a longer period after a disturbance.

Conditions for oscillations between major subsystems may depend upon many variable factors, e.g., loading of generators, load levels, and system voltages, which are difficult to predict with precision. The subharmonic frequencies can be reduced with the application of power system stabilizers (PSSs). When a PSS is not applied, a fast-acting or wider-band excitation system has greater chance of destabilizing local-machine system oscillations. The National Electric Regulatory Council (NERC) recommends power system stabilizer (PSS) for generators that exceed 30 MVA or a group of generators that exceeds 75 MVA.

The large rotor angle stability, or transient stability as it is normally termed, is concerned when the power systems are subjected to large perturbations, i.e., system faults. The resulting system response involves large excursions of rotor angle swings, and the system equations are nonlinear.

Classically, it was thought that due to insufficient damping torque, the instability occurs during the first swing (see Figure 15.1, curve a). In large power systems, transient instability may not occur during the first swing. It can be a result of superimposition of several oscillation modes, which may result in *larger* deviation of rotor angle in subsequent swings (see Figure 15.1, curve b). The curve c of Figure 15.1 shows the transient stability as the rotor angle swings damp out. The time frame of interest may be 1–3 s.

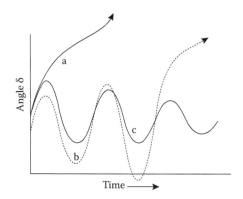

FIGURE 15.1
Rotor angle stability (curve a). Instability during the first swing (curve b). Instability due to larger rotor angle swings after a number of swings (curve c), decaying rotor angle swings depicting stability.

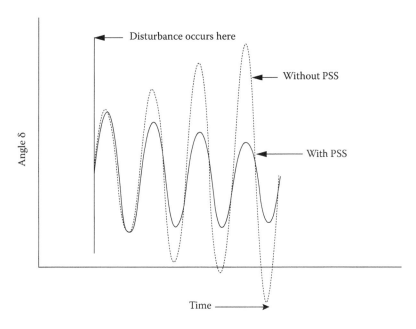

FIGURE 15.2
Dynamic instability: swings controlled by the application of a PSS.

The transient mode can sometimes lead to dynamic mode. The rotor angle may continue to swing at lower frequencies, i.e., interarea oscillations. Figure 15.2 shows dynamic instability as the oscillations increase with time. With the application of a PSS, these can be damped as shown in this figure.

15.1.2 Voltage Instability

Voltage instability can again be classified into two categories:

15.1.2.1 Large Disturbance Instability

The voltages in the system should be controlled following large disturbances, such as faults, loss of generation, or circuit contingencies, which may force the power flow through alternate routes of higher impedances. The generators may hit their reactive power capability limit. The loads have certain voltage tolerance limits for successful ride-through capability of the disturbance, before these will fall out of operation.

The analysis requires nonlinear dynamic performance of the system over a period of time which can extend to minutes. The rotor angle instability can lead to voltage instability. The gradual loss of synchronism, as the rotor angles between the machines depart more than 180°, would result in low voltages in the network close to the electrical center.

15.1.2.2 Small Disturbance Voltage Instability

The instability is concerned with system ability to control voltages following small perturbations, i.e., incremental changes in the system load. This is essentially of a steady-state nature. The reactive power flow in a mainly reactive tie circuit requires a difference of voltage at the tie ends, ΔV. A system is said to be stable if V–Q sensitivity is positive for every bus and unstable if V–Q sensitivity is negative for at least one bus.

The stability of synchronous and induction motors on voltage dips is discussed in Volume 2 of this series. Also, the concepts of stability of synchronous motors and calculations of pull out power factor are discussed in Chapter 10.

15.1.3 Static Stability

The term static stability may seem a misnomer for the dynamic nature of the electrical systems. Yet, it can be assumed that the response of a system to a gradually occurring change is without an oscillation—practically, this is not correct but forms the basis of some fundamental concepts.

The power–angle characteristic of a synchronous generator can be derived from the basic phasor diagram shown in Figure 15.3. All resistances and losses are neglected, and all parameters are as defined in Chapter 10. It can be shown that

$$P = \frac{EV \sin \delta}{X_d} + \frac{1}{2} V^2 \left(\frac{1}{X_q} - \frac{1}{X_d} \right) \sin 2\delta$$

$$Q = \frac{EV \cos \delta}{X_d} - V^2 \left(\frac{\cos^2 \delta}{X_d} + \frac{\sin^2 \delta}{X_q} \right) \tag{15.1}$$

The second terms in these equations disappear if saliency is neglected and $X_d = X_q$. The saliency gives some element of stability even when the excitation of the generator is removed.

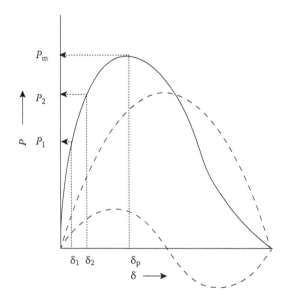

FIGURE 15.3
Power–angle characteristics of a salient pole synchronous generator.

This shows that for a gradual increase in the power output of the generator, the torque angle should increase, with phase of V fixed. In this way, the energy balance is maintained. We say that the generator is statically stable if

$$\frac{\partial P}{\partial \delta} > 0 \tag{15.2}$$

In fact, this ratio is called the synchronizing power and develops with every small variation of shaft power output, which results in a change in δ. From Equation 15.1, the synchronizing power is

$$P_s = EV \cos\delta - V^2 \left(\frac{1}{X_d} - \frac{1}{X_q} \right) \cos 2\delta \tag{15.3}$$

Whenever there is any perturbation to the steady-state operation of a synchronous machine, the synchronizing power is brought into play, tending to counteract the disturbance and bring the system to a new stable point.

15.2 Equal Area Concept of Stability

We discussed equal area criteria of stability in Chapter 10 in connection with stability of synchronous motors. Equation 15.1 for the power output can be plotted as a power–angle relation, as shown in Figure 15.3. The angle δ is called the torque angle. The second term

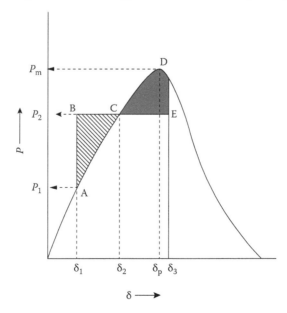

FIGURE 15.4
Torque angle swings on the application of a sudden disturbance, using equal area criteria of stability.

in the power output relation has a sin 2δ term, which makes the power–angle curve of the generator peaky in the first half cycle. Referring to Figure 15.4, if the output shaft power demands an increase, from P_1 to P_2, the torque angle increases from δ_1 to δ_2. The limit of δ_2 is reached at the peak of the torque angle curve, δ_p, then at δ_p synchronizing power is zero. If the load is increased very gradually and there are no oscillations due to a step change, then the maximum load that can be carried is given by the torque angle δ_p. This maximum limit is much higher than the continuous rating of the generator and the operation at this point will be very unstable—a small excursion of load will result in instability.

It is obvious that excitation plays an important role (Equation 15.1). The higher the E, the greater the power output. On a simplistic basis, torque angle curves of a synchronous machine can be drawn with varying E. Neglecting the second frequency term, these will be half sinusoids with varying maximum peak.

On a step variation of the shaft power (increase), the torque angle will overshoot to a point δ_3 (Figure 15.4), which may pass the peak of the stability curve, δ_p. It will settle down to δ_2 only after a series of oscillations. If these oscillations damp out, we say that the stability will be achieved. If these oscillations diverge, then the stability will be lost.

Note the following:

1. The inertia of the synchronous generators and prime movers is fairly large.
2. The speed and power–angle transients are much slower than electrical, current, and voltage transients.

In Figure 15.4, the area of triangle ABC translates into the acceleration area due to variation of the kinetic energy of the rotating masses:

$$A_{accelerating} = ABC = \int_{\delta_1}^{\delta_2} (T_{shaft} - T_e) d\delta \tag{15.4}$$

The area CDE is the deceleration area:

$$B_{decelerating} = CDE = \int_{\delta_2}^{\delta_3} (T_{shaft} - T_e) d\delta \tag{15.5}$$

In acceleration and deceleration, the machine passes the equilibrium point denoted by δ_2 giving rise to oscillations, which will either increase or decrease. If the initial impact is large enough, the machine will be instable in the very first swing.

If

$$A_{accelerating} = B_{decelerating} \tag{15.6}$$

then there are chances of the generator remaining in synchronism. The asynchronous torque produced by the dampers has been neglected in this analysis. Also the synchronizing power is assumed to remain constant during the disturbance. It is clear that at point C the accelerating power is zero, and assuming a generator connected to an infinite bus, the speed continues to increase. The speed is more than the speed of the infinite bus. At point E, the relative speed is zero and the torque angle ceases to increase, but the output is more

Power System Stability

than the input and the torque angle starts decreasing which results in deceleration of the rotor. But for damping, these oscillations can continue. This is the concept of "equal area criterion of stability."

15.2.1 Critical Clearing Angle

From the equal area criteria, it is easy to infer that the disturbance, say a fault, should be removed quickly enough for the machine stability. The critical clearing angle is defined as the maximum angle at which the faulty section must be isolated to maintain the stability of the system. Figure 15.5 shows three P/δ curves, which are under normal operation, during fault, and after fault clearance—curves marked A, C, and B, respectively. Note that during fault, some synchronizing power can be transmitted depending upon the nature of fault and the system configuration. For the condition that area $A1 = A2$ (same results if we make, rectangle $abcd = dfghbc$), because $abcd = A1 + kebc$ and $dfghbc = A2 + kebc$. This gives the following equation:

$$(\delta_m - \delta_0)P_s = \int_{\delta}^{\delta_c} r_1 P_m \sin\delta d\delta + \int_{\delta_c}^{\delta_m} r_2 P_m \sin\delta d\delta \tag{15.7}$$

Substituting

$$P_s = P_m \sin\delta \tag{15.8}$$

During fault, the maximum power P_m is reduced by a factor r_1, and after fault, it is $r_2 P_m$. This gives

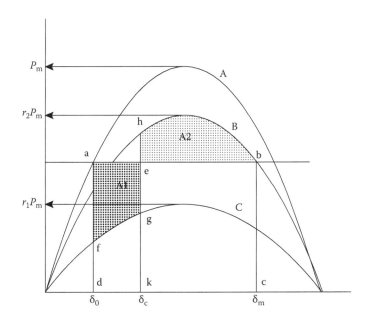

FIGURE 15.5
Critical fault clearing angle using equal area criteria of stability.

$$\cos\delta_c = \frac{(\delta_m - \delta_0)\sin\delta_0 - r_1\cos\delta_0 + r_2\cos\delta_m}{r_2 - r_1} \quad (15.9)$$

where δ_c is the critical clearing angle. If the actual clearing angle is greater than the critical clearing angle, the system will be unstable. This brings an important factor in transient stability, i.e., fast protective relaying and circuit breakers with lower interrupting time. The clearing time of the fault will be the relay operating time plus the breaker interrupting time.

From Figure 15.5

$$P_s = P_m \sin\delta_0 = r_2 P_m \sin\delta_m = r_2 P_m \sin(\pi - \delta_m)$$

Thus

$$\sin\delta_0 = r_2 \sin(\pi - \delta_m)$$

$$\delta_m = \pi - \sin^{-1}\left(\frac{\sin\delta_0}{r_2}\right) \quad (15.10)$$

Note that P_s, the synchronizing power, is assumed constant during and after the disturbance.

15.3 Factors Affecting Stability

The factors affecting stability are dependent upon the type of stability problem. The considerations for large rotor angle instability, voltage instability, interarea oscillations, and turbine generator torsional problems are quite different. The following is a general list:

1. Short-circuit ratio of the generator (see Volume 1).
2. Inertia constant H.
3. Transient reactance of the generator, though all other generator parameters impact to some extent, for example, single and double dampers.
4. Interconnecting system impedances between machines in a multimachine system; the postfault system reactance as seen from the generator.
5. The prior loading on the generator, which determines the initial torque angle, prior to disturbance. The higher the loading, the closer the generator to its P_m.
6. The type of fault or perturbation, a three-phase fault at the terminals of the machines, will be a worst-case scenario for the large rotor angle stability of the synchronous generator. The transient stability on such faults is sacrificed due to economical reasons. The fault types in decreasing order of severity will be two phase-to-ground, phase-to-phase, and phase-to-ground.
7. The nature of the loads served and the presence of rotating motor loads may profoundly impact the results.
8. Speed of protective relaying.

9. Interrupting time of the circuit breakers: one cycle synchronous breakers have been developed.
10. Excitation system and voltage regulator types and response; redundant voltage regulators with bumpless transfer on failure, PV, or constant kvar controls.
11. Fast load shedding (frequency dependent with undervoltage settings).
12. Control system response of turbines, prime movers, governing systems, boilers, PSSs, and voltage regulators. PSSs play a major role in the small-signal stability.
13. The system interconnections, the parameters of transmission lines, and characteristics and ratings of other machines relative to the stability of machine/machines under consideration.
14. Auto-reclosing: single-pole switching has been used to enhance transient stability (see Chapter 7). Only the faulted phase is tripped and the unfaulted phases support transfer of synchronizing power. This works well as approximately 70% of the faults in the transmission lines are of single phase-to-ground type. This may, however, impose negative-sequence loading on the generators in operation; it can excite 120 Hz torque oscillations and result in secondary arcing phenomena and overvoltages.
15. Series and shunt compensation of transmission lines, SVCs, and FACTs (see Chapter 1).
16. Steam turbine fast valving (applicable to thermal power plants) involves rapid opening and closing of the steam valves in a predetermined manner to reduce generator accelerating power following a severe transmission line fault. The reheat intercept valves in a turbine may control up to 70% of unit power and rapid opening and closing of these valves can significantly reduce the turbine output.
17. Fast tripping of a generator after it pulls out of step, as keeping it online, results in avoidable stresses to the generator itself and the system as it draws power from and supplies power into the system. While the unit is tripped from the system, the turbine is not tripped, turbine controls limit the over speed, and the unit can be again brought online quickly.
18. System separation and islanding the faulty section can prevent a major cascade and prevent propagating the disturbance to the rest of the system. Modern relays with adaptive controls have been applied. Much power is generated in dispersed generation and cogeneration facilities that run in synchronism with the utilities, and fast isolation is of importance.
19. Application of Narain Hingorani subsynchronous resonance suppressor (NH-SSR) (see Chapter 13).

Thus, the list of factors affecting power system stability is long.

15.4 Swing Equation of a Generator

The swing equation relates to the motion of the rotor, which can be written as

$$J\frac{d^2\omega_r}{dt^2} = T_m - T_e \tag{15.11}$$

where
  J is the total moment of inertia (kg-m²)
  ω_r is the angular displacement of mechanical rotor (rad)
  T_m and T_e are the mechanical and electrical torques (N-m).

This can be converted into a per unit form by noting the relation between H, the inertia constant, and J, the moment of inertia

$$2H\frac{d^2\omega_{ru}}{dt^2} = T_{mu} - T_{eu} \tag{15.12}$$

where

$$H = \frac{J\omega_{om}^2}{2VA_{base}} \text{ and } \omega_{ru} = \omega_r/\omega_0 \tag{15.13}$$

The subscript u indicates per unit values. H is the inertia constant, as defined in Equation 11.57. It is repeated here in terms of units prevalent in the United States

$$H = \frac{2.31 \times RPM^2 \times WR^2 (\text{lb}/\text{ft}^2) \times 10^{-10}}{\text{MVA rating of machine}} \text{ MW s/MVA} \tag{15.14}$$

H does not vary over a large limit for various machines. Typical data for steam and hydro units of various sizes are available in the current literature.

We are more interested in writing the swing equation in terms of angular position of the rotor in electrical degrees, δ, with reference to a synchronously rotating reference δ_0, at $t = 0$. Angle δ is easily interpreted from the phasor diagrams of the machine:

$$\delta = \omega_r t - \omega_0 t + \delta_0$$

$$\dot{\delta} = \omega_r - \omega_0 = \Delta\omega_r \tag{15.15}$$

$$\ddot{\delta} = \omega_0 \dot{\omega}_{ru}$$

Thus, we can write

$$\frac{2H}{\omega_0}\frac{d^2\delta}{dt^2} = T_{mu} - T_{eu} \tag{15.16}$$

We add another term to this equation to account for damping, which is proportional to the speed. Then, the equation becomes

$$\frac{2H}{\omega_0}\frac{d^2\delta}{dt^2} = T_{\text{mu}} - T_{\text{eu}} - K_D\omega_{\text{ru}} \tag{15.17}$$

Equation 15.17 is called the swing equation. It can be represented in terms of two differential equations of the first order:

$$\frac{d\omega_r}{dt} = \frac{1}{2H}(T_m - T_e - K_D\Delta\omega_r)$$
$$\frac{d\delta}{dt} = \omega_0\Delta\omega_r \tag{15.18}$$

The additional subscript u has been dropped and the equation is understood to be in pu.
Linearizing

$$\Delta\dot{\omega}_r = \frac{1}{2H}[\Delta T_m - K_1\Delta\delta - K_D\Delta\omega_r]$$
$$\Delta\dot{\delta} = \omega_0\Delta\omega_r \tag{15.19}$$

K_1 (pu ΔP/radian) = synchronizing coefficient.

The term $K_1\Delta\delta$ may be called synchronizing power, which acts to accelerate and decelerate the inertia to bring it to the stable operating point, if one exists. For small deviations, K_1 is the slope of the transient power–angle curve, at the particular steady-state operating point (see Figure 15.6):

$$K_1 = \frac{dP}{d\delta}\bigg|_{\delta_0} = \frac{VE'_q}{X'_d + X_e}\cos\delta_0 \tag{15.20}$$

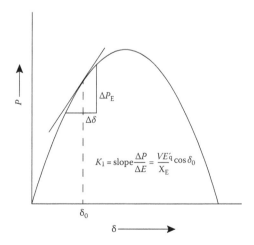

FIGURE 15.6
Power–angle curve showing derivation of synchronizing coefficient K.

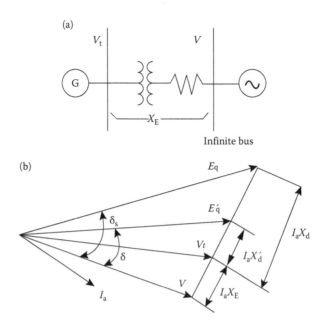

FIGURE 15.7
(a) A generator connected to an infinite bus through external impedance. (b) Phasor diagram of a generator on an infinite bus.

Referring to Figure 15.7 of a synchronous generator connected to an infinite bus through a reactance X_e and ignoring saliency

E'_q = internal voltage behind a transient reactance
E_q = internal voltage behind a synchronous reactance
V = infinite bus voltage
V_t = generator terminal voltage. And δ is the angle as shown between E'_q and V.

Equation 15.17 governs the dynamic response, having an oscillation frequency of approximately

$$\omega_n \approx \sqrt{\frac{K_1 \omega_s}{2H}} \text{ rad/s} \tag{15.21}$$

15.5 Classical Stability Model

In the classical treatment of stability, the following assumptions are made:

1. The generator model is *type 1* with all its assumptions, represented by an internal generator voltage behind a transient reactance.

2. The loads are represented as constant impedance loads.
3. The excitation systems, governing systems, are not modeled. The mechanical power remains constant.
4. Damping or asynchronous power is neglected.
5. The mechanical rotor angle of the machine coincides with the angle of voltage behind transient reactance.

These assumptions may not be always valid, but are illustrative of the methodology of simple stability calculations.

The swing equation reduces to

$$\frac{2H}{\omega_s}\frac{d^2\delta}{dt^2} = P_m - P_e \tag{15.22}$$

The general network equations are

$$\begin{vmatrix} \bar{I}_g \\ 0 \end{vmatrix} = \begin{vmatrix} \bar{Y}_{11} & \bar{Y}_{1j} \\ \bar{Y}_{j1} & \bar{Y}_{jj} \end{vmatrix} \begin{vmatrix} \bar{E} \\ \bar{V} \end{vmatrix} \tag{15.23}$$

where $\bar{I}_g$ is the vector $n \times 1$ of generator currents, $\bar{E}$ is the vector of internal generator voltages, $\bar{V}$ is the vector of bus voltages, $\bar{Y}_{11}$ is the $n \times n$ matrix, $\bar{Y}_{1j}$ is the $n \times m$ matrix, $\bar{Y}_{j1}$ is the $m \times n$ matrix, and $\bar{Y}_{jj}$ is the $m \times m$ matrix.

We eliminate all load nodes and are interested only in the generator currents.

From Equation 15.23

$$\bar{0} = \bar{Y}_{j1}\bar{E} + \bar{Y}_{jj}\bar{V}$$

or

$$\bar{V} = -\bar{Y}_{jj}^{-1}\bar{Y}_{j1}\bar{E} \tag{15.24}$$

Also

$$\bar{I}_g = \bar{Y}_{11}\bar{E} + \bar{Y}_{1j}\bar{V}$$

$$= \left(\bar{Y}_{11} - \bar{Y}_{1j}\bar{Y}_{jj}^{-1}\bar{Y}_{j1}\right)\bar{E} \tag{15.25}$$

$$= \bar{Y}\bar{E}$$

The $\bar{Y}$ matrix changes with the system conditions. In the prefault state, the load flow study is conducted to determine $\bar{E}$ and the angles δ. At the instant of fault, the system configuration changes and the matrix is modified and again modified after the fault is cleared. Thus, there are three distinct conditions in the solution.

15.6 Modern Transient Stability Methods

A modern transient stability study program shall model the following:

- The transient electrical parameters of the generators, including saturation and damping and also the mechanical system.
- The detailed control circuits of the excitation systems.
- Modeling of governing systems especially for dynamic stability studies.
- Modeling of prime mover, steam, or hydraulic governing.
- Dynamic models of loads such as motors with their transient electrical parameters and mechanical data.
- Modeling of PSS as required.

The time-domain methods described above for the analysis of stability problem may involve solving thousands of algebraic and nonlinear differential equations, complexity depending upon the models used and the power system studied. This may be fairly time-consuming and slow down the computing speed.

The time-domain methods analyze the prefault, faulted, and postfault systems. The direct methods integrate the faulted system only and determine without examining the postfault system, whether the system will be stable after fault clearance. The system energy, when the fault is cleared, is compared to a critical energy value, not discussed.

15.7 Excitation Systems

A fundamental block circuit control diagram of the excitation systems is shown in Figure 15.8. Here, no attempt is made to describe the fundamental characteristics of the excitation systems, their stability, saturation characteristics, ac and dc exciters, regulation characteristics, etc. The response of an exciter system is determined from the curve shown in Figure 15.9. Referring to this figure, the rate of increase or decrease of the excitation output voltage is given by line ac, so that area acd = area abd. This means that the rate of rise if maintained constant will develop the same voltage–time area as obtained from the curve for a specified period (0.5 s, shown in this figure). The *response ratio* is $(ce-ao)/\{(ao)0.5\}$ pu V/s. Nominal response is used as a figure of merit for comparing different types of excitation systems; misleading results can occur if different types of limiters or different values of inductances are permitted.

The response ratio is an approximate measure of the rate of rise of the exciter open-circuit voltage in 0.5 s, if the excitation control is adjusted suddenly in the maximum increase position. This can be considered more like a step input of sufficient magnitude to drive the exciter to its ceiling voltage, i.e., exciter operating under no-load conditions.

15.7.1 Fast Response Systems

The time of 0.5 s in IEEE standards was chosen because it approximates to one-half period of natural electromechanical oscillation of average power system. Modern fast systems

Power System Stability

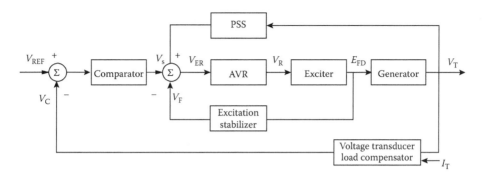

FIGURE 15.8
Basic control circuit block diagram of an excitation system.

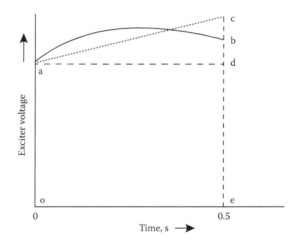

FIGURE 15.9
Determination of excitation system response according to IEEE Standards.

may reach ceiling voltage in much smaller time. Fast response systems are defined as the ones which reach ceiling voltage in 0.1 s or less and *oe* in Figure 15.9 is replaced (= 0.1 s).

15.7.2 Types of Excitation Systems

Earlier, IEEE report "Computer representation of excitation system" classified the excitation systems into four types: (1) type 1—represented continuously acting regulator and exciter, (2) type 2—represented rotating rectifier (brushless) system, (3) type 3—static with terminal potential and current supplies, and (4) type 4—noncontinuously acting regulator. These models are still in use. Currently, the types of excitation systems and AVR models can be described according to IEEE Std. 421.5, IEEE Recommended Practice for Excitation

System Models for Power System Stability Studies, which classifies these into three major categories, based upon the excitation power source:

1. DC excitation systems utilize a dc current generator with a commutator for the excitation power source (IEEE types DC1A, DC2A, and DC3A).
2. AC excitation systems use an alternator, and either stationary or rotating rectifiers (brushless) to produce the dc power (IEEE types AC1A, AC2A, AC3A, AC4A, AC5A, and AC6A).
3. ST type excitation systems use static controlled rectifiers for the dc power supply (IEEE types ST1A and ST2A).

The type designations in parenthesis are from the IEEE standard. Manufacturers specify the time constants and data based upon the IEEE models, and these data may differ from the specimen data included in IEEE standards. For any rigorous stability study, the manufacturer's data should be used. Furthermore, many models may be manufacturer's specific and may not exactly confirm to IEEE types, especially for equipment of European origin.

15.8 Transient Stability in a Simple Cogeneration System

A simple system for the study of transient stability is shown in Figure 15.10. Two in-plant generators each of 42.5 MW, 13.8 kV, and 0.85 power factor serve the plant loads and operate in synchronism with the utility source through 60 MVA step-down transformer. The

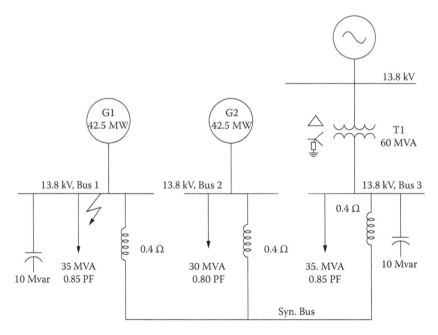

FIGURE 15.10
A simple system configuration for study of transient stability.

TABLE 15.1
Data for Figure 15.10

Equipment	Data
Utility source 138 kV	$Z+ = 0.009277 + j0.01392$ pu, $Z0 = 0.2003 + j0.03479$ pu 100 MVA base
Transformer	60 MVA, $Z = 9\%$, $X/R = 34.1$
Reactors	0.4 Ω, $X/r = 80$, 2000 A
Generators G1 and G2 42.5 MW, 0.85 PF, 13.8 kV	$X_d'' = 15\%, X_d' = 28\%, X_2 = 18\%, X_0 = 7\%, X_d = 155\%, X_{dv} = 165\%,$ $X_L = 15\%, X_q = 145\%, X_{qv} = 155\%, X_q' = 65\%, T_{do}' = 6.5$ s, $T_{qo}' = 1.25$ s, saturation, $S_{break} = 0.8, S_{100} = 1.07, S_{120} = 1.15,$ damping $= 0.9,$ total $H = 4.9, X/R = 70$
Exciter G1, ST2 Type	$KA = 250, KC = 0.001, KE = 1, KF = 0.06, TA = 0.03, TE = 1.25, TF = 1, TR = 0.005,$ $VR_{max} = 17.5, VR_{min} = -15.5, Efd_{max} = 6.6$
Exciter G2, ST1 Type	$KA = 52, KC = 0.05, KF = 0.114, TA = 0.01, TB = 0.92, TC = 0, TF = 0.6, TR = 0,$ $VI_{max} = 2.4, VI_{min} = -2.4, VR_{max} = 4.6, VR_{min} = 0$

protective relaying is not shown in this figure and the data for the transient system study model are shown in Table 15.1.

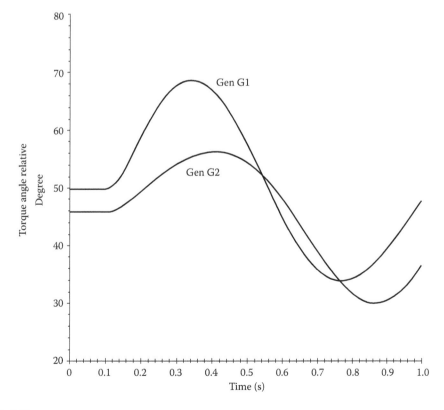

FIGURE 15.11
Generators torque angle swings, showing stability.

It is necessary to conduct short-circuit and load flow study before embarking on transient stability study. The two 10 Mvar capacitor banks are required to maintain system voltages when any source goes out of service. The short-circuit currents are limited to 1000 MVA, within the short-circuit ratings of the circuit breakers, not shown in this figure.

A three-phase fault occurs on bus 1 at 0.1 s and is cleared, say, by the differential relays in approximately four cycles, assuming three cycle rated circuit breakers. The stability is investigated in the following figures:

Figure 15.11: This shows torque angle swings of generators 1 and 2. Note that the torque angle of generator G1 swings from running level of 50° to approximately 68° and then falls back. This figure alone is good enough to illustrate that the system is stable.

Figure 15.12: This figure illustrates the bus voltage dips during fault and then its recovery. Note that the voltages on all three 13.8 kV buses dip severely. The voltage on the faulted bus 1 is zero during the fault. Though the voltages recover quickly after a fault clearance, the motor starter contactors can drop out even during the first cycle of the voltage dip. The contactors for essential loads can be stabilized, see Chapter 10.

Figures 15.13 and 15.14: These figures show the transients in the exciter currents and voltages.

Figures 15.15 and 15.16: These figures depict the transients in the active and reactive power of the generators. The faulted generator cannot supply any active or reactive power; its terminal voltage is zero during the fault.

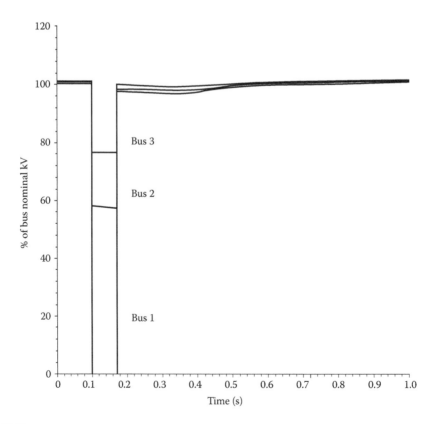

FIGURE 15.12
Bus recovery voltage profiles during fault and after fault clearance.

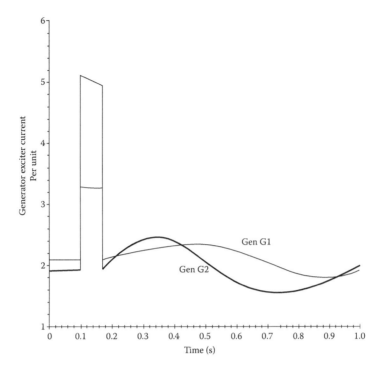

FIGURE 15.13
Transients in generators' exciter current.

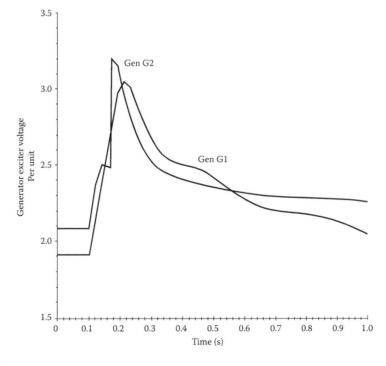

FIGURE 15.14
Transients in generators' exciter voltage.

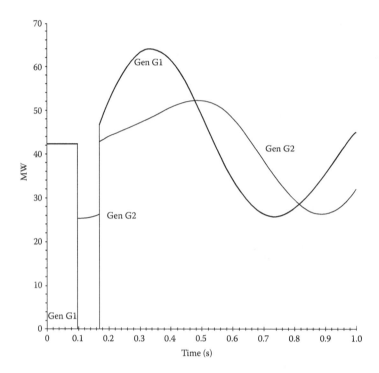

FIGURE 15.15
Generators' MW transients.

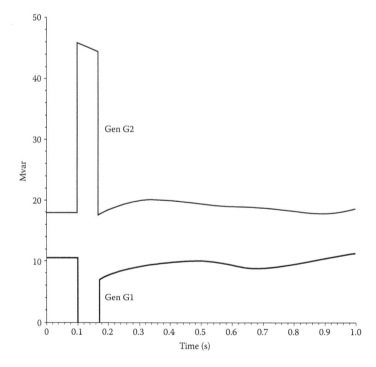

FIGURE 15.16
Generators' Mvar transients.

Power System Stability 613

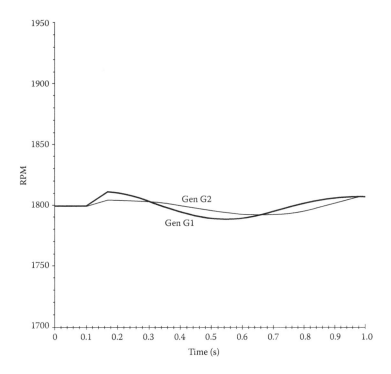

FIGURE 15.17
Generators' speed transients.

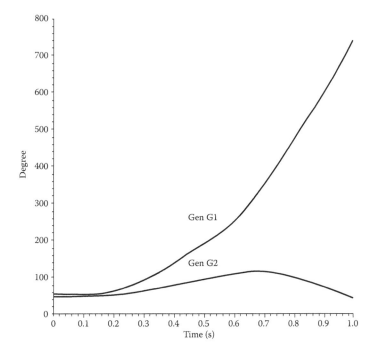

FIGURE 15.18
Generators' torque angle swings. Generator G1 goes out of synchronism at 0.4 s.

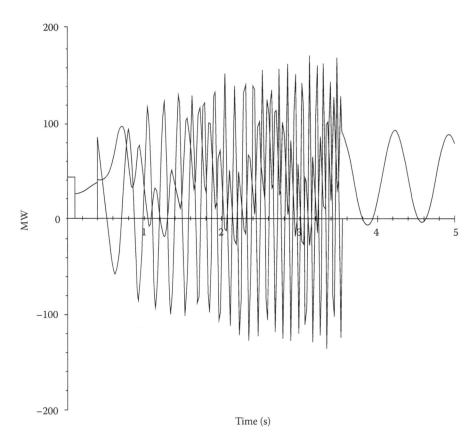

FIGURE 15.19
Violent swings in MW outputs of generators.

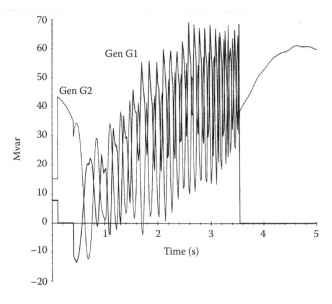

FIGURE 15.20
Violent swings in Mvar outputs of generators.

Figure 15.17: This figure illustrates the speed transients, which are small. In an industrial system, the frequency is held practically constant so long as the system is connected to the utility source. The relatively small in-plant generators cannot dictate the system frequency. Thus, underfrequency load shedding is not possible so long as the system is connected to utility source.

It can be documented that the system is stable for a bus fault cleared in about seven cycles with five cycle breakers. Thus, the importance of bus differential protection is well illustrated. If the fault clearance is prolonged beyond about 12 cycles, the system will become unstable. Figure 15.18 illustrates that the torque angle of generator G1 has diverged beyond 180° at 0.4 s and this generator has fallen out of step. Figures 15.19 and 15.20 show the violent oscillations of the active and reactive power of the generators, which will cause stresses and shaft fatigue. The out-of-step protection of the generators described in Chapter 11 is very useful for preventing these stresses.

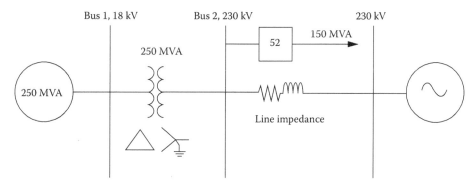

FIGURE 15.21
A system configuration for the application of PSS.

TABLE 15.2

Data for Figure 15.21

Equipment	Data
Utility source 230 kV	$Z+=0.0063+j0.00992$ pu, $Z0=0.02003+j0.003479$ pu 100 MVA base
Transformer	250 MVA, $Z=10\%$, $X/R=42$
Reactors	0.4 Ω, $X/r=80$, 2000 A
Generator G1 18 kV, 250 MVA, 0.85 PF	$X_d''=19\%, X_d'=32\%, X_2=21\%, X_0=10\%, X_d=200\%, X_{dv}=220\%,$ $X_L=19\%, X_q=195\%, X_{qv}=210\%, X_q'=75\%, T_{do}'=7.5$ s, $T_{qo}'=1.4$ s, saturation, $S_{break}=0.8, S_{100}=1.07, S_{120}=1.15,$ damping $=0.9$ total $H=5.5, X/R=105$
Exciter G1, AC1 Type	$KA=400, KC=0.2, KD=0.38, KE=1, KF=0.03, TA=0.02, TB=0, TC=0, TE=0.8,$ $TF=1, TR=0$ $VR_{max}=14.5, VR_{min}=-14.5, SE_{max}=0.1, SE75=0.03, Efd_{max}=4.18$
PSS Type PSS1A	V_{SI}=speed, $K_S=3.15, V_{STMAX}=0.9, V_{STMin}=-0.9, A_1=A_2=0, T_1=0.76, T_2=0.1,$ $T_3=0.76, T_4=0.1, T_5=1, T_6=0.1$

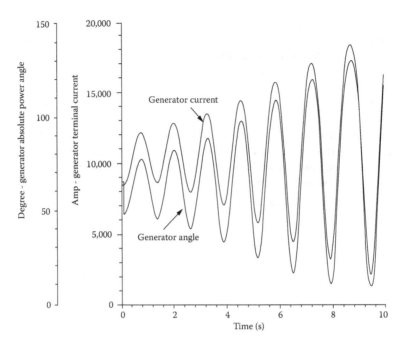

FIGURE 15.22
Generator torque angle and terminal current diverging transients.

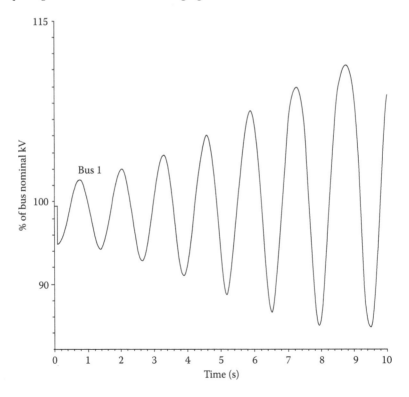

FIGURE 15.23
Bus 1 voltage increasing transients.

Power System Stability

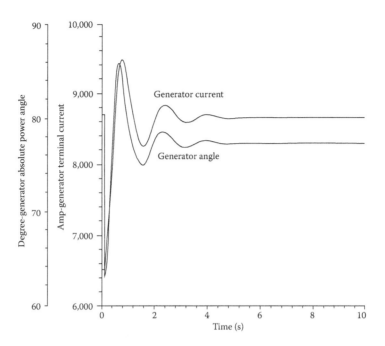

FIGURE 15.24
Damping of generator torque angle and terminal current transients with the application of PSS.

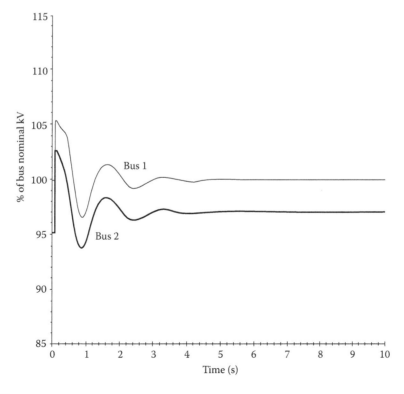

FIGURE 15.25
Damping of voltage transients at buses 1 and 2 with the application of PSS.

15.9 A System Illustrating Application of PSS

Figure 15.21 illustrates a large 250 MVA generator which supplies a load of 150 MVA, connected to GSU primary and the rest of the generator power is supplied into 230 kV system through line impedances. The data for transient stability modeling are shown in Table 15.2.

A stability simulation is carried out with sudden load dropping of 150 MVA. First, consider that no PSS is provided. Simulation is carried for 10 s.

The generator torque angle swings and terminal current transients are shown in Figure 15.22. Note that the transients are diverging. Soon the generator will fall out of step. Figure 15.23 shows the voltage swings at bus 1. Similar voltage swings occur at bus 2.

Next, all other conditions remaining identical apply a PSS, IEEE type PSS1A, the control circuit parameters as shown in Table 15.2.

Figure 15.24 illustrates how effectively PSS dampens the torque angle and generator terminal current oscillations, and Figure 15.25 illustrates damping of voltage transients.

The bibliography provides further reading.

Bibliography

1. PM Anderson, AA Fouad. *Power System Control and Stability*. IEEE Press, New Jersey, 1994.
2. P Kundar. *Power System Stability and Control*. McGraw Hill, New York, 1994.
3. EW Kimbark. *Power System Stability*, Vol. 3. John Wiley & Sons, New York, 1956.
4. CC Young, Equipment and system modeling for large scale stability studies. *IEEE Trans PAS-91*(1), 99–109, 1972.
5. CIGRE Task Force 38.02.05. Load modeling and dynamics. *Electra*, 1990.
6. SB Crary. *Power System Stability*, Vol. 2. John Wiley & Sons, New York, 1947.
7. VA Venikov. *Transient Phenomena in Electrical Power Systems*. Pergamon Press, Macmillan, New York, 1964.
8. K Ogata. *State-Space Analysis of Control Systems*. Prentice Hall, Englewood Cliffs, NJ, 1967.
9. PL Dandeno, RL Hauth, RP Schulz. Effects of synchronous machine modeling in large scale system studies. *IEEE Trans*, PAS-92, 926–933, 1973.
10. IEEE Committee Report. Supplementary definitions and associated tests for obtaining parameters for synchronous machines stability simulations. *IEEE Trans Power Syst*, PAS-99, 1625–1633, 1980.
11. IEEE Std. 1110, IEEE Guide for Synchronous Generator Modeling Practices and Applications in Power System Stability Analyses, 2002.
12. FP de Mello, JW Feltes, TF Laskowski, LJ Oppel. Simulating fast and slow dynamic effects in power systems. *IEEE Comput Appl Power*, 5(3), 33–38, 1992.
13. IEEE Committee Report. Dynamic stability assessment practices in North America. *IEEE Trans Power Syst*, 3(3), 1310–1321, 1988.
14. WT Carson. *Power System Voltage Stability*. McGraw Hill, New York, 1994.
15. Voltage Stability and Long Term Stability Working Group. Bibliography on voltage stability. *IEEE Trans Power Syst*, 13(1), 115–125, 1998.
16. DJ Hill. Nonlinear dynamic load models with recovery for voltage stability study. *IEEE Trans Power Syst*, 8(1), 166–176, 1993.
17. HD Chaing, I Dobson, RJ Thomas, JS Thorp, L Fekih-Ahmed. On the voltage collapse in the power systems. *IEEE Trans Power Syst*, 5(2), 601–611, 1990.

18. T Athay, R Podmore, S Virmani. A practical method for direct analysis of transient stability. *IEEE Trans*, PAS-98(2), 573–584, 1979.
19. MA Pai. *Energy Function Analysis for Power System Stability*. Kluwer Academic Publishers, Boston, MA, 1989.
20. HD Chiang, CC Chu, G Cauley. Direct stability analysis of electrical power systems using energy functions: Theory, applications and perspective. *Proc IEEE*, 83(11), 1497–1529, 1995.
21. HD Chiang, FF Wu, PP Varaiya. A BCU method for direct analysis of power system transient stability. *IEEE Trans Power Syst*, 8(3), 1194–1208, 1994.
22. EPRI. *Load Modeling for Power Flow and Transient Stability Computer Studies*. Report EL-5003, 1987.
23. CIGRE SC38-WG02 Report. State of the art in non-classical means to improve power system stability. *Electra*, (118), 88–113, 1988.
24. AJ Gonzales, GC Kung, C Raczkowski, CW Taylor, D Thonn. Effects of single- and three-pole witching and high speed reclosing on turbine generator shafts and blades. *IEEE Trans*, PAS-103, 3218–3228, 1984.
25. IEEE Working Group Report. Turbine fast valving to aid system stability: Benefits and other considerations. *IEEE Trans*, PWRS-1, 143–153, 1986.
26. RT Byerly, RJ Bennon, DE Sherman. Eigenvalue analysis of synchronizing power flow oscillations in large power systems. *IEEE Trans*, PAS-101, 235–243, 1982.
27. N Martins. Efficient eigenvalue and frequency response methods applied to power system small-signal stability studies. *IEEE Trans*, PWRS-1, 217–225, 1986.
28. EPRI Report EL-5798. The small signal stability program package, vol. 1. Final report of Project 2447-1, 1990.
29. P Kundur, GJ Rogers, DY Wong, L Wang, MG Lauby. A comprehensive computer program for small signal stability analysis of power systems. *IEEE Trans*, PWRS-5, 1076–1083, 1990.
30. ANSI/IEEE Std. 421.1, IEEE Standard Definitions of Excitation Systems for Synchronous Machines, 1986.
31. IEEE Std. 421.2, IEEE Guide for Identification, Testing, and Evaluation of Dynamic Performance of Excitation Control Systems, 1990.
32. IEEE Std. 421.5, IEEE Recommended Practice for Excitation System Models for Power System Stability Studies, 1992.
33. IEEE Committee Report. Computer representation of excitation system. *IEEE Trans*, PAS-87, 1460–1464, 1968.
34. JC Das. *Transients in Electrical Systems*. McGraw-Hill, New York, 2010.

16

Substation Automation and Communication Protocols Including IEC 61850

16.1 Substation Automation

During the last 30 years, the substation automation (SA) has taken large strides. Earlier the control of substations was based on discrete electronic or electromechanical elements, and several functions were carried out separately by specific subsystems. The microprocessor-based substation control systems were originally conceived as remote terminal unit (RTU)-centric architecture, and later on as local area network (LAN) architecture. Modern industrial and utility power systems are complex. From the view points of protection, communication, and control, a variety of intelligent electronic devices (IEDs) must be connected together and operate in synchronism with reliability. The automation requires high-speed data transfer and reliability for effective control and managements. Consider, for example, Modbus RTU protocol. It is a master–slave protocol, and can address up to 254 slaves, primarily defined on RS-485, but can be operated on Ethernet. The data are addressed via 2-byte registers and Modbus packet can transmit up to 120 registers per message.

16.2 System Functions

The SA system (SAS) must execute a number of functions. A simplified model of SAS is shown in Figure 16.1. The SA depends upon the interface between substation and its associated equipment to provide and maintain a high level of confidence demanded for the power system operation. It uses a variety of devices integrated into a functional package by a communication technology for monitoring and control. SA has one or more communication connections to the outside world, which include utility operation centers, maintenance offices, and engineering centers. Most SASs connect to a traditional SCADA system master station serving the real-time needs for operation.

16.3 Control Functions

The control functions include the following:

- Selecting, opening, and closing circuit breakers and switches
- Blocking and unblocking control commands

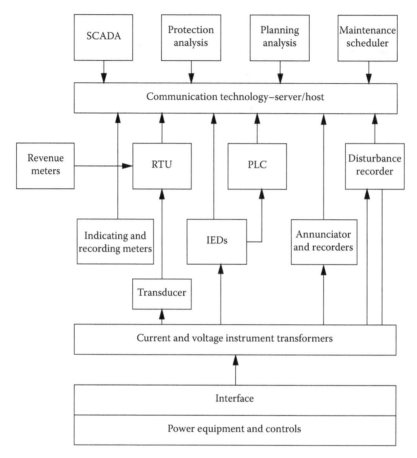

FIGURE 16.1
Power station SA functional diagram.

- Monitoring
- Display of substation configuration with circuit breaker and switch position indicators. The mechanical switches and auxiliary relays may mechanically bounce and the interpretation can be multiple operations of the primary device. Modern IEDs have a debounce algorithm to filter contact bounce
- Processing data from power transformers, reactors, and other power equipment
- Display of substation events, with prior and post change of conditions
- Alarming: Alarming to the operator regarding all adverse conditions and the parameters that exceed set points and represent a risk to substation integrity
- Measurements: Measurements of watts, vars, current, voltage power factor, MWH, and Mvar flows
- Protective relays
- Monitoring the health of the relays—hardware or software failures in MMPRs, event reports, pre- and postfault data capture
- Control and monitoring of auxiliary systems
- Displays

- Battery monitors
- Automatic transfer switches
- Alarms for abnormal conditions
- Voltage regulation
- ULTC remote monitoring and controls
- Step-voltage regulators
- Voltage and disturbance recorders

16.4 Wire Line Networks

16.4.1 Point-to-Point Networks

The communication link between IEDs and SA can be a point-to-point connection with IED connecting directly to SA controller. Many IEDs can interconnect to a data concentrator that serves as a communication hub. In earlier communications, these were RS-232 serial pathways. Often utilities use optical fiber links for these connections.

16.4.2 Point-to-Multipoint Networks

Most automation systems rely on point-to-multipoint connections for the IEDs. The IEDs that share a common protocol often support a party line communication pathway, where they share the channel. The SA controller uses it as a master–slave communication bus, where an SA controller controls the traffic. All devices on the channel must be addressable so that only one device communicates at one time. RS-485 is the most common point-to-multipoint bus. It may use a shielded copper pair, terminated at each end of the bus with a matching resistor equal to the characteristic impedance of the bus cable. RS-485 can support 32 devices on the channel, maximum channel length being typically 4000′. The longer the bus, the slower it runs. The maximum speed is 1.0 Mbps, though, practically, it runs much slower with a speed of 19.2 kbps. RS-485 bus is linear, and stubs will cause reflections. The TS-485 devices are wired in a daisy chain.

16.4.3 Peer-to-Peer Network

Here each device has equal access to the communication bus and can also message any other device. This is substantially different from a master–slave environment. A peer-to-peer network must prevent message collisions. In a control scheme called "token ring," a message is passed from one device to another through communication bus which gives authority to transmit messages. Different schemes control the access time. The buses can be RS-485 or higher speed coaxial cable arrangements. When the token is lost or device fails, the bus must restart.

Ethernet, IEEE Standard 802x is another scheme. Ethernet is widely used in the information technology environment and is finding its way into SA. The media can be coaxial cable or twisted-pair cable. Many utilities are extending their wide area networks (WANs) to substations, where it is becoming a pathway to SCADA and automation. Some utilities are using LANs to connect IEDs. A growing number of IEDs support

Ethernet communications over LANs, and communication interface modules (NIMs) are available to make transition.

16.5 System Architecture

16.5.1 Level 1: Field Devices

The choice of IEDs is important. These have internal memory to entire data such as analog values, status changes, sequence of events, and power quality. A number of events can be stored in first-in-first out.

16.5.2 Level 2: Substation Data Concentrator

The substation data concentrator polls each device for the stored data consistent with the utility's SCADA system. For example, status updates every 2 s, generator metering functions every 1.5 s, and so on.

16.5.3 Level 3: SCADA Systems

All acquired data should be communicated to SCADA system via a communication link from the data concentrator. All data acquired for nonoperational purposes are communicated to data warehouse via a communication link. A data warehouse supports the client–server architecture of data exchange between the system and the corporate users over WAN.

16.5.4 LAN Protocols

A substation LAN is a communication network of high speed within the substation and extending into the switchyard. It quickly transfers the data between IEDs. In 1996, the EPRI-sponsored utility substation communication initiative performed benchmark and simulation testing of different LAN technologies. The substation configuration tested had 47 IEDs. The response requirements were 4 ms for a protection event, 111 transactions per second for SCADA traffic, and 600 s to transmit a fault record (which is very slow compared to modern standards). The communication profiles tested were FMS/Profibus at 12 Mbps, MMS/Trim 7, Ethernet at 10 and 100 Mbps, and switched Ethernet. Initially, the testing was performed with four test-bed nodes using four 133 MHz Pentium computers. The four nodes simulated 47 IEDs. This was followed up with 20 nodes using 20 computers of 133 MHz. The tests showed that FMS/Profibus at 12 Mbps could not meet the trip time requirements for the protective devices. However, MMS/Ethernet did meet the requirements.

16.5.5 SCADA Communication Requirements

The communication requirements should meet the following criteria:

- Overall system topology, star or mesh, see sections to follow
- Identifications of communication traffic flows and end-system locations
- Device and processor capabilities

- Device addressing schemes
- Communication network traffic characteristics
- Performance requirements and timing issues
- Reliability and failure of a communication path or device
- Application data formats
- Operational requirements

In the mid-1980s, many standard organizations looked at the problems of proliferations of SCADA protocols.

16.5.5.1 Distributed Network Protocol

In 1993, the distributed network protocol (DNP) was developed by Harris, Distributed Automation Products in Calgary, Canada. The DNP user group is a forum of 300 users. The present version is DNP3, which is defined in three distinct levels. Level 1, DNP V3.00 implementation, DNP-L1, has the least functionality and is for simple IEDs, and DNP V3.00 Level 2 describes a subset of protocols slightly larger than level 1. DNP-L3 describes a set of protocols larger than level 2; it has the most functionality for SCADA master station-communication front-end processors. Some benefits are interoperability between multilayer devices, fewer protocols to support in the field, reduced software costs, and shorter delivery schedules. It is implemented by various vendors, is based upon IEC 870-5 standards, and is in the process of becoming an IEEE standard. It allows multiple masters, and layered protocols allow mix and match features. It has address capability of over 65,000 devices and 4×10^9 data points of each data type. Figure 16.1 is a basic block circuit diagram of SA.

16.5.5.2 IEEE 802.3 (Ethernet)

The standard defines a communication protocol which is called "carrier sense multiple access collision detect." It works under the broadcasting principles of carrying all delivered messages to all IEDs connected to LAN.

16.5.5.2.1 IEC 870-5

IEC TC57 WG# continued work on a telecontrol protocol and has issued several standards in the IEC 60870-5 series that collectively define an international consensus for telecontrol.

16.5.5.2.2 UCA 1.0

The EPRI UCA project published its initial results in 1991. The UCA 1.0 specifications outline communication architecture based on existing international standards. It specifies use of the Manufacturing Message Specifications (MMS: ISO 9506) in the application layer for substation communications.

16.5.5.2.3 Intercontrol Center Communications Protocol

The UCA 1.0 work became the basis for IEC 60870-6-503 entitled, "Telecontrol equipment and systems—Part6-503: Telecontrol compatible with ISO standards and ITU-T recommendations—TASE.2 Services and protocol." Also known as Intercontrol Center Communications Protocol (ICCP), this specification calls for the use of MMS and was designed to provide standardized communication services between centers, but it has been used to provide communication services between a control center and its associated substations.

16.5.5.2.4 UCA 2.0

Work continued to develop the substation and IED communication portions of UCA. This work resulted in issuance of UCA 2.0 report, which was published as an IEEE Technical Report 1550-1000 EPRI/UCA Communication Architecture, Version 2.0, 1999, IEEE Product No. SS1117-TBR, IEEE Standard No: TR 1550-1999, Nov. 1999.

16.6 IEC 61850 Protocol

IEC 61850 [1] is the international standard for substation/plant automation systems. It defines the communication between devices in a substation or facility and the related system requirements, and supports all automation functions and their engineering. The technical approach makes IEC 61850 flexible and adaptable to future needs. The scope of the standard addresses the following:

- Technically define communication methods and specify their attributes
- Recommendations for SAS testing and commissioning
- Guidelines for WAN control and monitoring
- Establish procedure for communications between substations
- Provide guidelines for SAS project management and network engineering

It has become synonymous with the following:

- Designator for the substation secondary systems with high degree of integration
- Reduced costs
- Greater flexibility
- Communication networks replacing hardwired connections
- Plug-and-play functionality
- Reduced construction and commissioning times

IEC 16850 is a large document consisting of 1850 pages, and experts from 20 countries participated in its development. A description of the various parts is shown in Table 16.1.
  The system requirements comprise the following reliability points:

- The system shall be operational even if one SAS component fails. Redundant schemes are designed so that single component failure does not result in loss of both redundant elements.
- Redundant SAS components are served from separate power sources.

TABLE 16.1
Description of IEC 16850 Part Numbers

Part Number	Description
1	Introduction and overview
2	Glossary
3	General requirements
4	System and project management
5	Communication requirements for functions and device models
6	Configuration language for electrical substation IEDs
ACSI	
7-1	Principles and models
7-2	ACSI
Data models	
7-3	Common data classes
7-4	Compatible logical node classes and data classes
Mapping to real communication networks (SCSM)	
8-1	Mapping to MMS and to ISO/IEC 8802-3
9-1	Sampled values over serial unidirectional multidrop point-to-point link
9-2	Sampled values over ISO 8802-3
Testing	
10	Conformance testing

- No failure mode should result in erratic trip signals or other malfunctions on the primary switchgear.
- Reliability classes R1, R2, and R3 are defined. The applicable class shall be agreed between a vendor and the user.
- The SAS vendor shall declare MTBF of SAS components and subsystems.
- The protection function shall be prioritized compared to other data exchange.

16.7 Modern IEDs

Developments in IED hardware design and developments of high-speed peer-to-peer communication protocols have resulted in a new generation of IEDs. These protective and control relays (described as MMPRs in previous chapters) have the capability of accepting multilevels of current and voltage inputs and of analyzing these values at increased speeds. The advantage of using these microprocessor-based IEDs is simplifications of device-to-device wiring, cost reduction, and reliability (see Chapter 4).

IEDs can be classified by their functions. Common types of IEDs include relays, circuit breaker controllers, voltage regulators, etc. An IED can perform more than one function taking advantage of microprocessor technology. An IED may have an operating system like Linux running in it.

An efficient way of reduction in device-to-device wiring is to use high-speed peer-to-peer IEC 61850 Generic Object Oriented Substation Event (GOOSE) messaging between protective relays. It is a fast, reliable, and interoperable device-to-device communication. GOOSE messages can be as follows:

- Time-critical data such as trips, blocks, and interlocks.
- The data transfer initiated only on the occurrence of an event, for example, a trip, closing or opening of a breaker, and change in the status of an arc-flash maintenance mode switch.
- Data can be sent periodically for self-test and reliability.
- Primarily it is local, but application to WAN is possible. A WAN may be defined as a computer network that covers a broad area. It can be a network whose communication links regional or cross-metropolitan boundaries, that is, utility networks. On the other hand, a LAN is a computer network covering a small physical area like a substation or a facility.
- It can also support virtual LANs that are created and configured to handle bandwidth more efficiently and provide additional network security.
- It is user defined: set of data that are "published" on the detection of a change in any of the contained data items.

16.8 Substation Architecture

A substation network is connected to the outside WAN via a secure gateway. Outside remote operators and control centers can use abstract communication service interface (ACSI) defined in part 7-2 of IEC 61850 to query and control devices in the substation. A substation bus is utilized as a medium bandwidth Ethernet, which carries all the ACSI requests and responses and substation event messages. There is another kind of bus which is called the process bus for communications in each bay. A process bus connects the IEDs to the traditional devices such as merge units and is realized as a high-bandwidth Ethernet network. A substation usually has one global substation bus but multiple process buses, one for each bay. Interactions in a substation fall under three categories:

- Data gathering and setting
- Data monitoring and reporting
- Event logging

In IEC standards, all enquires and control activities toward physical devices are modeled as getting or setting the values of the corresponding data attributes, and data monitoring/reporting provides an efficient way to track system status.

16.9 IEC 61850 Communication Structure

The IEC defines a rather complicated communication structure, see Figure 16.2. Five kinds of protocols are defined in the standard:

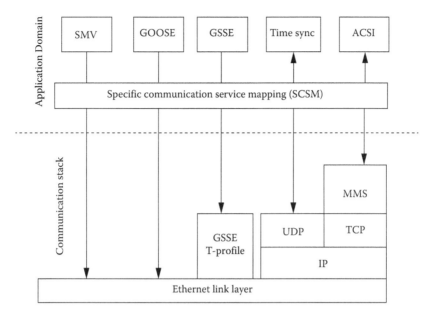

FIGURE 16.2
IEC 61850 communication profiles.

1. ACSI
2. GOOSE
3. Generic substation status event
4. Sampled measured value multicast profile
5. Time-synchronized profile

ACSI is the primary interface in IEC. It is the interface by which applications communicate with servers and is an important part of logical connections between two nodes. An object-oriented approach is adopted in the design of ACSI, which includes a hierarchal and comprehensive data model and a set of available services in each class in this data model. Though the data model is defined outside the scope of ACSI, actually it is a part of ACSI. Figure 16.3 shows the hierarchical data models of IEC. The server is the topmost component in this hierarchy. It serves as the common point of physical devices and logical objects. Each server hosts several files or logical devices. A logical device is the logical correspondence of a physical device—it is basically a group of logical nodes (LNs).

16.10 Logical Nodes

Information is exchanged between all IEDs which comprise the system and more precisely the data are exchanged between the functions and subfunctions resting in the devices. The smallest part of the function that exchanges data is called the logical node (LN) in IEC 61850. The LN groups, first letter listed, are shown in Table 16.2.

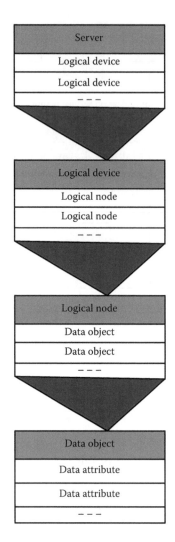

FIGURE 16.3
Hierarchy of IEC 61850 data model.

Data exchanged between LNs is modeled as data objects. An LN usually contains several data objects. Each data object is an instance of data class and has a common data type. Data attributes are typed and restricted by some functional constraints. Functional constraints provide a way to organize all data attributes in an LN by functions. Figure 16.4 shows the circuit breaker XCBR information tree. Figure 16.5 shows the anatomy of IEC 61850 object names.

16.11 Ethernet Connection

IEC 61850 uses an Ethernet connection as the physical medium of communication between IEDs. Logical I/O via Ethernet communication is used in place of traditional hardwired

TABLE 16.2

Logical Node Groups

Logical Node	Description
L	System LN (2)
P	Protection (28)
R	Protection related (10)
C	Control (5)
G	Generic (3)
I	Interfacing and archiving (4)
A	Automatic control (4)
M	Metering and measurements (8)
S	Sensor and monitoring (4)
X	Switchgear (2)
T	Instrument transformers (2)
Y	Power transformers (4)
Z	Further power system equipment (15)

Examples: PDIF, differential protection; *RBRF*, breaker failure; *MMXU*, measurement unit; *YPTR*, power transformer.

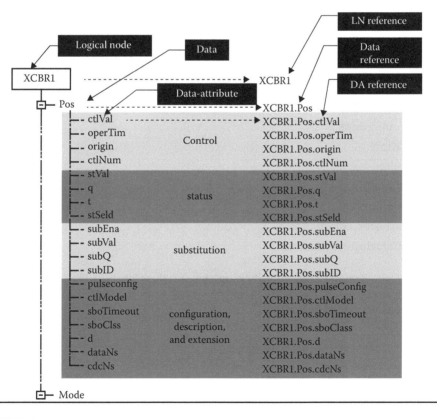

FIGURE 16.4
Circuit breaker model: XCBR information tree.

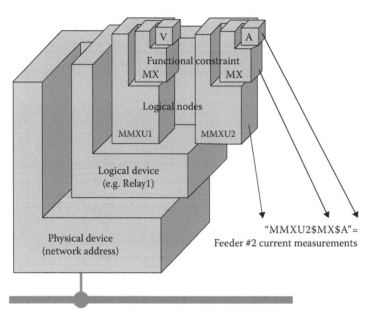

FIGURE 16.5
Anatomy of IEC 61850 object nomenclature.

systems to exchange information between protective IEDs. This may include connected device I/O, protective element statuses, and programmable logic states. The implementations are able to send messages between protective relays at a speed of 1–4 ms. Analog data can be exchanged through GOOSE messaging. Some applications are as follows:

- High-speed bus transfer schemes
- Switching set groups
- Load shedding
- Bus protection
- Breaker failure initiate
- Transfer tripping
- Remote start–stop commands
- Blocking and tripping schemes

As an example, consider the zone interlocking scheme described in Figure 6.30. If IEDs and Ethernet connections are used, the blocking signal from the feeder IEDs to the main breaker IED can be transmitted fast:

IEC 61850 100 Mbps LAN response time	2–4 ms
Feeder relay (IED) response time	2 ms
Main circuit breaker (IED) response time	2 ms
Input recognition time	0–2 ms
Output recognition time	2–4 ms
Total	8–14 ms

Substation Automation and Communication Protocols Including IEC 61850

This will further reduce the arc-flash incident energy release.

Also use of Ethernet allows simultaneous use of multiple protocols on the same hardware, for example, Modbus and DNP.

IEC introduced a common language (manufacturer independent) which can be used to exchange information. Export of IED's description into a common XML-based language is allowed, and the so called ICD file (IED capability description) contains all information about the IED, which allows the user to configure a GOOSE message. This configuration can be performed by IEC 61850 system configurator. All ICD files get imported into the IEC 61850 system configurator and the GOOSE message can be programmed by specifying the sender (publisher) and receivers (subscribers) of a message. The whole descriptions of the system, including description of GOOSE messages, get stored in Substation configuration description (SCD) file. Each proprietary tool is able to import this SCD file and extract the information needed for the IED.

Figure 16.6 shows that one device publishes information, and the subscriber devices receive it. The reaction of each subscriber depends on its configuration and functionality. This figure shows that the sender X trip command, "trip CB A," has no reaction on GOOSE receiver Y, but GOOSE receiver Z is configured to receive it and trip the breaker. Note that the figure shows only one receiver responding, but a GOOSE message can be received and used by many receivers.

Replacing hardwired connections with digital communications requires new understanding and practices—a transition from physical terminations to logical interconnections between IEDs which open up a vast opportunity of data communications and controls. It is fast, deterministic, and automatically connected to predefined logical terminations between the publisher and receivers. Experience indicates that over 50% of copper terminations, associated cost and labor, and potentialities for mistakes are eliminated. The key concepts are summarized as follows:

- *Interoperability*: The ability of system components to function effectively with other components including communication among multiple vendor components using GOOSE protocols.

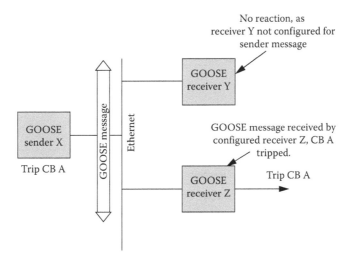

FIGURE 16.6
GOOSE messaging.

- *Interchangeability*: The ability to interchange a system component with another manufacturer's component which fulfills the same functions.
- *Vendor independent*: Selection of system components not dependent upon one specific supplier.
- *Substation architecture*: See Section 16.8.
- *Networking features*: Compliance with IEEE 1613 [2,3] and IEC 61850.

16.12 Networking Media

Networking media can be divided into two categories:

1. Twisted-pair copper
2. Fiber optic

The major considerations of selection between the two options are as follows:

- The "optical power budget" (OPB) which can be described as the maximum permissible attenuation of the light signal as it travels from the transmitter to the receiver, while still permitting reliable communication
- The distance involved, which is a major consideration
- The speed of communication, which is defined in megabits (MB) per second
- The cost and reliability

Ethernet over fiber is becoming the medium of choice where long distances are involved and immunity from electromagnetic interference is desired. The cost difference between copper and fiber cables is not much; however, the fiber Ethernet transceivers are more expensive. Table 16.3 provides some basic idea of the OPB and the distances. The terms used in this table are explained further in the following sections.

TABLE 16.3

Port Description, Typical Distance, and Power Budget

Port Type	Description	Typical Distance	Power Budget
10/100Base T	10/100 MB, RJ45 copper unshielded	100 m	
10/100Base T	10/100 MB, RJ45 copper shielded	150 m	
100Base FX	100 MB, multimode SC fiber optic (full duplex)	2 km	14 dB
100Base FX	100 MB, single-mode SC fiber optic	20 km	12.5 dB
100Base FX	100 MB, single-mode SC fiber optic	70 km	32.5 dB
1000Base FX	1 GB, single-mode 1550 LC fiber optic	70 km	20 dB

16.12.1 Copper Twisted Shielded and Unshielded

The interface is defined by speed, the modulation type (Base), and the physical interface. For example, T or TX is the twisted pair and FL or FX is the fiber. 10 Base T and 100 Base TX are the most common twisted-pair copper media standards. The cable can be shielded twisted pair (STP) or unshielded twisted pair (UTP). UTP has several categories 1–7, with their specific fields of applications. The cable consists of four pairs of wires, color coded and terminated in RJ45 connectors; the maximum permitted length is 100 m for the UTP cable and 150 m for the STP cable. In practical installations, the distance is a serious limitation and a copper twisted pair is not in much use.

16.12.2 Fiber-Optic Cable

The fiber-optic cable is classified into multimode and single mode. Obviously, the single-mode fiber can be used for much longer distances and gives less attenuation. Both the multimode and the single-mode fiber cables can support a wide range of wavelengths, notably 820, 1300, and 1550 nm. Figure 16.7 shows the single-mode and multimode fiber cables, and the outer clad is 120 μm in both cases; however, the cores are of vastly different diameters. This figure shows transmission through these two cable types. The differences are that for a range of light injection angles, there is a total reflection of the light which is being transmitted down the core—analogous to a perfect mirrored surface. Multimode has a larger core diameter and supports multiple injection angles resulting in a substantial amount of input pulse spreading. A single-mode fiber continuously focuses the light into the center. As a result, the single-mode fiber has less attenuation for a given wavelength of light.

The maximum distance can be calculated by worst case OPB. The OPB can be considered as the maximum permitted attenuation of light signal, while permitting reliable communication

$$\text{OPB}_{\text{worst}} = \text{OPB} - (1 + n)\text{dB}$$

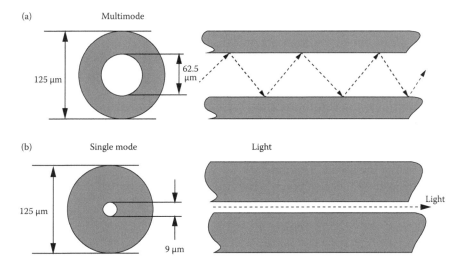

FIGURE 16.7
(a) Fiber-optic multimode cable cross section and light transmission. (b) Fiber-optic single-mode cable cross section and light transmission.

where n denotes the number of pairs, ndB gives insertion loss of n pairs, and 1 dB is the loss for LED aging. Then, the worst case OPB can be divided by cable loss; for example, for a 62.5/125 μm it will be of the order of 2.8 dB/km, while for 9/125 μm (single mode, 1550 nm) it is 0.2–0.25 dB/km.

A half-duplex fiber-optic system provides communication in both directions, but only one direction at a time. The transmitted signal must stop before a response can be sent back. A full-duplex or double-duplex allows simultaneous communications in both directions. IEDs can send and receive data simultaneously over the link [4].

16.13 Network Topologies

Some network topologies are as follows:

- Star
- Redundant star
- Mesh
- Ring

The port formed by connecting one Ethernet switch to another is often called an uplink port. Figure 16.8a and b shows a redundant star and ring configurations, respectively, using redundant ports on the IEDs. In a single star, a single point failure causes a loss of communication, additional Ethernet switches are required, and the network recovery time is 5–6 ms per Ethernet switch. A redundant star provides higher availability at the cost of additional hardware.

A ring architecture, Figure 16.8b, provides network redundancy, and, using proprietary techniques, has a failure recovery time of 5 ms per Ethernet switch. Integrated Ethernet switches can be provided in the IEDs. The IED is internally connected to the Ethernet switch through internal hardware. The last IED can be connected to the first IED to form a ring network. When one of the components fails, it should be detected and the communication path reconfigured. IEEE standard performs this function through Rapid Spanning Tree Protocol (RSTP). The protocol sends messages to the various nodes in the network to detect the broken path and reconfigure. The "link pulses" are sent when idle, and used to detect connectivity and communication capability. When a receiver loses a link, it indicates a problem and the communication path is routed in the other direction around the ring. The link loss function also allows IEDs to recover, where one of the two fiber-optic cables connected to the IEDs is damaged. Upon detection of the broken transmit fiber, IEDs will switch to the secondary port.

Architectures using Ethernet switches require one Ethernet switch for approximately 12 IEDs when using redundant fiber-optic communications.

16.13.1 Prioritizing GOOSE Messages

Quality of service signifies prioritizing the traffic, so that the critical data are processed first. Considering a trip command and metering data communications—the trip command must be prioritized. IEC 61850 GOOSE messaging provides a priority setting with

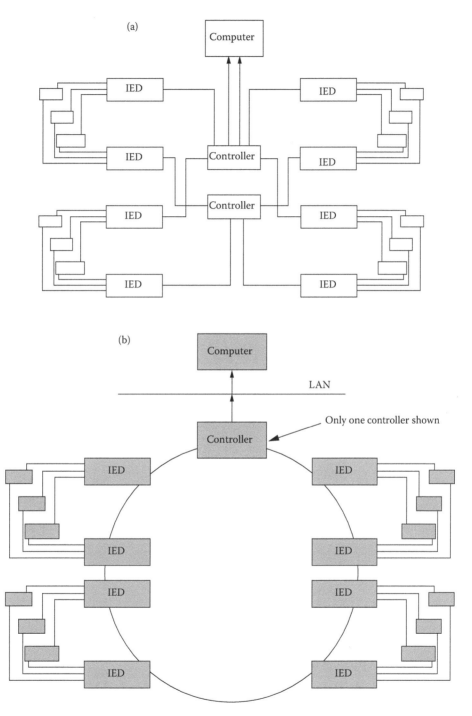

FIGURE 16.8
(a) Redundant star architecture using redundant ports on IEDs. (b) Ring architecture using redundant ports on IEDs.

eight levels of property. Processed in an Ethernet switch, the message with the highest priority is moved to the front of the queue. Reliability of critical messages is improved by using redundant Ethernet networks, redundant Ethernet ports, and power supplies.

IEEE Power System Relaying Committee Working Group WG119 [2,3] lays down the following recommendations when using IEC 61850 for critical applications:

- Multiple switches are connected in a ring, so that there are two paths for any switch port used by a relay to any other such switch port. Ethernet switches include RSTP by which the switches use a default path without calculating messages in the loop. One link in the loop is blocked. If the ring suffers a break or one switch fails, the switches can detect the path lose, unblock the spare path, and set up new routing.
- Many GOOSE-capable relays have primary and failover communication ports. Provide two switches in two groups: connect the relay primary port to one group and the failover port to another switch group.

16.13.2 Techno-Economical Justifications

- Cost reduction by replacing many control cables with fiber-optic links for alarm and control signals.
- Reduced construction for trenches and cable raceways for new substations.
- Maintenance cost reduction by taking advantage of multifunction relays to reduce the number of IEDs.
- Panel-to-panel cable reduction using extensive communication infrastructure and GOOSE messaging.
- Reduced control building and associated costs.
- Reduction and elimination of auxiliary relays reduce failure points, panel size, and wiring.
- Pretested systems, as a part of quality assurance requirements.
- Reduction of devices such as push buttons, mechanical relays, and selector switches.

Studies [5–9] provide further reading.

16.14 A Sample Application

Consider a large industrial distribution system, load demand of 90 MVA, served through three main 13.8 kV buses, A, B, C, located some distance apart. These buses have plant generation and utility interconnections and are interconnected through reactors connected to a synchronizing bus. There are a total of 150 substations in the distribution system, and there are 35 substation buildings. Assume that all substations are indoor type and a substation building may house substations connected to three 13.8 kV buses, see Figure 16.9. The substation buildings are spread over an area of 2 square miles.

Substation Automation and Communication Protocols Including IEC 61850   639

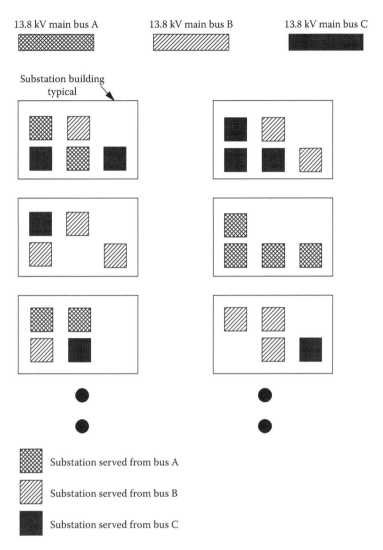

FIGURE 16.9
A schematic picture of substation layouts and connections in a large industrial distribution system.

Consider that there are 150 smart IEDs, one in each substation. The desired operation is that 150 IEDs should trip 13.8 kV breakers if a fault occurs. Also the metering data on the secondary of each substation is required to be communicated to a central control location along with status indications. The system can be implemented by hard-wired trips, transducers, and controllers, which will require considerable wiring, labor, and terminations. Also remote tripping of breakers over long distances may need auxiliary relays extending the power supplies or providing redundant batteries and chargers.

Alternatively, the system can be configured with IEC GOOSE network in ring formation, using about 35 managed Ethernet switches and a couple of controllers. The controllers will be programmed to poll the IEDs and data can be presented in a web-based HMI. A switch can be installed and programmed at the door of each substation so that it acts simultaneously on all the IEDs in a substation to bring alternate settings into operation and a worker can safely enter the substation. Bringing alternate settings on all the substations in a building by a mere flip of a switch ensures that arc-flash hazard as well the arc-flash boundary is reduced to safe levels. The status of any IED, its settings, metering data, and breaker status can be accessed on any computer connected to LAN. The system is expandable and new devices can be added, with their unique addresses and programming. Also see References [5–9].

References

1. IEC 61850 Ed. 1.0. Communication Networks and Systems in Substations-All Parts, 2011.
2. IEEE 1613. IEEE Standard. Environmental and Testing Requirements for Communications Networking Devices Installed in Electrical Power Substations, 2009.
3. IEEE 1613a. IEEE Standard. Environmental and Testing Requirements for Communications Networking Devices Installed in Electrical Power Substations, Amendment 1, 2011.
4. C Wester, M Adamiak. Practical applications of Ethernet in substations and industrial facilities. *Conferences Record, IEEE Pulp and Paper Industry Technical Conference*, June 2011, Nashville, TN.
5. IEEE PSRC Working Group. Redundancy considerations for protective relaying systems. *Conferences Record, Texas A and M Protective Relaying Conference*, April 2010.
6. KP Brand. The standard IEC 61850 as prerequisite for intelligent applications in substations. *IEEE Power Eng Soc Gen Meet*, 1, 714–718, 2004.
7. J McDonald. Substation automation and integration and availability of information. *IEEE Power Energy Mag*, 99(2), 22–31, 2003.
8. D Dolezilek, IEC 61850: What you need to know about functionality and practical implementation, Schweitzer Engineering Laboratories, Inc. 2004, 2005. www.selinc.com.
9. E Atienza. Testing and troubleshooting IEC 61850 GOOSE-based control and protection schemes, Schweitzer Engineering Laboratories, Inc. 2004, 2005. www.selinc.com.

17

Protective Relaying for Arc-Flash Reduction

17.1 Arc-Flash Hazard

In the past, industrial electrical systems in the USA have been designed considering prevalent standards, i.e., ANSI/IEEE, NEC, OSHA, UL, NESC, etc., and arc-flash hazard was not a direct consideration for the electrical system designs. This environment is changing fast (rather it has already considerably changed and there are a number of innovations and arc-flash reduction products available today), and the industry is heading toward innovations in the electrical system designs, equipment, and protection to limit the arc-flash hazard, as it is detrimental to the workers safety. This opens another chapter of the power system design, analysis and calculations, and protection (hitherto not required). There is a spate of technical literature and papers on arc-flash hazard, its calculation, and mitigation. Reference works [1–8] describe arcing phenomena and arc-flash calculations, sometimes commenting on the equations in IEEE Standard 1584 [9]. Due to this criticism, a joint venture between IEEE and NFPA is underway, retesting and revising existing IEEE Standard 1584. For example, one major criticism has been that the IEEE 1584 equations are empirical fit based on test rig up, which has vertical bus bars; however, the main bus bar orientation in most equipment is horizontal.

Arc flash is a dangerous condition associated with the unexpected release of tremendous amount of energy caused by an electric arc within electrical equipment [10]. This release is in the form of intense light, heat, and blast of arc products, which may consist of vaporized components of enclosure material—copper, steel, or aluminum. Intense sound and pressure waves also emanate from the arc flash that resembles a confined explosion. Arcing occurs when the insulation between the live conductors breaks down, due to aging, surface tracking, treeing phenomena, and human error while maintaining electrical equipment in the energized state. The insulation systems are not perfectly homogeneous, and voids form due to thermal cycling. In non-self-restoring insulations, the treeing phenomenon starts with a discharge in a cavity, which enlarges over a period of time and the discharge patterns resemble tree branches, hence, the name "treeing." As the treeing progresses, discharge activity increases and ultimately the insulation resistance may be sufficiently weakened, and breakdown occurs under electrical stress. The treeing phenomenon is of particular importance in XLPE and non-self-restoring insulations. Surface tracking occurs due to abrasion, irregularities, contamination, and moisture, which may lead to an arc formation between the line and the ground. An example will be a contaminated insulator under humid conditions. Though online monitoring and partial discharge measurements are being applied as diagnostic tools, the randomness associated with a fault and insulation breakdown is well recognized, and a breakdown can occur at any time, jeopardizing a worker's safety, who may be in close proximity of the energized equipment. Arc temperatures are of the order of 35,000°F, about four times the temperature on the surface of the sun. An arc flash can therefore cause serious fatal burns.

Although reference to electrical safety can be found as early as about 1888, it was only in 1982 that Ralph Lee [11] correlated arc flash and body burns with short-circuit currents.

The OSHA definition of a recordable injury, TRIR, for 1 year of exposure, is as follows:

$$\text{TRIR} = \frac{\text{Total number of recordable injuries and accidents}}{200,000\,\text{h}}. \tag{17.1}$$

Most insurance companies accept this parameter of definition because there is a cost associated with these incidents.

An arcing phenomenon is associated with other hazards, too, namely, arc blast, fire hazard, and shock hazard.

17.1.1 Arc Blast

Unlike arc flash, which is associated with thermal hazard and burns, arc blast is associated with extreme pressure and rapid pressure buildup. Consider a person positioned directly in front of an event and high pressure impinging upon the chest and close to the heart and the hazard associated with it.

A substance requires a different amount of physical space when it changes state, say from solid to vaporized particles. When the liquid copper evaporates, it expands 67,000 times. In documented instances, a motor terminal box exploded as a result of force created by the pressure buildup, parts flying across the room [12]. A pressure measurement of 2160 lbs/ft^2 around the chest area and a sound level of 141.5 dB at 2 ft have been made.

17.1.2 Fire Hazard and Electrical Shock

Fire hazard [13,14] and electrical shock are the other hazards. There are a number of ways by which the exposure to shock hazard occurs. The resistance of the contact point, the insulation of the ground under the feet, the flow of current path through the body, the body weight, and the system voltage level are all important. The threshold perception level is 0.0010–0.0007 A, and at about 40 mA, the shock duration of 1 s can cause ventricular fibrillation and can be fatal. On average, one person is electrocuted in work places *every day in the USA*.

17.1.3 Time Motion Studies

Of necessity and for the continuity of processes, maintenance of electrical equipment in energized state has to be allowed for. If all maintenance work could be carried out in de-energized state, a short circuit cannot occur when a worker is maintaining the equipment and therefore there is no risk of arc-flash hazard. However, in the continuous process plants, where the shutdown of a process can result in colossal amount of loss, downtime, and restarting, it becomes necessary to maintain the equipment in the energized state. Prior to the institution of arc-flash standards, this has been carried out for many years jeopardizing worker safety and there are documented cases of injuries including fatal burns.

The time/motion studies show that the human reaction time to sense, judge, and runaway from a hazardous situation varies from person to person. A typical time is of the order of 0.4 s (approximately 25 cycles). This means that 24 cycles is the shortest time in which a person can view a condition and *begin* to move or act. In all other conditions, it is not possible to see a hazardous situation and move away from it. As will be further demonstrated, this reaction time is too large for workers to move away and shelter themselves from an arc-flash hazard situation.

Thermal burns due to arc flash are only a part of the picture for overall worker safety, and for training and establishing sound maintenance procedures and guidelines.

17.2 Arc-Flash Hazard Analysis

Currently, there are following standards for arc-flash calculations:

1. NFPA 70E, revised in 2017 [15]
2. IEEE 1584, 2002, which will undergo revisions [9]
3. IEEE 1584a, 2004, amendment 1 [16]
4. IEEE P1584b/D5 [17]

NFPA 70E, year 2017, in Annex D, Table D.1 provides limitations of various calculation methods. This is reproduced in Table 17.1. *The standard does not express any preference, regarding which method should be used.* The work in Reference [17] recognizes use of knowledge and experience of those who have performed studies as a guide in applying the standard.

It is recognized that to construct an *accurate* mathematical model of the arcing phenomena is rather impractical. This is because of the spasmodic nature of the fault caused by arc elongation blowout effects, physical flexing of cables and bus bars under short circuits, possible arc reignition, turbulent flow of plasma, high-temperature gradients (the temperature at the core being of the order of 25,000 K while at the arc boundary of the order of 300–2000 K), and conductivity factors changing at a fast rate. IEEE 1584 equations are empirical equations based upon laboratory test results, though the standard includes some Lee's equations also.

TABLE 17.1

Limitations of Calculation Methods

Source	Limitation/Parameter
Ralph Lee Paper [11]	Calculates arc-flash protection boundary for arc in open air; conservative over 600 V and becomes more conservative as voltage rises
Doughty/Neal Paper [18]	Calculates incident energy for a three-phase arc on systems rated 600 V or below, applies to short-circuit currents between 16 kA and 50 kA
Ralph Lee Paper [11]	Calculates incident energy for a three-phase arc in open air on systems rated above 600 V, becomes more conservative as voltage rises
IEEE Std. 1584 [9]	Calculates incident energy and arc-flash boundary for 208 V to 15 kV, three-phase 50–60 Hz; 700–106,000 A short-circuit currents, and 13–152 mm conductor gaps[a]
ANSI/IEEE C2 [19]	Calculates incident energy for open-air phase-to-ground arcs 1–500 kV for live line work

Source: NFPA 70E-2017.

[a] Note that IEEE 1584 contains a theoretically derived model applicable for any voltage.

17.2.1 Ralph Lee's Equations

Ralph Lee's equations from Reference [11] are as follows:
The maximum power in a three-phase arc is

$$P = \text{MVA}_{bf} \times 0.707^2 \, \text{MW} \tag{17.2}$$

where MVA_{bf} is the bolted fault MVA.

The distance (in feet) of a person from the arc source for a just curable burn, i.e., skin temperature remains less than 80°C, is

$$D_c = (2.65 \text{MVA}_{bf} t)^{1/2} \tag{17.3}$$

where t is the time of exposure in seconds.

The equation for the incident energy produced by a three-phase arc in open air on systems rated above 600 V is given by

$$E = \frac{793 F V t_A}{D^2} \, \text{cal/cm}^2 \tag{17.4}$$

where

D = distance from the arc source (inches)
F = bolted fault short-circuit current (kA)
V = system phase-to-phase voltage (kV)
t_A = arc duration (s)

For the low-voltage systems of 600 V or below and for an arc in the open air, the estimated incident energy is

$$E_{MA} = 5271 D_A^{-1.9593} t_A \left[0.0016 F^2 - 0.0076 F + 0.8938 \right] \tag{17.5}$$

where E_{MA} is the maximum open-air incident energy in cal/cm², F is the short-circuit current in kA (range 16–50 kA), and D_A is the distance from arc electrodes in in. (for distances 18 in. or greater).

The estimated energy for an arc in a cubic box of 20 in., open on one side, is given by

$$E_{MB} = 1038.7 D_B^{-1.4738} t_A \left[0.0016 F^2 - 0.0076 F + 0.8938 \right] \tag{17.6}$$

where E_{MB} is the incident energy and D_B is the distance from arc electrodes (in.) (for distances 18 in. or greater).

17.2.2 IEEE 1584 Equations

The IEEE equations are applicable for the electrical systems operating at 0.208–15 kV, three phase, and 50 Hz or 60 Hz. The available short-circuit current range is 700–106,000 A, and the conductor gap 13–152 mm. For three-phase systems in open-air substations, open-air transmission systems, a theoretically derived model is available. For system voltage *below* 1 kV, the following equation is solved:

$$\log I_a = K + 0.662 \log_{10} I_{bf} + 0.0966V + 0.000526G + 0.5588V(\log_{10} I_{bf}) - 0.00304G(\log_{10} I_{bf})$$

(17.7)

where

I_a = arcing current (kA)
G = conductor gap, typical conductor gaps are specified in Reference [9]
K = −0.153 for open-air arcs, and −0.097 for arc in a box
V = system voltage (kV)
I_{bf} = bolted three-phase fault current (kA, rms symmetrical).

For systems of 1 kV or higher, solve the following equation:

$$\log_{10} I_a = 0.00402 + 0.983 \log_{10} I_{bf}.$$

(17.8)

This expression is valid for arcs both in open air and in a box. Use $0.85 I_a$ to find a second arc duration. This second arc duration accounts for variations in the arcing current and the time for the overcurrent device to open. Calculate incident energy using both $0.85 I_a$ and I_a and use the higher value.

Incident energy at working distance, empirically derived equation, is given by

$$\log_{10} E_n = K_1 + K_2 + 1.081 \log_{10} I_a + 0.0011G.$$

(17.9)

The equation is based upon data normalized for an arc time of 0.2 s, where

E_n = incident energy (J/cm²) normalized for time and distance
K_1 = 0.792 for open air and −0.555 for arcs in a box
K_2 = 0 for ungrounded and high-resistance grounded systems and −0.113 for grounded systems. Note that resistance grounded, HR grounded, and ungrounded systems are all considered ungrounded for the purpose of calculation of incident energy.
G = conductor gap (mm), see Table 17.2.

TABLE 17.2

Classes of Equipment and Typical Bus Gaps

Classes of Equipment	Typical Bus Gap (mm)
15 kV switchgear	152
5 kV switchgear	104
Low-voltage switchgear	32
Low-voltage MCCs and panel boards	25
Cable	13
Other	Not required

Conversion from normalized values gives the equation

$$E = 4.184 C_f E_n \left(\frac{t}{0.2}\right)\left(\frac{610^x}{D^x}\right) \quad (17.10)$$

where

E = incident energy (J/cm²)
C_f = calculation factor which is equal to 1.0 for voltages above 1 kV and 1.5 for voltages at or below 1 kV.
t = arcing time (s)
D = distance from the arc to the person, working distance (see Table 17.3)
x = distance exponent as given in Reference [9] and reproduced in Table 17.4.

A theoretically derived equation can be applied for voltages above 15 kV or when the gap is outside the range in Table 17.2 from Reference [9]:

$$E = 2.142 \times 10^6 VI_{bf}\left(\frac{t}{D^2}\right). \quad (17.11)$$

The arc-flash protection boundary, defined further, empirically derived equation, is

$$D_B = \left[4.184 C_f E_n \left(\frac{t}{0.2}\right)\left(\frac{610^x}{E_B}\right)\right]^{1/x} \quad (17.12)$$

where E_B is the incident energy in J/cm² at the distance of arc-flash protection boundary. For the empirically derived equation:

$$D_B = \left[2.142 \times 10^6 VI_{bf}\left(\frac{t}{E_B}\right)\right]^{1/2}. \quad (17.13)$$

TABLE 17.3

Classes of Equipment and Typical Working Distances

Classes of Equipment	Typical Working Distance (mm)
15 kV switchgear	910
5 kV switchgear	910
Low-voltage switchgear	610
Low-voltage MCCs and panel boards	455
Cable	455
Other	To be determined in field

Source: IEEE, *IEEE Guide for Performing Arc-Flash Hazard Calculations*, Standard 1584, 2002.

TABLE 17.4

Factors for Equipment and Voltage Classes

System Voltage (kV)	Equipment Type	Typical Gap between Conductors	Distance x Factor
0.208–1	Open air	10–40	2.000
	Switchgear	32	1.473
	MCC and panels	25	1.641
	Cable	13	2.000
>1–5	Open air	102	2.000
	Switchgear	13–102	0.973
	Cable	13	2.000
>5–15	Open air	13–153	2.000
	Switchgear	153	0.973
	Cable	13	2.000

Source: IEEE, *IEEE Guide for Performing Arc-Flash Hazard Calculations,* Standard 1584, 2002.

Due to the complexity of IEEE equations, the arc-flash analysis is run on digital computers. Most commercially available programs analyze arc-flash hazard as a subroutine to short-circuit calculations. It is obvious that the incident energy release and the consequent hazard depend upon the following:

- The available three-phase rms symmetrical short-circuits current. In low-voltage systems, the arc-flash current will be 50%–60% of the bolted three-phase current, due to arc voltage drop. In medium- and high-voltage systems, it will be only slightly lower than the bolted three-phase current. The short-circuit currents are accompanied by a dc component, whether it is the short circuit of a generator, motor, or utility source. However, for arc-flash hazard calculations the *dc component is ignored*. Also any unsymmetrical fault currents, such as line-to-ground fault currents, need not be calculated. As is evident from the cited equations, only three-phase bolted fault current needs to be calculated.
- The time duration for which the event lasts. This is obviously the sum of protective relay (or any other protection device) operating time and the opening time of the switching device. For example, if the relay operating time is 20 cycles, and the interrupting time of the circuit breaker is 5 cycles, then the arc-flash time is 25 cycles.
- The type of equipment, that is, switchgear or MCC and the operating voltage.
- The system grounding.

17.3 Hazard/Risk Categories

NFPA Table 130.7(C) (11) describes the personal protective equipment (PPE) characteristics for hazard/risk category of 0, and 1 through 4. These are shown in Table 17.5. There is a slight revision with respect to the 2004 edition.

The Standard ASTM F1506 [18] calls for every flame resistance garment to be labeled with an arc energy rating, arc thermal performance exposure value (ATPV). The rating of

TABLE 17.5

Protective Clothing Characteristics

Hazard Category	Clothing Description	Arc Rating of PPE (cal/cm²)
0	Nonmelting, flammable materials, i.e., untreated cotton, wool, rayon, or silk, or blends of these materials, with a fabric weight of 4.5 oz/yd²	N/A
1	Arc-treated FR shirt and FR pants or FR overall	4
2	Arc-treated FR shirt and FR pants or FR overall	8
3	Arc-treated shirt and pants or FR overall, and arc suit selected so that the system rating meets the required minimum	25
4	Arc-treated shirt and pants or FR overall, and arc-flash suit selected so that the system rating meets the required minimum	40

Source: NFPA 70E-Electrical Safety in Workplace, 2017.

the garment is matched with the calculated incident energy release level. The test method of determining the ATPV states that the incident energy on a multilayer system of materials which results in 50% probability of sufficient heat transfer through the test specimen is predicted to cause onset of second-degree skin burn injury [15].

The maximum incident energy for which PPE is specified is 40 cal/cm² (167.36 J/cm²). It is not unusual to encounter energy levels much higher than 40 cal/cm² in actual electrical systems. Standards do not provide guidelines for higher incident energy levels. Incident energy reduction techniques can be applied; otherwise, it is prudent not to maintain such equipment in energized state.

A category 4 PPE outfit looks like a space suit with face hood, eye shields, cover, and gloves. It restricts the mobility of a worker to perform delicate tasks, for example, maintenance work on terminals and wiring.

Thus, not only an accurate calculation of the incident energy level but its reduction in the planning and design stage and the selection of appropriate protection and relaying of electrical systems are gaining importance [19,20].

17.3.1 Hazard Boundaries

The boundaries are defined in the NFPA, and the following synopsis is relevant here (see Figure 17.1).

The flash protection boundary is the distance at which threshold of second-degree burns can occur and the incident energy release is 1.2 cal/cm² (5.0 J/cm²). This is the boundary that is calculated by computer-based programs. Inside the boundary, the energy level will be higher. This boundary should not be crossed by anyone, including a qualified person, without wearing the required PPE (see Table 17.5). The PPE outfits are designed to minimize the risk of sustaining energy greater than 1.2 cal/cm². *That is, threshold of second-degree burns can still occur even with appropriate PPE, and these burns are considered curable.*

Unqualified persons, that is, those not specifically trained to carry out the required tasks, are safe when they stay away from the energized part a certain distance, which is the limited approach boundary. They should not cross the limited approach boundary and arc-flash boundary unless escorted by a qualified person.

Crossing restricted approach boundary means that special shock prevention techniques and equipment are required and an unqualified person is not allowed to cross this boundary.

Protective Relaying for Arc-Flash Reduction

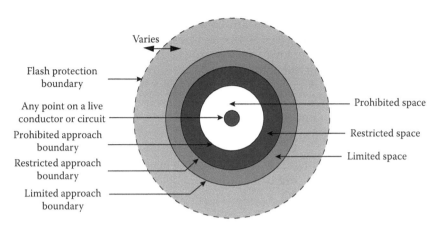

FIGURE 17.1
Arc-flash boundaries.

Finally, the prohibited approach boundary establishes the space that can be crossed only, as if a live contact with exposed energized conductors or circuits was planned.

Working distance is defined as the closest distance to a worker's body excluding hands and arms. IEEE Standard 1584 [9] specifies required working distances, see Table 17.3. For example, for a 15 kV switchgear it is 36 in., while for a 480 V MCC it is 18 in. A larger working distance reduces the incident energy and, therefore, the HRC. Recommendations of IEEE for the working distances are followed.

The limited approach, restricted approach, and prohibited approach boundaries are all defined based upon the system voltage. *No calculations are required for establishing these boundaries.*

17.4 System Grounding: Impact on Incident Energy

Equation 17.9 for the calculation of incident energy has a factor K_2 for system grounding. It is zero for ungrounded and high-resistance grounded systems and −0.113 for the solidly grounded systems. Thus, the incident energy release is higher for the ungrounded or high-resistance grounded systems compared to solidly grounded systems. The physical phenomenon leading to release of incident energy for solidly grounded and ungrounded systems is not explicitly stated in the standard.

Lee's equations do not consider system grounding and the incident energy will be the same, whether the system is grounded or ungrounded.

However, it is not the intent that solidly grounded systems should be adopted to reduce arc-flash hazard. Much lower equipment damage and continuity of processes can be achieved with high-resistance grounded systems. For low-voltage distributions, generally, the solidly grounded systems are not applied and the ungrounded systems are extinct (no more used). The solidly grounded systems versus high-resistance grounding systems are amply discussed in the current literature, see Chapter 2. Approximately 70% of the faults in the electrical systems are line-to-ground faults. Sometimes these may be self-clearing and of transient nature (for example, in OH line systems) or may evolve into three-phase faults over a period of time.

Thus, the probability of a worker being subject to arc flash due to ground faults is much higher. The IEEE equations take a safe stance and state that if the hazard level is calculated for three-phase faults, it will be lower for the ground faults. In a single-phase-to-ground fault, the arc tends to extinguish at natural current zero. In three-phase systems, the phases are displaced 120 electrical degrees and natural current zeros are similarly displaced. In industrial systems, the line-to-ground fault current in solidly grounded systems can even exceed three-phase bolted fault current (see Chapter 2). A footnote in the NFPA-90 reads as follows:

> ..., high resistance grounding of low-voltage and 5 kV (nominal) systems, current limitations, and specifications of covered bus within equipment are techniques available to reduce the hazard of the system.

Example 17.1

Calculate the incident energy using IEEE and Ralph Lee's equations for a 30 kA three-phase bolted fault current in a 13.8 kV metal-clad switchgear, resistant grounded. The arcing time is 30 cycles for the arc-fault current through the protective device.

Using IEEE Equation 17.8 calculate the arcing current:

$$\log I_a = 0.00402 + 0.983 \log I_{bf}.$$

This gives $I_a = 28.578$ kA. Note that the arc-flash time is calculated based upon the arcing fault current through the protective device and not the bolted fault current. A further calculation for arc-flash time is required at $0.85 I_a$. Let us assume for this example that 30 cycles are applicable. Then, calculate normalized incident energy from Equation 17.9:

$$\log E_n = (-0.555) + (-0.113) + 1.081 \log(28.578) + 0.0011(153).$$

Here $K_1 = -0555$ (arc in a box), $K_2 = -0.113$ (resistance grounded system), and $G = 153$ mm (see Table 17.3). This gives $\log E_n = 1.074$. Calculate incident energy in J/cm² using (Equation 17.10):

$$E = 4.184(1)(11.858)\left(\frac{0.5}{0.2}\right)\left(\frac{610^{0.973}}{910^{0.973}}\right).$$

Here, t is the arcing time, which is equal to 0.5 s, the distance x is given in Table 17.4 and for 15 kV it is 0.973, and D, the working distance from Table 17.3, is 910 mm.

This gives $E = 84.4$ J/cm² = 20.08 cal/cm².

Calculate the arc-flash boundary (Equation 17.12):

$$D_B = \left[4.184(1)(11.858)\left(\frac{0.5}{0.2}\right)\left(\frac{610^{0.973}}{5}\right)\right]^{1/0.973}.$$

Here, $C_f = 1$, E_n as calculated before is 11.858 J/cm², and $E_B = 5$ J/cm² by definition. Thus $D_B = 16{,}541$ mm = 54.267 ft. This is rather a large distance at which a worker can get threshold of second-degree burns.

Lee's Equations

The incident energy release given by Equation 17.4 is 126.6 cal/cm² and from Equation 17.3 the distance for a curable burn is 30.82 ft. There is a vast difference in these calculations. Note that Lee's equations do not consider system grounding. IEEE equations are normally used for arc-flash evaluations.

17.5 Maximum Duration of an Arc-Flash Event

A maximum duration of 2 s for the total fault clearance time of an arc-flash event is considered, though, in some cases, the fault clearance time can be higher.

Tables 17.6 and 17.7 show the arc-flash boundary calculations according to IEEE 1584 equations for a bolted three-phase fault current of 30 kA in a 13.8 kV switchgear, 13.8 kV system is resistance grounded, and also in a 480 V MCC, 480 V system high-resistance grounded.

For a 30 kA, 2 s fault in 13.8 kV switchgear, the incident energy boundary is 3539 in., which is equal to 295 ft. For 1 s fault duration, it is 144.6 ft. This is rather a large distance. The variation of the incident energy boundary at which a worker can be exposed to 1.2 cal/cm² of incident energy and sustain threshold of second-degree burns seems to be very large.

The working space, in an electrical room switchgear installation in front of electrical equipment, is limited and may be as low as 5–6 ft, and must meet NEC requirements.

TABLE 17.6

Arc-Flash Boundary and Incident Energy Release for 30 kA of Bolted Fault Current (28.58 kA rms Arc-Flash Current) in 13.8 kV Switchgear, 13.8 kV System Resistance Grounded, Working Distance = 36 in., Gap = 153 mm

Arc Duration (s)	Arc-Flash Boundary (in.)	Incident Energy
0.058	74	3.0
0.5	851	26
1.0	1736	52
1.5	2633	78
2	3539	104

TABLE 17.7

Arc-Flash Boundary and Incident Energy Release for 30 kA of Bolted Fault Current (16.76 kA rms Arc-Flash Current) in 480 V MCC, 480 V System High-Resistance Grounded, Working Distance = 18 in., Gap = 25 mm

Arc Duration (s)	Arc-Flash Boundary (in.)	Incident Energy
0.050	36	3.8
0.5	147	38
1.0	225	75
1.5	288	113
2	343	151

Many companies, as a policy, keep the electrical rooms locked and a worker must have the required PPE outfit before entering the electrical rooms.

A distribution system may be operated in various operating modes, that is, a tie line or a generator switched in or switched out of service. Bolted short-circuit currents in each of these operating modes are required to be calculated and the arc-flash hazard is calculated for a variation between 100% and 85% of arcing current in each scenario.

17.5.1 Equipment Labeling

The equipment can be, then, labeled with respect to the maximum HRC calculated in each of the operating modes. Generally, the label contains equipment identification, system voltage, PPE required (Category 1, 2, 3, or 4), incident energy in cal/cm^2 or J/cm^2, arc-flash boundary, restricted and prohibitive approach boundary, and the identification of the upstream protective device that cleared the fault. Such a label must be provided on each electrical equipment based upon the arc-flash analysis study [15].

17.6 Protective Relaying and Coordination

For arc-flash analysis, ground fault protection and its coordination need not be considered, as the calculations are based upon three-phase bolted faults, which will be cleared by the phase overcurrent protection relays, differential relays, arc-flash sensing relays, or other unit protection schemes. We have amply discussed these protection schemes in this volume, except arc-flash sensing relays.

17.7 Arc Protection Relays

Arc-flash detection relays provide even faster fault detection times compared to the differential relays. With modern AFD technology, this arc-fault detection time is of the order of ¼ cycles or even lower, based on 60 Hz. High-speed insulated gate bipolar transistors (IGBTs) are used to provide fully rated trip outputs rather than the relatively slow conventional dry contacts. Thus, AFDs form a technically acceptable and cost-effective way to reduce the arc-flash hazard.

17.7.1 Principle of Operation

The light emitted during an arc-flash event is significantly brighter than the normal lighting background. This radiant energy can be easily detected using light sensors.

Though relatively new in North America, the arc-flash relays have been around since the early 1990s. These earlier relays used only single-point light receptacles, which are called lens sensors. One or more lens sensors are located in a compartment, where a potential arc hazard can occur. Each sensor is connected radially to electronics, through clad fibers. Because of opaque cladding, only the extreme ends of the radial fibers are light sensitive.

Protective Relaying for Arc-Flash Reduction

A limited number of sensors may be attached to an electronic package. Multiple electronic packages are then interconnected to provide the required coverage.

During the year 2000, a second generation of AFD relays was introduced. In addition to the lens sensors, a long unclad fiber optic that could absorb light throughout its length was developed. This had several advantages, as it reduced the cost of installation. A single optic fiber sensor can be as long as 200 ft and can cover the same zone of protection as bus differential protection, but at a reduced cost (see Chapter 8). The concern for any shadows from the internal structures that could block the light is eliminated. Also if the fiber sensor is configured in a loop, which is normally done in the current installation, it can provide regular self-checking features of integrity and breakage of fiber loop, producing an alarm if a problem is detected in the integrity of the sensor.

Thousands of arc-flash detection systems have been installed in 36 countries worldwide over the past 6 years. Optically based arc-flash relaying is the fastest protection available for arc-flash hazard reduction.

Arc flash produces a bright light. The light produced has been measured to be 100,000–250,000 lux at a distance of 3 m. A camera flash can produce 234,000 lux at 18 in. and the direct sunlight at any distance is 100,000 lux. High-intensity LED flash light can produce 28,000 lux. The intensity of normal substation lighting is 200–300 lux.

Visible light consists of light spectrum ranging from 400 to 700 nm wavelengths. The arc-flash tests show that most of the radiated energy has a wavelength of 200–600 nm. Consequently, AFDs are designed to operate at the lower end of visible light spectrum and slightly including ultraviolet light.

Figure 17.2 shows the photometric data for an arc current of 20 kA, based upon the test results. As the intensity of light decreases very fast after an arc-flash event, the response time of the AFD system must be small.

The light captured by the sensor is amplified and compared to a preselected light reference level. Once it exceeds the set level, a light signal is activated. AFD relays may be provided with a selector switch to select auto or manual light reference level. In the manual mode, the light reference level can be set through a front potentiometer. Figure 17.3 shows the sensitivity of lens sensors of a manufacturer at various backlight compensation settings [5,6].

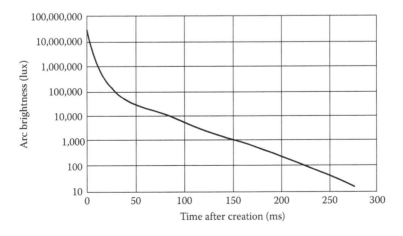

FIGURE 17.2
Arc brightness versus time after a 20 kA fault.

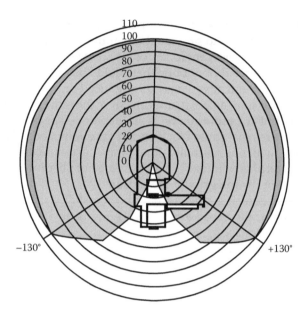

FIGURE 17.3
Relative sensitivity of a lens sensor from different angles of lighting.

17.7.2 Light Sensor Types

The common types of sensors are lens-point sensors and bare fiber-optic sensors. Figure 17.4 is a depiction of sensitivity of sensors. The normal operating sector is 130°. Practically, the light is also reflected from the compartment walls. For fiber sensors, the incident angle of lighting is not applicable. The bare fiber sensor consists of a high-quality plastic fiber-optic cable without a jacket. Figure 17.5a,b shows a lens (point sensor) and Figure 17.5c depicts a fiber sensor. Special termination and splice kits are required, which are not discussed. A bare fiber sensor makes possible the detection in large areas. The detecting reach depends upon several factors:

- Light source energy
- Fiber length
- Reflectances
- Backlight compensation settings

The length of the sensor fiber for a switchgear compartment is selected according to the available short-circuit or ground fault currents. Table 17.8 lists these selection criteria.
The table has been compiled based upon:

- Copper bus bars
- Arc length = 10 cm
- Surrounding light = 400 lux
- No reflecting surfaces
- Light reference level set one scale mark to the right from the minimum

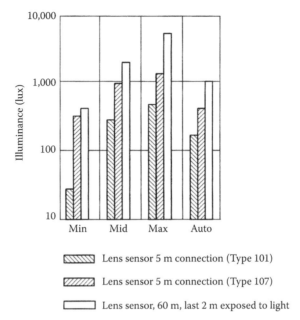

FIGURE 17.4
Sensitivity of sensors at various compensation settings.

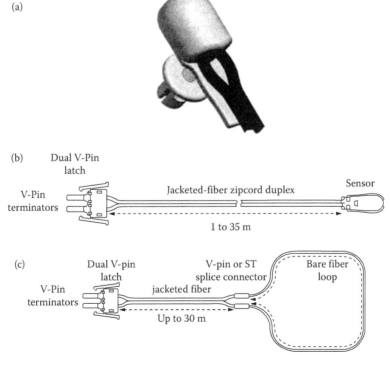

FIGURE 17.5
(a,b) A lens sensor and (c) a fiber sensor.

TABLE 17.8

Minimum Length of the Exposed Sensor Fiber per One Switchgear Compartment

Fault Current (rms)	Distance between Sensor and Arc			
	100 cm	200 cm	300 cm	400 cm
0.5 kA	20	Note	Note	Note
0.7 kA	20	70	210	280
1.4 kA	20	20	20	140
2.2 kA	20	20	20	20

Note: Not operational.

A manufacturer will place the sensors to detect arc-flash radiant light properly in an air-insulated switchgear to be protected. Some guidelines for the installation of the sensors are as follows:

- Each sensor fiber is routed to cover the specific protection zone, that is, bus bar cable, circuit breaker, and PT compartments, etc.
- The best location is near the top and rear of circuit breaker cell and away from the breaker. For bus bars, the best location is usually along the back wall, centrally located with respect to bus bars.
- The high-temperature surfaces are to be avoided. Temperatures above 60°C can decrease the performance over a period of time. The fibers are not secured directly to the bus.
- The fiber is not installed in conduit or in a raceway that will shield it from the light emitted by an arc-flash event. Where exposure to the light is not required, protective tubing is installed.
- The fiber is secured using nylon ratchet clamps or similar fiber management products. The purpose is to keep the fiber away from moving parts and prevent snagging the fiber when racking the breakers in and out.
- Protective rubber grommets are used when routing sensors through metal walls. The hole size required is small, which is of the order of 10 mm.
- Sharp bends are avoided. The minimum permanent bending radius of the sensor fiber is 2 in. The fiber can be broken or damaged if this minimum bending radius is not maintained.
- Correct installation tools are required. There is a limit to the number of splices that can be installed in a given length of sensor fiber to maintain reliable operation. The following rules are advocated by one manufacturer:

    Maximum total length without splices or with one splice: 160 m
    With two splices: 50 m
    With three splices: 40 m

- When the existing equipment is to be retrofitted with AFD protection, additional care in the installation and testing is required.

In a radial unit installation system, the breaks cannot be detected.

17.7.3 Other Hardware

The routing of fibers in a 13.8 kV, single-high metal-clad switchgear lineup is shown in Figure 17.6a,b. This shows two feeder breakers on each side of the tie breaker and two main breakers. The sensing fibers are run through the main bus compartment, circuit breaker compartments, PT compartment, cable compartment, and bus section breaker compartment. In the bus section breaker compartment, the fibers overlap much like differential zones of protection. The fiber runs through lower main and feeder breakers and returns through the bus compartments and through the PT compartment to device A. The right and left sides are symmetrical (details not completely shown in Figure 17.6a,b). The fiber in the cable compartment is routed through device B, only one relay shown. *One B device per circuit breaker is used.* The device functionality is described below:

A: Central unit
B: High-speed trip unit or AFD relay
C: Fiber extension unit
D: Lens sensor extension unit [5,6].

17.7.4 Selective Tripping

With the arrangement shown in Figure 17.7a, selective trip arrangement is obtained. For a fault in the cable compartment, only feeder circuit breaker can be tripped through device B. The main fiber loop does not enter the cable compartments. This ensures that only the respective feeder circuit breaker is tripped for a fault which is downstream of the feeder circuit breaker. Similarly for the faults on the right and left sides of the tie circuit breaker, only the circuit breakers on that side plus the tie circuit breaker are tripped. Regardless of where the flash occurs along the fiber, the same circuit breakers will be tripped. Thus, in order to achieve selectivity separate fibers are used, as shown in this figure. For a bus differential protection, the cable compartment faults remain outside the differential zone of protection. With AFDs the faults in the cable or PT compartments can be easily detected.

The light is transmitted from the sensor to the detector located remotely in the relay. For reliability, this is done through a fiber-optic loop. For lens sensors, each lens has an input and output connection. The input is connected to the transmitter in the relay and the output to the detector in the relay. The loop connection allows periodic testing of the system by injecting light from the transmitter through the loop and back to the detector. This can work with either the lens sensor or the bare fiber sensor.

About 200 ft of optical fiber loop is adequate to cover most applications. Extension units can be daisy chained and up to approximately 20 additional fiber loops of 200 ft each can be connected for a total effective sensor length of 4000 ft.

A simple tripping scheme is shown in Figure 17.7a. For a fault in any one phase, both primary and secondary circuit breakers are tripped. Compare this with Figure 17.7b, which provides independent arc-flash detection for the downstream faults. The extension unit (Ext. unit in Figure 17.7b) will trip its associated feeder circuit breaker. Simultaneously, it communicates to the central unit that a downstream trip has occurred. If the fault is not cleared within the programmed time, the central unit will trip its associated circuit breakers, thereby providing coordinated arc-flash backup protection and selective fault clearing.

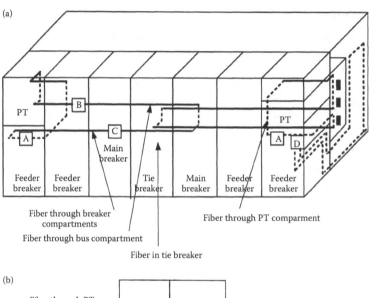

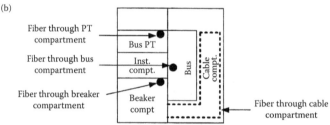

FIGURE 17.6
(a,b) Routing of fibers through bus compartment, breaker compartment, PT and cable compartment, and bus tie breaker compartment.

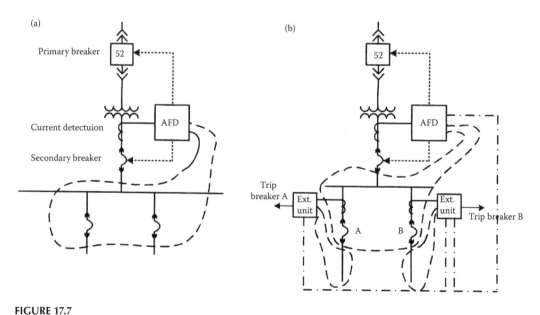

FIGURE 17.7
(a) Routing of fibers for nonselective AFD protection and (b) routing of fibers for selective AFD protection.

17.7.5 Supervision with Current Elements

Figure 17.8 shows a schematic diagram of the AFD system. This shows that the trip can be based upon the light sensing only, or it can be supervised through a high-speed current sensing. An obvious advantage is that low levels of phase and ground fault currents are sensed, limiting the equipment damage as well as arc-flash hazard. In addition, the security of the system is enhanced. Figure 17.8 shows a selector switch to enable or disable the current supervision. The set current level can be even below the load current. Conventional current elements have a response time of the order of 6–20 ms. This delay is unacceptable and for arc-flash detection, high-speed overcurrent elements should act as fast as the arc detection.

17.7.6 Applications

AFD systems can be applied to:

- Metal-clad and cubical-type switchgear
- Load interrupter switchgear
- Medium-voltage MCCs.

There has been some concern that low-voltage circuit breakers produce enough ambient light while operating, yet AFD technology has been applied to low-voltage circuit breakers and MCCs and commercial products are available. A light sensing system should not be triggered by light emissions of properly interrupting low-voltage circuit breakers. Filtering the light is an easily workable solution. Light filters can be designed that block the light from interrupting circuit breakers as well ambient conditions.

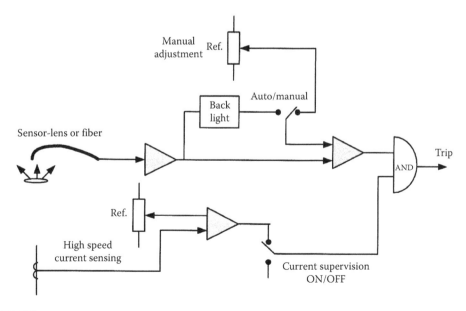

FIGURE 17.8
A logic circuit diagram of AFD protection.

17.7.7 Self-Testing of Sensors

Both point and fiber sensors can be tested by performing a loop-back test. An LED couples the test light to the loop or point sensor. For the loop sensor, the light travels through the fiber-optic cable to the sensor in the relay. For a point sensor, the light travels through fiber-optic cable and is emitted by the sensor dome. This emitted light by the sensor dome is picked up by the adjacent fiber-optic cable and returned to the sensor in the relay. For either type of sensors, the light sensed during the test is compared to the set limits, thus verifying the functionality of the light sensor relay and measuring the optic path attenuation. A test result outside the test limits can be alarmed. The fiber cuts, kinks, or disconnections can be detected. The self-test features are important to ensure the integrity of the AFD systems. Currently, there are no standards for testing the performance of AFD systems. The self-tests can be made so that the test signals and an arc-flash event are distinguishable and no nuisance trip occurs, if an arc-flash event occurs during the testing [8].

17.8 Accounting for Decay in Short-Circuit Currents

The application of short-circuit currents for calculations of switching devices' short-circuit duties is quite different from their applications to arc-flash calculations.

First consider the fault decrement curve of the generator. Note that the first cycle and interrupting duty ac components of the current are identical. For the calculation of duties on a circuit breaker, these currents are considered with appropriate multiplying factors till the interruption takes place. Similarly, the motor contributions will decay fast with in a couple of cycles depending on the motor rating. It will be erroneous to consider these currents nondecaying for the duration of arcing time.

Consider a system bus which has multiple sources connected to it, which will contribute to the short-circuit currents. For example, Figure 17.9 shows a 13.8 kV bus, resistance grounded system, which has a 40 MVA generator and a utility interconnection through a 45 MVA transformer. The generator operates in synchronism with a 138 kV utility source. A number of transformers carrying rotating ac motor load (induction and synchronous motors) and some static loads, not shown in Figure 17.9, are served from this bus. This is a typical industrial distribution cogeneration bus. The load distributions from this bus can be equivalent Thévenin impedances, as shown in this figure.

The calculated short-circuit current for a fault on the bus is 38 kA rms symmetrical. With respect to the calculations of short-circuit currents for arc-flash analysis, using available arc-flash calculation software programs, the options are as follows:

1. No algorithm to account for the decay from the generators or motors is available. A user can select either the first cycle (momentary) or interrupting (1.5–5 cycle) symmetrical currents.
2. The program facilitates knocking out the motor contribution after a user selectable time delay and similarly reduce the generator short-circuit current, after a time delay.
3. Some programs may summate the arc-flash energy calculated in small time strips.

Protective Relaying for Arc-Flash Reduction

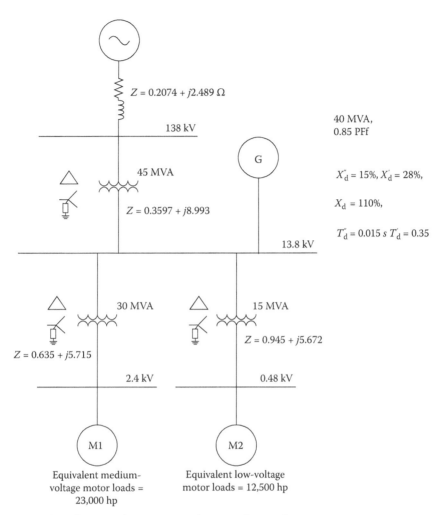

FIGURE 17.9
A 13.8 kV bus with multiple sources of short-circuit contributions.

Consider a fault on 13.8 kV bus of Figure 17.9. With no decay from the generators or motors, the incident energy accumulation profile is shown in Figure 17.10a. In this figure, for a bus fault the generator breaker trips prior to opening of the utility tie breaker. The motor contributions continue for the entire period. This results in energy accumulation as shown.

Now, assume that the motor contributions are dropped in six cycles and generator contribution is reduced to 300% of its full load current in 0.5 s. Then this situation is depicted in Figure 17.10b. The total energy accumulated is reduced, as shown in the shaded area, compared to the situation shown in Figure 17.10a.

Yet, the total energy accumulations shown in Figure 17.10b are not accurate. In the real-world situation, it is not the step reduction but decay at a certain rate given by the transient parameters of the generators and motors (see Figure 17.10c).

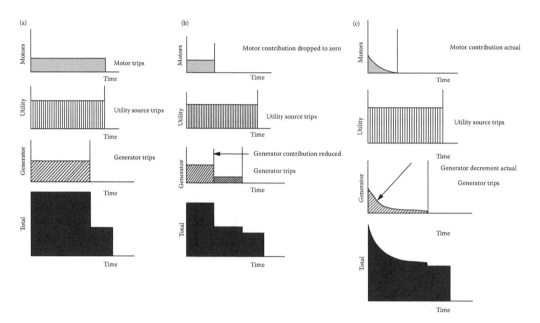

FIGURE 17.10
(a through c) Accumulation of incident energy, see text.

Therefore, the incident energy calculated using the three figures will be
$E_{\text{fig 22a}} > E_{\text{fig 22b}} > E_{\text{fig 22c}}$.

In other words, the calculations are overly conservative. In Figure 17.10b, it is only guesswork when the step change should be made.

There is no computer software that can simulate the results of Figure 17.10c. Theoretically, trapezoidal rule of integration or other step-by-step numerical techniques can be used. IEC standard [21] recommends integration over the period of short circuit to calculate the accumulated energy.

Example 17.2

Consider that the fault on bus A, fed from the generator and utility tie, is removed simultaneously in 0.5 s. This is arbitrary, to show the difference in calculations using the methodology shown in Figure 17.10a through c.

Calculation 1 (Figure 17.10a): No decay of short-circuit current from the generators or motors. The calculated results are shown in the first row of Table 17.9. Incident energy release = 36 cal/cm² and HRC = 4.

Calculation 2 (Figure 17.10b): Motor contribution knocked out in 8 cycles and generator contribution reduced to 350% in 15 cycles. Incident energy release = 30 cal/cm² and HRC = 4 (see row 2 of Table 17.9).

Calculation 3 (Figure 17.10c): First plot/calculate the overall current–time decrement curve of the short-circuit current. Utility source is considered nondecaying and motor and generator short-circuit contributions decay.

Plot decrement curves of the generators and motors. The symmetrical component of the generator short-circuit current, generator at no load at the rated voltage of the generator, using the generator parameters from Figure 17.9 can be plotted, as shown in Figure 17.11. The dc component is ignored. Similarly, the decrement curves of motors

TABLE 17.9

Variations in the Calculations of Incident Energy and HRC, 13.8 kV Switchgear, Resistance Grounded System, Trip Delay 0.5 s, Breaker Opening Time 0.080 s (Five-Cycle Breaker). Row 1, No Decay, Row 2, AC Motor and Generator Current Decay, Step Change, Row 3, Accurate Calculations

Cal. No.	Bolted Current (kA rms)	Arcing Current (kA rms)	Incident Energy (cal/cm^2)	HRC
1	37.90	35.96	36	4
2	37.90	35.96	30	4
3			21.11	3

can be plotted. This gives the overall decrement curve by summation of the three components. The plot is extended to time 0.001 s (0.06 cycle).

To calculate the incident energy, divide the 0.5 s interval into three parts, arbitrary chosen for reasonable accuracy:

$$0.001 \text{ to } 0.01 \text{ s}$$
$$0.01 \text{ s to } 0.1 \text{ s}$$
$$0.1 \text{ s to } 0.5 \text{ s}.$$

The average current in each interval and its time duration are shown in Table 17.10, with corresponding energy release. The summation gives 21.11 cal/cm^2 and HRC=3. There is a vast difference in the results obtained with calculating the total current decay, dividing it to a number of sections, taking the average in each section, calculating the energy release in each section, and then summing it up.

This shows that accounting for decay of short-circuit currents makes much difference in the ultimate results. Practically, with the hand calculations, it will be very time consuming to plot the decay at each bus, divide it into a number of segments, and then calculate the incident energy and summate, though the calculation algorithm can be computerized.

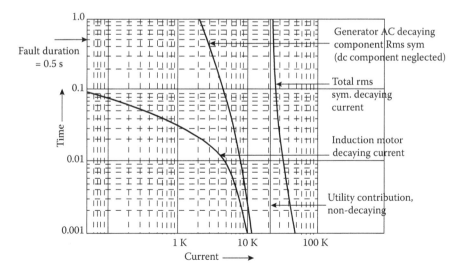

FIGURE 17.11
Calculations of short-circuit current decay profiles.

TABLE 17.10

Calculations of Incident Energy by Plotting Overall Fault Decrement Curve

Bolted Current (kA)	Arcing Current (kA)	Duration in (s)	Incident Energy (cal/cm^2)
35	33.25	0.01	0.61
29.2	27.64	0.09	4.5
23.5	22.48	0.40	16

17.8.1 Reducing Short-Circuit Currents

To reduce the arc-flash hazard, the short-circuit levels in the electrical systems can be reduced by the following:

1. The relative magnitudes of 33 kA at 2.4 kV and 19 kA at 4.16 kV, for the same equipment ratings, bring about a voltage selection criterion for arc-flash reduction—that is, at the design stage select 4.16 kV rather than 2.4 kV for medium-voltage secondary distributions. This will have cost impacts depending upon the ratings, especially for induction motors.
2. For low-voltage systems of the same ratings and configurations, the short-circuit currents in a 575/600 V system will be approximately 83% of those in 480 V systems. (Low-voltage cables of 1 or 2 kV voltage grade may be required for drive systems.)
3. Rating of the transformers plays another major role in reducing the short-circuit levels. For example, two 3.75 MVA medium-voltage transformers can be used compared to a single 7.5 MVA unit. Similarly for low-voltage distributions two 1.0 MVA transformers can be used versus a single 2.0 MVA transformer.
4. Double-ended substations with autoclosing of normally open bus section circuit breaker can be used. This configuration is not so popular, yet it is possible to transfer rotating loads (induction motors) without loss of processes (see Chapter 9).
5. Current-limiting reactors are a possibility, but careful analysis and study are required. These may have adverse effect on voltage profiles and reactive power loss, and are not recommended simply to lower short-circuit currents for the purpose of reducing HRC.

17.9 Arc-Resistant Switchgear

An arc-resistant switchgear for medium-voltage and low-voltage applications will give zero incident energy outside the enclosure at the working distance. It must withstand an internal arcing fault and meet the requirements of IEEE C37.20.7-2007 [22].

Assembly type 1 has an arc-resistant design for the front of the equipment only. Type 2 has freely accessible exterior for front back and sides. Accessibility suffixes A, B, C, and D further categorize the equipment, with respect to accessibility. For example, suffix "B" means that under normal operation of the equipment opening of the doors or cover of compartments specifically designed as low-voltage control or instrumentation compartments is permissible. This requires additional testing [22]. If any cover or panel is opened, which is not intended to be opened under normal operation, the equipment is no longer arc resistant.

The arc-resistant switchgear is specified for the maximum short circuit duration and IEEE standard [22] recommends a time duration of 0.5 s. Practically, the time duration can be reduced when fast-differential or light-sensing protective features are provided. The arc-flash duration withstand time has a major impact on the cost of the equipment.

The switchgear may be provided with pressure relief devices to exhaust the overpressure. Plenums may be run on top or rear of the assembly to conduct arc-flash products outside the switchgear room. It is not the intention to lay down the specifications for applications, suffice to state that this is one means to reduce arc-flash hazard. Nevertheless, all maintenance tasks cannot be carried out simply because the equipment is arc resistant. NFPA 70E-2017 [15] specifies an HRC of 4 for the arc-resistant switchgear, type 1 or 2, with clearing time of <0.5 s for insertion or removal of circuit breaker from cubicles with door open. (This should be analyzed case by case depending upon the protection and system fault levels.) Of course, rarely, the circuit breakers will be locally racked out in modern work environment. Industrial establishments are resorting to remote racking, and the presence of a worker near the equipment is not required except to engage the remote racking mechanisms.

17.10 Arc-Flash Calculations

Table 17.11 shows the maximum arc-flash time, which is the sum of relay operating time and breaker interrupting time for limiting the HRC to level 2 (8 cal/cm^2) and also level 4 (40 cal/cm^2) in the medium-voltage systems. This table shows that in order to limit HRC to 2, a 30 kA bolted short-circuit fault in a 13.8 kV system must be cleared in 0.15 s, and a 40 kA fault must be cleared in 0.11 s. Deducting five cycles for the breaker operating time, the time available for the downstream coordination is 0.027 and 0.06 s, respectively. This is too small for any downstream time–current coordination. *Thus, time–current coordination, howsoever implemented, cannot reduce the hazard level to 2 or lower for 13.8 kV distributions.*

Table 17.12 shows the calculations with differential relays and light-sensing relays. For an older version of differential relays, an operating time of 1.5 cycles is considered while for newer microprocessor-based multifunction relays (MMPRs) with current differential an operating time of ¾ of a cycle is considered. An operating time of ¼ cycle is considered

TABLE 17.11

Maximum Arcing Time for Limiting HRC to 2 and 4 for 13.8, 4.16, and 2.4 kV Systems, Resistance Grounded, Working Distance 36 in.

System Voltage (kV)	SC Current, (kA rms sym.)	FCT (s)	Incident Energy (cal/cm^2)
13.8	40	0.11	8
		0.57	40
	30	0.15	8
		0.75	40
2.4	33	0.16	8
		0.78	40
4.16	19	0.28	8
	19	1.43	40

TABLE 17.12

Arc-Flash Hazard Reduction in Medium-Voltage Systems through Differential and Arc-Flash Light-Sensing Relays

System Voltage (kV)	Bus Bolted Fault (kA rms sym.)	Arc Fault (kA rms sym.)	Breaker Interrupting Time Cycles	Device Operating Time Cycles	Gap (mm)	Arc-Flash Boundary (in.)	Incident Energy (cal/cm^2)	HRC	Remarks
13.8	40	37.92	5	1.5	153	241	7.6	2	1.5 cycles diff.
				0.75		169	6.8	2	3/4 cycle diff.
				0.25		155	6.2	2	1/4 cycle arc flash
			3	1.5		131	5.3	2	1.5 cycles diff.
				0.75		110	4.4	2	3/4 cycle diff.
				0.25		94	3.8	1	1/4 cycle arc flash
13.8	30	28.58	5	1.5	153	176	5.6	2	1.5 cycles diff.
				0.75		124	5	2	3/4 cycle diff.
				0.25		113	4.6	2	1/4 cycle arc flash
			3	1.5		96	3.9	1	1.5 cycles diff.
				0.75		80	3.3	1	3/4 cycle diff.
				0.25		68	2.8	1	1/4 cycle arc flash
4.16	19 (see text)	18.24	5	1.5	104	94	3	1	1.5 cycles diff.
				0.75		66	2.7	1	3/4 cycle diff.
				0.25		60	2.5	1	1/4 cycle arc flash
			3	1.5		51	2.1	1	1.5 cycles diff.
				0.75		43	1.8	1	3/4 cycle diff.
				0.25		37	1.5	1	1/4 cycle arc flash
2.4	33 (see text)	31.38	5	1.5	104	172	5.5	2	1.5 cycles diff.
				0.75		121	4.9	2	3/4 cycle diff.
				0.25		110	4.5	2	1/4 cycle arc flash
			3	1.5		94	3.8	1	1.5 cycles diff.
				0.75		78	3.2	1	3/4 cycle diff.
				0.25		67	2.7	1	1/4 cycle arc flash

for light-sensing relays supervised by current. These relays operate with either three-cycle or five-cycle circuit breakers. This table shows that even for a 40 kA short-circuit current in a 13.8 kV resistance grounded system HRC level of 1 can be obtained with light-sensing relays. The calculated HRC levels do not exceed 2 anywhere in this table.

17.10.1 Reduction of HRC through a Maintenance Mode Switch

With the proper application of an MMPR, with alternate protective settings modified/brought in operation through *maintenance mode switch* can reduce the hazard level, sometimes equivalent to that obtained by differential relays. The premise is *that a fault may not coincide with the maintenance tasks and, therefore, during maintenance the selective coordination is sacrificed to some extent.* If a fault does occur, the worker as well as the equipment is protected, but a wider area may be shutdown. In an MMPR changing or bringing into service, a protection function is performed by forming an AND gate with the switch position and the protective settings, which actuate a discrete output. This can be wired in the trip circuit.

The authors of Reference [23] state that annual exposures for low-voltage MCCs are 365; for other low-voltage equipment 52; and for rest of the equipment at HV (>34 kV), MV (1–34 kV), and MV MCCs, it is the total of 30. Therefore, the reduction of HRC for low-voltage MCCs is important. It also implies that LV MCCs are the most frequently maintained equipment while energized.

According to NEC Article 240.85, for the low-voltage transformer protection it is not necessary to provide a secondary main breaker, and protection with primary overcurrent protection devices is acceptable provided certain conditions stated in this article are met. Table 450-3(A) of NEC permits this omission of secondary protection device in supervised locations. Table 17.15 shows that for the arc-flash protection this leads to a hazardous condition for a bus fault on the low-voltage switchgear, which is directly connected to the transformer secondary. The arc-fault current as seen by the current-limiting fuse, on the primary side, is reduced in the transformation ratio, that is by a factor of 28.75, and the fuse takes a long time to operate resulting in release of large amount of incident energy. Even if the maximum arcing time is reduced to 2 s, the incident energy release is much above 40 cal/cm^2.

Before we discuss this situation and the measures to reduce arc-flash hazards in the low-voltage systems, there have been many product innovations in the industry for controlling the arc-flash hazard in low-voltage distribution systems. Some of these and relevant aspects can be enumerated below.

i. *Zone interlocking*. This is discussed in Chapter 6.

ii. *CPU-based protection and control*: Low-voltage power circuit breakers with redundant CPU-based protection, monitoring, control, and diagnostic functions, which replace discrete devices and hard wiring, are available. These provide enhanced reliability and arc-flash reduction. Synchronization between two CPUs is maintained through a hardwired sync clock and the CPUs and critical controls are powered through in-built redundant UPS systems. For arc-fault reduction, bus differential algorithm with zone-based overcurrent protection is used. For work near the equipment, reduced energy let-through mode can be selected through a remotely located switch. This enables more sensitive protection settings to lower the arc-flash incident energy. The maintenance arc-flash reduction system is being called with a variety of names; essentially it acts upon the trip settings to modify these during maintenance operation.

iii. *Current-limiting fuses:* A current-limiting fuse is designed to reduce equipment damage by interrupting the rising fault current before it reaches its peak value. Within its current-limiting range, the fuse operates within 1/4 to 1/2 cycles. By limiting the rising fault current, the I^2t let-through to the fault is reduced because of two counts: (1) high speed of fault clearance in 1/4 cycle typically in the current-limiting range and (2) fault current limitation. This reduces the fault damage [24]. Practically (1) a large size fuse may be provided to achieve maximum coordination, and (2) some lack of coordination must be accepted.

iv. *Short time bands of LVPCBs*: This is discussed in Chapter 5 (see Table 5.10).

v. *Coordination between instantaneous settings*: This is discussed in Chapter 6.

vi. *Innovations*, such as closed-door remote racking, insulated bus bars, system redundancy, through the door voltage indicators, incorporation of safety shutters, arc-flash protected infra-red scanning windows, online insulation testing systems, etc., are strategies for worker's safety.

17.11 System Configuration for Study

A radial distribution system is shown in Figure 17.12. A single 1200 A 13.8 kV circuit breaker serves four substation transformers: three transformers of 2000/2240 kVA, 480 V and one of 7500/9375 kVA, 2.4 kV secondary. All transformers have 13.8 kV windings in delta connection and the secondary windings in wye connection. The 480 V transformer wye windings are high-resistance grounded and 2.4 kV transformer wye windings are low-resistance grounded through a 200-A resistor. Downstream distribution from one 2000 kVA low-voltage transformer TX1 and 7500 kVA, 2.4 kV transformer TX2 is shown. The 480 V transformer serves a lineup of low-voltage switchgear provided with LVPCBs having electronic trip programmers with LSI functions. At first consider that *there are no main secondary circuit breakers either on 480 V or 2.4 kV transformers*, that is, circuit breakers BK5 and BK6 are not provided. The low-voltage switchgear serves a number of low-voltage motor control centers; one typical MCC is shown. For coordination purposes, we need to consider the largest motor/feeder on any MCC. In the industrial distribution system, generally, the 200 hp motor rating is the maximum that is applied and higher rated motors are connected to medium-voltage distributions. The 2.4 kV transformer serves as a lineup of 2.4 kV metal-clad switchgear, which in turn serves some medium-voltage MCCs. The major ratings of the equipment are shown in this figure. The interconnecting cable sizes are detailed in Table 17.13.

The 7500/9375 kVA transformer TX2 is protected by relay R1 on 13.8 kV circuit breaker BK1. The 2000/2240 kVA transformer TX1 is protected with 150E current-limiting fuse. The medium-voltage MCC is served from a relayed feeder circuit breaker, BK4. The 2000 hp motor has a NEMA 2 fused starter, with 700 A vacuum contactor and 36R fuse. The system ground fault protection is not shown. All switching devices are applied well within their short-circuit ratings. The three-phase bolted short-circuit level at 13.8 kV switchgear is 30 kA rms symmetrical.

This radial distribution system meets the requirements of NEC and is "well protected" with the protection devices and loads required to be served. Many industrial distribution systems implemented on this basis are in service. *However in the analysis to follow, it is shown that this system is very deficient from the arc-flash hazard mitigation consideration.*

17.11.1 Coordination Study and Observations

Let us consider fault locations shown from F1 through F12. For arc-flash evaluations, faults in any location of the equipment assembly must be considered. Table 17.14 shows the fault locations and the protective devices which will clear the fault, when main circuit breakers are not provided.

Protective Relaying for Arc-Flash Reduction

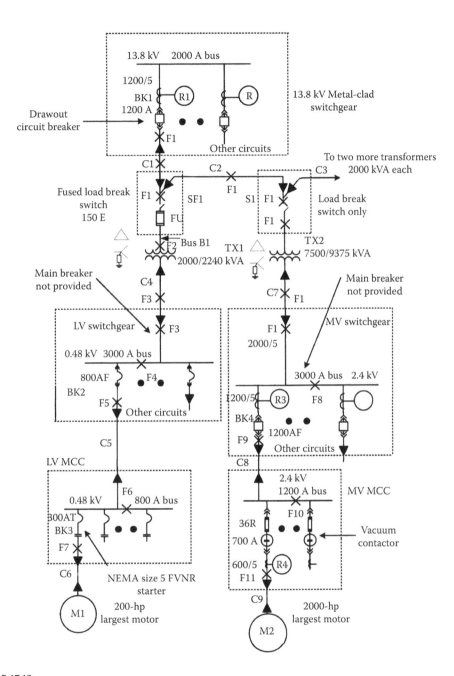

FIGURE 17.12
A medium-voltage and low-voltage distribution system for coordination and arc-flash analysis.

Figure 17.13 shows the TCC plot for the 2000/2240 kVA transformer protective devices and Figure 17.14 shows the TCC plot for the 7500/9500 kVA transformer. The following observations are of interest.

- The transformer ANSI frequent fault curves are protected.
- The system is well coordinated.

TABLE 17.13

Cable Sizes and Lengths (Figure 17.12)

Cable Designation	Cable Description	Number in Parallel per Phase
C1, C2	15 kV grade, 3/C, 500 kcmil, 130% insulation level, 90°C temperature, XLPE	2
C3	15 kV grade, 3/C, 500 kcmil, 130% insulation level, 90°C temperature, XLPE	1
C7	5 kV grade, 1/C, 500 kcmil, 130% insulation level, 90°C temperature, XLPE	4
C8	5 kV grade, 1/C, 500 kcmil, 130% insulation level, 90°C temperature, XLPE	2
C9	5 kV grade, 1/C, 500 kcmil, 130% insulation level, 90°C temperature, XLPE	2
C4	600 V grade, 3/C, 500 kcmil, THHW, NEC 90°C temperature	7
C5	600 V grade, 3/C, 500 kcmil, THHW, NEC 90°C temperature	2
C6	600 V grade, 3/C, 500 kcmil, THHW, NEC 90°C temperature	1

TABLE 17.14

Faults at Various Locations in Figure 17.12, and the Protective Devices Clearing These Faults, When Main Secondary Breakers Are Not Provided

Fault	Description	Fault Clearing Device
F1, F8	On the secondary side of the 13.8 kV feeder breaker in the 13.8 kV switchgear itself, in the cable circuits C1 and C2, in the fused load break switch upstream of the 150E fuse, in the 7500/9375 kVA transformer load break switch, cable C7 to the 2.4 kV switchgear, in the primary cable terminations in the 2.4 kV switchgear, and also in the 2.4 kV switchgear bus, as no main secondary beaker is present.	Relay R1, feeder breaker BK1
F2	Load side of the fuse and up to the primary windings of 2000/2240 kVA transformer.	Fuse FS1
F3, F4	On the secondary side of 2000/2240 kVA transformer, on the cable terminations in the low-voltage switchgear, and also on the low-voltage switchgear bus, fault location F4; as there is no main secondary breaker.	Fuse FS1
F5, F6	On the secondary side of the feeder beaker in low-voltage switchgear, in the incoming cable terminations at low-voltage MCC, and on the MCC bus, fault location 5.	Feeder breaker BK2
F7	On the secondary side of the motor feeder in the low-voltage MCC.	Motor starter breaker BK3
F9, F11	On the load side of the feeder circuit breaker to medium-voltage MCC, on the incoming terminations of the medium-voltage MCC, and also on the MCC bus, fault location F11.	Relay R3, breaker BK4
F12	On the load side of the 2000 hp motor starter	Motor starter fuse 36R

- The 150E fuse clears the transformer inrush point. There is a slight overlap with the feeder short-time delay setting, but this is acceptable. The transformer rated current at 13.8 kV is 93.7 A at its full fan cooled rating of 2240 kVA. Thus, a lower rated class E fuse, for example 125E, could be used, but the overlap with feeder short-time characteristics will increase.

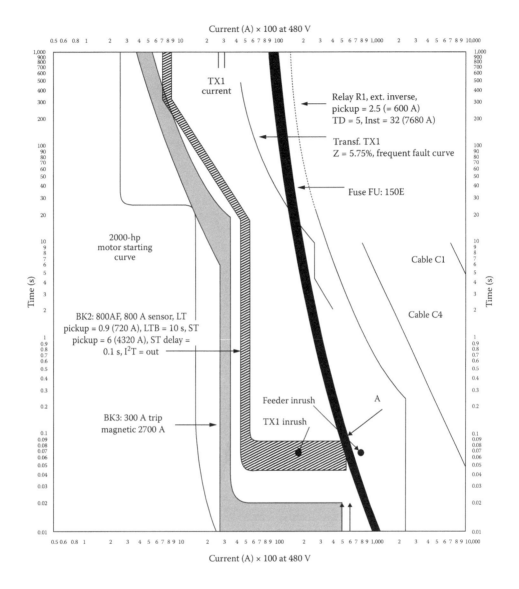

FIGURE 17.13
Time–current coordination for 480 V, 2000 kVA transformer protective devices as shown in Figure 17.12.

- To coordinate with 200 A MCCB, motor circuit breaker BK3, a setting higher than the minimum short-time delay setting available on the trip programmer of circuit breaker BK2 is not required.
- The motor starting curves are drawn for the actual motor starting time, driving its load inertia.
- The 2000 hp motor thermal curve is plotted based upon the manufacturer's data, and the multifunction relay R4 shown in Figure 17.14 admirably protects it.
- The settings on relay R1 protect the 7500/9375 kVA transformer as per NEC guidelines and allow the maximum system load equivalent to the total installed kVA rating of the transformer to be carried on a continuous basis.

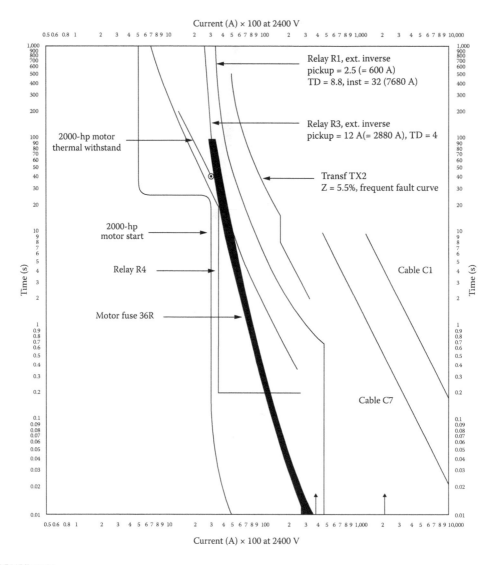

FIGURE 17.14
Time–current coordination for 2.4 kV, 7500 kVA transformer protective devices as shown in Figure 17.12, no main breaker.

- The instantaneous settings on relay R1 take into account the total feeder inrush; that is, all the connected transformers will take their inrush currents on switching. This total inrush current is calculated as 6500 A at 13.8 kV.
- Curves for only a few of the cables are plotted as specimens.

17.11.2 Arc-Flash Calculations in Figure 17.12

Table 17.15 shows the results of arc-flash hazard calculations according to IEEE Guide. The working distance, gap in mm according to equipment type, and the system grounding are shown in this table. The following observations are of interest:

TABLE 17.15

Arc-Flash Hazard Analysis (Figure 17.12), and TCC Plots (Figures 17.13 and 17.14)

Equipment Faulted	Voltage (kV)	Upstream Trip Device	Ground	Air Gap (mm)	Bolted Fault (kA)	Arc Fault (kA)	Trip Time (s)	Opening Time (s)	Arc Time (s)	Arc-Flash Boundary (in.)	Working Distance (in.)	Incident Energy (cal/cm²)	PPE
SF1	13.8	R1	No	153	35.02	33.26	0.016	0.083	0.099	152.5	36	6.1	2
S1	13.8	R1	No	153	33.96	32.28	0.016	0.083	0.099	147.5	36	5.9	2
Bus duct B1	13.8	FU	No	153	35.16	33.41	0.01	0	0.01	14.4	36	0.6	0
MV switchgear	2.4	R1	No	102	4.83	4.61	0.673	0.083	0.757	1089	36	33.1	4
MV MCC	2.4	R3	No	102	26.69	25.46	0.437	0.083	0.521	710	36	21.8	3
Fault F12, MV MCC	2.4	36R Fuse	No	102	26.47	25.25	0.023	0	0.023	22.9	36	1	0
LV switchgear	0.48	FU	No	32	1.37	0.58	35.77	0	35.77	3346	24	1729	Danger
LV MCC	0.48	BK2	No	25	36.06	19.57	0.15	0	0.15	78.9	18	13.6	3
LV MCC Fault F7	0.48	BK3	No	25	36.06	19.57	0.017	0	0.017	18.2	18	1.5	1

- The currents shown are the ones that flow in the device which clears the fault. For example, a fault on low-voltage switchgear is cleared by the transformer primary fuse, and the current through the fuse is only 1.37 kA. It is reduced in the transformation ratio; in this case, it is reduced by a factor of 28.75. Further, $I_a=85\%$ is used for the calculation of arcing time.
- The operating times of the devices such as fuses, low-voltage trip programmers, and MCCBs are built into the curves published by the manufacturer (see Chapter 1). For relayed circuit breaker, the trip time and the circuit breaker opening time are shown separately. We have five cycle breakers so the opening time is 0.083 s.
- The disadvantage of not having a main secondary circuit breaker is demonstrated by these calculations. As the secondary short-circuit currents are reduced in the ratio of transformation, when reflected on the primary side, the primary protection device takes a long time to clear the fault. In Table 17.15, the transformer primary fuse takes 35.76 s to clear a fault on the low-voltage switchgear bus, releasing an immense amount of incident energy (1729 cal/cm^2). This is not acceptable. Even if the arcing time is limited to 2 s, the incident energy release exceeds 40 cal/cm^2.
- The PPE required at medium-voltage switchgear is 4, while at medium-voltage MCC and low-voltage MCC, it is 3.
- Multiple PPE levels can exist on the same equipment. For example, the medium-voltage switchgear has a PPE of 4 for a bus fault, but for a fault in the load-side terminals or outgoing cable connections, the PPE is 3.

Though from TCC plots in Figures 17.13 and 17.14, the protection seems to be adequate, but not so from arc-flash reduction considerations. It becomes, therefore, necessary to reduce the incident energy by faster fault clearance times, which means additional protection devices have to be provided.

Next, we will demonstrate how the incident energy can be reduced so that a PPE of >2 is not required anywhere in the system.

17.11.3 Reducing HRC Levels with Main Secondary Circuit Breakers

In these calculations, main secondary circuit breakers, BK5 and BK6, are provided in Figure 17.15. This coordination is shown in Figures 17.16 and 17.17. Provision of 2000 A circuit breaker BK6 requires shifting of the circuit breaker BK1 relay R1 curve. The arc-flash calculations are shown in Table 17.16. In this table, only the arc-flash hazard for the faulted equipment undergoing a change due to provision of main circuit breakers is documented. The following is noteworthy:

- The incident energy at the medium-voltage switchgear slightly increases, while at low-voltage switchgear it is much reduced from the extreme hazard to PPE3.
- A fault at F3 in the incoming cable terminations of the low-voltage circuit breaker BK5 will still be cleared by the primary fuse FU. Thus, though the incident energy is reduced on the rest of the switchgear, the main secondary circuit breaker is exposed to extreme hazard and cannot be maintained in the energized state.
- A similar situation is depicted for fault at F1 on the incoming side of the 2.4 kV circuit breaker BK6.

Protective Relaying for Arc-Flash Reduction

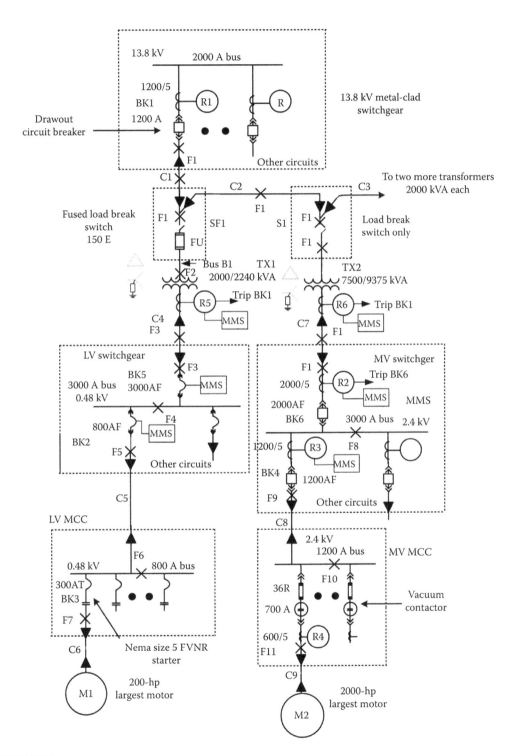

FIGURE 17.15
Modified distribution system of Figure 17.12 for arc-flash analysis.

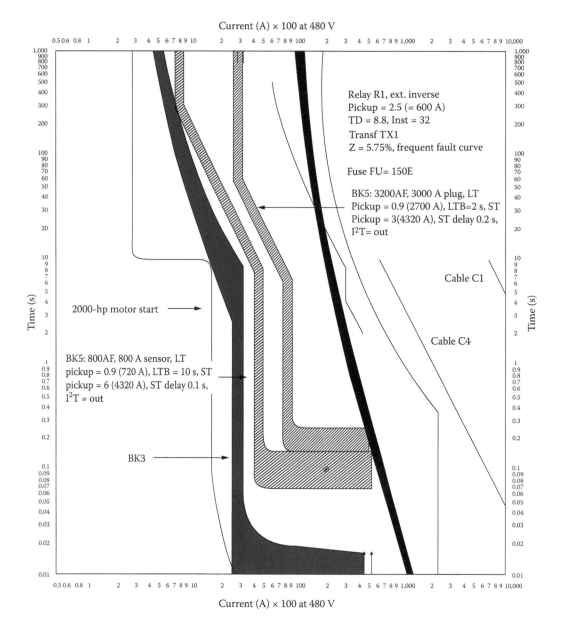

FIGURE 17.16
Time–current coordination for 480 V, 2000 kVA transformer protective devices as shown in Figure 17.15, with main secondary breaker.

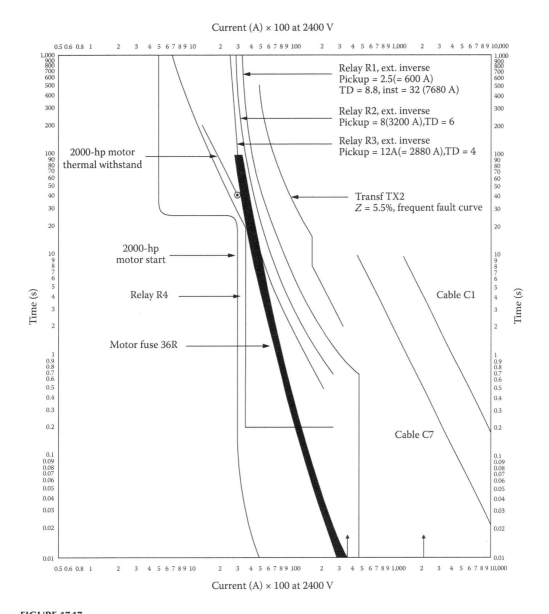

FIGURE 17.17
Time–current coordination for 2.4 kV, 7500 kVA transformer protective devices as shown in Figure 17.15, with main secondary breaker.

TABLE 17.16
Arc-Flash Hazard Analysis (Figure 17.15), with Main Secondary Breakers, TCC Plots (Figures 17.16 and 17.17)

Equipment Faulted	Voltage (kV)	Upstream Trip Device	Ground	Air Gap (mm)	Bolted Fault (kA)	Arc Fault (kA)	Trip Time (s)	Opening Time (s)	Arc Time (s)	Arc-Flash Boundary (in.)	Working Distance (in.)	Incident Energy (cal/cm²)	PPE
MV switchgear	2.4	R2	No	102	27.81	26.51	0.722	0.083	0.805	1161	36	35.2	4
Breaker BK6	2.4	R1	No	102	4.83	4.61	0.673	0.083	0.757	1089	36	33.1	4
LV switchgear	0.48	BK4	No	32	39.37	19.68	0.25	0	0.25	129.7	24	14.4	3
Breaker BK5	0.48	FU	No	32	1.37	0.58	35.77	0	35.77	3346	24	1729	Danger

Thus, the installation of the main circuit breakers has much reduced the arc-flash hazard on the main buses, but for a fault on the main circuit breaker itself, which is cleared by the primary protection, the incident energy release is not reduced and these circuit breakers cannot be maintained in the energized state.

17.11.4 Maintenance Mode Switches on Low-Voltage Trip Programmers

The trip programmers for low-voltage circuit breakers can be provided with maintenance mode switch. The coordination with instantaneous settings on low-voltage trip programmers is shown in Figure 17.18, and Figure 17.19 shows the instantaneous settings activated through maintenance mode switch on relays R2 and R3. Table 17.17 shows the HRC levels. These are reduced to below 2 for all fault locations except for a fault on the main circuit breakers BK5 and BK6 (see also [25–27]).

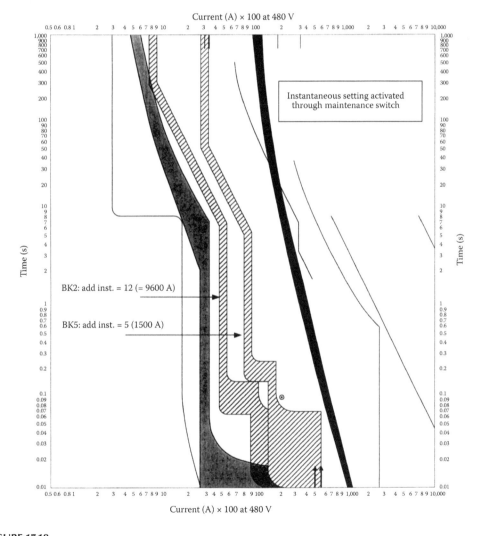

FIGURE 17.18
Time–current coordination for 480 V, 2000 kVA transformer protective devices as shown in Figure 17.15, with main secondary breaker and instantaneous settings activated through maintenance mode switch.

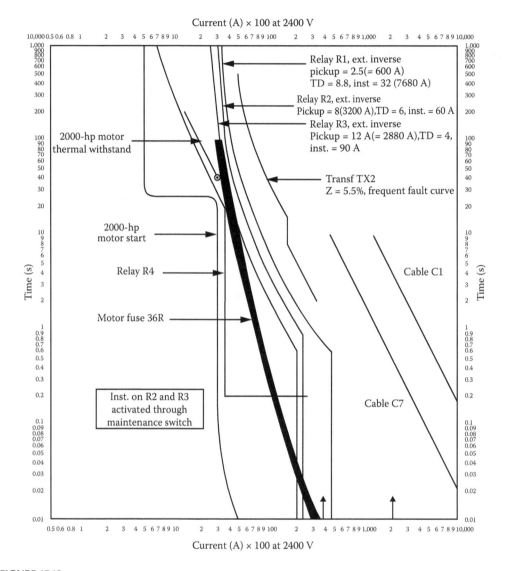

FIGURE 17.19
Time–current coordination for 2.4 kV, 7500 kVA transformer protective devices as shown in Figure 17.15, with main secondary breaker and instantaneous settings activated through maintenance mode switch.

TABLE 17.17
Arc-Flash Hazard Analysis, with Main Secondary Breakers Provided with Instantaneous Settings, TCC Plots (Figures 17.18 and 17.19)

Equipment Faulted	Voltage (kV)	Upstream Trip Device	Ground	Air Gap (mm)	Bolted Fault (kA)	Arc Fault (kA)	Trip Time (s)	Opening Time (s)	Arc Time (s)	Arc-Flash Boundary (in.)	Working Distance (in.)	Incident Energy (cal/cm²)	PPE
MV switchgear	2.4	R4	No	102	27.81	26.51	0.016	0.083	0.099	107.4	36	4.3	2
MV MCC	2.4	R3	No	102	26.69	25.46	0.916	0.083	0.099	102.9	36	4.2	2
LV switchgear	0.48	BK2	No	32	39.37	19.68	0.099	0	0.099	52.9	24	4.8	2
LV MCC	0.48	BK2	No	25	36.06	19.57	0.07	0	0.07	43.3	18	6.3	2

References

1. WA Brown, R Shapiro. Incident energy reduction techniques. *IEEE Ind Mag*, 15(3), 53–61, 2009.
2. T Gammon, J Matthews. Conventional and recommended arc power and energy calculations and arc damage assessment. *IEEE Trans Ind Appl*, 39(3), 197–203, 2003.
3. T Gammon, J Matthews. Instantaneous arcing fault models developed for building system analysis. *IEEE Trans Ind Appl*, 37(1), 197–203, 2001.
4. VV Terzija, HJ Koglin. On the modeling of long arc in still air and arc resistance calculations. *IEEE Trans Power Delivery*, 19(3), 1012–1017, 2004.
5. AD Stokes, DK Sweeting. Electrical arcing burn hazards. *IEEE Trans Ind Appl*, 42(1), 134–142, 2006.
6. HB Land, III. Determination of the case of arcing faults in low-voltage switchboards. *IEEE Trans Ind Appl*, 44(2), 430–436, 2008.
7. HB Land, III. The behavior of arcing faults in low-voltage switchboards. *IEEE Trans Ind Appl*, 44(2), 437–444, 2008.
8. R Wilkins, M Allison, M Lang. Effects of electrode orientation in arc flash testing, In *IEEE IAS Annual Meeting, Hong Kong*, pp. 459–465, 2005.
9. IEEE. *IEEE Guide for Performing Arc-Flash Hazard Calculations*. Standard 1584, 2002.
10. JC Das. *Arc Flash Hazard*, McGraw-Hill Year Book of Science and Technology, 18–20, 2008.
11. R Lee. The other electrical hazard: Electrical arc blast burns. *IEEE Trans Ind Appl*, 1A-18(3), 246–251, 1982.
12. MG Droouet, F Nadeau. Pressure waves due to arcing faults in a substation. *IEEE Trans Power Appar Syst*, PAS-98(5), 1632–1635, 1979.
13. Society of Fire Protection Engineers (SFPE) Task Force, Engineering Guide to Predicting First and Second Degree Skin Burns, Boston, MA, 2000.
14. NFPA 921, Guide for Fire and Explosion Investigations.
15. NFPA 70E, Electrical Safety in Workplace, 2017.
16. IEEE 1584a, IEEE Guide for Performing Arc-Flash Hazard Calculations—Amendment 1, 2004.
17. IEEE P 1584b/D5, Draft 5, IEEE Guide for Specifications of Scope and Deliverable Requirements for an Arc-Flash Hazard Study in Accordance with 1584, 2013.
18. ASTM F1506, Specifications for Textile Materials for Wearing Apparel for Use by Electrical Workers Exposed to Momentary Electrical Arc and Related Thermal Hazards, 1998.
19. JC Das. Design aspects of industrial distribution systems to limit arc flash hazard. *IEEE Trans Ind Appl*, 41(6), 1467–1482, 2005.
20. JC Das. Protection planning and system design to reduce arc flash incident energy in a multi-voltage level distribution system to 8 cal/cm^2 or less—part II analysis. *IEEE Trans Ind Appl*, 47, 408–420, 2011.
21. IEC 60909-0, Short-Circuit Currents in Three-Phase AC Systems, Part-0, Calculation of Currents, 2001, Also IEC 60909-1, Factors for Calculation.
22. ANSI/IEEE C37.20.7. IEEE Guide for Testing Metal Enclosed Switchgear Rated up to 38 kV for Internal Arcing Faults, 2007.
23. DR Doan, JK Slivka, CJ Bohrer. A summary of arc flash hazard assessment and safety improvements. *IEEE Trans Ind Appl*, 45(4), 1210–1216, 2009.
24. RL Doughty, TE Neal, TL Macalady, V Saportia, K Borgwald. The use of low-voltage current-limiting fuses to reduce arc-flash energy. *IEEE Trans Ind Appl*, 36(6), 1741–1749, 2000.
25. JC Das. *Arc Flash Hazard Analysis and Mitigation*, IEEE Press, Piscataway, NJ, 2012.
26. RL Doughty, TE Neal, HL Floyd, II. Predicting incident energy to better manage the electrical arc hazard on 600-V power distribution systems. *IEEE Trans Ind Appl*, 36(1), 257–269, 2000.
27. WJ Lee, M Sahni, K Methaprasyoon, C Kwan, Z Ren, M Sheeley. A novel approach for arcing fault detection for medium-voltage/low-voltage switchgear. *IEEE Trans Ind Appl*, 45(4), 2009.

Appendix A: Device Numbers according to IEEE C37.2

Device Numbers according to IEEE C37.2

Device Number	Description
1	Mater element, generally used for hand-operated devices
2	Time delay setting or closing relay, except device functions 48, 62, and 79
3	Checking or interlocking relay
4	Master contactor
5	Stopping device
6	Starting circuit breaker
7	Rate-of-rise relay
8	Control power disconnecting device
9	Reversing device
10	Unit sequence switch
11	Multifunction device
12	Over-speed device
13	Synchronous-speed device
14	Under-speed device
15	Speed-or frequency-matching device
16	Data communication device
17	Shunting or discharge device
18	Accelerating or de-accelerating device
19	Starting-to-run transition contactor
20	Electricity operated valve
21	Distance relay
22	Equalizer circuit breaker
23	Temperature control device
24	Volts/Hz relay
25	Synchronizing or synchronizing-check relay
26	Apparatus thermal device
27	Undervoltage relay
28	Flame detector
29	Isolating contactor
30	Annunciator relay
31	Separate excitation device
32	Directional power relay
33	Position switch
34	Master sequence device
35	Brush-operating or slip-ring short-circuiting device
36	Polarity or polarizing voltage device
37	Undercurrent or underpower relay
38	Bearing protection device

(Continued)

(*Continued*)

Device Numbers according to IEEE C37.2

Device Number	Description
39	Mechanical condition monitor
40	Field relay
41	Field circuit breaker
42	Running circuit breaker
43	Manual transfer or selector device
44	Unit sequence starting relay
45	Atmospheric condition monitor
46	Reverse-phase or phase-balance relay
47	Phase-sequence or phase-balance voltage relay
48	Incomplete sequence relay
49	Machine or transformer thermal relay
50	Instantaneous overcurrent relay
51	AC time delay relay
52	AC circuit breaker
53	Exciter or dc generator relay
54	Turning gear engaging device
55	Power factor relay
56	Field application relay
57	Short-circuiting or grounding device
58	Rectification failure relay
59	Overvoltage relay
60	Voltage or current balance relay
61	Density switch or sensor
62	Time-delay stopping or opening relay
63	Pressure switch
64	Ground detector relay
65	Governor
66	Notching or jogging device
67	AC directional overcurrent relay
68	Blocking relay
69	Permissive control device
70	Rheostat
71	Level switch
72	DC circuit breaker
73	Load-resistor contactor
74	Alarm relay
75	Position changing mechanism
76	DC overcurrent relay
77	Telemetering device
78	Phase-angle measurement or out-of-step relay
79	AC reclosing relay
80	Flow switch
81	Frequency relay
82	DC load-measuring reclosing relay

(*Continued*)

(*Continued*)

Device Numbers according to IEEE C37.2

Device Number	Description
83	Automatic selector control or transfer relay
84	Operating mechanism
85	Carrier or pilot-wire receiver relay
86	Lockout relay
87	Differential protection relay
88	Auxiliary motor or motor generator
89	Line switch
90	Regulating device
91	Voltage directional relay
92	Voltage and power directional relay
93	Field-changing contactor
94	Tripping or trip-free relay
95–99	For specific applications where other numbers are not suitable

In addition, suffix letters as per this standard are used for clarity. Generally, the meaning of suffix letters is clearly spelled out on single-line diagrams.

Reference

IEEE C37.2

Index

Page numbers followed by "*f*" indicate figures; those followed by "*t*" indicate tables.

A

Abstract communication service interface (ACSI), 628, 629
AC pilot relaying, 561
Adaptive protection, 17–18
    and redundancy, 18*f*
Adaptive relaying, 101
Additive polarity, 65*f*, 66
Adjustable speed drives (ASDs), 373
    grounding of, 249, 251*f*, 252
Alpha plane, 565–566, 567*f*
ANSI curves, 26–27
Antifriction ball/roller bearings, 361
Application layer, of utility of future, 9*f*
    customer interaction, 11
    grid and customer analytics, 11
    real-time awareness and control, 11
Arc blast, 642
Arc-flash detection (AFD) relays, 652–660
Arc-flash event
    equipment labeling, 652
    maximum duration of, 651–652
Arc-flash hazards, 641–643
    analysis, 643–647
        calculations, 665–668, 672–674
        HRC reduction through maintenance mode switch, 666–668
        HRC reduction with main secondary circuit breakers, 674, 675–677*f*, 678*t*, 679
        and short-circuit currents, 660–664
        system configuration for study, 668–679
        trip programmers for low-voltage circuit breakers, 679, 679–680*f*, 681*t*
    arc blast, 642
    fire hazard and electrical shock, 642
    ground fault protection and coordination, 652
    hazard boundaries, 648–649
    reduction of, 641–681
    risk categories, 647–649
    standards for calculations, 643
    system grounding, 649–651
    time motion studies, 642–643
Arc protection relays
    applications, 659
    light sensor types, 654–656
    other hardware, 657, 658*f*
    principle of operation, 652–653, 654*f*
    selective tripping, 657, 658*f*
    self-testing of sensors, 660
    supervision with current elements, 659
Arc resistance, 528–530
Arc-resistant switchgear, 664–665
Arc thermal performance exposure value (ATPV), 647–648
ASTM F1506, 647–648
Attenuation mode, 589, 590*t*
Audio tone channels, 588, 589*t*
Aurora vulnerability, 20*t*
Autoclosing, for transmission lines, 510
Automatic resynchronization, 336
Automatic transfer switches (ATSs), 145–147
Automatic voltage regulator (AVR), 551
Autotransfer, of synchronous motors, 331–333
Autotransformer, 483, 495*f*
Axial vibrations, 364–365

B

Back-up protection, 7
    for generator using differential relays, 410–411
Ball and roller bearings, 361
Banding, 364
Bare fiber-optic sensors, 654
Batteries, sizing, 143–149
    automatic transfer switches, 145–147
    battery chargers, 147–148
    DC load profile for, 145*f*
    short-circuit and coordination considerations, 148–149
    standards for, 144
    system configurations, 144, 146–147*f*
Battery chargers
    as battery eliminator, 148
    equalizing charge, 148
    floating operation, 147–148
    switch mode operation, 148

Battery systems
    battery monitoring system, 141–142
    lead acid batteries, 138–139
        pasted plate type, 140, 141
        plante type, 139–140
        tubular plate type, 140, 141
        valve-regulated lead acid, 140–141
    nickel–cadmium batteries, 142
    pocket plate nickel–cadmium batteries, 142–143
    types of batteries, 138–139
Bearing protection
    antifriction ball and roller bearings, 361
    bearing failures
        excessive axial or thrust loading, 362
        mechanical problems, 361–362
    devices, 362
    end play and rotor float for coupled sleeve bearings horizontal motors, 362, 363t
    sleeve bearings, 361
Bimetallic thermal trip device, 163, 164f
Bleeder valve, 502
Blinder, 525f
Blocking pilot protection, 574–575
Blocking schemes, 561, 570
Boric acid fuses, 161
Breaker and half scheme, 322, 323f
Breaker failure relay, 45–46
Breaker monitor, 125, 127t, 128f
Brushless-type excitation system, 377–378
Brushless-type of generators, 419–420, 421f
Brush-type controllers, 376–377
Buchholz relay, 501
Bulk electric system (BES), 20–21
Bus-bar protection and autotransfer of loads
    bus differential relays, 301
    bus faults, 301
    bus transfer schemes, 326–329
    differential protection of bus configurations, 318–326
    direction comparison bus protection, 316
    high-impedance differential relays, 301–308
    low-impedance current differential relays, 308–316
    momentary paralleling, 330–333
    reclosing, 326
    using linear couplers, 317
Bus differential relays, 301
    low-voltage bus bars, 301
Bus faults, 301
Bushing current transformer, 56
Bushing flashovers, 445
Bus transfer schemes, 326–329
    fast bus transfer, 327
    in-phase transfer, 329
    residual voltage transfer, 327–328
Bypass switches, 322

C

Cable selection, for HR grounded systems, 275–276
Capacitive trip devices, 149
Capacitor-coupled voltage transformers, 84–88
    applications to distance relay protection, 88
    transient performance, 85–88
Capacitor protective gaps, 546
"Carrier sense multiple access collision detect," 625
Carrier transmitters, 587t
Cast coil transformers, 499
Channel impairment, 564
CIGRE working group, 560, 561f
Circuit breakers, 326, 477
    for substation transformers, 448
    XCBR information tree, 630, 631f
Class E controllers, ratings of, 349t
Class E fuses, for transformer protection, 159, 160f
Classical stability model, 604–605
Close logic, 118
Coaxial cable, 585
Coefficient of grounding (COG), 233
Coercive force, 75
Cogeneration system, coordination in, 197, 199, 201f, 202f, 203t, 204f
Combined bus differential zones, 322, 324f
Combustible gas relay, 504
Combustion turbine generators (CTGs), 423
Commercial low-voltage switchgear, 3
Communication-assisted tripping, 530–531
Communication methods, selection of, 562t
Conductor-cooled generators, 404
Conservator tank design, 501
Constant field current, 379
Constant power factor, 379
Constant reactive power output, 379
Constant voltage controller, 379
Consumer load profile, with/without solar generation, 12f
Cooling time constants, 345
Coordinating time interval (CTI), 184, 510
    guidelines for selecting, 185t
Coordination timers, 545–546
Core-balance CTs, for ground fault protection, 298–299

Core monitor, 404
Corpals benchmark model, 548
Coupling capacitor, 584
Coupling loss, 591
CPU-based protection and control, 667
Critical clearing angle, 599–600
Cross-compounded generators, 408f
Cross-connected generators, 413f
Current-limiting fuses, 460, 463t, 667
Current-limiting molded case circuit breakers, 164–168, 199
Current transformers (CTs)
    application
        considerations, 66–70, 71f
        future directions, 77–79
        practicality, 75–76
    connections, and phase angle compensation, 486–488
    constructional features of, 56, 57f
    CT ratio
        and class C accuracy, 62–64, 65f
        and phase angle errors, 59–62
    equivalent circuit of, 73f
    for low-resistance grounded medium-voltage systems, 76–77
    metering accuracies, 53, 54t
    polarity of, 64–66
    and PT test switches, 151, 153–154f
    relaying accuracies, 53–55
    saturation, 312–313, 314f, 567–568
    secondary terminal voltage rating, 56–59
    series and parallel connections of, 70, 72f
    transient performance of, 70–75
Cyber security, 18–19, 20t

D

DC control schematics, of close and trip circuit of generator breaker, 137, 138f
Dedicated circuit breakers, for substation transformers, 448
Delta grounded systems, corner of, 242–243
Delta–wye transformer, 481, 483, 486, 491
Demand metering, 120–122
Deterioration, 357
Device 21
    application of, 395–396
    locked rotor protection using, 347–348
Device 51V, 391f, 395
Device numbers according to IEEE C37.2, 683–685
Differential protection
    bus configurations
        breaker and half scheme, 322, 323f
        combined bus differential zones, 322, 324f
        double bus double breaker, 319, 320f
        double bus single breaker with bus tie, 319–322
        ground fault bus differential protection, 325–326
        main and transfer bus, 319, 320f
        ring bus, 322, 324f
        sectionalized bus, 318–319
        single bus, 318, 319f
    generator protection
        backup protection using differential relays, 410–411
        characteristics, 411–415
        generator winding connections, 407–408, 409–410
        protection provided by differential relays, 408
        self-balancing differential scheme, 409
        tripping modes, 415
        for unit-connected generator, 412f
    motor protection
        flux balancing current differential, 371, 373f
        full differential protection with ground fault differential, 372, 374f
    pilot protection
        alpha plane, 565–567
        CT saturation, 567–568
        enhanced current differential characteristics, 569–571
        three-terminal protection, 568–569
        of two-terminal and three-terminal lines, 564–565
    transformer protections, 477–489
        autotransformer protection, 483
        electromechanical transformer differential relays, 480–482
        with grounding transformer, 483
        harmonic restraint, 483
        microprocessor-based transformer differential relays, 483–489
        sensitive ground fault protection, 489–491, 497f
        three-winding transformers protection, 483
        zigzag grounding transformer protection, 491
Differential relays
    low-impedance relays, settings, 315–316
    overcurrent differential protection, 29, 32–33f, 34

Differential relays (*cont.*)
  overlapping the zones of protection, 34, 35–36f
  partial differential schemes, 33f, 34
  percent differential relays, 34, 36, 37f
Digital microwave, 564
Digital relays, 24, 471
Direct fiber connections, 564
Directional comparison blocking (DCB), 530
  trip scheme, 545
  and unblocking scheme, 575–576
Directional comparison unblocking (DCBU), 530
Directional ground fault relays, 280–284
Directional overcurrent relays, 37–41
  application of, 517–519
Directional relays, 316, 518–519, 531
Direction comparison bus protection, 316
Direct/nonpermissive underreaching transfer trip, 571, 572f
Direct overreaching transfer trip, 572–573
Discrete Fourier transform (DFT), 15
Distance protection, 522–530
  distance relay characteristics, 524–528
  effect of arc fault resistance, 528–530
  operating time in first zone, 528
  zoned distance relays, 523–524
Distance relays, 50–52, 400
  characteristics, 524–528
Distance time blocking scheme, 574f
Distributed energy resources (DERs), 7
Distributed network protocol (DNP), 625
Distribution lines, 509
Double bus double breaker, 319, 320f
Double bus single breaker with bus tie, 319–322
Double-frequency tuning, 90
Double-line-to-ground fault, 285
Drain coils, 584
Dropout of motor contactors, 330–331
Dry-type reactors, 494, 497f, 498f
Dry-type transformers, 454, 455f, 499
Dual-level volts/Hz sensing, 422, 423f
Dual-phase comparison blocking, 579, 580f
Dynamic bus replica, 309, 314–315
Dynamic contingency analysis (DCA), 16
Dynamic CT ratio corrections, 488
Dynamic stability, *see* Small rotor angle stability

E

East Coast Blackout (2003), 13
Electrically reset lockout relay, 149, 150f

Electrical shock, 642
Electrical system coordination (article 240.12 from NEC), 211
Electrical utility business models and regulatory framework, drivers for, 7, 8
Electromechanical relays, 24–26
  construction of directional overcurrent, 40f
  metallic pilot wire protection using, 562–564
  percentage differential relay, 36
  relay overtravel for, 184
  transformer differential relays, 480–482
Electronic power fuses, 175–176, 177–178f
Embedded temperature detectors, 359–360
EMTP simulations, 549–551
  for ground fault in HRG system, 287–294, 295f
Enabling layer, of utility of future, 9f
  enabling infrastructure, 11
  incremental intelligence, 11
End users, 11
"Equal area criterion of stability," 383, 597–600
Equipment grounding, 231
Equipment protection, 4
Equivalent circuit of multiple grounded systems, 246, 247f
Ethernet, 623–624
  connection, 630, 632–634
  IEEE 802.3, 625–626
Events analyzes, relay operations, 130–131, 132–133f, 132t
Excitation systems
  determination of, 606, 607f
  fast response systems, 606–607
  fundamental block circuit control diagram, 606, 607f
  types of, 607–608
External transformer faults
  overloads, 443
  overvoltage and underfrequency, 444
  short-circuit faults, 443
Extraneous voltages, 563

F

Fast bus transfer, 327
Fault arc resistance, effect of, 528–530
Fault conditions, 330
Fault current, loop system with one source of, 520–522
Fault decrement curve, 393
Fault detection method, 316, 559, 578
Fault induced error, 535
Fault resistance, 527

Index

Federal Energy Regulatory Commission (FERC), 2
Federal Information Processing Standard Publication 200, 19
Ferroresonance, 461
　damping, 84
Fiber optics, 562
　cable, 635–636
Field controllers, types of, 378–379
Field-to-ground fault protection, of motors, 46f
51V (voltage-controlled and voltage-restraint protection), 390–397
Fire hazard, 642
First benchmark model (FBM), 547, 554f
Fixed mounted primary circuit breaker, for substation transformers, 448
Fixed-percentage differential relay, 36, 37
Foundational layer, of utility structure, 9f
　business and regulatory, 10–11
　foundational infrastructure and resources, 9–10
　organization and processes, 10
　standards and models, 10
FPX switchboard type of relay, 502, 503f
Frequency relays, 47–48, 49–50f
Frequency shift carrier, 587–588
Full-cosine filter, 306
Fuses, 151, 154–161
　medium-voltage fuses, 155–156, 159, 160f
　selection for medium-voltage applications, 159, 161
　semiconductor fuses, 161–162, 163t

G

Gate turn-off (GTO) inverter, 249, 252
Generator protection
　antimotoring protection, 433–434
　breaker failure, 431–432
　differential protection, 407–415
　functionality of modern MMPR for, 388–389
　generator stator ground fault protection, 415–418
　generator thermal overload, 404–406
　inadvertent energization of generator, 429–431
　loss of excitation protection, 400–403
　loss of potential, 434–435
　negative-sequence protection, 397–399
　out-of-step protection, 427–429
　over- and underfrequency protection, 423–427
　protection of industrial generators, 387–388
　ratings of synchronous generators, 387
　rotor ground fault protection, 419–420
　synchronization, 436–437
　tripping schemes, 437–440
　under- and overvoltage protection, 436
　voltage-controlled and voltage-restraint protection, 390–397
　volts per hertz protection, 420–422
Generator step-up (GSU) transformer, 395
Generator thermal overload, 404–406
Generator winding connections, 407–408
　application depending on, 409–410
Global positioning systems (GPSs), 14
GOOSE (Generic Object Oriented Substation Event) messages, 627–628, 633
　prioritizing, 636, 638
Great North East Blackout (1965), 13
Greenhouse emissions, reduction of, 7
Ground directional relays (GDRs), 282
Ground distance relay, 531, 536
Ground fault coordination and protection, 196–197, 200f, 201t, 372–373, 418, 473
　bus differential protection, 325–326
　detectors, 316
　directional ground fault relays, 280–284
　effect of nonhomogeneous system on reactance elements, 534–535
　high-resistance fault coverage and remote infeed, 536–538
　in high-resistance grounded systems, 271–276
　of industrial bus-connected generators, 277–280
　in low-resistance grounded medium-voltage systems, 266–268
　monitoring of grounding resistors, 299
　operating logic selection for directional elements, 285
　quadrilateral ground distance and mho ground distance characteristics, 533–534
　remote tripping, 268
　requirements of, 469
　in resonant grounded systems, 276–277
　selective high-resistance grounding systems, 285–299
　in solidly grounded systems, 257–265
　studies in practical systems, 277–285
　in ungrounded systems, 268–271
　zero-sequence mutual coupling, 535–536
　zero-sequence overcurrent, 531–533
Grounding in mine installations, 253–254

Grounding of adjustable speed drive (ASD) systems, 249, 251f, 252
Grounding systems, study of, 231
Grounding transformer, 483
Group feed system, for substation transformers, 448

H

Half-cosine filter, 306
Hand reset lockout relays, 149, 150f
Hard-wired remote trips, 150–151, 153f
Harmonic restraint
　percentage, 36
　in relay, 483
High-impedance differential relays
　for bus protection, 301–308
　comparison with, 316
　high-impedance MMPRs, 305–308
　versus low-impedance relays, 316
　open-circuited CT, 308
　sensitivity for internal faults, 305
High-resistance fault coverage and remote infeed, 536–538
High-resistance grounded (HRG) systems, 233, 237–239
　ground fault protection in
　　insulation stresses and cable selection for, 275–276
　　nondiscriminatory alarms and trips, 271, 272f
　　protection against second ground fault, 274–275
　　protection of motors, 273–274
　　pulsing-type ground fault detection system, 272–273, 274f
　　selective ground fault clearance, 271–272, 273f
　for selective ground fault protection, 285–299
　　accuracy of low current pickup settings, 299
　　EMTP simulation, 287–294, 295f
　　generator 100% stator winding protection, 294–299
High-voltage circuit breakers, 135–138
　current-limiting fuse characteristics, 136f
　DC control schematics, 137, 138f
　failure modes of circuit breakers, 137t
High-voltage fuses, 155
HV transmission lines, subsynchronous resonance in, 551–555
Hybrid grounding system
　for generator, 285, 286f
　for industrial bus-connected generators, 248–249, 250–251f
Hydro generators, 400
Hysteresis loop, of magnetic material, 74

I

IC cooling designations, 366, 368t
ICD file (IED capability description), 633
IEC 61850 protocols
　communication profiles, 628–629
　description of part numbers, 627t
　Ethernet connection, 630, 632–634
　hierarchical data models, 629, 630f
　logical nodes (LNs), 629–630, 631–632f, 631t
　for substation automation (SA) systems, 626–627
IEC 870-5 standards, 625
IEC curves, 28, 29f, 30–31f, 32f
IEEE 1584 equations, for electrical systems, 644–647
IEEE 802.3 (Ethernet) standards
　IEC 870-5, 625
　Intercontrol Center Communications Protocol (ICCP), 625–626
　UCA 1.0, 625
　UCA 2.0, 626
Impedance locus, 400
Impedance relays, *see* Distance relays
Impulse margin time, *see* Relay overtravel
Incident energy, 649–651, 662
Induction cup relays, 48
Induction disk relay, 48
Induction generator effect, 547
Induction motors, 358t
Industrial bus-connected generators
　ground fault protection of, 277–280
　hybrid grounding system for, 248–249, 250–251f
Industrial generators, protection of, 387–388
Innovation layer, for utilities, 9f
　pilot and demonstration projects, 12
　research and development, 12
In-phase transfer, 329
Instrument transformers
　capacitor-coupled voltage transformers, 84–88
　current transformers (CTs), *see* Current transformers (CTs)
　line (wave) traps, 88–91
　transducers, 91
　voltage transformers, 79–84
Insulated case circuit breakers (ICCBs), 168

Index

Insulation stresses, for HR grounded systems, 275–276
Intelligent electronic devices (IEDs), 627–628
Intercontrol Center Communications Protocol (ICCP), 625–626
Interior metal water piping system, 248
Internal faults, sensitivity of relay for, 305
Internal transformer faults
    bushing flashovers, 445
    interturn winding faults, 444–445
    part-winding resonance, 445
    short circuit in windings, 444
    in ULTC and off–load tap changing equipment, 445
Intertripping, 559
Interturn faults, 496
Interturn winding faults, 444–445
Inverse volts/Hz sensing, 422, 423f
IP enclosure designations, 366, 367t
Iron-core reactor, 494
IT and CS cyber security approaches, 19t

K

Knee point, 54
    voltage, 55
Krondroffer starter, control circuit of, 353
kV distribution, for coordination, 192–195, 196f, 197f, 198f

L

Large disturbance instability, 595
Large rotor angle stability, 594
Large-sized generators, 387–388
Latch bits, 123–124
Lead acid batteries, 138–139
    pasted plate type, 140, 141t
    plante type, 139, 140t
    tubular plate type, 140, 141t
    valve-regulated lead acid, 140–141
Lead antimony, 140
Lead calcium, 140
Lead selenium, 140
Lens-point sensors, 654
Lens sensors, 652–655
Less-flammable liquids, for transformers, 466–468
Let-through characteristics
    of current-limiting MCCB, 165, 166f
    of semiconductor fuses, 161, 162f
    of types of fuses, 154, 158f
Light sensor types, 654–656

Linear couplers, bus protection using, 317
Lines protection, 509–555
Line (wave) traps, 88–91
    in power line carrier, 584, 586f
Line tuner, 584–585
Liquid-filled transformers
    conservator tank design, 501
    positive pressure inert gas, 500
    sealed tank construction, 500
Liquid level gauge, 502
Liquid temperature indicator, 503
Load encroachment, 543
    logic, 530–531
Load profiles, 12
    consumer load with/without solar panels, 12f
    voltage regulation, 13f
Load shedding, 424, 427
Local area network (LAN) protocols, 624
Locked-rotor current of motors, 188, 190t
Locked rotor protection, 347–348
Lockout relays, 149–150
Logarithmic decrement, 548
Logical nodes (LNs), 629–630, 631–632f, 631t
Loss of potential (LOP), 114
    generator protection
        automatic synchronizing, 437
        manual synchronizing, 436–437
Low- and medium-voltage contactors, 176, 178t, 179
Low arc-flash circuit breaker design, 168f
Lower current ratio, 63
Low-impedance current differential relay
    for bus protection, 308–316
    CT saturation, 312–313, 314f
    differential settings, 315–316
    dynamic bus replica, 314–315
    high-impedance relays, comparison, 316
Low-resistance grounded systems, 236
    for ground fault protection and coordination, medium-voltage, 266–268
Low-voltage circuit breakers
    current-limiting MCCBs, 164–168
    insulated case circuit breakers (ICCBs), 168
    low-voltage power circuit breakers (LVPCBs), 168–169
    molded case circuit breakers (MCCBs), 162–164
    motor circuit protectors (MCPs), 171–172, 173f, 174t
    other pertinent data of, 172–173
    short-time bands of LVPCBs' trip programmers, 169–170, 171f, 172f

Low-voltage fuses types, 154–155, 157t
Low-voltage motors, 336–337
Low-voltage power circuit breakers (LVPCBs), 165, 168–170
    short-time bands of, 169–170, 171f, 172f, 668
Low-voltage radial distribution system, for coordination, 186–192, 193f
Low-voltage relayed power circuit breakers, 185t

M

Machine field ground fault relay, 46–47
Magnetic contactors, 350–351
Magnetic instantaneous trip device, 163, 164f
Magnitude, 14
Main and transfer bus, 319, 320f
Maintenance mode switches
    HRC reduction through, 666–668
    on low-voltage trip programmers, 679
Mark frequency, 579
Medium- and low-voltage distribution system for coordination and arc-flash analysis, 668, 669f
Medium-sized generator protection, 387
Medium-voltage distribution system, for coordination, 193, 194f, 218, 219f
Medium-voltage fuses, 155–156, 159, 160f
    variations in fuse time–current characteristics, 159, 160f
Medium-voltage induction motors, 335–336
Medium-voltage low-resistance grounded system
    for ground fault protection and coordination, 266–268
Medium-voltage motor starters, 348–351
    class E1, 349
    class E2, 349–350
    coordination, 193, 196f
    low-voltage magnetic contactors, 350–351
Medium-voltage synchronous motors, 336
Metal-clad switchgear, 34, 35, 64
Metallic pilot wire protection using electromechanical relays, 562–564
Metal oxide varistor (MOV) supervision, 306
Metering accuracies, 53, 54t
Mho, 525f, 527
    ground distance characteristics, 533–534
    ground distance element reach, 542
    phase distance element reach, 540
    unit, 51
Microprocessor-based multifunction relays (MMPRs), 1, 24, 56, 488, 564
    analyzing events, 130–132
        aliases, 131
        sequential event recorder, 131
        triggering, 131, 132–133f, 132t
    breaker monitor, 125, 127t, 128f
    close logic, 118
    demand metering, 120–122
    dimensions, 95, 96f
    environmental compatibility, 95t
    frequency settings, 114, 115t, 116f
    front panel, 94–95, 127–130
    functionality, 93–94
        communications and controls, 94
        for generator protection, 388–389
        monitoring features, 94
        protection features, 93
        voltage-based protections, 93–94
    global settings, 124–125
    high-impedance, 305–308
    latch bits, 123–124
    logical settings, 122
    loss of potential (LOP), 114
    low-impedance, 309, 310f
    port settings, 125, 125–127t
    power elements, 113
    for protection of medium-voltage motors, 335–336
    reclose supervision logic, 119–120, 121f
    relay bit words, 106, 107–108f
    relay settings, 132–133
    settings, 101–106
    specifications, 95, 97–99t, 99–101
    time delay overcurrent protection, 106, 108–111, 112f
    trip logic, 114–115, 117
    voltage-based elements, 111–113
Microprocessor-based transformer differential relays, 483–489
    CT connections and phase angle compensation, 486–488
    dynamic CT ratio corrections, 488
    security under transformer magnetizing currents, 488–489
Microwave channels, 589
Microwave frequencies, 562
Mine installations, grounding in, 253–254
Modal analysis, of transmission lines, 589–590
Modern intelligent electronic devices (IEDs), 627–628
Modern protective relaying, *see* Protective relays/relaying
Modern transient stability methods, 606
Mode shapes, of torsional motion, 548

Molded case circuit breakers (MCCBs), 155, 162–164, 187
   current-limiting, 164–168
Momentary paralleling
   autotransfer of synchronous motors, 331–333
   dropout of motor contactors, 330–331
   fault conditions, 330
Motor circuit protectors (MCPs), 155, 171–172, 173f, 174t, 187
Motor contactors, dropout of, 330–331
Motor protection
   axial vibrations, 364–365
   bearing protection, 361–362, 363t
   and coordination study, 337–338
   coordination with motor thermal damage curve, 338–346
   differential protection, 371–372
   embedded temperature detectors, 359–360
   ground fault protection, 273–274, 372–373
   Krondroffer starter, control circuit, 353
   locked rotor protection, 347–348
   low-voltage motors, 336–337
   medium-voltage induction motors, 335–336
   medium-voltage motor starters, 348–351
   medium-voltage synchronous motors, 336
   motor characteristics, 335
   motor enclosure, 365–366
   motor insulation classes and temperature limits, 357–360
   NEC and OSHA requirements, 354–357
   negative-sequence currents, effect of, 367–371
   NEMA standards for insulation temperature rise limits, 357–358, 359t
   relative shaft vibrations, 363–364, 365
   resistance temperature detector (RTD)
      biasing, 347–348
      polling, 360
   rotor ground fault protection, 385
   stability concepts of synchronous motors, 379–384
   and synchronization, 373–379
   two-wire and three-wire controls, 352–353
   undervoltage protection, 354
   variable-speed motor protection, 373
   vibrations, 363–365
Motor thermal damage curve, coordination with, 338–346
Multiple grounded systems, 245–246
   equivalent circuit of, 246, 247f
Multiple isolated generators, on medium-voltage systems, 387

Multiratio bushing current transformer, 56
Multirestraint differential systems, 309

N

Narain Hingorani subsynchronous resonance suppressor (NH-SSR), 551–552
NEC requirements
   and NESC requirements system grounding, 247–248
   and OSHA requirements for motor protection, 354–357
   for overcurrent protection of transformers, 445–446
   relating to overcurrent selectivity, 211–215, 216f, 217f, 218t
   for solidly grounded systems, 257–262
Negative-sequence currents
   effect of, 367–371
   phase balance protection, 369–370
   time delay and instantaneous negative-sequence protection, 370–371
Negative-sequence differential settings, 567–568
Negative-sequence impedance, 533, 537
Negative-sequence protection, 397–399
NEMA standards
   for insulation temperature rise limits, 357–358, 359t
Networking media, 634–636
   fiber-optic cable, 635–636
   twisted-pair copper, 635
Network topologies, 636–638
   prioritizing GOOSE messages, 636, 638
   techno-economical justifications, 638
Neutral conductor, 245–246
Neutral grounding resistors, 299
Neutralizing reactor, 563
Neutrals, artificial, 243–244
Nickel–cadmium batteries, 138, 142
North American Electric Reliability Corporation (NERC), 2
   and CIP standards, 20–22
North East Blackout (1965), 13
Nuisance trips, 525f
Numerical relay, 24

O

Off–load tap changing equipment, faults in, 445
Off-set mho, 525f
Oil-immersed reactors, 497–498

100% stator winding ground fault protection, 294–299
On-load tap changing (OLTC), 471
On–off carrier, 586–587
Open-circuited CT, 308
Operating time, in first zone, 528
Optical power budget (OPB), 634
Out-of-step protection, 427–429
Overall transformer protection
  addition of secondary relay, 477
  coordination study and observations, 474–477
  system configuration, 471–474
Overcurrent differential protection, 29, 32–33f, 34
Overcurrent protections
  art of compromise, 215–221, 222f, 223t, 223f
  computer-based coordination, 183
  coordinating time interval (CTI), 184, 185t
  coordination in cogeneration system, 197, 199, 201f, 202f, 203t, 204f
  coordination on instantaneous basis, 199–211
  coordination settings for, 203t
  data for coordination study, 182–183
  fundamental considerations, 181
    for coordination, 185–186, 187f
  ground fault coordination, 196–197, 200f, 201t
  initial analysis, 184
  kV distribution, 192–195, 196f, 197f, 198f
  low-voltage distribution system, 186–192, 193f
  NEC requirements of selectivity, 211–215, 216f, 217f, 218t
  UPS systems, protection and coordination of, 227–228, 229f
  zone selective interlocking, 221–222, 224–227
Overcurrent relays, 341, 342–343f, 517–519
  ANSI curves, 26–27
  IEC curves, 28, 29f, 30–31f, 32f
  for transformer primary protection, 463–465
Overexcitation, 422, 493
Overfrequency protection, 423–427
Overfrequency relays, 48, 49f
Overlapping zones, of protection, 6f, 34, 35–36f
Overloads, 443
  capability of generators, 404f
Overreaching, 536
Overvoltages, 354, 444, 496
  protection, 41

P

Partial differential schemes, 33f, 34
Part-winding resonance, 445

Pasted plate batteries, 140, 141t
Peer-to-peer wire line network, 623–624
Percent differential relays, 34, 36, 37f
Permissive overreaching transfer tripping (POTT), 530
Permissive underreaching transfer trip, 571, 573f
Personal protective equipment (PPE)
  characteristics, for hazard/risk category, 647, 648t
Phase angle, 14
  compensation, 486–488
  errors, CT ratio and, 59–62
Phase balance protection, 369–370
Phase bus transfer, 329f
Phase comparison schemes, 577–582
  dual-phase comparison blocking, 579, 580f
  segregated phase comparison, 579–582
  single-phase comparison blocking, 578–579
Phase fault detectors, 316
Phase velocities, 590t
Phasor data concentrators (PDCs), 15
Phasor measurement units (PMUs), 14–16
  hierarchical system of, 16f
Pilot protection
  direct blocking scheme, 574–575
  directional comparison blocking and unblocking scheme, 575–576
  direct overreaching transfer trip, 572–573
  direct underreaching transfer trip, 571, 572f
  metallic pilot wire protection, 562–564
  modal analysis, 589–590
  modern line current differential protection, 564–571
  other losses, 591
  permissive underreaching transfer trip, 571
  phase comparison schemes, 577–582
  pilot systems, 559–561
  power line carrier, 583–589
  signal frequencies, 561–562
Pilot signaling, 575
Pilot wire protection, 36–37
Plante batteries, 139, 140t
Pocket plate nickel–cadmium batteries, 142–143
Point-to-multipoint wire line networks, 623
Point-to-point wire line networks, 623
Polarity of instrument transformers, 64–66
Polarizing quantity, 282
Pole disagreement protection, 499f
Positive pressure inert gas, 500
Positive temperature coefficient (PTC) thermistors, 359
Potential transformers (PTs)
  connections, 83–84

Index

limits of accuracy classes for, 62f
Power elements, 113
Power line carrier, 583–589
    audio tone channels, 588
    coaxial cable, 585
    coupling capacitor and drain coils, 584
    frequencies, 561
    line tuner, 584–585
    microwave channels, 589
    transmitter/receiver, 585–588
Power system stabilizer (PSS), 551
    classical stability model, 604–605
    classification, 593–597
        rotor angle stability, 594–595
        static stability, 596–597
        voltage instability, 595–596
    critical clearing angle, 599–600
    equal area concept of stability, 597–600
    excitation systems, 606–608
    list of factors affecting, 600–601
    modern transient stability methods, 606
    swing equation of generator, 601–604
    system illustrating application of, 615–617f, 615t, 618
    transient stability in simple cogeneration system, 608–615
Pressure-relief devices, 468, 502
Pressure rise relay, 502
Pressure–vacuum gauge, 502
Primary selective system, for substation transformers, 448
"Product type" of relay, 490
Protection standards, for power facilities, 1–2
Protective relays/relaying, 1, 510–522
    for arc-flash reduction, 641–681
    breaker failure relay, 45–46
    cyber security, 18, 19, 20t
    design aspects and reliability, 1–2
    design criteria of protective systems, 3–4
    differential relays, 28–36
    directional overcurrent relays, 37–41
    directional overcurrent relays application, 517–519
    distance relays, 50–52
    electromechanical relays, 24–26
    equipment and system protection, 4–5
    frequency relays, 47–48, 49–50f
    fundamental power system knowledge, 2–3
    loop system with one source of fault current, 520–522
    machine field ground fault relay, 46–47
    overcurrent relays, 26–28
    pilot wire protection, 36–37
    reclosing relays, 45
    relay types
        classification of, 23–24
        other, 52
    smart grids, *see* Smart grids
    unit protection systems, 5–7
    voltage relays, 41–44
Pull-out (PO) protection, 336
    calculation of PO power factor, 383–384
Pulsing-type high-resistance grounding system, 272–273, 274f

Q

Quadrilateral, for radial system, 525f
Quadrilateral ground distance, characteristics, 533–534

R

Radial distribution system, 471
    for substation transformers, 446–448
Radiofrequencies, 561
Ralph Lee's equations, 644, 651
Rapid Spanning Tree Protocol (RSTP), 636
Rated primary voltage and ratios, 79–82
Ratio correction factor (RCF), 60
Reach of the distance relay, 526–527
Reactance, 525f
    elements, nonhomogeneous system effect on, 534–535
    error, 534
    grounding, 241–242
Reclose logic and supervision, 119–120, 121f
Reclosing
    for bus faults, 326
    relays, 45
Redundancy, 1–2
    adaptive protection and, 18
Redundant battery system, with automatic transfer switch (ATS), 144, 146–147
Redundant star architecture, 636, 637f
Relative shaft vibrations, 363–365
    limits of, 365
    special machines, 364
    standard machines, 363, 364t
Relay bit words, 106, 107–108f
Relayed medium-voltage circuit breakers, 185t
Relaying accuracies
    C classification, 53–55, 54f
    ratio errors for relaying class CTs, 56, 58t
    T classification, 55, 55f
    X classification, 55

Relay overtravel, 184
Reliability, 1–2, 4
Reluctance torque, 380
Remote trips, 150–151, 152–153f, 268
Research and development, 12
Residual voltage transfer, 327–328
Resistance temperature detector (RTD)
    biasing, 347–348
    polling, 360
Resistive fault coverage, 536–538
Resonant grounding system, 242
    ground fault protection in, 276–277
Restricted earth fault (REF) protection, 489
Resynchronization, 374
Revenue metering instrument transformers, 53
Reverse directional check, 544, 545
RF hybrid loss, 591
Ring architecture, 636, 637f
Ring bus, 322, 324f
Robustness, 564, 571
Roller bearings, 361
Rotor angle stability, 594–595
Rotor ground fault protection
    brushless type of generators, 419–420
    slip ring machines, 419
    slip ring-type synchronous motors, 384f, 385
RS-485 bus, 623
RTs (Remote Terminals), 11
Run time constant (RTC), 345

S

Saturation detector, 312–313
Saturation factor, 58–59
Saturation voltage, 57–58
Sealed (valve-regulated) lead acid batteries, 140–141
Sealed tank construction, for transformers, 500
Secondary selective system, for substation transformers, 448–449
Secondary terminal voltage rating, 56–59
    saturation factor, 58–59
    saturation voltage, 57–58
Second benchmark model (SBM), 548
Second ground fault, protection against, 274–275
Sectionalized bus, 318–319
Segregated phase comparison, 579–582
Self-balancing differential scheme, 409
Self-extinguishing ground faults, 262f, 263
Self-monitoring, 564
Self-testing of sensors, 660
Self-tuned line traps, 91

Semiconductor fuses, 161–162, 163t
Sensitive ground fault differential protection, for transformer, 489–491
Sequential event recorder (SER), 131
Sequential tripping, 438–439, 440f
Series and parallel connections of current transformers, 70, 72f
Series compensated lines, 546–551
    analysis, 548
    example with EMTP simulations, 549–551
    models, 547–548
    steady-state excitation, 547
    subsynchronous resonance, 547
Shielded twisted pair (STP) cable, 635
Short-circuit and coordination considerations, 148–149
Short-circuit currents
    and arc-flash analysis calculations, 660–664
    reducing, 664
    transformer faults, 443
    in windings, 444
Shunt capacitor bank protection, 507
Shunt loss, 591
Shunt reactor protection, 494–498
Signal extension, blocking, 546
Signal frequencies, 561–562
Single bus, 318, 319f
Single-frequency tuning, 90
Single-line-to-ground fault, 280–281, 285
Single-phase comparison blocking, 578–579
Single phasing and ferroresonance, 461
Sinusoidal waveform, phasor representation of, 14f
Sizing the batteries, see Batteries, sizing
Sleeve bearings, 361
Slip ring machines, 419
Slip ring-type synchronous motors, 385
Small disturbance voltage instability, 595–596
Small generators, 387
Small rotor angle stability, 594, 595f
Smart grids
    application layer, 9f, 11
    drivers for utilities, 8f
    enabling layer, 9f, 11
    foundational layer, 9–11, 9f
    framework for, 7–12
    innovation layer, 9f, 12
    modern technologies leading to, 13–18
        adaptive protection, 17–18
        system integrity protection schemes (SIPSs), 16–17
        WAMSs and PMUs, 14–16
    voltage–var control, concept of, 12, 13f

Index

Solidly grounded systems, 232–236
   hazards in, 235–236
   improving coordination in low-voltage systems, 263–265
   protection and coordination in, 257–265
Source impedance ratio (SIR), 539
Space frequency, 579
Square waves, 577
Static frequency relay, 48
Static relay, 24
Static stability, 596–597
Stator ground fault protection schemes, for generator stator windings
   generator ground fault differential scheme, 418
   subharmonic voltage injection, 416–418
   third harmonic differential scheme, 415–416
Steady-state impedance, 552
Steam turbine generators, 400
Subharmonic voltage injection, 416–418
Substation automation (SA), 621
   control functions, 621–623
   IEC 61850 protocols, 626–627
     communication profiles, 628–629, 630f
     Ethernet connection, 630, 632–634
     logical nodes (LNs), 629–630, 631t, 631f, 632f
   modern IEDs, 627–628
   networking media, 634–636
   network topologies, 636–638
   sample application, 638–640
   substation architecture, 628
   system architecture
     LAN protocols, 624
     level 1: field devices, 624
     level 2: substation data concentrator, 624
     level 3: SCADA systems, 624
     SCADA communication requirements, 624–626
   system functions, 621, 622f
   wire line networks, 623–624
Substation configuration description (SCD) file, 633
Subsynchronous resonance (SSR)
   mitigation in HV transmission lines, 551–555
     NGH-SSR, 551–552
     supplemental excitation damping controls (SEDCs), 554–555
     thyristor-controlled series capacitor (TCSC), 552–554
     torsional relay, 555
Subtractive polarity, 64–66
Subtransmission lines, 510

Supervisory control and data acquisition systems (SCADAs), 14, 15
   communication requirements, 624–626
     distributed network protocol (DNP), 625
     IEEE 802.3 (Ethernet), 625–626
Supplemental excitation damping controls (SEDCs), 554
Surge arrester, 91
Surge protection, 506
Swing equation, of generator, 601–604
Switchboard-type fault-pressure relay, 502, 503f
Synchronizing power, 598, 603
Synchronous digital hierarchy (SDH), 564
Synchronous generators
   connected to infinite bus, 604
   power–angle characteristic of, 596
   ratings of, 387
Synchronous motors, 358–359t, 373–379
   autotransfer of, 331–333
   brushless-type excitation system, 377–378
   brush-type controllers, 376–377
   PO power factor calculation, 383–384
   rotor ground fault protection, 385
   stability concepts of, 379–384
   types of field controllers, 378–379
Synchronous optical network (SONET), 564
Synchrophasors, 14
System grounding, 231
   artificially derived neutrals, 243–244
   corner of delta grounded systems, 242–243
   grounding in mine installations, 253–254
   grounding of adjustable speed drive (ASD) systems, 249, 251f, 252
   high-resistance grounded systems, 237–239
   hybrid grounding system for industrial bus-connected generators, 248–249, 250–251f
   low-resistance grounded systems, 236
   methods of, 232f
   multiple grounded systems, 245–246, 247f
   NEC and NESC requirements, 247–248
   reactance grounding, 241–242
   resonant grounding, 242
   solidly grounded systems, 232–236
   ungrounded systems, 239–241
System integrity protection schemes (SIPSs), 5, 16–17
System protection, 4–5

T

Thermal burdens, 83
Thermal damage curve, of motor, 338–346

Thermal overloads, 404–406
Thermal relays, 503
Thermistors, 359
Thermocouples, 359
Third harmonic differential scheme, for generator stator windings, 415–416
32QG reverse directional check, 544
345 kV tapped transmission line, protection of, 538–546
Three-phase relays, 526
Three-subsystem segregated scheme, 581, 582f
Three-terminal protection, 568–569
Three-winding transformers protection, 483
Three-wire control circuit, 351f, 352–353
Through-fault current withstand capability
   categories, 450–453
   dry-type transformers, 454
   observation on faults, 454
   through-fault curve
      analytical construction of, 455–458
      protection with respect to, 458–460
   of transformer, 449–454
Through-fault protection
   primary fuse protection, 470
   primary relay protection, 471
Thyristor-controlled series capacitor (TCSC), 552–554
Thyristor switch, 551–552
Time–current characteristics
   of fuses, 154, 156f, 159, 160f
   of low-voltage trip programmer, 165, 167f
   of MCCB, 170, 172f
   of MCP, 171, 173f
   of OCPDs, 215f, 216f, 217f
   of overcurrent relays, 29f, 30–31f
Time–current coordination curves (TCCs), 199
Time delay overcurrent protection, 106, 108–111, 112f
Time motion studies, 642–643
Timer, 316
Token ring, 623
Torque angle swings of generators, 609f, 610
Torsional interaction, 547
Torsional mode damping, 548
Torsional relay, 555
Total correction factor (TCF), 61
Transducers, 91
Transfer bus, 319, 320f
Transfer trip scheme, 561
Transformer accessories
   combustible gas relay, 504
   liquid level gauge, 502
   liquid temperature indicator, 503
   pressure-relief device, 502
   pressure–vacuum gauge and bleeder valve, 502
   rapid pressure rise relay, 502
   surge protection, 506
   underload tap changing equipment, 504–505
   winding temperature indicator (thermal relays), 503
Transformer connections
   system configurations, 447f
      dedicated circuit breakers, 448
      fixed mounted primary circuit breaker, 448
      group feed system, 448
      primary selective system, 448
      radial system of distribution, 446–448
      secondary selective system, 448–449
Transformer enclosures, 498–501
   dry-type and cast coil transformers, 499
   liquid-filled transformers, 500–501
Transformer faults
   external faults
      overloads, 443
      overvoltage and underfrequency, 444
      short-circuit faults, 443
   internal faults
      bushing flashovers, 445
      interturn winding faults, 444–445
      part-winding resonance, 445
      short circuit in windings, 444
      in ULTC and off–load tap changing equipment, 445
Transformer primary fuse protection
   considerations, 461–463
   overcurrent relays, 463–465
   single phasing and ferroresonance, 461
   variations in fuse characteristics, 460–461
Transformers
   differential protection, 477–489
   ground fault protection, 297–298
   listing less-flammable liquids for, 466–468
   overall transformer protection, 471–477
   protection of parallel running transformers, 491–492
   shunt reactor protection, 494–498
   through-fault current withstand capability of, 449–454
   through-fault protection, 470–471
   volts per hertz protection, 493
   winding connections, effect of, 468–469
Transient performance

Index 701

of capacitor-coupled voltage transformers, 85–88
of current transformers (CTs), 70–75
Transients in grounding systems, 231
Transient stability, 510; *see also* Large rotor angle stability
  in simple cogeneration system, 608–615
Transient torques, 547
Transmission lines, 509–510, 526*f*, 531
Transmitter/receiver, 585–588
  frequency shift carrier, 587–588
  on–off carrier, 586–587
Treeing phenomenon, 641
Trip/restraint decisions, 569
Trip logic, 114–115, 117
Tripping modes, 415
Tripping schemes, 437–440
Trip programmers for low-voltage circuit breakers, 679, 679–680*f*, 681*t*
TRIR (recordable injury), 642
Tubular plate batteries, 140, 141*t*
Tuning methods, of line traps, 90–91
Twisted-pair copper networking media, 635
Two-subsystem segregated scheme, 581, 583*f*
Two-wire control circuit, 351*f*, 352–353

U

UCA 1.0 communication architecture, 625
UCA 2.0, 626
Unbalanced faults, 544
Unblocking pilot protection, 575–576
Underfrequency, 444
  protection, 423–427
  relays, 48, 49*f*
Underload tap changing (ULTC) equipment, 504–505, 506*f*
  faults in, 445
Undervoltage protection, 41
  of motors, 354
Ungrounded systems, 239–241
  ground fault protection in, 268–270
  nondiscriminatory alarms and trips, 270, 271*f*
Unit auxiliary transformer (UAT), 454, 458, 459–460
Unit protection systems, 3, 5–7
  back-up protection, 7
  overlapping zones of differential protection, 6*f*
Unit transformer (UT), 458–460
Unshielded twisted pair (UTP) cable, 635

Uplink port, 636
UPS (uninterruptible power supply) system, protection and coordination of, 227–228, 229*f*

V

Valve-regulated lead acid batteries, 140–141
Variable-percentage differential relay, 36, 37
Variable-speed motor protection, 373, 375*t*
Var–volt control, 12, 13*f*
Vibrations
  axial vibrations, 364–365
  relative shaft vibrations, 363–364, 365
Volta, Allesandro, 138
Voltage-based elements, 111–113
Voltage-controlled device, 390–397
  51V settings on, 395, 397*f*
Voltage instability
  large disturbance instability, 595
  small disturbance voltage instability, 595–596
Voltage relays, 41–44
Voltage-restraint device, 390–397
  51V settings on, 395, 396*f*
Voltage transformers
  accuracy rating, 82
  ferroresonance damping, 84
  PT connections, 83–84
  rated primary voltage and ratios, 79–82
  thermal burdens, 83
Voltage unbalance, 370
Volta Pile, 138
Volts per hertz protection, 420–422, 423*f*, 493

W

Water ingress, 498
Waveform recognition (WFR), 209
Wide-area control framework, 18*f*
Wide-area measurement systems (WAMSs), 14–16
Wideband tuning, 91
Winding temperature indicator (thermal relays), 503
Window-type current transformers, 56, 57
Wire line networks
  peer-to-peer network, 623–624
  point-to-multipoint networks, 623
  point-to-point networks, 623
Working distances, 649
Wound-type current transformer, 56

Z

Zero-sequence circuit, for fault, 281, 282f
Zero-sequence currents, on ground fault, 243–244
Zero-sequence mutual coupling, 535–536
Zero-sequence overcurrent, 531–533

Zigzag grounding transformer protection, 491
Zigzag transformer, 243–244, 483
Zoned distance relays, 523–524
Zone interlocking, 667
Zone selective interlocking, 173–175, 221–222, 224–227